Stochastic Mechanics
Random Media
Signal Processing and Image Synthesis
Mathematical Economics and Finance
Stochastic Optimization
Stochastic Control
Stochastic Models in Life Sciences

Stochastic Modelling and Applied Probability

(Formerly: Applications of Mathematics)

29

Robert J. Elliott · Lakhdar Aggoun ·
John B. Moore

Hidden Markov Models

Estimation and Control

Springer

Authors

Robert J. Elliott

Haskayne School of Business
Scurfield Hall
University of Calgary
2500 University Drive NW
Calgary, Alberta
T2N 1N4, Canada
relliott@ucalgary.ca

John B. Moore

Department of Systems Engineering
and Cooperative Research
Centre for Robust and Adaptive Systems
Research School of Physical Sciences and Engineering
Australian National University
Canberra, ACT 0200, Australia

Lakhdar Aggoun

Department of Mathematics and Statistics
College of Science
Sultan Qaboos University
P.O.BOX 36 Al-Khoud
Sultanate of Oman
laggoun@squ.edu.om

Managing Editors

B. Rozovskiĭ

Division of Applied Mathematics
Brown University
182 George St
Providence, RI 02912, USA
rozovsky@dam.brown.edu

G. Grimmett

Statistical Laboratory
Centre for Mathematical Sciences
University of Cambridge
Wilberforce Road
Cambridge CB3 0WB, United Kingdom
g.r.grimmett@statslab.cam.ac.uk

ISBN 978-0-387-94364-0

e-ISBN 978-0-387-84854-9

DOI 10.1007/978-0-387-84854-9

Stochastic Modelling and Applied Probability ISSN 0172-4568

Library of Congress Cataloging-in-Publication Data
Elliott, Robert James.
Hidden Markov models : estimation and control / Robert J. Elliott,
Lakhdar Aggoun, John B. Moore.
p. cm. — (Applications of mathematics ; vol. 29)
Includes bibliographical references and indexes.
ISBN 0-387-94364-1
1. Markov processes. I. Aggoun, Lakhdar. II. Moore, John B.
(John Barratt), 1941– . III. Title. IV. Series: Applications of
mathematics ; 29.
QA274.7.E45 1994
519.2′33–dc20 94-28643

Mathematics Subject Classification (2000): 93E11, 93E20, 60G35, 60H30

Corrected, 3rd printing 2008

Cover design: WMXDesign GmbH, Heidelberg

Printed on acid-free paper

9 8 7 6 5 4 3 2 1

springer.com

Preface to the Second Edition

Applications and interest in Hidden Markov Models continues to grow. The authors have been pleased with the reception of the first edition of this book.

Following comments and feedback from friends, students and others working with Hidden Markov Models the second edition contains clarifications, improvements and some new material.

In Chapter 2 the derivation of the basic filters related to the Markov chain are each presented explicitly, rather than as special cases of one general filter. Furthermore, equations for smoothed estimates are given.

The dynamics for the Kalman filter are derived explicitly as special cases of our general results and new expressions for a Kalman smoother are given. The Chapters on control of Hidden Markov Chains are expanded and clarified.

Preface

This work is aimed at mathematics students in the area of stochastic dynamical systems and at engineering graduate students in signal processing and control systems. First-year graduate-level students with some background in systems theory and probability theory can tackle much of this material, at least once the techniques of Chapter 2 are mastered (with reference to the Appendices and some tutorial help). Even so, most of this work is new and would benefit more advanced graduate students. Familiarity with the language of the general theory of random processes and measure-theoretic probability will be a help to the reader. Well-known results such as the Kalman filter and Wonham filter, and also H^2, H^∞ control, emerge as special cases. The motivation is from advanced signal processing applications in engineering and science, particularly in situations where signal models are only partially known and are in noisy environments. The focus is on optimal processing, but with a counterpoint theme in suboptimal, adaptive processing to achieve a compromise between performance and computational effort.

The central theme of the book is the exploitation, in novel ways, of the so-called reference probability methods for optimal estimation and control. These methods supersede, for us at least, the more familiar innovation and martingale representation methods of earlier decades. They render the theory behind the very general and powerful estimation and control results accessible to the first-year graduate student. We claim that these reference probability methods are powerful and, perhaps, comprehensive in the context of discrete-time stochastic systems; furthermore, they turn out to be relevant for systems control. It is in the nature of mathematics that these methods were first developed for the technically more demanding area of continuous time stochastic systems, starting with the theorems of Cameron and Martin (1944), and Girsanov (1960). The reference probability approach to optimal filtering was introduced in continuous-time in Duncan (1967), Mortensen (1966) and Zakai (1969). This material tends to be viewed as inaccessible to graduate students in engineering. However, apart from contributions in Boel (1976), Brémaud and van Schuppen (1976), di Masi and Runggaldier (1982), Segall (1976b), Kumar and Varaiya (1986b) and Campillo and le Gland (1989), there has been little work on discrete-time filtering and control using the measure change approach.

An important feature of this book is the systematic introduction of new, equivalent probability measures. Under the new measure the variables of the observation process, and at times the state process, are independent, and the computations are greatly simplified, being no more difficult than processing for linear models. An inverse change of measure returns the variables to the "real world" where the state influences the observations. Our methods also apply in continuous time, giving simpler proofs of known theorems together with new results. However, we have chosen to concentrate on models whose state is a noisily observed Markov chain. We thus avoid much of the delicate mathematics associated with continuous-time diffusion processes.

The signal models discussed in this text are, for the main part, in discrete time and, in the first instance, with states and measurements in a discrete set. We proceed from discrete time to continuous time, from linear models to nonlinear ones, from completely known models to partially known models, from one-dimensional signal processing to two-dimensional processing, from white noise environments to colored noise environments, and from general formulations to specific applications.

Our emphasis is on recent results, but at times we cannot resist the temptation to provide "slicker" derivations of known theorems.

This work arose from a conversation two of the authors had at a conference twenty years ago. We talked about achieving adaptive filter stability and performance enhancement using martingale theory. We would have been incredulous then at what we have recently achieved and organized as this book. Optimal filtering and closed-loop control objectives have been attained for quite general nonlinear signal models in noisy environments. The optimal algorithms are simply stated. They are derived in a systematic manner with a minimal number of steps in the proofs.

Of course, twenty years ago we would have been absolutely amazed at the power of supercomputers and, indeed, desktop computers today, and so would not have dreamt that *optimal* processing could actually be implemented in applications except for the simplest examples. It is still true that our simply formulated optimal algorithms can be formidable to implement, but there are enough applications areas where it is possible to proceed effectively from the foundations laid here, in spite of the dreaded curse of dimensionality.

Our work starts with discrete-time signal models and with states and measurements belonging to a discrete set. We first apply the change-of-measure technique so that the observations under a probability measure are independent and uniformly distributed. We then achieve our optimization objectives, and, in a final step, translate these results back to the real world. Perhaps at first glance, the work looks too mathematical for the engineers of today, but all the results have engineering motivation, and our pedagogical style should allow an engineer to build the mathematical tools without first taking numerous mathematics courses in probability theory and stochastic systems. The advanced mathematics student may find later chapters immediately accessible and see earlier chapters as special cases. However, we believe many of the key insights are right there in the first technical chapter. For us, these first results were the key to most of what follows, but it must be admitted that only by tackling the harder, more general problems did we develop proofs which we now use to derive the first results.

Actually, it was just two years ago that we got together to work on hidden Markov model (HMM) signal processing. One of us (JBM) had just developed exciting application studies for such models in biological signal processing. It turns out that ionic channel currents in neuron cell membranes can now be observed using Nobel prize winning apparatus measuring femto (10^{-15}) amps. The noise is white and Gaussian but dominates the signals. By assuming that the signals are finite-state Markov chains, and adaptively estimating transition probability and finite state values, much information can be obtained about neural synapses and the synaptic response to various new drug formulations. We believed that the on-line biological signal processing techniques which we developed could be applied to communication systems involving fading channels, such as mobile radio communications.

The key question for us, two years ago, was how could we do all this signal processing, with uncertain models in noisy environments, *optimally*? Then, if this task was too formidable for implementation, how could we achieve a reasonable compromise between computational effort and performance? We believed that the martingale approach would be rewarding, and it was, but it was serendipitous to find just how powerful were the reference probability methods for discrete-time stochastic systems. This book has emerged somewhat as a surprise.

In our earlier HMM studies, work with Ph.D. student Vikram Krishnamurthy and postdoctoral student Dr. Lige Xia set the pace for adaptive HMM signal processing. Next, work with Ph.D. student Hailiang Yang helped translate some continuous-time domain filtering insights to discrete time. The work of some of our next generation of Ph.D. students, including Iain Collings, features quite significantly in our final manuscript. Also, discussions with Matt James, Alain Bensoussan, and John Baras have been very beneficial in the development of the book. We wish to acknowledge to seminal thinking of Martin Clarke in the area of nonlinear filtering and his influence on our work. Special thanks go to René Boel for his review of the first version of the book and to N. Krylov for supplying corrections to the first printing.

The support of the Cooperative Research Centre for Robust and Adaptive Systems, the Boeing Commercial Airplane Company, and the NSERC Grant A7964 are gratefully acknowledged. We acknowledge the typing support of Shelley Hey, and Marita Rendina, and LaTeX programming support of James Ashton.

Contents

Part I
Introduction

Chapter 1
Hidden Markov Model Processing

1.1 Models, Objectives, and Methods

The term *hidden Markov model* (HMM) is now quite familiar in the speech signal processing community and is gaining acceptance for communication systems. It is perhaps a little less daunting, and yet more mysterious, than the term *partially observed stochastic dynamical system model*, which is a translation familiar to people in systems theory, or its applications areas of estimation and control theory. The term HMM is frequently restricted to models with states and measurements in a discrete set and in discrete time, but here we allow relaxation of these restrictions. We first work with the more restricted HMM class, termed here a *discrete* HMM, and then show how to cope with the more general stochastic dynamical systems.

The term *estimation* is used to cover signal filtering, model parameter identification, state estimation, signal smoothing, and signal prediction. *Control* refers to selecting actions which effect the signal-generating system in such a way as to achieve certain control *objectives*. The control actions can be based on on-line signal processing to achieve *feedback control* or by off-line calculations to achieve *feedforward* or *open-loop control*.

The term *reference probability methods* refers to a procedure where a probability measure change is introduced to reformulate the original estimation and control task in a fictitious world, but so that well-known results for identically and independently distributed (i.i.d.) random variables can be applied. Then the results are reinterpreted back to the real world with the original probability measure.

1.2 Book Outline

In the next part of this book, we first work with discrete-time, discrete-state HMMs to achieve optimal estimation algorithms via reference probability methods, and then repeat the discussion with HMMs of increasing complexity. Continuous-range states and continuous-time HMM models are studied, and indeed two-dimensional (image) estimation is developed. In the second part of the book, the focus is on

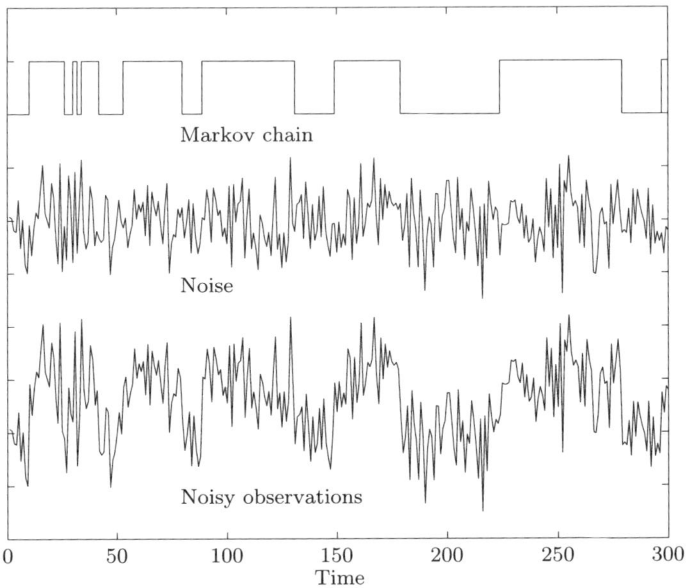

Fig. 2.1 Binary Markov chain in noise

optimal control algorithms. At times, certain application tasks are studied to give a realistic measure of the significance of the results. We have sought to inspire both the interest of engineering students in the mathematics, and the curiosity of the mathematics students in the engineering or science applications.

Consider for example the situation depicted in Figure 2.1. Here, a binary message sequence $X_k, k = 1, 2, \ldots$ consists of ones and zeros. Perhaps the binary signal X (a Markov chain) is transmitted on a noisy communications channel such as a radio channel, and so when it is detected at the receiver, the resultant signal is Y_k (quantized to 15 levels in this case). The middle trace of Figure 2.1 depicts the additive noise in the channel, and the lower trace gives the received signal quantized to 15 levels.

In the first instance, we develop optimal estimation algorithms for discrete HMMs which are discrete in time, in the state, and in the measurement space. They have state space models in terms of processes X_k and Y_k defined for all $k \in \mathbb{N}$, the set of positive integers, with dynamics:

$$X_{k+1} = AX_k + V_{k+1}, \tag{2.1}$$

$$Y_{k+1} = CX_k + W_{k+1}. \tag{2.2}$$

Here the states X_k and measurements Y_k are *indicator functions* as

$$X_k \in S_X = \{e_1, e_2, \ldots, e_N\}, \qquad Y_k \in S_Y = \{f_1, f_2, \ldots, f_M\},$$

where e_i (resp. f_i) is the unit vector with unity in the ith position and zeros elsewhere. The matrices A, C consist of transition probabilities and so have elements a_{ij}, c_{ij} in $\mathbb{R}^+$ and are such that $\sum_{i=1}^N a_{ij} = \sum_{i=1}^M c_{ij} = 1$, or equivalently, so that with $\underline{1} = (1, 1, \dots, 1)'$, then

$$\begin{aligned} \underline{1}'A &= \underline{1}, & \underline{1}'X_k &= 1, & \underline{1}'V_k &= 0, \\ \underline{1}'C &= \underline{1}, & \underline{1}'Y_k &= 1, & \underline{1}'W_k &= 0. \end{aligned}$$

Here the prime denotes transposition. The random *noise* terms of the model are V_k and W_k; these are martingale increment processes. We term V_k the driving noise and W_k the measurement noise. If $Y_k = X_k$, then the state is no longer hidden.

Actually, the HMM model above is more general than might first appear. Consider a process $\mathscr{X}_k$ with its state space being an arbitrary finite set $S_{\mathscr{X}} = \{s_1, \dots, s_N\}$, which are polytope vertices, as depicted in Figure 2.2 for the case $N = 3$. By considering the "characteristic" or "indicator" functions $\phi_k(s_i)$, defined so that $\phi_k(s_i) = 0$ if $i \neq k$ and $\phi_k(s_k) = 1$, and writing $X_k := (\phi_1(\mathscr{X}_k), \dots, \phi_N(\mathscr{X}_k))$ we see that at any time k just one component of X_k is one and the others are zero. Therefore, we can consider the process X_k, derivative from $\mathscr{X}_k$, whose state space is the set $S_X = \{e_1, \dots, e_N\}$ of unit (column) vectors $e_i = (0, \dots, 1, 0, \dots, 0)'$ of $\mathbb{R}^N$, which are simplex vertices, as depicted in Figure 2.2 for the case $N = 3$. So without loss of generality, the state space of X can be taken to be the set of unit vectors e_i which has all elements zero, save unity in the ith position. Similarly, the state space of the finite-state process Y can be taken to be a set of standard unit vectors $S_Y = \{f_1, \dots, f_M\}$. For those indicator functions, it turns out that expectations are probabilities in that $P(X = e_i) = E[\langle X, e_i \rangle]$ where $\langle X, e_i \rangle$ is the ith element X^i of X. Also, nonlinear operations on indicator functions X are linear (affine) in X.

The *estimation* task of prime interest for discrete HMMs is to estimate the sequence $\{X_k\}$ from the measurement sequence $\{Y_k\}$ in some optimal way. *Filtered estimates*, also termed *forward estimates*, are denoted $\hat{X}_k$ and are estimates at time k of X_k based on processing past and present measurements $\{Y_1, Y_2, \dots, Y_k\}$. The elements of $\hat{X}_k$ are conditional probabilities so that $\underline{1}'\hat{X}_k = 1$. Thus, $\hat{X}_k$ lies within the simplex Δ_N depicted in Figure 2.2. The estimates $\hat{X}_k$ are frequently calculated

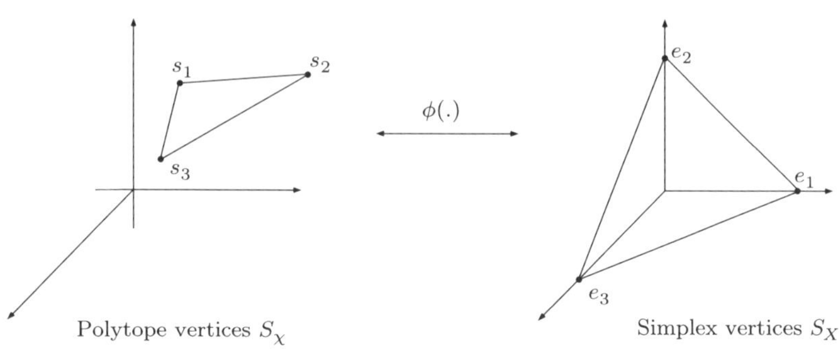

Fig. 2.2 Depiction of state sets

(often in *unnormalized* form), as a forward recursion. *Smoothed estimates* are estimates at time k of X_k based on processing past, present, and future measurements $\{Y_1, \ldots, Y_k, \ldots, Y_m\}$ with $m > k$. *Backward estimates* are those based only on the future measurements. These are usually calculated as a backward recursion from the end of the batch of measurement data. Of course, forward and backward estimates can be combined to yield smooth estimates, where estimation is based on past, present, and future measurements. The simplest situation is when the parameters a_{ij}, c_{ij} are assumed known.

Related estimation problems concern the expected number of jumps (transitions) $\mathscr{J}_k$ and state occupation times $\mathscr{O}_k$. These, in turn, allow a re-estimation of signal model parameters a_{ij}, c_{ij}, should these not be known precisely in advance but only be given prior estimates.

Actually, the schemes that generate the filtered and smoothed estimates work in the first instance with *conditional state estimates*, i.e., estimates of the probability that a particular Markov chain state occurs at each time instant given the measurements. *Forward (filtered)* estimates are illustrated for our example above in Figure 2.3, *backward estimates* in Figure 2.4, and *smoothed* estimates in Figure 2.5.

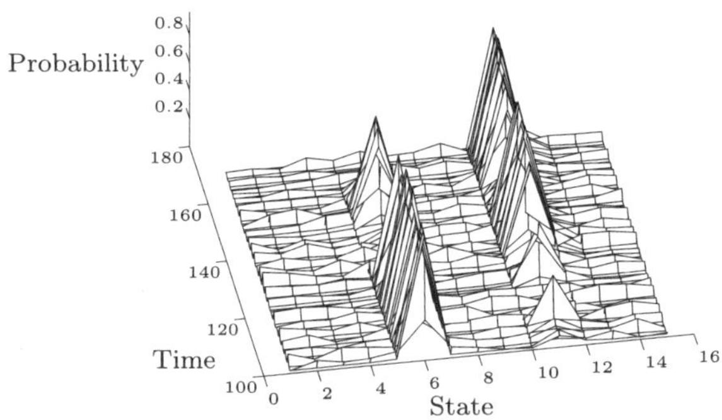

Fig. 2.3 Evolution of forward (filtered) estimate

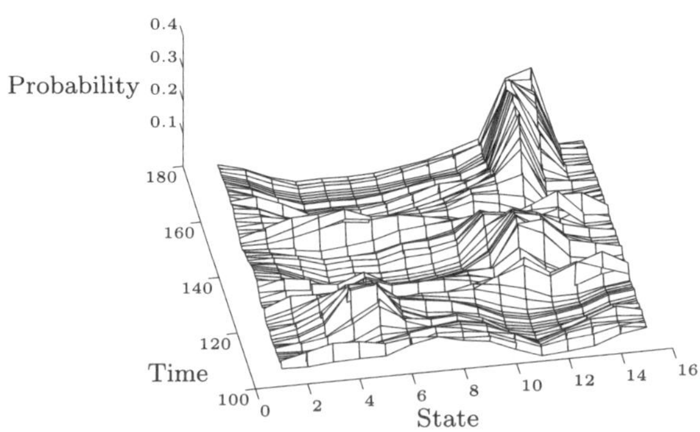

Fig. 2.4 Evolution of backward estimates

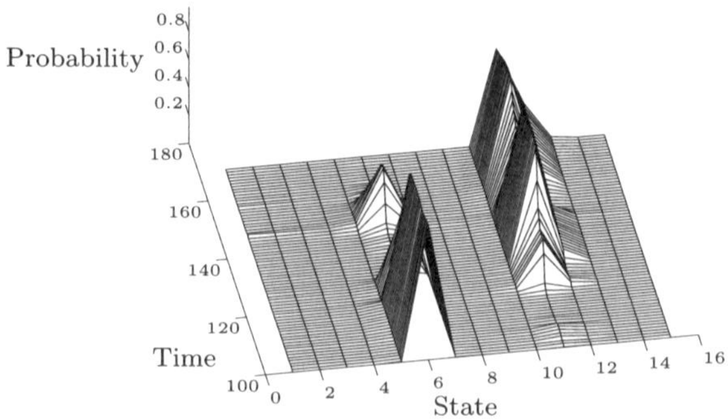

Fig. 2.5 Evolution of smoothed estimates

The *maximum a posteriori probability* (MAP) *state estimates* are the states at which the a posteriori estimates are maximized, i.e., they follow the peaks in the figures.

Such *parameter estimation* can be achieved in a *multipass estimation* procedure in which the batch of data is processed based on the previous estimates to give improved estimates. The familiar *expectation maximization* (EM) algorithm arises in this context. Also studied is *recursive*, or *on-line estimation*, where improved estimates are calculated as each new measurement arrives.

The advantage of working with discrete HMMs is that the optimal estimation algorithms are finite-dimensional with the dimension independent of the length of the measurement sequence.

A basic technique used throughout the book is a change-of-probability measure. This is a discrete-time version of Girsanov's Theorem (Elliott and Yang 1992); see also Appendix A. A new probability $\overline{P}$ is defined such that under $\overline{P}$ the observations are independent (and often identically distributed) random variables. Calculations take place in the *mathematically ideal* world of $\overline{P}$ using Fubini's Theorem which allows interchange of expectations and summations; see Loève (1978) and also Appendix A. They are then related to the *real world* by an inverse change of measure. The situation is depicted in Figure 2.6, and contrasts the direct optimal filter derivation approach of Figure 2.7.

As this book seeks to demonstrate, in discrete time this Girsanov approach brings with it many rewards. In continuous time, where stochastic integrals are involved, similar techniques can be used, though they require careful and detailed analysis.

To illustrate this reference measure approach in a very simple situation, consider a coin for which the probability of heads is p and the probability of tails is q. It is well known that the simple probability space which can be used to describe the one throw of such a coin is $\Omega = \{H, T\}$, with a probability measure P such that $P(H) = p$, $P(T) = q = 1 - p$. (Here we suppose p is neither 0 nor 1.) Suppose now we wish to adjust our statistics in coin tossing experiments to that of a fair coin. We can achieve this mathematically by introducing a new probability measure $\overline{P}$ such that $\overline{P}(H) = 1/2 = \overline{P}(T)$. This implies the event $\{H\}$ had been weighted by a factor $\overline{P}(H)/P(H) = 1/2p$, and the event $\{T\}$ has been weighted by a factor

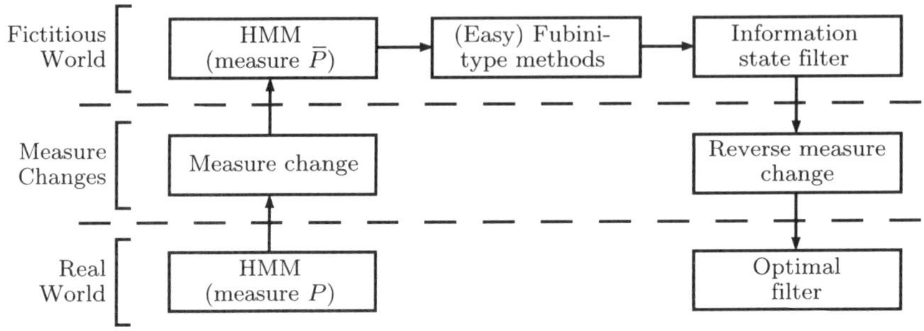

Fig. 2.6 Reference probability optimal filter derivation

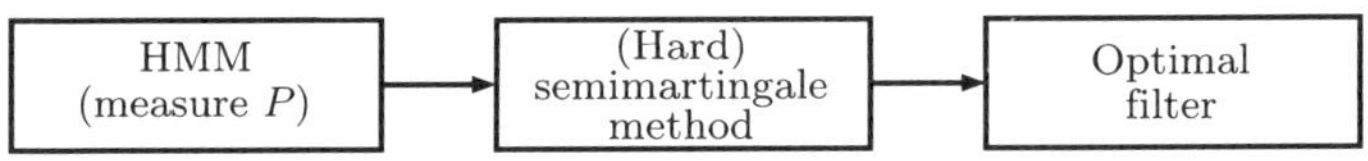

Fig. 2.7 Direct optimal filter derivation

$\overline{P}(T)/P(T) = 1/2q$. The function $\overline{P}(.)/P(.)$ is the *Radon-Nikodym derivative* of the fair (uniform) (1/2, 1/2) $\overline{P}$-measure against the (p,q), P-measure; in fact, the function $\overline{P}(.)/P(.)$ can be used to define $\overline{P}$ because clearly

$$\overline{P}(.) = \frac{\overline{P}(.)}{P(.)}P(.).$$

One can work in the *fictitious world* with the probability $\overline{P}$ to achieve various mathematical objectives, and then reinterpret these results back in the *real world* with a measure change back to P via the inverse Radon-Nikodym derivative.

The reference probability approach to estimation is first discussed for the simplest of discrete HMM models studied in Chapter 2. In Chapter 3, we consider the case where the observations are not restricted to a finite set, but have a continuous range.

In Chapters 4 and 5 HMMs are studied with states and measurements in a continuous range so we deal with the general discrete-time models

$$\begin{aligned} x_{k+1} &= a(x_k) + v_{k+1}, \\ y_k &= c(x_k) + w_k, \end{aligned}$$

where $x_k \in \mathbb{R}, a(.), c(.)$ are nonlinear functions, and v_k, w_k are noise disturbances in a continuous range. Even more general HMMs are also considered. The well-known *Kalman filter* emerges as a special case of the results. As a convention throughout the book, we denote discrete range variables by uppercase variables.

In Chapter 6, more attention is focused on asymptotically optimum model parameter estimation and suboptimal filters to achieve practical adaptive schemes such as might be used in communication systems. A key idea is to work with the so-called

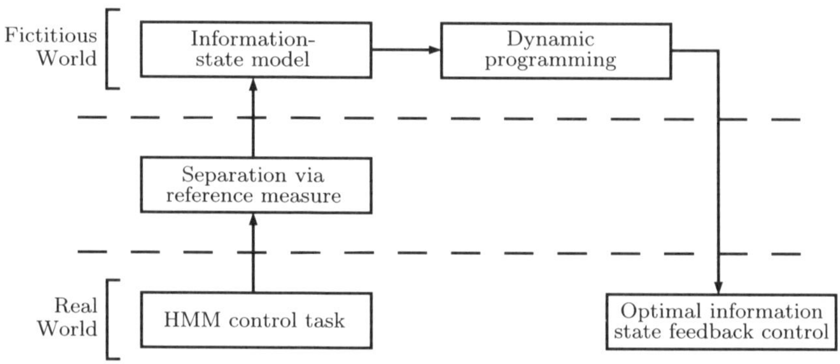

Fig. 2.8 Separation principle

information-state signal models derived using HMM estimation theory, and then apply conditional Kalman filters.

In Chapters 7 and 8, continuous-time HMMs are considered to illustrate how the discrete-time theory can be generalized to the continuous-time framework. Again, reference probability methods are seen to simplify the technical approach.

In Chapter 9, the estimation techniques are developed for two-dimensional image processing. The reference measure method is applied here in a *hidden Markov random field* situation. This is a two-dimensional version of the Markov chain framework. Our estimators and filters are no longer recursive in the usual sense for scanned image data, but we develop techniques to obtain and update them in an optimal manner. The results, which are a natural extension of those in earlier chapters, are presented in an open-ended manner, since at this stage it is not clear if they will provide any significant contributions to the area of image processing.

In the last part of the book concerning *optimal control* our key objective is to show that *optimal feedback* control laws can be formulated for quite general HMMs. These are well-known ideas discussed, for example, in Kumar and Varaiya (1986b) and Bertsekas (1987). Both risk neutral and risk-sensitive situation are considered. In the risk-sensitive case, the feedback is in terms of an *information-state estimate* which takes account of the cost, rather than in terms of estimates of the states themselves. The information-state estimates give the total information about the model states available in the measurements.

The feedback optimal control results arise quite naturally from the optimal estimation results and associated methodologies of the first part of the book together with appropriate applications of the *principle of optimality* and *dynamic programming*, so familiar in optimal control theory. Indeed, the control results are a triumph of the probability reference methods, which allows a representation as depicted in Figures 2.8, 2.9, and 2.10, and in turn validates this approach for optimal estimation theory. Moreover, once stochastic control problems have been discussed, we find that optimality, and indeed robustness, results for deterministic models can be achieved by certain limiting operations; see Figure 2.11.

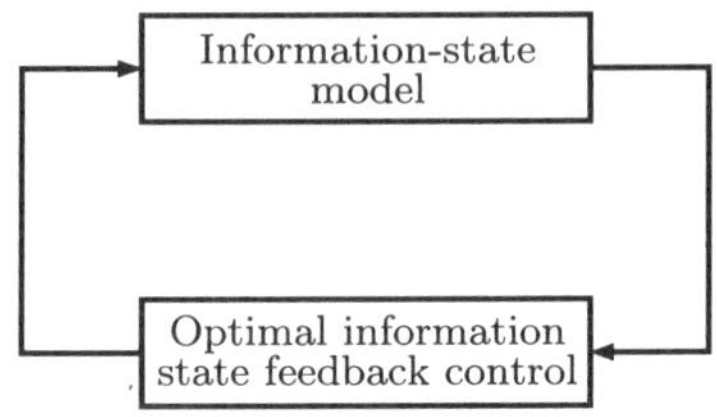

Fig. 2.9 Information-state feedback control

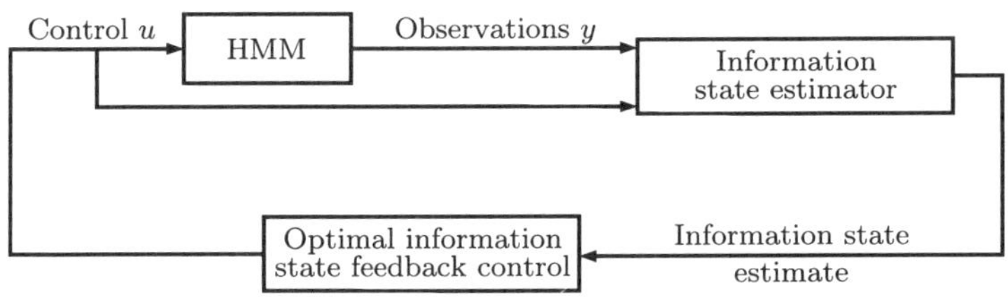

Fig. 2.10 Information-state estimate feedback control

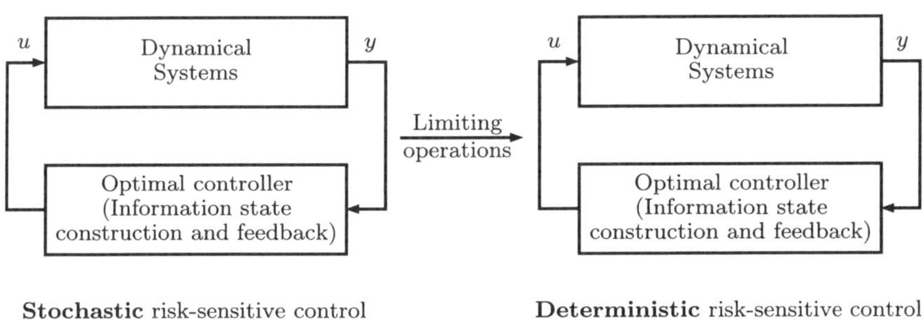

Fig. 2.11 Nonlinear risk-sensitive (H^2, H^∞) control

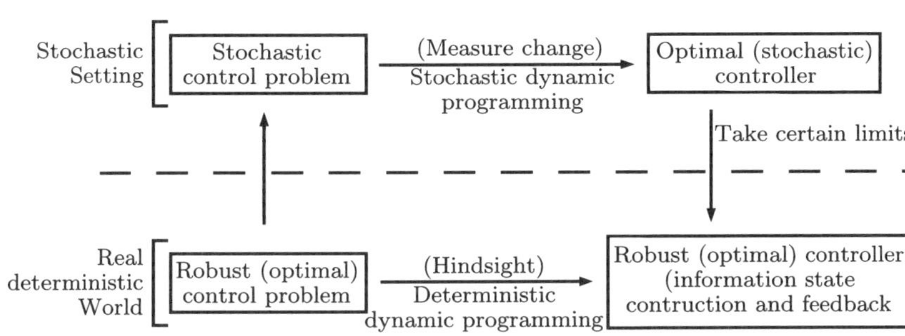

Fig. 2.12 Application of stochastic theory to deterministic problems

Indeed, a hindsight approach allows a new optimal/robust feedback nonlinear deterministic control theory to develop; see Figure 2.12. This material is mentioned in Chapter 11, but is not fleshed out in detail. Furthermore, the task of achieving practical finite-dimensional feedback controllers for nonlinear stochastic (and thereby deterministic) systems, could be tackled via an optimal theory for nonlinear models linearized with respect to the states.

Part II
Discrete-Time HMM Estimation

Chapter 2
Discrete States and Discrete Observations

2.1 Introduction

In this chapter, we deal with signals denoted by $\{X_k\}$, $k \in \mathbb{N}$ in *discrete time*. These signals are further restricted to a discrete set and are thus termed *discrete-state* signals. They transit between elements in this set with transition probabilities dependent only on the previous state, and so are *Markov chains*. The *transition probabilities* are independent of time, and so the Markov chains are said to be *homogeneous*. The Markov chain is not observed directly; rather there is a discrete-time, finite-state observation process $\{Y_k\}, k \in \mathbb{N}$, which is a noisy function of the chain. Consequently, the Markov chain is said to be *hidden* in the observations.

Our objective is to estimate the state of the chain, given the observations. Our preference is to achieve such estimation on-line in an optimal recursive manner, using what we term optimal estimators. The term *estimator* covers the special cases of *on-line filters*, where the estimates are calculated as the measurements are received, *on-line predictors* where there is a prediction at a fixed number of discrete time instants in the future, and *on-line smoothers* where there is improved estimation achieved by using a fixed number of future measurements as well as the previous ones. We also seek recursive filters and smoothers for the number of jumps from one state to another, for the occupation time of a state, and for a process related to the observations.

In the first instance, we assume that the equations describing the HMM are known. However, if this is not the case, it is possible to estimate the parameters also on-line and so achieve *adaptive* (or *self-tuning*) estimators. Unfortunately, it is usually not practical to achieve optimal adaptive estimators. In seeking practical suboptimal schemes, a *multipass scheme* is to update the parameters estimates only after processing a large data set, perhaps the entire data set. At the end of each pass through this data set, the parameter estimates are updated, to yield improved parameter estimates; see, for example, the so-called expectation maximization (EM) scheme; see Dempster, Laird and Rubin (1977). Our approach requires only a forward pass through the data to achieve parameter updates, in contrast to earlier so-called *forward-backward* algorithms of the Baum-Welch type (Baum and Petrie 1966).

Hidden Markov models have been found useful in many areas of probabilistic modeling, including speech processing; see Rabiner (1989). We believe our model is of wide applicability and generality. Many state and observation processes of the form (2.14) arise in the literature. In addition, certain time-series models can be approximated by HMMs.

As mentioned in the introduction, one of the fundamental techniques employed throughout this book is the discrete-time *change of measure*. This is a version of *Girsanov's Theorem* (see Theorem A.1.2). It is developed for the discrete-state HMM in Section 2.3 of this chapter.

A second basic observation is the *idempotent property* of the indicator functions for the state space of the Markov chain. With X one of the unit (column) vectors e_i, $1 \le i \le N$, prime denoting transpose, and using the inner product notation $\langle a,b\rangle = a'b$, this idempotent property allows us to write the square XX' as $\sum_{i=1}^{N} \langle X, e_i\rangle e_i e_i'$ and so obtain *closed (finite-dimensional), recursive filters* in Sections 2.4–2.9. More generally, any real function $f(X)$ can be expressed as a linear functional $f(X) = \langle f, X\rangle$ where $\langle f, e_i\rangle = f(e_i) = f_i$ and $f = (f_1, \dots, f_N)$. Thus with $X^i = \langle X, e_i\rangle$,

$$f(X) = \sum_{i=1}^{N} f(e_i) X^i = \sum_{i=1}^{N} f_i X^i. \tag{1.1}$$

For the vector of indicator functions X, note that from the definition of expectations of a simple random variable, as in Appendix A,

$$E[\langle X, e_i\rangle] = \sum_{j=1}^{N} \langle e_j, e_i\rangle P(X = e_j) = P(X = e_i). \tag{1.2}$$

Section 2.10 of this chapter discusses similar estimation problems for a discrete-time, discrete-state hidden Markov model in the case where the noise terms in the Markov chain X and observation process Y are not independent. A test for independence is given. This section may be omitted on a first reading.

2.2 Model

All processes are defined initially on a *probability space* $(\Omega, \mathscr{F}, P)$. Below, a new probability measure $\overline{P}$ is defined. See Appendix A for related background in probability theory.

A system is considered whose state is described by a finite-state, homogeneous, discrete-time Markov chain X_k, $k \in \mathbb{N}$. We suppose X_0 is given, or its distribution known. If the state space of X_k has N elements it can be identified without loss of generality, with the set

$$S_X = \{e_1, \dots, e_N\}, \tag{2.1}$$

where e_i are unit vectors in $\mathbb{R}^N$ with unity as the ith element and zeros elsewhere.

Write $\mathscr{F}_k^0 = \sigma\{X_0,\dots,X_k\}$, for the *σ-field* generated by $X_0,\dots,X_k$, and $\{\mathscr{F}_k\}$ for the *complete filtration* generated by the $\mathscr{F}_k^0$; this augments $\mathscr{F}_k^0$ by including all subsets of events of probability zero. Again, see Appendix A for related background in probability theory. The *Markov property* implies here that

$$P(X_{k+1} = e_j \mid \mathscr{F}_k) = P(X_{k+1} = e_j \mid X_k).$$

Write

$$a_{ji} = P(X_{k+1} = e_j \mid X_k = e_i), \qquad A = (a_{ji}) \in \mathbb{R}^{N\times N} \tag{2.2}$$

so that using the property (1.2), then

$$E[X_{k+1} \mid \mathscr{F}_k] = E[X_{k+1} \mid X_k] = AX_k. \tag{2.3}$$

Define

$$V_{k+1} := X_{k+1} - AX_k. \tag{2.4}$$

So that

$$X_{k+1} = AX_k + V_{k+1}. \tag{2.5}$$

This can be referred to as a *state equation.*

Now observe that taking the *conditional expectation* and noting that $E[AX_k \mid X_k] = AX_k$, we have

$$E[V_{k+1} \mid \mathscr{F}_k] = E[X_{k+1} - AX_k \mid X_k] = AX_k - AX_k = 0,$$

so $\{V_k\}$, $k \in \mathbb{N}$, is a sequence of martingale increments.

The state process X is not observed directly. We suppose there is a function $c(.,.)$ with finite range and we observe the values

$$Y_{k+1} = c(X_k, w_{k+1}), \qquad k \in \mathbb{N}. \tag{2.6}$$

The w_k in (2.6) are a sequence of independent, identically distributed (i.i.d.) random variables, with V_k, w_k being mutually independent.

$\{\mathscr{G}_k^0\}$ will be the σ-field on Ω generated by $X_0, X_1, \dots, X_k$ and $Y_1, \dots, Y_k$, and $\mathscr{G}_k$ its completion. Also $\{\mathscr{Y}_k^0\}$ will be the σ-field on Ω generated by $Y_1, \dots, Y_k$ and $\mathscr{Y}_k$ its completion. Note $\mathscr{G}_k \subset \mathscr{G}_{k+1} \subset \cdots$ and $\mathscr{Y}_k \subset \mathscr{Y}_{k+1} \subset \cdots$. The increasing family of σ-fields is called a *filtration*. A function is $\mathscr{G}_k^0$-measurable if and only if it is a function of $X_0, X_1, \dots, X_k, Y_1, \dots, Y_k$. Similarly, for $\mathscr{Y}_k^0$, $\mathscr{Y}_k$. See also Appendix A.

The w_k in (2.6) are a sequence of independent, identically distributed (i.i.d.) random variables, with V_k, w_k being mutually independent. The pair of processes (X_k, Y_k), $k \in \mathbb{N}$, provides our first, basic example of a hidden Markov model, or HMM. This term is appropriate because the Markov chain is not observed directly but, instead, is hidden in the noisy observations Y. In this HMM the time parameter is discrete and the state spaces of both X and Y are finite (and discrete). Note that there is a unit delay between the state X at time k and its measurement Y at time $k+1$. A *zero delay observation model* is discussed later in this chapter.

Suppose the range of $c(.,.)$ consists of M points. Then we can identify the range of $c(.,.)$ with the set of unit vectors

$$S_Y = \{f_1, \dots, f_M\}, \qquad f_j = (0, \dots, 1, \dots, 0)' \in \mathbb{R}^M, \tag{2.7}$$

where the unit element is the jth element.

We have assumed that $c(.,.)$ is independent of the time parameter k, but the results below are easily extended to the case of a nonhomogeneous chain X and a time-dependent $c(.,.)$.

Now (2.6) implies

$$P(Y_{k+1} = f_j \mid X_0, X_1, \dots, X_k, Y_1, \dots, Y_k) = P(Y_{k+1} = f_j \mid X_k).$$

Write

$$C = (c_{ji}) \in \mathbb{R}^{M \times N}, \qquad c_{ji} = P(Y_{k+1} = f_j \mid X_k = e_i) \tag{2.8}$$

so that $\sum_{j=1}^M c_{ji} = 1$ and $c_{ji} \geq 0$, $1 \leq j \leq M$, $1 \leq i \leq N$. We have, therefore,

$$E[Y_{k+1} \mid X_k] = CX_k. \tag{2.9}$$

If $W_{k+1} := Y_{k+1} - CX_k$, then taking the conditional expectation and noting $E[CX_k \mid X_k] = CX_k$ we have

$$\begin{aligned} E[W_{k+1} \mid \mathscr{G}_k] &= E[Y_{k+1} - CX_k \mid X_k] \\ &= CX_k - CX_k = 0, \end{aligned}$$

so W_k is a $(P, \mathscr{G}_k)$ martingale increment and

$$Y_{k+1} = CX_k + W_{k+1}. \tag{2.10}$$

Equation (2.10) can be thought of as an *observation equation*. The case where, given $\mathscr{G}_k$, the noise terms W_k in the observations Y_k are possibly correlated with the noise terms V_k in the Markov chain will be considered in Section 2.10.

Notation 2.1 *Write* $Y_k^i = \langle Y_k, f_i \rangle$ *so* $Y_k = \left(Y_k^1, \dots, Y_k^M\right)'$, $k \in \mathbb{N}$. *For each* $k \in \mathbb{N}$, *exactly one component is equal to* 1, *the remainder being* 0.

Note $\sum_{i=1}^M Y_k^i = 1$. Write $c_{k+1}^i = E\left[Y_{k+1}^i \mid \mathscr{G}_k\right] = \sum_{j=1}^N c_{ij} \langle e_j, X_k \rangle$ and $c_{k+1} = (c_{k+1}^1, \dots, c_{k+1}^M)'$. Then

$$c_{k+1} = E[Y_{k+1} \mid \mathscr{G}_k] = CX_k. \tag{2.11}$$

We shall suppose initially that $c_k^i > 0$, $1 \leq i \leq M$, $k \in \mathbb{N}$. (See, however, the construction of P from $\overline{P}$ in Section 2.3). Note $\sum_{i=1}^M c_k^i = 1$, $k \in \mathbb{N}$. We shall need the following result in the sequel.

Lemma 2.2 *With* diag(z) *denoting the diagonal matrix with vector* z *on its diagonal, we have*

$$\begin{aligned} V_{k+1}V'_{k+1} &= \operatorname{diag}(AX_k) + \operatorname{diag}(V_{k+1}) - A \operatorname{diag} X_k A' \\ &\quad - AX_k V'_{k+1} - V_{k+1}(AX_k)' \end{aligned} \tag{2.12}$$

and

$$\begin{aligned} \langle V_{k+1} \rangle &:= E\left[V_{k+1}V'_{k+1} \mid \mathscr{F}_k\right] \\ &= E\left[V_{k+1}V'_{k+1} \mid X_k\right] \\ &= \operatorname{diag}(AX_k) - A \operatorname{diag} X_k A'. \end{aligned} \tag{2.13}$$

Proof From (2.4)

$$X_{k+1}X'_{k+1} = AX_k(AX_k)' + AX_kV'_{k+1} + V_{k+1}(AX_k)' + V_{k+1}V'_{k+1}.$$

However, $X_{k+1}X'_{k+1} = \operatorname{diag}(X_{k+1}) = \operatorname{diag}(AX_k) + \operatorname{diag}(V_{k+1})$. Equation (2.12) follows. The terms on the right side of (2.12) involving V_{k+1} are martingale increments; conditioning on X_k we see

$$\langle V_{k+1} \rangle = E\left[V_{k+1}V'_{k+1} \mid X_k\right] = \operatorname{diag}(AX_k) - A \operatorname{diag} X_k A'.$$

□

Similarly, we can show that

$$\langle W_{k+1} \rangle := E\left[W_{k+1}W'_{k+1} \mid \mathscr{G}_k\right] = \operatorname{diag}(CX_k) - C \operatorname{diag} X_k C'.$$

In summary then, we have the following state space signal model for a Markov chain hidden in noise with discrete measurements.

Discrete HMM *The discrete HMM under P has the* state space equations

$$\boxed{\begin{aligned} X_{k+1} &= AX_k + V_{k+1}, \\ Y_{k+1} &= CX_k + W_{k+1}, \qquad k \in \mathbb{N}, \end{aligned}} \tag{2.14}$$

where $X_k \in S_X$, $Y_k \in S_Y$, A and C are matrices of transition probabilities given in (2.2) *and* (2.8). *The entries satisfy*

$$\sum_{j=1}^{N} a_{ji} = 1, \qquad a_{ji} \geq 0, \tag{2.15}$$

$$\sum_{j=1}^{M} c_{ji} = 1, \qquad c_{ji} \geq 0. \tag{2.16}$$

V_k and W_k are martingale increments satisfying

$$\begin{aligned} E\left[V_{k+1} \mid \mathscr{F}_k\right] = 0, &\qquad E\left[W_{k+1} \mid \mathscr{G}_k\right] = 0, \\ \langle V_{k+1} \rangle := E\left[V_{k+1}V'_{k+1} \mid X_k\right] &= \operatorname{diag}(AX_k) - A \operatorname{diag} X_k A', \\ \langle W_{k+1} \rangle := E\left[W_{k+1}W'_{k+1} \mid X_k\right] &= \operatorname{diag}(CX_k) - C \operatorname{diag} X_k C'. \end{aligned}$$

2.3 Change of Measure

The idea of introducing new probability measures, as outlined in the previous chapter, is now discussed for the observation process Y. This measure change concept is the key to many of the results in this and the following chapters.

We assume, for this measure change, $c_\ell^i > 0$, $1 \le i \le M$, $\ell \in \mathbb{N}$. This assumption, in effect, is that given any $\mathscr{G}_k$, the observation noise is such that there is a nonzero probability that $Y_{k+1}^i > 0$ for all i. This assumption is later relaxed to achieve the main results of this section. Define

$$\lambda_\ell = \sum_{i=1}^{M} \left(\frac{M^{-1}}{c_\ell^i} \right) \langle Y_\ell, f_i \rangle, \tag{3.1}$$

and

$$\Lambda_k = \prod_{\ell=1}^{k} \lambda_\ell. \tag{3.2}$$

Note that $Y_\ell^i = 1$ for only one i at each ℓ, and $Y_\ell^i = 0$ otherwise, so that λ_ℓ is merely the product of unity terms and one nonunity term. Consequently, since λ_k is a nonlinear function of Y_k, then property (1.1) tells us that $\lambda_k = \lambda_k(Y_k) = \sum_{i=1}^{M} Y_k^i / M c_k^i$.

Lemma 3.1 *With the above definitions*

$$E[\lambda_{k+1} \mid \mathscr{G}_k] = 1. \tag{3.3}$$

Proof Applying the properties (1.1) and (1.2),

$$\begin{aligned}
E[\lambda_{k+1} \mid \mathscr{G}_k] &= E\left[\sum_{i=1}^{M} \frac{1}{M c_{k+1}^i} Y_{k+1}^i \,\middle|\, \mathscr{G}_k \right] \\
&= \frac{1}{M} \sum_{i=1}^{M} \frac{1}{c_{k+1}^i} P\left(Y_{k+1}^i = 1 \mid \mathscr{G}_k\right) \\
&= \frac{1}{M} \sum_{i=1}^{M} \frac{1}{c_{k+1}^i} \cdot c_{k+1}^i = 1.
\end{aligned}$$

Here as in many places, we interchange expectations and summations, for a simple random variable. This is permitted, of course, by a special case of Fubini's Theorem; see Loève (1978) and Appendix A. □

We now define a new probability measure $\overline{P}$ on $(\Omega, \bigvee_{\ell=1}^{\infty} \mathscr{G}_\ell)$ by putting the restriction of the Radon-Nikodym derivative $d\overline{P}/dP$ to the σ-field $\mathscr{G}_k$ equal to Λ_k. Thus

$$\left. \frac{d\overline{P}}{dP} \right|_{\mathscr{G}_k} = \Lambda_k. \tag{3.4}$$

[The existence of $\overline{P}$ follows from *Kolmogorov's Extension Theorem* (Kolmogorov 1933)]; see also Appendix A. This means that, for any set $B \in \mathscr{G}_k$,

$$\overline{P}(B) = \int_B \Lambda_k \, dP.$$

Equivalently, for any $\mathscr{G}_k$*-measurable* random variable ϕ

$$\overline{E}[\phi] = \int \phi \, d\overline{P} = \int \phi \frac{d\overline{P}}{dP} dP = \int \phi \Lambda_k \, dP = E[\Lambda_k \phi], \tag{3.5}$$

where $\overline{E}$ and E denote expectations under $\overline{P}$ and P, respectively. In the discrete-state case under consideration, $d\overline{P}/dP$ reduces to the ratio $\overline{P}/P$ and the integrations reduce to sums. This equation exhibits the basic idea of the change of measure; for most of the results in this book a big challenge is to determine the appropriate forms for λ and Λ. It is not straightforward to give insight into this process other than to illustrate by examples and present hindsight proofs. Perhaps the measure changes of Chapter 3 are the most transparent, and more discussion is given for these.

We now give a conditional form of *Bayes' Theorem* which is fundamental for the results that follow. The result relates conditional expectations under two different measures. Recall that ϕ is *integrable* if $E|\phi| < \infty$. First we shall consider a simple case.

Consider the experiment of throwing a die. The set of outcomes is $\Omega = \{1, 2, \ldots, 6\}$. Suppose the die is not necessarily balanced, so that the probability of i showing is $P(i) = p_i$, $p_1 + \cdots + p_6 = 1$.

The σ-field $\mathscr{F}$ associated with this experiment is the collection of all subsets of Ω, including the empty set ϕ. The sets in $\mathscr{F}$ are the *events*. (See also Appendix A.) The probability of the event "odd number," for instance, is $P\{1,3,5\} = p_1 + p_3 + p_5$. Consider the sub-$\sigma$-field $\mathscr{G}$ of $\mathscr{F}$ defined by $\mathscr{G} = \{\Omega, \phi, \{1,3,5\}, \{2,4,6\}\}$.

Now suppose ϕ is a real random variable on $(\Omega, \mathscr{F})$, that is, $\phi(i) \in \mathbb{R}$ for $i = 1, 2, \ldots, 6$. The mean, or expected, value of ϕ is then $E[\phi] = \sum_{i=1}^{6} \phi(i) p_i$.

The conditional expected value of ϕ, given $\mathscr{G}$, $E[\phi \mid \mathscr{G}]$, is then a function which is constant on the smallest, nonempty sets of $\mathscr{G}$. That is,

$$E[\phi \mid \mathscr{G}](i) = \frac{\phi(1)p_1 + \phi(3)p_3 + \phi(5)p_5}{p_1 + p_3 + p_5}, \qquad \text{if } i \in \{1,3,5\},$$

$$E[\phi \mid \mathscr{G}](i) = \frac{\phi(2)p_2 + \phi(4)p_4 + \phi(6)p_6}{p_2 + p_4 + p_6}, \qquad \text{if } i \in \{2,4,6\}$$

We note that $\psi = E[\phi \mid \mathscr{G}]$ can be considered a function on $(\Omega, \mathscr{F})$ and that then $E[E[\phi \mid \mathscr{G}]] = E[\phi]$.

Suppose we now rebalance the die by introducing weights $\Lambda(i)$ on the different faces. Note that Λ is itself, therefore, a random variable on $(\Omega, \mathscr{F})$.

Write $\overline{p}_i = \Lambda(i) p_i = \overline{P}(i)$, $i = 1, \ldots, 6$, for the new balance proportion assigned to the ith face. Then, because $\overline{P}$ is to be a probability measure, $E[\Lambda] = \overline{p}_1 + \cdots + \overline{p}_6 = \Lambda(1) p_1 + \cdots + \Lambda(6) p_6 = 1$.

We have the following expressions:

$$E[\Lambda\phi \mid \mathscr{G}](i) = \frac{\phi(1)\Lambda(1)p_1 + \phi(3)\Lambda(3)p_3 + \phi(5)\Lambda(5)p_5}{p_1 + p_3 + p_5}, \qquad \text{if } i \in \{1,3,5\},$$

$$E[\Lambda\phi \mid \mathscr{G}](i) = \frac{\phi(2)\Lambda(2)p_2 + \phi(4)\Lambda(4)p_4 + \phi(6)\Lambda(6)p_6}{p_2 + p_4 + p_6}, \qquad \text{if } i \in \{2,4,6\}.$$

Similarly,

$$E[\Lambda \mid \mathscr{G}](i) = \frac{\Lambda(1)p_1 + \Lambda(3)p_3 + \Lambda(5)p_5}{p_1 + p_3 + p_5}, \qquad \text{if } i \in \{1,3,5\},$$

$$E[\Lambda \mid \mathscr{G}](i) = \frac{\Lambda(2)p_2 + \Lambda(4)p_4 + \Lambda(6)p_6}{p_2 + p_4 + p_6}, \qquad \text{if } i \in \{2,4,6\}.$$

However, with $\overline{E}$ denoting expectation under the new probability $\overline{P}$:

$$\overline{E}[\phi \mid \mathscr{G}](i) = \frac{\phi(1)\overline{p}_1 + \phi(3)\overline{p}_3 + \phi(5)\overline{p}_5}{\overline{p}_1 + \overline{p}_3 + \overline{p}_5}, \qquad \text{if } i \in \{1,3,5\},$$

$$\overline{E}[\phi \mid \mathscr{G}](i) = \frac{\phi(2)\overline{p}_2 + \phi(4)\overline{p}_4 + \phi(6)\overline{p}_6}{\overline{p}_2 + \overline{p}_4 + \overline{p}_6}, \qquad \text{if } i \in \{2,4,6\}.$$

Consequently, $\overline{E}[\phi \mid \mathscr{G}] = E[\Lambda\phi \mid \mathscr{G}]/E[\Lambda \mid \mathscr{G}]$.

We now prove this result in full generality. For background on conditional expectation see Elliott (1982b).

Theorem 3.2 (Conditional Bayes Theorem) *Suppose $(\Omega, \mathscr{F}, P)$ is a probability space and $\mathscr{G} \subset \mathscr{F}$ is a sub-σ-field. Suppose $\overline{P}$ is another probability measure absolutely continuous with respect to P and with Radon-Nikodym derivative $d\overline{P}/dP = \Lambda$. Then if ϕ is any $\overline{P}$ integrable random variable*

$$\overline{E}[\phi \mid \mathscr{G}] = \psi \quad \textit{where} \quad \psi = \frac{E[\Lambda\phi \mid \mathscr{G}]}{E[\Lambda \mid \mathscr{G}]} \quad \textit{if } E[\Lambda \mid \mathscr{G}] > 0$$
$$\textit{and} \quad \psi = 0 \quad \textit{otherwise.}$$

Proof Suppose B is any set in $\mathscr{G}$. We must show

$$\int_B \overline{E}[\phi \mid \mathscr{G}]\, d\overline{P} = \int_B \frac{E[\Lambda\phi \mid \mathscr{G}]}{E[\Lambda \mid \mathscr{G}]}\, d\overline{P}.$$

Define $\psi = E[\Lambda\phi \mid \mathscr{G}]/E[\Lambda \mid \mathscr{G}]$ if $E[\Lambda \mid \mathscr{G}] > 0$ and $\psi = 0$ otherwise. Then $\overline{E}[\phi \mid \mathscr{G}] = \psi$.

Suppose A is any set in $\mathscr{G}$. We must show $\int_A \overline{E}[\phi \mid \mathscr{G}]\, d\overline{P} = \int_A \psi\, d\overline{P}$. Write $G = \{\omega : E[\Lambda \mid \mathscr{G}] = 0\}$, so $G \in \mathscr{G}$. Then $\int_G E[\Lambda \mid \mathscr{G}]\, dP = 0 = \int_G \Lambda\, dP$ and $\Lambda \geq 0$ a.s.

So either $P(G)=0$, or the restriction of Λ to G is 0 a.s. In either case, $\Lambda=0$ a.s. on G.

Now $G^c=\{\omega : E[\Lambda \mid \mathcal{G}]>0\}$. Suppose $A\in\mathcal{G}$; then $A=B\cup C$ where $B=A\cap G^c$ and $C=A\cap G$. Further,

$$\begin{aligned}\int_A \overline{E}[\phi \mid \mathcal{G}]\, d\overline{P} &= \int_A \phi\, d\overline{P} = \int_A \phi\Lambda\, dP \\ &= \int_B \phi\Lambda\, dP + \int_C \phi\Lambda\, dP. \end{aligned} \tag{3.6}$$

Of course, $\Lambda=0$ a.s. on $C\subset G$, so

$$\int_C \phi\Lambda\, dP = 0 = \int_C \psi\, d\overline{P}, \tag{3.7}$$

by definition.

Now

$$\begin{aligned}\int_B \psi\, d\overline{P} &= \int_B \frac{E[\Lambda\phi \mid \mathcal{G}]}{E[\Lambda \mid \mathcal{G}]}\, d\overline{P} \\ &= \overline{E}\left[I_B \frac{E[\Lambda\phi \mid \mathcal{G}]}{E[\Lambda \mid \mathcal{G}]}\right] \\ &= E\left[I_B\Lambda \frac{E[\Lambda\phi \mid \mathcal{G}]}{E[\Lambda \mid \mathcal{G}]}\right] \\ &= E\left[E\left[I_B\Lambda \frac{E[\Lambda\phi \mid \mathcal{G}]}{E[\Lambda \mid \mathcal{G}]} \mid \mathcal{G}\right]\right] \\ &= E\left[I_B E[\Lambda \mid \mathcal{G}] \frac{E[\Lambda\phi \mid \mathcal{G}]}{E[\Lambda \mid \mathcal{G}]}\right] \\ &= E[I_B E[\Lambda\phi \mid \mathcal{G}]] \\ &= E[I_B\Lambda\phi].\end{aligned}$$

That is

$$\int_B \Lambda\phi\, dP = \int_B \psi\, d\overline{P}. \tag{3.8}$$

From (3.6), adding (3.7) and (3.8) we see that

$$\begin{aligned}&\int_C \Lambda\phi\, dP + \int_B \Lambda\phi\, dP = \int_A \Lambda\phi\, dP \\ &= \int_A \overline{E}[\phi \mid \mathcal{G}]\, d\overline{P} = \int_A \psi\, d\overline{P},\end{aligned}$$

and the result follows. □

A sequence $\{\phi_k\}$ is said to be $\mathcal{G}$*-adapted* if ϕ_k is $\mathcal{G}_k$-measurable for every k. Applying Theorem 3.2 result to the P and $\overline{P}$ of (3.4) we have the following:

Lemma 3.3 *If $\{\phi_k\}$ is a $\mathcal{G}$-adapted integrable sequence of random variables, then*

$$\overline{E}[\phi_k \mid \mathcal{Y}_k] = \frac{E[\Lambda_k \phi_k \mid \mathcal{Y}_k]}{E[\Lambda_k \mid \mathcal{Y}_k]}.$$

Lemma 3.4 *Under $\overline{P}$, $\{Y_k\}$, $k \in \mathbb{N}$, is a sequence of i.i.d. random variables each having the uniform distribution that assigns probability $\frac{1}{M}$ to each point f_i, $1 \le i \le M$, in its range space.*

Proof With $\overline{E}$ denoting expectation under $\overline{P}$, using Lemma 3.1, Theorem 3.2 and properties (1.1) and (1.2), then

$$\begin{aligned}
\overline{P}\left(Y_{k+1}^j = 1 \mid \mathcal{G}_k\right) &= \overline{E}\left[\langle Y_{k+1}, f_j\rangle \mid \mathcal{G}_k\right] \\
&= \frac{E\left[\Lambda_{k+1}\langle Y_{k+1}, f_j\rangle \mid \mathcal{G}_k\right]}{E\left[\Lambda_{k+1} \mid \mathcal{G}_k\right]} \\
&= \frac{\Lambda_k E\left[\lambda_{k+1}\langle Y_{k+1}, f_j\rangle \mid \mathcal{G}_k\right]}{\Lambda_k E\left[\lambda_{k+1} \mid \mathcal{G}_k\right]} \\
&= E\left[\lambda_{k+1}\langle Y_{k+1}, f_j\rangle \mid \mathcal{G}_k\right] \\
&= E\left[\prod_{i=1}^{M}\left(\frac{1}{Mc_{k+1}^i}\right)^{Y_{k+1}^i}\langle Y_{k+1}, f_j\rangle \,\middle|\, \mathcal{G}_k\right] \\
&= E\left[\sum_{i=1}^{M}\left(\frac{1}{Mc_{k+1}^i}\right)Y_{k+1}^i Y_{k+1}^j \,\middle|\, \mathcal{G}_k\right] \\
&= \frac{1}{Mc_{k+1}^j} E\left[Y_{k+1}^j \mid \mathcal{G}_k\right] \\
&= \frac{1}{Mc_{k+1}^j} c_{k+1}^j = \frac{1}{M} = \overline{P}\left(Y_{k+1}^j = 1\right),
\end{aligned}$$

a quantity independent of $\mathcal{G}_k$ which finishes the proof. □

Now note that $\overline{E}[X_{k+1} \mid \mathcal{G}_k] = E[\Lambda_{k+1}X_{k+1} \mid \mathcal{G}_k]/E[\Lambda_{k+1} \mid \mathcal{G}_k] = E[\lambda_{k+1}X_{k+1} \mid \mathcal{G}_k] = AX_k$ so that under $\overline{P}$, X remains a Markov chain with transition matrix A.

A Reverse Measure Change

What we wish to do now is start with a probability measure $\overline{P}$ on $(\Omega, \bigvee_{n=1}^{\infty}\mathcal{G}_n)$ such that

1. the process X is a finite-state Markov chain with transition matrix A and
2. $\{Y_k\}$, $k \in \mathbb{N}$, is a sequence of i.i.d. random variables and

$$\overline{P}\left(Y_{k+1}^j = 1 \mid \mathcal{G}_k\right) = \overline{P}\left(Y_{k+1}^j = 1\right) = \frac{1}{M}.$$

Suppose $C=(c_{ji})$, $1\le j\le M$, $1\le i\le N$ is a matrix such that $c_{ji}\ge 0$ and $\sum_{j=1}^{M} c_{ji}=1$.

We shall now construct a new measure P on $(\Omega, \bigvee_{n=1}^{\infty}\mathscr{G}_n)$ such that under P, (2.14) still holds and $E\left[Y_{k+1} \mid \mathscr{G}_k\right]=CX_k$. We again write

$$c_{k+1}=CX_k$$

and $c^i_{k+1}=\langle c_{k+1}, f_i\rangle=\langle CX_k, f_i\rangle$, so that

$$\sum_{i=1}^{M} c^i_{k+1}=1. \tag{3.9}$$

The construction of P from $\overline{P}$ is inverse to that of $\overline{P}$ from P. Write

$$\overline{\lambda}_\ell=\prod_{i=1}^{M}\left(Mc^i_\ell\right)^{Y^i_\ell}, \qquad \ell\in\mathbb{N}, \tag{3.10}$$

and

$$\overline{\Lambda}_k=\prod_{\ell=1}^{k}\overline{\lambda}_\ell. \tag{3.11}$$

Lemma 3.5 *With the above definitions*

$$\overline{E}\left[\overline{\lambda}_{k+1} \mid \mathscr{G}_k\right]=1. \tag{3.12}$$

Proof Following the proof of Lemma 3.5

$$\begin{aligned}\overline{E}\left[\overline{\lambda}_{k+1} \mid \mathscr{G}_k\right] &= \overline{E}\left[\prod_{i=1}^{M}\left(Mc^i_{k+1}\right)^{Y^i_{k+1}} \mid \mathscr{G}_k\right]\\ &= M\sum_{i=1}^{M} c^i_{k+1}\overline{P}\left(Y^i_{k+1}=1 \,\middle|\, \mathscr{G}_k\right)\\ &= M\sum_{i=1}^{M}\frac{c^i_{k+1}}{M}=\sum_{i=1}^{M} c^i_{k+1}=1,\end{aligned}$$

□

This time set

$$\left.\frac{dP}{d\overline{P}}\right|_{\mathscr{G}_k}=\overline{\Lambda}_k. \tag{3.13}$$

[The existence of P follows from Kolmogorov's Extension Theorem (Kolmogorov 1933); see also Appendix A.]

Lemma 3.6 *Under P,*

$$E\left[Y_{k+1} \mid \mathscr{G}_k\right]=CX_k.$$

Proof Using Theorem 3.2 and the now familiar properties (1.1) and (1.2), then

$$\begin{aligned}
P\left(Y_{k+1}^{j}=1 \mid \mathscr{G}_k\right) &= E\left[\left\langle Y_{k+1}, f_j\right\rangle \mid \mathscr{G}_k\right] \\
&= \frac{\overline{E}\left[\overline{\Lambda}_{k+1}\left\langle Y_{k+1}, f_j\right\rangle \mid \mathscr{G}_k\right]}{\overline{E}\left[\overline{\Lambda}_{k+1} \mid \mathscr{G}_k\right]} \qquad (\text{case } \overline{\Lambda} \neq 0) \\
&= \frac{\overline{E}\left[\overline{\lambda}_{k+1}\left\langle Y_{k+1}, f_j\right\rangle \mid \mathscr{G}_k\right]}{\overline{E}\left[\overline{\lambda}_{k+1} \mid \mathscr{G}_k\right]} \\
&= \overline{E}\left[\prod_{i=1}^{M}\left(Mc_{k+1}^{i}\right)^{Y_{k+1}^{i}}\left\langle Y_{k+1}, f_j\right\rangle \,\middle|\, \mathscr{G}_k\right] \\
&= M\overline{E}\left[c_{k+1}^{j}\left\langle Y_{k+1}, f_j\right\rangle \mid \mathscr{G}_k\right] = c_{k+1}^{j}.
\end{aligned}$$

In case $\overline{\Lambda}_{k+1}=0$ we take $\dfrac{0}{0}=1$, and the result follows. □

2.4 Unnormalized Estimates and Bayes' Formula

Recall our discrete HMM of Section 2.2; recall also that $\mathscr{Y}_k$ is the complete σ-field generated by knowledge of $Y_1,\dots,Y_k$ and $\mathscr{G}_k$ is the complete σ-field generated by knowledge of X_0, $X_1,\dots,X_k$ and $Y_1,\dots,Y_k$. We suppose there is a probability $\overline{P}$ on $(\Omega, \bigvee_{n=1}^{\infty}\mathscr{G}_n)$ such that, under $\overline{P}$, $X_{k+1}=AX_k+V_{k+1}$, where V_k is a $\left(\overline{P},\mathscr{G}_k\right)$ martingale increment. That is, $\overline{E}\left[V_{k+1} \mid \mathscr{G}_k\right]=0$ and the $\{Y_k\}$ are i.i.d. with $\overline{P}\left(Y_k^j=1\right)=\frac{1}{M}$, and the Y_k are conditionally independent of V_k, given $\mathscr{G}_k$, under both P and $\overline{P}$. We also have via the double expectation property listed in Appendix A,

$$\begin{aligned}
\overline{E}\left[V_{k+1} \mid \mathscr{Y}_{k+1}\right] &= \overline{E}\left[\overline{E}\left[V_{k+1} \mid \mathscr{G}_k, \mathscr{Y}_{k+1}\right] \mid \mathscr{Y}_{k+1}\right] \\
&= \overline{E}\left[\overline{E}\left[V_{k+1} \mid \mathscr{G}_k\right] \mid \mathscr{Y}_{k+1}\right] = 0.
\end{aligned} \tag{4.1}$$

The measure P is then defined using (3.13). Recall from Lemma 3.3 that for a $\mathscr{G}$-adapted sequence $\{\phi_k\}$,

$$E\left[\phi_k \mid \mathscr{Y}_k\right] = \frac{\overline{E}\left[\overline{\Lambda}_k \phi_k \mid \mathscr{Y}_k\right]}{\overline{E}\left[\overline{\Lambda}_k \mid \mathscr{Y}_k\right]}. \tag{4.2}$$

Remark 4.1 This identity indicates why the unnormalized conditional expectation $\overline{E}\left[\overline{\Lambda}_k \phi_k \mid \mathscr{Y}_k\right]$ is investigated. ■

Write $q_k(e_r)$, $1 \le r \le N$, $k \in \mathbb{N}$, for the unnormalized, conditional probability distribution such that

$$\overline{E}\left[\overline{\Lambda}_k \left\langle X_k, e_r\right\rangle \mid \mathscr{Y}_k\right] = q_k(e_r).$$

Note that an alternative standard notation for this unnormalized conditional distribution is α; this is used in later chapters for a related distribution.

Now $\sum_{i=1}^{N}\langle X_k, e_i\rangle = 1$, so

$$\sum_{i=1}^{N} q_k(e_i) = \overline{E}\left[\overline{\Lambda}_k \sum_{i=1}^{N}\langle X_k, e_i\rangle \mid \mathscr{Y}_k\right] = \overline{E}\left[\overline{\Lambda}_k \mid \mathscr{Y}_k\right].$$

Therefore, from (4.2) the normalized conditional probability distribution

$$p_k(e_r) = E\left[\langle X_k, e_r\rangle \mid \mathscr{Y}_k\right]$$

is given by

$$p_k(e_r) = \frac{q_k(e_r)}{\sum_{j=1}^{k} q_k(e_j)}.$$

To conclude this section with a basic example we obtain a recursive expression for q_k. Recursive estimates for more general processes will be obtained in Section 2.5.

Notation 4.2 *To simplify the notation we write* $c_j(Y_k) = M\prod_{i=1}^{M} c_{ij}^{Y_k^i}$.

Theorem 4.3 *For* $k \in \mathbb{N}$ *and* $1 \le r \le N$, *the recursive filter for the unnormalized estimates of the states is given by*

$$\boxed{q_{k+1} = A \operatorname{diag} c(Y_{k+1}) \cdot q_k.} \tag{4.3}$$

Proof Using the independence assumptions under $\overline{P}$ and the fact that $\sum_{j=1}^{N}\langle X_k, e_j\rangle = 1$, as well as properties (1.1) and (1.2), we have

$$\begin{aligned}
q_k(e_r) &= \overline{E}\left[\langle X_{k+1}, e_r\rangle \overline{\Lambda}_{k+1} \mid \mathscr{Y}_{k+1}\right] \\
&= \overline{E}\left[\langle AX_k + V_{k+1}, e_r\rangle \overline{\Lambda}_k \prod_{i=1}^{M} \left(Mc_{k+1}^i\right)^{Y_{k+1}^i} \,\middle|\, \mathscr{Y}_{k+1}\right] \\
&= M\overline{E}\left[\langle AX_k, e_r\rangle \overline{\Lambda}_k \prod_{i=1}^{M} \left(\langle CX_k, f_i\rangle\right)^{Y_{k+1}^i} \,\middle|\, \mathscr{Y}_{k+1}\right]
\end{aligned}$$

[because V_{k+1} is a martingale increment with (4.1) holding]

$$\begin{aligned}
&= M\sum_{j=1}^{N} \overline{E}\left[\langle X_k, e_j\rangle a_{rj}\overline{\Lambda}_k \mid \mathscr{Y}_{k+1}\right] \prod_{i=1}^{M} c_{ij}^{Y_{k+1}^i} \\
&= M\sum_{j=1}^{N} \overline{E}\left[\langle X_k, e_j\rangle a_{rj}\overline{\Lambda}_k \mid \mathscr{Y}_k\right] \prod_{i=1}^{M} c_{ij}^{Y_{k+1}^i}
\end{aligned}$$

(because y_k is i.i.d. under $\overline{P}$)

$$= M\sum_{j=1}^{N} q_k(e_j)\, a_{rj} \prod_{i=1}^{M} c_{ij}^{Y_{k+1}^i}.$$

Using Notation 4.2 the result follows. □

Remark 4.4 This unnormalized recursion is a discrete-time form of Zakai's Theorem (Zakai 1969). This recursion is linear. ■

2.5 A General Unnormalized Recursive Filter

We continue to work under measure $\overline{P}$ so that

$$X_{k+1} = AX_k + V_{k+1} \tag{5.1}$$

and the Y_k are independent random variables, uniformly distributed over $f_1, \ldots, f_M$.

Notation 5.1 *If* $\{H_k\}$, $k \in \mathbb{N}$, *is any integrable sequence of random variables we shall write*

$$\gamma_k(H_k) = \overline{E}\left[\overline{\Lambda}_k H_k \mid \mathscr{Y}_k\right]. \tag{5.2}$$

Note this makes sense for vector processes H.

Using Lemma 3.3 we see that

$$E[H_k \mid \mathscr{Y}_k] = \frac{\overline{E}\left[\overline{\Lambda}_k H_k \mid \mathscr{Y}_k\right]}{\overline{E}\left[\overline{\Lambda}_k \mid \mathscr{Y}_k\right]} = \frac{\gamma_k(H_k)}{\gamma_k(1)}. \tag{5.3}$$

Consequently $\gamma_k(H_k)$ is an unnormalized conditional expectation of H_k given $\mathscr{Y}_k$. We shall take $\gamma_0(X_0) = E[X_0]$; this provides the initial value for later recursions.

Now suppose $\{H_k\}$, $k \in \mathbb{N}$, is an integrable (scalar) sequence. With $\Delta H_{k+1} = H_{k+1} - H_k$, $H_{k+1} = H_k + \Delta H_{k+1}$, then

$$\gamma_{k+1}(H_{k+1}) = \overline{E}\left[\overline{\Lambda}_{k+1} H_k \mid \mathscr{Y}_{k+1}\right] + \overline{E}\left[\overline{\Lambda}_{k+1} \Delta H_{k+1} \mid \mathscr{Y}_{k+1}\right].$$

Consider the first term on the right. Then, using the now familiar properties (1.1) and (1.2),

$$\begin{aligned}
\overline{E}\left[\overline{\Lambda}_{k+1} H_k \mid \mathscr{Y}_{k+1}\right] &= \overline{E}\left[\overline{\Lambda}_k H_k \overline{\lambda}_{k+1} \mid \mathscr{Y}_{k+1}\right] \\
&= \overline{E}\left[\overline{\Lambda}_k H_k M \prod_{i=1}^{M} \langle CX_k, f_i \rangle^{Y^i_{k+1}} \;\middle|\; \mathscr{Y}_{k+1}\right] \\
&= \sum_{j=1}^{N} \overline{E}\left[\overline{\Lambda}_k H_k \langle X_k, e_j \rangle \mid \mathscr{Y}_k\right] M \prod_{i=1}^{M} c_{ij}^{Y^i_{k+1}} \\
&= \sum_{j=1}^{N} c_j(Y_{k+1}) \langle \gamma_k(H_k X_k), e_j \rangle.
\end{aligned}$$

In this way the estimate for $\gamma_{k+1}(H_{k+1})$ introduces $\gamma_k(H_k X_k)$. A technical trick is to investigate the recursion for $\gamma_{k+1}(H_{k+1}X_{k+1})$. A similar discussion to that above then introduces the term $\gamma_k\left(H_k X_k X_k'\right)$; this can be written $\sum_{i=1}^N \langle \gamma_k(H_k X_k), e_i\rangle e_i e_i'$. Therefore, the estimates for $\gamma_{k+1}(H_{k+1}X_{k+1})$ can be recursively expressed in terms of $\gamma_k(H_k X_k)$ (together with other terms). Writing $\underline{1}$ for the vector $(1,1,\dots,1)' \in \mathbb{R}^N$ we see $\langle X_k, \underline{1}\rangle = \sum_{i=1}^N \langle X_k, e_i\rangle = 1$, so

$$\langle \gamma_k(H_k X_k), \underline{1}\rangle = \gamma_k(H_k \langle X_k, \underline{1}\rangle) = \gamma_k(H_k). \tag{5.4}$$

Consequently, the unnormalized estimate $\gamma_k(H_k)$ is obtained by summing the components of $\gamma_k(H_k X_k)$. Furthermore, taking $H_k = 1$ in (5.4) we see

$$\gamma_k(1) = \langle \gamma_k(X_k), \underline{1}\rangle = \overline{E}\left[\overline{\Lambda}_k \mid \mathscr{Y}_k\right] = \sum_{i=1}^N q_k(e_i)$$

using the notation of Section 2.4. Therefore, the normalizing factor $\gamma_k(1)$ in (5.3) is obtained by summing the components of $\gamma_k(X_k)$.

We now make the above observations precise by considering a more specific, though general, process H.

Suppose, for $k \geq 1$, H_k is a scalar process of the form

$$\begin{aligned} H_{k+1} &= \sum_{\ell=1}^{k+1} \left(\alpha_\ell + \langle \beta_\ell, V_\ell\rangle + \langle \delta_\ell, Y_\ell\rangle\right) \\ &= H_k + \alpha_{k+1} + \langle \beta_{k+1}, V_{k+1}\rangle + \langle \delta_{k+1}, Y_{k+1}\rangle. \end{aligned} \tag{5.5}$$

Here $V_\ell = X_\ell - AX_{\ell-1}$ and α_ℓ, β_ℓ, δ_ℓ are $\mathscr{G}$-predictable processes of appropriate dimensions, that is, α_ℓ, β_ℓ, δ_ℓ are $\mathscr{G}_{\ell-1}$ measurable, α_ℓ is scalar, β_ℓ is N-dimensional, and δ_ℓ is M-dimensional.

Notation 5.2 *For any process ϕ_k, $k \in \mathbb{N}$, write*

$$\gamma_{m,k}(\phi_m) = \overline{E}\left[\overline{\Lambda}_k \phi_m X_k \mid \mathscr{Y}_k\right]. \tag{5.6}$$

Theorem 5.3 *For $1 \leq j \leq M$ write $c_j = Ce_j = \left(c_{1j}, \dots, c_{Mj}\right)'$ for the jth column of $C = (c_{ij})$ and $a_j = Ae_j = \left(a_{1j}, \dots, a_{Nj}\right)'$ for the jth column of $A = (a_{ij})$. Then*

$$\begin{aligned} &\gamma_{k+1,k+1}(H_{k+1}) \\ &\quad = \sum_{j=1}^N c_j(Y_{k+1})\left\{\left\langle \gamma_{k,k}(H_k) + \gamma_{k+1,k}\left(\alpha_{k+1} + \langle \delta_{k+1}, Y_{k+1}\rangle\right), e_j\right\rangle a_j \right. \\ &\qquad\qquad \left. + \left[\operatorname{diag}(a_j) - a_j a_j'\right] \overline{E}\left[\left\langle \overline{\Lambda}_k X_k, e_j\right\rangle \beta_{k+1} \mid \mathscr{Y}_{k+1}\right]\right\}. \end{aligned} \tag{5.7}$$

Proof

$$\begin{aligned} &\gamma_{k+1,k+1}(H_{k+1}) \\ &\quad = \overline{E}\left[X_{k+1} H_{k+1} \overline{\Lambda}_{k+1} \mid \mathscr{Y}_{k+1}\right] \end{aligned}$$

$$
\begin{aligned}
&= \overline{E}\big[(AX_k + V_{k+1})(H_k + \alpha_{k+1} + \langle \beta_{k+1}, V_{k+1}\rangle + \langle \delta_{k+1}, Y_{k+1}\rangle) \\
&\qquad \times \overline{\Lambda}_k \overline{\lambda}_{k+1} \mid \mathscr{Y}_{k+1}\big] \\
&= \overline{E}\big[((H_k + \alpha_{k+1} + \langle \delta_{k+1}, Y_{k+1}\rangle) AX_k + \langle V_{k+1}\rangle \beta_{k+1}) \\
&\qquad \times \overline{\Lambda}_k \overline{\lambda}_{k+1} \mid \mathscr{Y}_{k+1}\big],
\end{aligned}
$$

$\{$because, as in Lemma 2.2,

$$
\begin{aligned}
&\overline{E}\big[\overline{\Lambda}_k \overline{\lambda}_{k+1} V_{k+1} V'_{k+1} \mid \mathscr{Y}_k\big] \\
&\qquad = \overline{E}\big[\overline{E}\big[\overline{\Lambda}_k \overline{\lambda}_{k+1} V_{k+1} V'_{k+1} \mid X_0, X_1, \dots, X_k, \mathscr{Y}_k\big] \mid \mathscr{Y}_k\big] \\
&\qquad = \overline{E}\big[\big\langle \overline{\Lambda}_k \overline{\lambda}_{k+1} V_{k+1}\big\rangle \mid \mathscr{Y}_k\big]\big\}
\end{aligned}
$$

$$
\begin{aligned}
&= \sum_{j=1}^{N} c_j(Y_{k+1}) \overline{E}\big[((H_k + \alpha_{k+1} + \langle \delta_{k+1}, Y_{k+1}\rangle) a_j \\
&\qquad\qquad + \langle V_{k+1}\rangle \beta_{k+1}) \overline{\Lambda}_k \big\langle X_k, e_j\big\rangle \mid \mathscr{Y}_{k+1}\big].
\end{aligned}
$$

Finally, because the Y are i.i.d. this final conditioning is the same as conditioning on $\mathscr{Y}_k$. Using Lemma 2.2 and Notation 5.2 the desired result follows. □

2.6 States, Transitions, and Occupation Times

Estimators for the State

Take $H_{k+1} = H_0 = \alpha_0 = 1$, $\alpha_\ell = 0$, $\ell \geq 1$, $\beta_\ell = 0$, $\ell \geq 0$ and $\delta_\ell = 0$, $\ell \geq 0$. Applying Theorem 5.3 we have again the *unnormalized filter* Equation (4.3) for $q_k = (q_k(e_1), \dots, q_k(e_N))$ in vector form:

$$
\boxed{q_{k+1} = \sum_{j=1}^{N} c_j(Y_{k+1}) \big\langle q_k, e_j\big\rangle a_j.} \tag{6.1}
$$

with normalized form

$$
p_k = q_k \langle q_k, \underline{1}\rangle^{-1}. \tag{6.2}
$$

This form is similar to that given by Aström (1965) and Stratonovich (1960). We can also obtain a recursive form for the unnormalized conditional expectation of $\big\langle X_m, e_p\big\rangle$ given $\mathscr{Y}_{k+1}$, $m < k+1$. This is the *unnormalized smoother*. For this we take $H_{k+1} = H_m = \big\langle X_m, e_p\big\rangle$, $m < k+1$, $1 \leq p \leq N$, $\alpha_\ell = 0$, $\beta_\ell = 0$ and $\delta_\ell = 0$. Applying Theorem 5.3 we have

$$
\boxed{\overline{E}\left[\overline{\Lambda}_{k+1} \langle X_m, e_p\rangle \mid \mathscr{Y}_{k+1}\right] = \sum_{j=1}^{N} c_j(Y_{k+1}) \big\langle \gamma_{m,k}(\langle X_m, e_p\rangle), e_j\big\rangle a_j.} \tag{6.3}
$$

We see that Equation (6.3) is indeed a recursion in k; this is why we consider $H_k X_k$. Taking the inner product with $\underline{1}$ and using Notation 5.1 gives the smoothed, unnormalized estimate

$$\gamma_k\left(\left\langle X_m, e_p\right\rangle\right)=\overline{E}\left[\overline{\Lambda}_k\left\langle X_m, e_p\right\rangle \mid \mathscr{Y}_k\right].$$

Estimators for the Number of Jumps

The number of jumps from state e_r to state e_s in time k is given by

$$\mathscr{J}_k^{rs}=\sum_{\ell=1}^k\left\langle X_{\ell-1}, e_r\right\rangle\left\langle X_\ell, e_s\right\rangle .$$

Using $X_\ell=A X_{\ell-1}+V_\ell$ this is

$$\begin{aligned}
& =\sum_{\ell=1}^k\left\langle X_{\ell-1}, e_r\right\rangle\left\langle A X_{\ell-1}, e_s\right\rangle+\sum_{\ell=1}^k\left\langle X_{\ell-1}, e_r\right\rangle\left\langle V_\ell, e_s\right\rangle \\
& =\sum_{\ell=1}^k\left\langle X_{\ell-1}, e_1 r\right\rangle a_{sr}+\sum_{\ell=1}^k\left\langle X_{\ell-1}, e_r\right\rangle\left\langle V_\ell, e_s\right\rangle .
\end{aligned}$$

Applying Theorem 5.3 with $H_{k+1}=\mathscr{J}_{k+1}^{rs}$, $H_0=0$, $\alpha_\ell=\left\langle X_{\ell-1}, e_r\right\rangle a_{sr}$, $\beta_\ell=\left\langle X_{\ell-1}, e_r\right\rangle e_s$, $\delta_\ell=0$ we have

$$\begin{aligned}
& \gamma_{k+1,k+1}\left(\mathscr{J}_{k+1}^{rs}\right) \\
&\quad =M \sum_{j=1}^N\left(\prod_{i=1}^M c_{ij}^{Y_{k+1}^i}\right)\Big\{\left\langle\gamma_{k,k}\left(\mathscr{J}_k^{rs}\right)+\gamma_{k,k}\left(\left\langle X_k, e_r\right\rangle a_{sr}\right), e_j\right\rangle a_j \\
&\qquad\qquad +\left[\operatorname{diag}\left(a_j\right)-a_j a_j^{\prime}\right] \\
&\qquad\qquad \times \overline{E}\left[\left\langle\overline{\Lambda}_k X_k, e_j\right\rangle\left\langle X_k, e_r\right\rangle e_s \mid \mathscr{Y}_{k+1}\right]\Big\} \\
& =M \sum_{j=1}^N\left(\prod_{i=1}^M c_{ij}^{Y_{k+1}^i}\right)\left\langle\gamma_{k,k}\left(\mathscr{J}_k^{rs}\right), e_j\right\rangle a_j \\
&\quad +M\left\langle q_k, e_r\right\rangle\left(\prod_{i=1}^M c_{ir}^{Y_{k+1}^i}\right)\left[a_{sr} a_r+e_s \operatorname{diag}\left(a_r\right)-e_s\left(a_r a_r^{\prime}\right)\right]
\end{aligned}$$

that is, using Notation 4.2,

$$\boxed{\begin{aligned}
\gamma_{k+1,k+1}\left(\mathscr{J}_{k+1}^{rs}\right)= & \sum_{j=1}^N c_j\left(Y_{k+1}\right)\left\langle\gamma_{k,k}\left(\mathscr{J}_k^{rs}\right), e_j\right\rangle a_j \\
& +c_r\left(Y_{k+1}\right)\left\langle q_k, e_r\right\rangle a_{sr} e_s .
\end{aligned}} \tag{6.4}$$

Together with the recursive Equation (6.1) for q_k we have in (6.4) a recursive estimator for $\gamma_{k,k}\left(\mathscr{J}_k^{rs}\right)$. Taking its inner product with $\underline{1}$, that is, summing its components, we obtain $\gamma_k\left(\mathscr{J}_k^{rs}\right) = \overline{E}\left[\overline{\Lambda}_k \mathscr{J}_k^{rs} \mid \mathscr{Y}_k\right]$.

Taking $H_{k+1} = H_m = \mathscr{J}_m^{rs}$, $\alpha_\ell = 0$, $\ell > m$, $\beta_\ell = 0$, $\ell \geq 0$, $\delta_\ell = 0$, $\ell \geq 0$, and applying Theorem 5.3 we obtain for $k > m$, the unnormalized smoothed estimate $\overline{E}\left[\overline{\Lambda}_{k+1} \mathscr{J}_m^{rs} X_{k+1} \mid \mathscr{Y}_{k+1}\right]$

$$\boxed{\gamma_{m,k+1}\left(\mathscr{J}_m^{rs}\right) = \sum_{j=1}^{N} c_j\left(Y_{k+1}\right)\left\langle \gamma_{m,k}\left(\mathscr{J}_m^{rs}\right), e_j\right\rangle a_j.} \tag{6.5}$$

Again, by considering the product $\mathscr{J}_m^{rs} X_k$ a recursive form has been obtained. Taking the inner product with $\underline{1}$ gives the smoothed unnormalized estimate $\overline{E}\left[\overline{\Lambda}_k \mathscr{J}_m^{rs} \mid \mathscr{Y}_k\right]$.

Estimators for the Occupation Time

The number of occasions up to time k for which the Markov chain X has been in state e_r, $1 \leq r \leq N$, is

$$\mathscr{O}_{k+1}^r = \sum_{\ell=1}^{k+1} \langle X_{\ell-1}, e_r \rangle .$$

Taking $H_{k+1} = \mathscr{O}_{k+1}^r$, $H_0 = 0$, $\alpha_\ell = \langle X_{\ell-1}, e_r\rangle$, $\beta_\ell = 0$, $\delta_\ell = 0$ and applying Theorem 5.3 we have

$$\begin{aligned}\gamma_{k+1,k+1}\left(\mathscr{O}_{k+1}^r\right) = M \sum_{j=1}^{N} \prod_{i=1}^{M} c_{ij}^{Y_{k+1}^i} &\left(\left\langle \gamma_{k,k}\left(\mathscr{O}_k^r\right), e_j\right\rangle\right.\\ &\left. + \left\langle \gamma_{k,k}\left(\langle X_k, e_r\rangle\right), e_j\right\rangle\right) a_j.\end{aligned}$$

That is

$$\begin{aligned}\gamma_{k+1,k+1}\left(\mathscr{O}_{k+1}^r\right) &= \sum_{j=1}^{N} c_j\left(Y_{k+1}\right)\left\langle \gamma_{k,k}\left(\mathscr{O}_k^r\right), e_j\right\rangle a_j\\ &\quad + c_r\left(Y_{k+1}\right)\langle q_k, e_r\rangle a_r.\end{aligned} \tag{6.6}$$

Together with (6.1) for q_k this equation gives a recursive expression for $\gamma_{k,k}\left(\mathscr{O}_k^r\right)$. Taking the inner product with $\underline{1}$ gives $\gamma_k\left(\mathscr{O}_k^r\right) = \overline{E}\left[\mathscr{O}_k^r \mid \mathscr{Y}_k\right]$. For the related smoother take $k > m$, $H_{k+1} = H_m = \mathscr{O}_m^r$, $\alpha_\ell = 0$, $\beta_\ell = 0$, $\delta_\ell = 0$ and apply Theorem 5.3 to obtain

$$\boxed{\gamma_{m,k+1}\left(\mathscr{O}_m^r\right) = \sum_{j=1}^{N} c_j\left(Y_{k+1}\right)\left\langle \gamma_{m,k}\left(\mathscr{O}_m^r\right), e_j\right\rangle a_j.} \tag{6.7}$$

Estimators for State to Observation Transitions

In estimating the parameters of our model in the next section we shall require estimates and smoothers of the process

$$\mathscr{T}_k^{rs} = \sum_{\ell=1}^{k} \langle X_{\ell-1}, e_r \rangle \langle Y_\ell, f_s \rangle$$

which counts the number of times up to time k that the observation process is in state f_s given the Markov chain at the preceding time is in state e_r, $1 \le r \le N$, $1 \le s \le M$. Taking $H_{k+1} = \mathscr{T}_{k+1}^{rs}$, $H_0 = 0$, $\alpha_\ell = 0$, $\beta_\ell = 0$, $\delta_\ell = \langle X_{\ell-1}, e_r \rangle f_s$ and applying Theorem 5.3

$$\gamma_{k+1,k+1}\left(\mathscr{T}_{k+1}^{rs}\right) = M \sum_{j=1}^{N} \prod_{i=1}^{M} c_{ij}^{Y_{k+1}^i} \left(\left\langle \gamma_{k,k}\left(\mathscr{T}_k^{rs}\right), e_j \right\rangle + \left\langle \gamma_{k,k}\left(\langle X_k, e_r \rangle \langle Y_{k+1}, f_s \rangle\right), e_j \right\rangle \right) a_j.$$

That is, using Notation 4.2,

$$\boxed{\begin{aligned}\gamma_{k+1,k+1}\left(\mathscr{T}_{k+1}^{rs}\right) = {} & \sum_{j=1}^{N} c_j(Y_{k+1}) \left\langle \gamma_{k,k}\left(\mathscr{T}_k^{rs}\right), e_j \right\rangle a_j \\ & + M \langle q_k, e_r \rangle \langle Y_{k+1}, f_s \rangle c_{sr} a_r.\end{aligned}}$$

Together with Equation (6.1) for q_k we have a recursive expression for $\gamma_{k,k}\left(\mathscr{T}_k^{rs}\right)$. To obtain the related smoother take $k+1 > m$, $H_{k+1} = H_m = \mathscr{T}_m^{rs}$, $\alpha_\ell = 0$, $\beta_\ell = 0$, $\delta_\ell = 0$ and apply Theorem 5.3 to obtain

$$\boxed{\gamma_{m,k+1}\left(\mathscr{T}_m^{rs}\right) = \sum_{j=1}^{N} c_j(Y_{k+1}) \left\langle \gamma_{m,k}\left(\mathscr{T}_m^{rs}\right), e_j \right\rangle a_j.} \tag{6.9}$$

This is recursive in k.

Remark 6.1 Note the similar form of the recursions (6.1), (6.4), (6.6), and (6.8). ∎

2.7 Parameter Reestimation

In this section we show how, using the expectation maximization (EM) algorithm, the parameters of the model can be estimated. In fact, it is a conditional pseudo *log-likelihood* that is maximized, and the new parameters are expressed in terms of the recursive estimates obtained in Section 2.6. We begin by describing the EM algorithm.

The basic idea behind the EM algorithm is as follows (Baum and Petrie 1966). Let $\{P_\theta, \theta \in \Theta\}$ be a family of probability measures on a measurable space $(\Omega, \mathscr{F})$ all absolutely continuous with respect to a fixed probability measure P_0 and let $\mathscr{Y} \subset \mathscr{F}$. The likelihood function for computing an estimate of the parameter θ based on the information available in $\mathscr{Y}$ is

$$L(\theta) = E_0\left[\frac{dP_\theta}{dP_0} \mid \mathscr{Y}\right],$$

and the maximum likelihood estimate (MLE) is defined by

$$\hat{\theta} \in \operatorname*{argmax}_{\theta \in \Theta} L(\theta).$$

The reasoning is that the most likely value of the parameter θ is the one that maximizes this conditional expectation of the density.

In general, the MLE is difficult to compute directly, and the EM algorithm provides an iterative approximation method:

Step 1. Set $p = 0$ and choose $\hat{\theta}_0$.
Step 2. (E-step) Set $\theta^* = \hat{\theta}_p$ and compute $Q(\cdot, \theta^*)$, where

$$Q(\theta, \theta^*) = E_{\theta^*}\left[\log \frac{dP_\theta}{dP_{\theta^*}} \mid \mathscr{Y}\right].$$

Step 3. (M-step) Find

$$\hat{\theta}_{p+1} \in \operatorname*{argmax}_{\theta \in \Theta} Q(\theta, \theta^*).$$

Step 4. Replace p by $p+1$ and repeat beginning with Step 2 until a stopping criterion is satisfied.

The sequence generated $\{\hat{\theta}_p, p \geq 0\}$ gives nondecreasing values of the likelihood function to a local maximum of the likelihood function: it follows from Jensen's Inequality, see Appendix A, that

$$\log L\left(\hat{\theta}_{p+1}\right) - \log L\left(\hat{\theta}_p\right) \geq Q\left(\hat{\theta}_{p+1}, \hat{\theta}_p\right),$$

with equality if $\hat{\theta}_{p+1} = \hat{\theta}_p$. We call $Q(\theta, \theta^*)$ a conditional pseudo-log-likelihood. Finding a set of parameters which gives a (local) maximum of the expected log-likelihood function gives an optimal estimate.

Our model (2.14) is determined by the set of *parameters*

$$\theta := (a_{ji},\ 1 \leq i, j \leq N,\ c_{ji},\ 1 \leq j \leq M,\ 1 \leq i \leq N)$$

which are also subject to the constraints (2.15) and (2.16). Suppose our model is determined by such a set θ and we wish to determine a new set

$$\hat{\theta} = (\hat{a}_{ji}(k),\ 1 \leq i, j \leq N,\ \hat{c}_{ji}(k),\ 1 \leq j \leq M,\ 1 \leq i \leq N)$$

which maximizes the conditional pseudo-log-likelihoods defined below. Recall $\mathscr{F}_k$ is the complete σ-field generated by $X_0, X_1, \dots, X_k$. Consider first the parameters a_{ji}. To replace the parameters a_{ji} by $\hat{a}_{ji}(k)$ in the Markov chain X we define

$$\Lambda_k = \prod_{\ell=1}^{k} \left(\sum_{r,s=1}^{N} \left[\frac{\hat{a}_{sr}(k)}{a_{sr}} \right] \langle X_\ell, e_s \rangle \langle X_{\ell-1}, e_r \rangle \right).$$

In case $a_{ji} = 0$, take $\hat{a}_{ji}(k) = 0$ and $\hat{a}_{ji}(k)/a_{ji} = 0$. Set

$$\left. \frac{dP_{\hat{\theta}}}{dP_\theta} \right|_{\mathscr{F}_k} = \Lambda_k.$$

To justify this we establish the following result.

Lemma 7.1 *Under the probability measure $P_{\hat{\theta}}$ and assuming $X_k = e_r$, then*

$$E_{\hat{\theta}} \left[\langle X_{k+1}, e_s \rangle \mid \mathscr{F}_k \right] = \hat{a}_{sr}(k).$$

Proof

$$\begin{aligned} E_{\hat{\theta}} \left[\langle X_{k+1}, e_s \rangle \mid \mathscr{F}_k \right] &= \frac{E \left[\langle X_{k+1}, e_s \rangle \Lambda_{k+1} \mid \mathscr{F}_k \right]}{E \left[\Lambda_{k+1} \mid \mathscr{F}_k \right]} \\ &= \frac{E\left[\langle X_{k+1}, e_s \rangle \frac{\hat{a}_{sr}(k)}{a_{sr}} \mid \mathscr{F}_k \right]}{E\left[\sum_{r=1}^{N} \left[\frac{\hat{a}_{sr}(k)}{a_{sr}} \right] \langle X_{k+1}, e_s \rangle \mid \mathscr{F}_k \right]} \\ &= \frac{\frac{\hat{a}_{sr}(k)}{a_{sr}} a_{sr}}{\sum_{r=1}^{N} \frac{\hat{a}_{sr}(k)}{a_{sr}} a_{sr}} \\ &= \hat{a}_{sr}(k). \end{aligned}$$

□

Notation 7.2 *For any process ϕ_k, $k \in \mathbb{N}$, write $\hat{\phi}_k = E[\phi_k \mid \mathscr{Y}_k]$ for its $\mathscr{Y}$-optional projection. In discrete time this conditioning defines the $\mathscr{Y}$-optional projection.*

Theorem 7.3 *The new estimates of the parameter $\hat{a}_{sr}(k)$ given the observations up to time k are given, when defined, by*

$$\boxed{\hat{a}_{sr}(k) = \frac{\hat{\mathscr{J}}_k^{rs}}{\hat{\mathscr{O}}_k^{r}} = \frac{\gamma_k\left(\mathscr{J}_k^{rs}\right)}{\gamma_k\left(\mathscr{O}_k^{r}\right)}.} \tag{7.1}$$

We take $\frac{0}{0}$ to be 0.

Proof

$$\log \Lambda_k = \sum_{r,s=1}^{N} \sum_{\ell=1}^{k} \langle X_\ell, e_s \rangle \langle X_{\ell-1}, e_r \rangle [\log \hat{a}_{sr}(k) - \log a_{sr}]$$
$$= \sum_{r,s=1}^{N} \mathscr{J}_k^{rs} \log \hat{a}_{sr}(k) + R(a),$$

where $R(a)$ is independent of $\hat{a}$. Therefore,

$$E[\log \Lambda_k \mid \mathscr{Y}_k] = \sum_{r,s=1}^{N} \hat{\mathscr{J}}_k^{rs} \log \hat{a}_{sr}(k) + \hat{R}(a). \tag{7.2}$$

Now the $\hat{a}_{sr}(k)$ must also satisfy the analog of (2.15)

$$\sum_{s=1}^{N} \hat{a}_{sr}(k) = 1. \tag{7.3}$$

Observe that

$$\sum_{s=1}^{N} \mathscr{J}_k^{rs} = \mathscr{O}_k^r \tag{7.4}$$

and in conditional form

$$\sum_{s=1}^{N} \hat{\mathscr{J}}_k^{rs} = \hat{\mathscr{O}}_k^r. \tag{7.5}$$

We wish, therefore, to choose the $\hat{a}_{sr}(k)$ to maximize (7.2) subject to the constraint (7.3). Write λ for the Lagrange multiplier and put

$$L(\hat{a}, \lambda) = \sum_{r,s=1}^{N} \hat{\mathscr{J}}_k^{rs} \log \hat{a}_{sr}(k) + \hat{R}(a) + \lambda \left(\sum_{s=1}^{N} \hat{a}_{sr}(k) - 1 \right).$$

Differentiating in λ and $\hat{a}_{sr}(k)$, and equating the derivatives to 0, we have the optimum choice of $\hat{a}_{sr}(k)$ is given by the equations

$$\frac{1}{\hat{a}_{sr}(k)} \hat{\mathscr{J}}_k^{rs} + \lambda = 0, \tag{7.6}$$
$$\sum_{s=1}^{N} \hat{a}_{sr}(k) = 1. \tag{7.7}$$

From (7.5)–(7.7) we see that $\lambda = -\hat{\mathscr{O}}_k^r$ so the optimum choice of $\hat{a}_{sr}(k)$, $1 \le s, r \le N$, is

$$\hat{a}_{sr}(k) = \frac{\hat{\mathscr{J}}_k^{rs}}{\hat{\mathscr{O}}_k^r} = \frac{\gamma_k(\mathscr{J}_k^{rs})}{\gamma_k(\mathscr{O}_k^r)}. \tag{7.8}$$

□

Note that the unnormalized conditional expectations in (7.8) are given by the inner product with $\underline{1}$ of (6.4) and (6.6).

Consider now the parameters c_{ji} in the matrix C. To replace the parameters c_{sr} by $\hat{c}_{sr}(k)$ we must now consider the Radon-Nikodym derivative

$$\widetilde{\Lambda}_k = \prod_{\ell=1}^{k} \left(\sum_{r=1}^{N} \sum_{s=1}^{M} \left[\frac{\hat{c}_{sr}(k)}{c_{sr}} \right] \langle X_{\ell-1}, e_r \rangle \langle Y_\ell, f_s \rangle \right).$$

By analogy with Lemma 3.1 we introduce a new probability by setting

$$\left. \frac{dP_{\hat{\theta}}}{dP_{\theta}} \right|_{\mathscr{G}_k} = \widetilde{\Lambda}_k.$$

Then $E_{\hat{\theta}}[\langle Y_{k+1}, f_s \rangle \mid X_k = e_r] = \hat{c}_{sr}(k)$.

Then

$$E\left[\log \widetilde{\Lambda}_k \mid \mathscr{Y}_k \right] = \sum_{r=1}^{N} \sum_{s=1}^{M} \mathscr{T}_k^{rs} \log \hat{c}_{sr}(k) + \widetilde{R}(c), \tag{7.9}$$

where $\widetilde{R}(c)$ is independent of $\hat{c}$. Now the $\hat{c}_{sr}(k)$ must also satisfy

$$\sum_{s=1}^{M} \hat{c}_{sr}(k) = 1. \tag{7.10}$$

Observe that

$$\sum_{s=1}^{M} \mathscr{T}_k^{rs} = \mathscr{O}_k^r$$

and conditional form

$$\sum_{s=1}^{M} \widehat{\mathscr{T}}_k^{rs} = \widehat{\mathscr{O}}_k^r. \tag{7.11}$$

We wish, therefore, to choose the $\hat{c}_{sr}(k)$ to maximize (7.9) subject to the constraint (7.11). Following the same procedure as above we obtain:

Theorem 7.4 *The maximum log likelihood estimates of the parameters $\hat{c}_{sr}(k)$ given the observation up to time k are given, when defined, by*

$$\boxed{\hat{c}_{sr}(k) = \frac{\gamma_k\left(\mathscr{T}_k^{rs}\right)}{\gamma_k\left(\mathscr{O}_k^r\right)}}. \tag{7.12}$$

We take $\frac{0}{0}$ *to be* 0.

Together with the estimates for $\gamma_k\left(\mathscr{T}_k^{rs}\right)$ given by the inner product with $\underline{1}$ of Equation (6.8) and the estimates for $\gamma_k\left(\mathscr{O}_k^r\right)$ given by taking the inner product with $\underline{1}$

of Equation (6.6) we can determine the optimal choice for $\hat{c}_{sr}(k)$, $1 \le s \le M-1$, $1 \le r \le N$. However, $\sum_{s=1}^{M} \hat{c}_{sr}(k) = 1$ for each r, so the remaining $\hat{c}_{Mr}(k)$ can also be found.

Remarks 7.5 The revised parameters $\hat{a}_{sr}(k)$, $\hat{c}_{sr}(k)$ determined by (7.8) and (7.12) give new probability measures for the model. The quantities $\gamma_k(\mathcal{J}_k^{rs})$, $\gamma_k(\mathcal{T}_k^{rs})$, $\gamma_k(\mathcal{O}_k^{r})$ can then be reestimated using the new parameters and perhaps new data, together with smoothing equations.

■

2.8 Recursive Parameter Estimation

In Section 2.7 we obtained estimates for the a_{ji} and the c_{ji}. However, these are not recursive, that is, the estimate at time k is not expressed as the estimate at time $(k-1)$ plus a correction based on new information. In this section we derive *recursive estimates* for the parameters. Unfortunately, these recursions are not in general finite-dimensional. Recall our discrete HMM signal model (2.14) is parametrized in terms of a_{ji}, c_{ji}. Let us collect these parameters into a parameter vector θ, so that we can write $A = A(\theta)$, $C = C(\theta)$. Suppose that θ is not known a priori. Let us estimate θ in a recursive manner, given the observations $\mathcal{Y}_k$. We assume that θ will take values in some set $\Theta \in \mathbb{R}^p$.

Let us now write $\mathcal{G}_k$ for the complete σ-field generated by knowledge of $X_0, X_1, \dots, X_k$, $Y_1, \dots, Y_k$, together with θ. Again $\mathcal{Y}_k$ will be the complete σ-field generated by knowledge of $Y_1, \dots, Y_k$. With this enlarged $\mathcal{G}_k$ the results of Sections 2.2 and 2.3 still hold. We suppose there is a probability $\overline{P}$ on $(\Omega \times \Theta, \bigvee_{\ell=1}^{\infty} \mathcal{G}_\ell)$ such that, under $\overline{P}$, the $\{Y_\ell\}$ are i.i.d. with $\overline{P}(Y_\ell^j = 1) = \frac{1}{M}$, and $X_{k+1} = AX_k + V_{k+1}$, where V_k is a $(\overline{P}, \mathcal{G}_k)$ martingale increment. Write $q_k^r(\theta)$, $1 \le r \le N$, $k \in \mathbb{N}$, for an unnormalized, conditional density such that

$$\overline{E}\left[\overline{\Lambda}_k \langle X_k, e_r \rangle I(\theta \in d\theta) \mid \mathcal{Y}_k\right] = q_k^r(\theta)\, d\theta.$$

Where $d\theta$ is Lebesgue measure on $\Theta \in \mathbb{R}^p$.

Here, $I(A)$ is the indicator function of the set A, that is, the function that is 1 on A and 0 otherwise. The existence of $q_k^r(\theta)$ will be discussed below. Equalities in the variable θ can be interpreted almost surely.

The normalized conditional density $p_k^r(\theta)$, such that

$$p_k^r(\theta)\, d\theta = E\left[\langle X_k, e_r \rangle I(\theta \in d\theta) \mid \mathcal{Y}_k\right],$$

is then given by

$$p_k^r(\theta) = \frac{q_k^r(\theta)}{\sum_{j=1}^{N} \int_\Theta q_k^j(u)\, du}.$$

We suppose an initial distribution $p_0(.) = \left(p_0^1(.), \ldots, p_0^N(.)\right)$ is given. This is further discussed in Remark 8.2. A recursive expression for $q_k^r(\theta)$ is now obtained:

Theorem 8.1 *For $k \in \mathbb{N}$, and $1 \le r \le N$, then the recursive estimates of an unnormalized joint conditional distribution of X_k and θ are given by*

$$\boxed{q_{k+1}^r(\theta) = a'_{r,(.)}\,\mathrm{diag}\left(q_k(\theta)\right) c_{(.)}\left(Y_{k+1}\right).} \tag{8.1}$$

Proof Suppose g is any real-valued *Borel function* on Θ. Then

$$\begin{aligned}
&\overline{E}\left[\langle X_{k+1}, e_r\rangle g(\theta)\overline{\Lambda}_{k+1} \mid \mathscr{Y}_{k+1}\right] \\
&\quad = \int_\Theta q_{k+1}^r(u)\, g(u)\, du \qquad (8.2) \\
&\quad = \overline{E}\left[\langle AX_k + V_{k+1}, e_r\rangle g(\theta)\overline{\Lambda}_k \sum_{i=1}^M M \langle CX_k, f_i\rangle \langle Y_{k+1}, f_i\rangle \;\Big|\; \mathscr{Y}_{k+1}\right] \\
&\quad = M\overline{E}\left[\langle AX_k, e_r\rangle g(\theta)\overline{\Lambda}_k \sum_{i=1}^M \langle CX_k, f_i\rangle \langle Y_{k+1}, f_i\rangle \;\Big|\; \mathscr{Y}_{k+1}\right] \\
&\quad = M\sum_{s=1}^N \overline{E}\left[\langle X_k, e_s\rangle a_{rs} g(\theta)\overline{\Lambda}_k \mid \mathscr{Y}_k\right] \prod_{i=1}^M c_{is}^{Y_{k+1}^i} \\
&\quad = M\int_\Theta \sum_{s=1}^N a_{rs} q_k^s(u)\, g(u)\, du \prod_{i=1}^M c_{is}^{Y_{k+1}^i}. \qquad (8.3)
\end{aligned}$$

As g is arbitrary, from (8.2) and (8.3) we see

$$q_{k+1}^r(u) = M\sum_{s=1}^N \left(a_{rs} q_k^s(u) \prod_{i=1}^M c_{is}^{Y_{k+1}^i}\right).$$

Using Notation 4.2 the result follows. □

Compared with Theorem 4.3 the new feature of Theorem 8.1 is that it updates recursively the estimate of the parameter.

Remark 8.2 Suppose $\pi = (\pi_1, \ldots, \pi_N)$, where $\pi_i = P(X_0 = e_i)$ is the initial distribution for X_0 and $h(\theta)$ is the prior density for θ. Then

$$q_0^r(\theta) = \pi_r h(\theta),$$

and the updated estimates are given by (8.1).

■

If the prior information about X_0 is that, say, $X_0 = e_i$, then the dynamics of X, (2.4) will move the state around and the estimate is given by (8.1). If the prior information about θ is that θ takes a particular value, then $h(\theta)$ (or a factor of h)

is a delta function at this value. No noise or dynamics enters into θ, so the equations (8.1) just continue to give the delta function at this value. This is exactly to be expected. The prior distribution h taken for θ must represent the a priori information about θ; it is not an initial guess for the value of θ.

Time-varying dynamics for θ could be incorporated in our model. Possibly $\theta_{k+1} = A_\theta \theta_k + v_{k+1}$, where v_{k+1} is the noise term. However, the problem then arises of estimating the terms of the matrix A_θ.

Finally, we note the equations (8.1) are really just a family of equations parametrized by θ. In particular, if θ can take one of finitely many values $\theta_1, \theta_2, \dots, \theta_p$ we obtain p equations (8.1) for each possible θ_i. The prior for θ is then just a distribution over $\theta_1, \dots, \theta_p$.

2.9 Quantized Observations

Suppose now the signal process $\{x_\ell\}$ is of the form

$$x_{k+1} = Ax_k + v_{k+1},$$

where $x_k \in \mathbb{R}^d$, $A = (a_{ji})$ is a $d \times d$ matrix and $\{v_\ell\}$, $\ell \in \mathbb{N}$, is a sequence of i.i.d. random variables with density function ψ. (Time-varying densities or nonlinear equations for the signal can be considered.) We suppose x_0, or its distribution, is known. The observation process is again denoted by Y_ℓ, $\ell \in \mathbb{N}$. However, the observations are *quantized*, so that the range space of Y_ℓ is finite. Here, also, we shall identify the range of Y_ℓ with the unit vectors $f_1, \dots, f_M$, $f_j = (0, \dots, 1, \dots, 0)' \in \mathbb{R}^M$, for some M. Again suppose some parameters $\theta \in \Theta$ in the model are not known. Write $\mathscr{G}_k$ for the complete σ-field generated by $x_0, x_1, \dots, x_k$, $Y_1, \dots, Y_k$ and θ; $\mathscr{Y}_k$ is the complete σ-field generated by $Y_1, \dots, Y_k$. If $Y_\ell^i = \langle Y_\ell, f_i \rangle$, $1 \le i \le M$, then $Y_\ell = \left(Y_\ell^i, \dots, Y_\ell^M\right)'$ and $\sum_{i=1}^M Y_\ell^i = 1$. Write

$$c_\ell^i = E\left[\langle Y_\ell, f_i \rangle \mid \mathscr{G}_{\ell-1}\right] = P\left(Y_\ell = f_i \mid \mathscr{G}_{\ell-1}\right).$$

We shall suppose

$$P\left(Y_\ell = f_i \mid \mathscr{G}_{\ell-1}\right) = P\left(Y_\ell = f_i \mid x_{\ell-1}\right), \qquad 1 \le i \le M, \ell \in \mathbb{N}.$$

In this case we write $c_\ell^i(x_{\ell-1})$. Suppose $c_\ell^i(x_{\ell-1}) > 0$, $1 \le i \le M$, $\ell \in \mathbb{N}$. Write

$$\Lambda_k = \prod_{\ell=1}^k \left(\sum_{i=1}^M \left[\frac{1}{M c_\ell^i(x_{\ell-1})} \right] \langle Y_\ell, f_i \rangle \right).$$

Defining $\overline{P}$ by setting

$$\left.\frac{d\overline{P}}{dP}\right|_{\mathscr{G}_k} = \Lambda_k$$

gives a measure such that

$$\overline{E}\left[\langle Y_\ell, f_i\rangle \mid \mathscr{G}_{\ell-1}\right] = \frac{1}{M}.$$

Suppose the parameter θ takes values in $\mathbb{R}^d$ and is random and unknown.

Suppose we start with a measure $\overline{P}$ on $\left(\Omega \times \mathbb{R}^d, \bigvee_{\ell=1}^{\infty}\mathscr{G}_\ell\right)$ such that

$$\overline{E}\left[\langle Y_\ell, f_i\rangle \mid \mathscr{G}_{\ell-1}\right] = \frac{1}{M}$$

and $x_{k+1} = Ax_k + v_{k+1}$. Write

$$\overline{\Lambda}_k = \prod_{\ell=1}^{k}\left(\sum_{i=1}^{M} M\left[c_\ell^i\left(x_{\ell-1}\right)\right]\langle Y_\ell, f_i\rangle\right).$$

[Note this no longer requires $c_{k+1}^i\left(x_k\right) > 0$.]

Introduce P by putting

$$\left.\frac{dP}{d\overline{P}}\right|_{\mathscr{G}_k} = \overline{\Lambda}_k.$$

Suppose f is any Borel function on $\mathbb{R}^d$ and g is any Borel function on Θ, and write $q_k(z,\theta)$ for an unnormalized conditional density such that

$$\overline{E}\left[\overline{\Lambda}_k I\left(x_k \in dz\right) I\left(\theta \in d\theta\right) \mid \mathscr{Y}_k\right] = q_k(z,\theta)\,dz\,d\theta.$$

Then

$$\overline{E}\left[f\left(x_{k+1}\right) g(\theta)\overline{\Lambda}_{k+1} \mid \mathscr{Y}_{k+1}\right] = \iint f(\xi)\, g(u)\, q_{k+1}(\xi,u)\, d\xi\, d\lambda(u). \tag{9.1}$$

The right-hand side is also equal to

$$= M\overline{E}\left[f\left(Ax_k + v_{k+1}\right) g(\theta)\overline{\Lambda}_k \prod_{i=1}^{M} c_{k+1}^i\left(x_k\right)^{Y_{k+1}^i} \mid \mathscr{Y}_{k+1}\right]$$
$$= M\iiint f(Az+v)\, g(u)\left[\prod_{i=1}^{M} c_{k+1}^i(z)^{Y_{k+1}^i}\right]\psi(v)\, q_k(z,u)\, dv\, dz\, d\lambda(u).$$

Write $\xi = Az + v$, so $v = \xi - Az$. The above is

$$= M\iiint f(\xi)\, g(u)\left(\prod_{i=1}^{M} c_{k+1}^i(z)^{Y_{k+1}^i}\right)\psi(\xi - Az)\, q_k(z,u)\, dz\, d\xi\, d\lambda(u). \tag{9.2}$$

Comparing (9.1) and (9.2) and denoting

$$c_{k+1}(Y_{k+1},z) = M\sum_{i=1}^{M} c_{k+1}^{i}(z)\langle Y_{k+1},f_i\rangle$$

we have the following result:

Theorem 9.1 *The recursive estimate of an unnormalized joint conditional density of the signal x and the parameter θ satisfies:*

$$\boxed{q_{k+1}(\xi,u) = \int_{\mathbb{R}^d} c_{k+1}(Y_{k+1},z)\,\psi(\xi - Az)\,q_k(z,u)\,dz.}$$

Example

In Kulhavy (1990) the following simple situation is considered. Suppose $\theta \in \mathbb{R}$ is unknown. $\{v_\ell\}$, $\ell \in \mathbb{N}$, is a sequence of i.i.d. $N(0,\sigma^2)$ random variables. The real line is partitioned into M disjoint intervals,

$$I_1 = (-\infty,\alpha_1),\, I_2 = [\alpha_1,\alpha_2),\ldots,I_{M-1} = [\alpha_{M-2},\alpha_{M-1}),\, I_M = [\alpha_M,\infty).$$

The signal process is $x_\ell = \theta + v_\ell$, $\ell \in \mathbb{N}$. The observation process Y_ℓ is an M-dimensional unit vector such that $Y_\ell^i = 1$ if $x_\ell \in I_i$. Then

$$\begin{aligned}
c_\ell^i &= P\big(Y_\ell^i = 1 \mid \mathscr{G}_{\ell-1}\big)\\
&= P\big(Y_\ell^i = 1 \mid \theta\big) = P\big(\alpha_{i-1} \le Y_\ell^i < \alpha_i \mid \theta\big)\\
&= \big(2\pi\sigma^2\big)^{-1/2}\int_{\alpha_{i-1}-\theta}^{\alpha_i-\theta} \exp\big(-x^2/2\sigma^2\big)dx\\
&= c_\ell^i(\theta), \qquad 1 \le i \le M.
\end{aligned}$$

Measure $\overline{P}$ is now introduced. Write $q_k(\theta)$ for the unnormalized conditional density such that

$$\overline{E}\big[\overline{\Lambda}_k I(\theta \in d\theta) \mid \mathscr{Y}_k\big] = q_k(\theta)\,d\theta.$$

Then, for an arbitrary Borel function g,

$$\begin{aligned}
\overline{E}\big[g(\theta)\overline{\Lambda}_{k+1} \mid \mathscr{Y}_{k+1}\big] &= \int_{\mathbb{R}} g(\lambda)\,q_{k+1}(\lambda)\,d\lambda\\
&= M\overline{E}\left[g(\theta)\overline{\Lambda}_k \sum_{i=1}^{M} c_{k+1}^i(\theta)\langle Y_{k+1},f_i\rangle \mid \mathscr{Y}_{k+1}\right]\\
&= M\int_{\mathbb{R}} g(\lambda)\left[\sum_{i=1}^{M} c_{k+1}^i(\lambda)\langle Y_{k+1},f_i\rangle\right] q_k(\lambda)\,d\lambda.
\end{aligned}$$

We, therefore, have the following recursion formula for the unnormalized conditional density of θ:

$$\boxed{q_{k+1}(\lambda) = \left(\prod_{i=1}^{M} c_{k+1}^{i}(\lambda)^{Y_{k+1}^{i}}\right) q_k(\lambda).} \tag{9.3}$$

The conditional density of θ given $\mathscr{Y}_k$ is then

$$p_k(\lambda) = \frac{q_k(\lambda)}{\int_{\mathbb{R}} q_k(\xi)\, d\xi}.$$

2.10 The Dependent Case

The situation considered in this section, (which may be omitted on a first reading), is that of a hidden Markov Model for which the "noise" terms in the state and observation processes are possibly dependent. An elementary prototype of this situation, for which the observation process is a single point process, is discussed in Segall (1976b). The filtrations $\{\mathscr{F}_k\}$, $\{\mathscr{G}_k\}$ and $\{\mathscr{Y}_k\}$ are as defined in Section 1.2. The semimartingale form of the Markov chain is, as in Section 2.2,

$$X_{k+1} = AX_k + V_{k+1}, \qquad k \in \mathbb{N},$$

where V_k is an $\{\mathscr{F}_k\}$ martingale increment, $a_{ji} = P(X_{k+1} = e_j \mid X_k = e_i)$ and $A = (a_{ji})$. Again the Markov chain is not observed directly; rather we suppose there is a finite-state observation process Y. The relation between X and Y can be given as $P(Y_{k+1} = f_r \mid \mathscr{G}_k) = P(Y_{k+1} = f_r \mid X_k)$ so that

$$Y_{k+1} = CX_k + W_{k+1}, \qquad k \in \mathbb{N},$$

where W_k is an $\{\mathscr{G}_k\}$ martingale increment, $c_{ji} = P(Y_{k+1} = f_j \mid X_k = e_i)$ and $C = (c_{ji})$. We initially assume c_{ji} positive for $1 \le i \le N$ and $1 \le j \le M$.

However, the noise, or martingale increment, terms V_k and W_k are not independent. In fact, the joint distribution of Y_k and X_k is supposed, given by

$$Y_{k+1}X_{k+1}' = SX_k + \Gamma_{k+1}, \qquad k \in \mathbb{N},$$

where $S = (s_{rji})$ denotes a $MN \times N$ matrix, or tensor, mapping $\mathbb{R}^N$ into $\mathbb{R}^M \times \mathbb{R}^N$ and

$$s_{rji} = P(Y_k = f_r,\, X_k = e_j \mid X_{k-1} = e_i) \qquad 1 \le r \le M,\ 1 \le i, j \le N.$$

Again Γ_{k+1} is a martingale increment, so $E[\Gamma_{k+1} \mid \mathscr{G}_k] = 0$.

If the terms are independent

$$SX_k = CX_k (AX_k)'.$$

In this dependent case, recursive estimates are derived for the state of the chain, the number of jumps from one state to another, the occupation time of the chain in any state, the number of transitions of the observation process into a particular state, and the number of joint transitions of the chain and the observation process. Using the expectation maximization algorithm optimal estimates are obtained for the elements a_{ji}, c_{ji} and s_{rji} of the matrices A, C, and S, respectively. Our model is again, therefore, adaptive or "self-tuning." In the independent case our results specialize to those of Section 2.5.

Dependent Dynamics

We shall suppose

$$P(Y_{k+1}=f_r, X_{k+1}=e_j \mid \mathscr{G}_k) = P(Y_{k+1}=f_r, X_{k+1}=e_j \mid X_k) \tag{10.1}$$

and write

$$s_{rji} = P(Y_{k+1}=f_r, X_{k+1}=e_j \mid X_k=e_i), \qquad 1 \le r \le M,\ 1 \le i,j \le N.$$

Then $S=(s_{rji})$ denotes a $MN \times N$ matrix, or tensor, mapping $\mathbb{R}^N$ into $\mathbb{R}^M \times \mathbb{R}^N$. From this hypothesis we have immediately:

$$Y_{k+1}X'_{k+1} = SX_k + \Gamma_{k+1}, \qquad k \in \mathbb{N}, \tag{10.2}$$

where Γ_{k+1} is a $(P, \mathscr{G}_k)$, $\mathbb{R}^M \times \mathbb{R}^N$ martingale increment.

Remark 10.1 Our model, therefore, involves the three sets of parameters (a_{ji}), (c_{ri}), and (s_{rji}).

■

Write $\underline{1} = (1,1,\dots,1)'$ for the vector, in $\mathbb{R}^M$ or $\mathbb{R}^N$ according to context, all components of which are 1.

Lemma 10.2 *For $\underline{1} \in \mathbb{R}^M$, then*

$$\langle \underline{1}, SX_k \rangle = AX_k. \tag{10.3}$$

For $\underline{1} \in \mathbb{R}^N$, then

$$\langle SX_k, \underline{1} \rangle = CX_k. \tag{10.4}$$

Proof In each case $\langle \underline{1}, \Gamma_k \rangle$ and $\langle \Gamma_k, \underline{1} \rangle$ are martingale increments. Taking the inner product of (10.2) with $\underline{1}$ the left side is, respectively, either $\langle \underline{1}, Y_{k+1}X'_{k+1} \rangle = X_k$ or $\langle Y_{k+1}X'_{k+1}, \underline{1} \rangle = Y_{k+1}$. Therefore, the result follows from the unique decompositions of the special semimartingales X_k and Y_k. □

In contrast to the independent situation, we have here $P\left[X_{k+1}=e_j \mid \mathscr{F}_k, \mathscr{Y}_{k+1}\right] = P\left[X_{k+1}=e_j \mid X_k, \mathscr{Y}_{k+1}\right]$. This is not, in general, equal to $P\left[X_{k+1}=e_j \mid X_k\right]$ so that knowledge of $\mathscr{Y}_k$, or in particular Y_k, now gives extra information about X_k.

Write

$$\alpha_{jir} = \frac{s_{rji}}{c_{ri}};$$

(recall the c_{ri} are positive). We then have the following:

Lemma 10.3 *With $\tilde{A}$ the $N \times (N \times M)$ matrix (α_{jir}), $1 \le i,j \le N$, $1 \le r \le M$,*

$$X_{k+1} = \tilde{A}\left(X_k Y'_{k+1}\right) + \tilde{V}_{k+1},$$

where

$$E\left[\tilde{V}_{k+1} \mid \mathscr{F}_k, \mathscr{Y}_{k+1}\right] = 0. \tag{10.5}$$

Proof

$$\begin{aligned} P&\left[X_{k+1}=e_j \mid X_k = e_i, Y_{k+1} = f_r\right] \\ &= \frac{P\left[Y_{k+1}=f_r, X_{k+1}=e_j \mid X_k = e_i\right]}{P\left[Y_{k+1}=f_r \mid X_k = e_i\right]} \\ &= \frac{s_{rji}}{c_{ri}} = \alpha_{jir}. \end{aligned}$$

With $\tilde{A} = (\alpha_{jir})$, $1 \le i,j \le N$, $1 \le r \le M$, we define $\tilde{V}_k$ by putting

$$X_{k+1} = \tilde{A}\left(X_k Y'_{k+1}\right) + \tilde{V}_{k+1}. \tag{10.6}$$

Then

$$\begin{aligned} E\left[\tilde{V}_{k+1} \mid \mathscr{F}_k, \mathscr{Y}_{k+1}\right] &= E\left[X_{k+1} \mid \mathscr{F}_k, \mathscr{Y}_{k+1}\right] - \tilde{A}\left(X_k Y'_{k+1}\right) \\ &= \tilde{A}\left(X_k Y'_{k+1}\right) - \tilde{A}\left(X_k Y'_{k+1}\right) = 0. \end{aligned}$$

□

In summary then, we have the following.

Dependent Discrete HMM The dependent discrete HMM is

$$\boxed{\begin{aligned} X_{k+1} &= \tilde{A}\left(X_k Y'_{k+1}\right) + \tilde{V}_{k+1} \\ Y_{k+1} &= C X_k + W_{k+1}, \qquad k \in \mathbb{N}, \end{aligned}} \tag{10.7}$$

where $X_k \in S_X$, $Y_k \in S_Y$, $\tilde{A}$ and C are matrices of transition probabilities given in Lemmas 10.3 and (2.8). The entries of $\tilde{A}$ satisfy

$$\sum_{j=1}^{N} \alpha_{jir} = 1, \qquad \alpha_{jir} \geq 0. \tag{10.8}$$

$\tilde{V}_k$ is a martingale increment satisfying

$$E\left[\tilde{V}_{k+1} \mid \mathscr{F}_k, \mathscr{Y}_{k+1}\right] = 0.$$

Next, we derive filters and smoothers for various processes.

The State Process

We shall be working under a probability measure $\overline{P}$ as discussed in Sections 2.3 and 2.4, so that the observation process is a sequence of i.i.d. random variables, uniformly distributed over the set of standard unit vectors $\{f_1, \ldots, f_M\}$ of $\mathbb{R}^M$.

Here Λ_k is as defined in Section 2.3. Using Bayes' Theorem we see that

$$\begin{aligned}
\overline{P}[X_{k+1} = e_j \mid \mathscr{F}_k, \mathscr{Y}_{k+1}] &= \overline{E}\left[\langle X_{k+1}, e_j\rangle \mid \mathscr{F}_k, \mathscr{Y}_{k+1}\right] \\
&= \frac{E\left[\langle X_{k+1}, e_j\rangle \Lambda_{k+1} \mid \mathscr{F}_k, \mathscr{Y}_{k+1}\right]}{E\left[\Lambda_{k+1} \mid \mathscr{G}_k, \mathscr{Y}_{k+1}\right]} \\
&= \frac{\Lambda_{k+1} E\left[\langle X_{k+1}, e_j\rangle \mid \mathscr{F}_k, \mathscr{Y}_{k+1}\right]}{\Lambda_{k+1}} \\
&= P[X_{k+1} = e_j \mid \mathscr{F}_k, \mathscr{Y}_{k+1}] \\
&= P[X_{k+1} = e_j \mid X_k, Y_{k+1}].
\end{aligned}$$

Therefore under $\overline{P}$, the process X satisfies (10.7). Write $\tilde{q}_k$, $k \in \mathbb{N}$, for the unnormalized conditional probability distribution such that

$$\overline{E}\left[\overline{\Lambda}_k X_k \mid \mathscr{Y}_k\right] := \tilde{q}_k.$$

Also write

$$\tilde{A}\left(e_j f_r'\right) = \alpha_{\cdot jr} = \left(\alpha_{1jr}, \alpha_{2jr}, \ldots, \alpha_{Njr}\right) \text{ and } s_{r \cdot j} = \left(s_{r1j}, \ldots, s_{rNj}\right).$$

Lemma 10.4 *A recursive formula for $\tilde{q}_{k+1}$ is given by*

$$\boxed{\tilde{q}_{k+1} = M \sum_{r=1}^{M} \sum_{j=1}^{N} \langle \tilde{q}_k, e_j\rangle \langle Y_{k+1}, f_r\rangle s_{r \cdot j} = MS\tilde{q}_k Y_{k+1}'.} \tag{10.9}$$

Proof

$$
\begin{aligned}
\tilde{q}_{k+1} &= \overline{E}\left[\overline{\Lambda}_{k+1}X_{k+1} \mid \mathscr{Y}_{k+1}\right] \\
&= \overline{E}\left[\overline{\Lambda}_k \prod_{r=1}^{M} \left(M\left\langle CX_k, f_r\right\rangle\right)^{Y^r_{k+1}} \overline{E}\left[X_{k+1} \mid \mathscr{F}_k, \mathscr{Y}_{k+1}\right] \mid \mathscr{Y}_{k+1}\right] \\
&= \overline{E}\left[\overline{\Lambda}_k \prod_{r=1}^{M} \left(M\left\langle CX_k, f_r\right\rangle\right)^{Y^r_{k+1}} \tilde{A}X_kY'_{k+1} \mid \mathscr{Y}_{k+1}\right] \\
&= M\sum_{r=1}^{M}\sum_{j=1}^{N} \left\langle \tilde{q}_k, e_j\right\rangle \left\langle Y_{k+1}, f_r\right\rangle c_{rj}\alpha_{\cdot jr} \\
&= M\sum_{r=1}^{M}\sum_{j=1}^{N} \left\langle \tilde{q}_k, e_j\right\rangle \left\langle Y_{k+1}, f_r\right\rangle s_{r\cdot j} \\
&= MS\tilde{q}_kY'_{k+1}.
\end{aligned}
$$

□

Remark 10.5 If the noise terms in the state X and observation Y are independent, then

$$
\begin{aligned}
SX_k &= E\left[Y_{k+1}X'_{k+1} \mid \mathscr{G}_k\right] \\
&= CX_k\left(AX_k\right)' \\
&= \sum_{i=1}^{N} \left\langle X_k, e_i\right\rangle c_i a'_i,
\end{aligned}
$$

where $c_i = Ce_i$ and $a_i = Ae_i$.

■

A General Recursive Filter

Suppose H_k is a scalar $\mathscr{G}$-adapted process such that H_0 is $\mathscr{F}_0$ measurable. With $\Delta H_{k+1} = H_{k+1} - H_k$, $H_{k+1} = H_k + \Delta H_{k+1}$. For any $\mathscr{G}$-adapted process ϕ_k, $k \in \mathbb{N}$, write $\tilde{\gamma}_{m,k}\left(\phi_m\right) = \overline{E}\left[\overline{\Lambda}_k\phi_mX_k \mid \mathscr{Y}_k\right]$. Then

$$
\begin{aligned}
&\tilde{\gamma}_{k+1,k+1}\left(H_{k+1}\right) \\
&\quad= \overline{E}\left[\overline{\Lambda}_{k+1}H_kX_{k+1} \mid \mathscr{Y}_{k+1}\right] + \overline{E}\left[\overline{\Lambda}_{k+1}\Delta H_{k+1}X_{k+1} \mid \mathscr{Y}_{k+1}\right] \\
&\quad= \overline{E}\left[\overline{\Lambda}_kH_k\tilde{A}\left(X_kY'_{k+1}\right)\overline{\lambda}_{k+1} \mid \mathscr{Y}_{k+1}\right] + \overline{E}\left[\overline{\Lambda}_{k+1}\Delta H_{k+1}X_{k+1} \mid \mathscr{Y}_{k+1}\right] \\
&\quad= M\sum_{r=1}^{M}\sum_{j=1}^{N} \left\langle \tilde{\gamma}_{k,k}\left(H_k\right), e_j\right\rangle \left\langle Y_{k+1}, f_r\right\rangle s_{r\cdot j} \\
&\qquad\qquad + \overline{E}\left[\overline{\Lambda}_{k+1}\Delta H_{k+1}X_{k+1} \mid \mathscr{Y}_{k+1}\right] \\
&\quad= MS\tilde{\gamma}_{k,k}\left(H_k\right)Y'_{k+1} + \overline{E}\left[\overline{\Lambda}_{k+1}\Delta H_{k+1}X_{k+1} \mid \mathscr{Y}_{k+1}\right]. \qquad (10.10)
\end{aligned}
$$

For the smoother at time $m < k+1$, we have

$$\begin{aligned}\tilde{\gamma}_{m,k+1}(H_m) &= M \sum_{r=1}^{M} \sum_{j=1}^{N} \langle \tilde{\gamma}_{m,k}(H_m), e_j \rangle \langle Y_{k+1}, f_r \rangle s_{r \cdot j} \\ &= MS\tilde{\gamma}_{m,k}(H_m) Y'_{k+1}. \qquad (10.11)\end{aligned}$$

Remark 10.6 The use of the product $H_{k+1}X_{k+1}$ and $H_m X_{k+1}$ is explained in Section 2.5. Specializing (10.10) and (10.11), estimates and smoothers for various processes of interest are now obtained.

■

The State Process

Here $H_{k+1} = H_0 = 1$ and $\Delta H_{k+1} = 0$. Denoting $\tilde{\gamma}_{k,k}(1)$ by $\tilde{q}_k$ we have from (10.10) and (10.11)

$$\boxed{\tilde{q}_{k+1} = MS\tilde{q}_k Y'_{k+1}} \qquad (10.12)$$

which we have already obtained in Lemma 10.4. For $m < k+1$ we have the smoothed estimate

$$\boxed{\tilde{\gamma}_{m,k+1}(\langle X_m, e_p \rangle) = MS\tilde{\gamma}_{m,k}(\langle X_m, e_p \rangle) Y'_{k+1}.} \qquad (10.13)$$

The Number of Jumps

Here $H_{k+1} = \mathscr{J}_{k+1}^{pq} = \sum_{n=1}^{k+1} \langle X_{n-1}, e_q \rangle \langle X_n, e_p \rangle$ and $\Delta H_{k+1} = \langle X_k, e_p \rangle \times \langle X_{k+1}, e_q \rangle$. Substitution of these quantities in (10.10) and (10.11) gives the estimates and smoothers for the number of jumps:

$$\boxed{\tilde{\gamma}_{k+1,k+1}(\mathscr{J}_{k+1}^{pq}) = M\left(S\tilde{\gamma}_{k,k}(\mathscr{J}_k^{pq}) Y'_{k+1} + \langle \tilde{q}_k, e_p \rangle \langle Y_{k+1}, s_{\cdot qp} \rangle e_q\right)} \qquad (10.14)$$

and for $m < k+1$ we have the smoothed estimate

$$\boxed{\tilde{\gamma}_{m,k+1}(\mathscr{J}_m^{pq}) = MS\tilde{\gamma}_{m,k}(\mathscr{J}_m^{pq}) Y'_{k+1}.} \qquad (10.15)$$

The Occupation Time

Here $H_{k+1} = \mathscr{O}_{k+1}^{p} = \sum_{n=1}^{k+1} \langle X_n, e_p \rangle$ and $\Delta H_{k+1} = \langle X_k, e_p \rangle$. Using again (10.10) and (10.11) we have the estimates

$$\boxed{\tilde{\gamma}_{k+1,k+1}(\mathscr{O}_{k+1}^{p}) = M\left(S\tilde{\gamma}_{k,k}(\mathscr{O}_k^{p}) Y'_{k+1} + \langle \tilde{q}_k, e_p \rangle \langle Y_{k+1}, s_{\cdot\cdot p} \rangle\right),} \qquad (10.16)$$

where $\langle Y_{k+1}, s_{\cdot\cdot p}\rangle = \sum_{r=1}^{M} \langle Y_{k+1}, f_r\rangle s_{r\cdot p}$, and the smoothers for $m < k+1$

$$\boxed{\tilde{\gamma}_{m,k+1}\left(\mathscr{O}_m^p\right) = MS\tilde{\gamma}_{m,k}\left(\mathscr{O}_m^p\right) Y'_{k+1}.} \tag{10.17}$$

The Process Related to the Observations

Here $H_{k+1} = \mathscr{T}_{k+1}^{ps} = \sum_{\ell=1}^{k+1} \langle X_{\ell-1}, e_p\rangle \langle Y_\ell, f_s\rangle$ and $\Delta H_{k+1} = \langle X_k, e_p\rangle \langle Y_{k+1}, f_s\rangle$. Again, substitution in (10.10) and (10.11) gives

$$\boxed{\tilde{\gamma}_{k+1,k+1}\left(\mathscr{T}_{k+1}^{ps}\right) = M\left(S\tilde{\gamma}_{k,k}\left(\mathscr{T}_k^{ps}\right) Y'_{k+1} + \langle \tilde{q}_k, e_p\rangle \langle Y_{k+1}, f_s\rangle s_{s\cdot p}\right)} \tag{10.18}$$

and for $m < k+1$ we have the smoothed estimate

$$\boxed{\tilde{\gamma}_{m,k+1}\left(\mathscr{T}_m^{ps}\right) = MS\tilde{\gamma}_{m,k}\left(\mathscr{T}_m^{ps}\right) Y'_{k+1}.} \tag{10.19}$$

The Joint Transition

In the dependent situation a new feature is the joint transition probabilities. Here $H_{k+1} = \mathscr{L}_{k+1}^{tqp} = \sum_{\ell=1}^{k+1} \langle Y_\ell, f_t\rangle \langle X_\ell, e_q\rangle \langle X_{\ell-1}, e_p\rangle$ and $\Delta H_{k+1} = \langle Y_{k+1}, f_t\rangle \langle X_{k+1}, e_q\rangle \times \langle X_k, e_p\rangle$. Estimates and smoothers for the joint transitions are obtained using again (10.10) and (10.11). These are:

$$\boxed{\tilde{\gamma}_{k+1,k+1}\left(\mathscr{L}_{k+1}^{tqp}\right) = M\left(S\tilde{\gamma}_{k,k}\left(\mathscr{L}_k^{tqp}\right) Y'_{k+1} + \langle \tilde{q}_k, e_p\rangle \langle Y_{k+1}, f_t\rangle s_{tqp} e_q\right)} \tag{10.20}$$

and

$$\boxed{\tilde{\gamma}_{m,k+1}\left(\mathscr{L}_m^{tqp}\right) = MS\tilde{\gamma}_{m,k}\left(\mathscr{L}_m^{tqp}\right) Y'_{k+1}.} \tag{10.21}$$

Parameter Estimation

Our hidden Markov model is described by the equations:

$$\begin{aligned} X_{k+1} &= AX_k + V_{k+1} \\ Y_{k+1} &= CX_k + W_{k+1} \\ Y_{k+1}X'_{k+1} &= SX_k + \Gamma_{k+1}, \qquad k \in \mathbb{N}. \end{aligned}$$

The parameters in the model are, therefore, given in a set

$$\begin{aligned} \theta = \big\{ & a_{ji}, 1 \le i, j \le N; \\ & c_{ji}, 1 \le j \le M,\ 1 \le i \le N; \\ & s_{rji}, 1 \le r \le M,\ 1 \le i, j \le N \big\}. \end{aligned}$$

These satisfy

$$\sum_{j=1}^{N} a_{ji} = 1, \qquad \sum_{j=1}^{M} c_{ji} = 1, \qquad \sum_{r=1}^{M} \sum_{j=1}^{N} s_{rji} = 1. \tag{10.22}$$

Suppose such a set θ is given and we wish to determine a new set $\hat{\theta} = \{(\hat{a}_{ji}(k)), (\hat{c}_{ji}(k)), (\hat{s}_{rji}(k))\}$ which maximizes the log-likelihood function defined below. Consider the parameters $(s_{rji}, 1 \le r \le M, 1 \le i, j \le N)$. To replace the joint transitions s_{rji} by $\hat{s}_{rji}(k)$ consider the Radon-Nikodym derivatives

$$\left.\frac{d\hat{P}}{dP}\right|_{\mathscr{G}_k} = \prod_{\ell=1}^{k} \prod_{r=1}^{M} \prod_{i,j=1}^{N} \left[\frac{\hat{s}_{rji}(k)}{s_{rji}(k)}\right]^{\langle Y_\ell, f_r\rangle \langle X_\ell, e_j\rangle \langle X_{\ell-1}, e_i\rangle}.$$

Therefore

$$E\left[\log \left.\frac{d\hat{P}}{dP}\right|_{\mathscr{G}_k} \mid \mathscr{Y}_k\right] = \sum_{r=1}^{M} \sum_{i,j=1}^{N} \hat{\mathscr{L}}_k^{rji} \log \hat{s}_{rji}(k) + \hat{R}(s), \tag{10.23}$$

where $\hat{R}(s)$ is independent of $\hat{s}$. Now observe that

$$\sum_{r=1}^{M} \sum_{j=1}^{N} \mathscr{L}_k^{rji} = \mathscr{O}_k^i. \tag{10.24}$$

Conditioning (10.24) on $\mathscr{Y}_k$ we have:

$$\sum_{r=1}^{M} \sum_{j=1}^{N} \hat{\mathscr{L}}_k^{rji} = \hat{\mathscr{O}}_k^i. \tag{10.25}$$

Now the $\hat{s}_{rji}(k)$ must also satisfy:

$$\sum_{r=1}^{M} \sum_{i=1}^{N} \hat{s}_{rji}(k) = 1. \tag{10.26}$$

We wish, therefore, to choose the $\hat{s}_{rji}(k)$ to maximize the conditional log-likelihood (10.23) subject to the constraint (10.26). Write λ for the *Lagrange multiplier* and put

$$F(\hat{s}, \lambda) = \sum_{r=1}^{M} \sum_{i,j=1}^{N} \hat{\mathscr{L}}_k^{rji} \log \hat{s}_{rji}(k) + \hat{R}(s) + \lambda \left(\sum_{r=1}^{M} \sum_{j=1}^{N} \hat{s}_{rji}(k) - 1 \right).$$

Equating the derivatives of F in $\hat{s}_{rji}(k)$ and λ to zero we have that the optimum choice of $\hat{s}_{rji}(k)$ is given, when defined, by

$$\boxed{\hat{s}_{rji}(k) = \frac{\hat{\mathscr{L}}_k^{rji}}{\hat{\mathscr{O}}_k^i} = \frac{\tilde{\gamma}_k\left(\mathscr{L}_k^{rji}\right)}{\tilde{\gamma}_k\left(\mathscr{O}_k^i\right)}.} \tag{10.27}$$

Similarly, as in Section 2.7 the optimal choice for $\hat{a}_{ji}(k)$ and $\hat{c}_{ji}(k)$ given the observations are, respectively, when defined

$$\boxed{\hat{a}_{ji}(k) = \frac{\tilde{\gamma}_k\left(\mathscr{J}_k^{ij}\right)}{\tilde{\gamma}_k\left(\mathscr{O}_k^i\right)}} \tag{10.28}$$

and

$$\boxed{\hat{c}_{ji}(k) = \frac{\tilde{\gamma}_k\left(\mathscr{T}_k^{ij}\right)}{\tilde{\gamma}_k\left(\mathscr{O}_k^i\right)}.} \tag{10.29}$$

Remark 10.7 We have found recursive expressions for $\tilde{\gamma}_k(\mathscr{O}_k^i)$, $\tilde{\gamma}_k(\mathscr{L}_k^{rji})$, $\tilde{\gamma}_k(\mathscr{J}_k^{ij})$ and $\tilde{\gamma}_k(\mathscr{T}_k^{ij})$. The revised parameters $\hat{\theta} = ((\hat{a}_{ji}(k)), (\hat{c}_{ji}(k)), (\hat{s}_{rji}(k)))$, are then determined by (10.27), (10.28), and (10.29). This procedure can be iterated and an increasing sequence of likelihood ratios obtained.

■

A Test for Independence

Taking inner products with $\underline{1} \in \mathbb{R}^N$, (10.16) and (10.20) provide estimates for $\tilde{\gamma}_k\left(\mathscr{O}_k^i\right)$ and $\tilde{\gamma}_k\left(\mathscr{L}_k^{rji}\right)$, respectively; an optimal estimate for $\hat{s}_{rji}(k)$ is then obtained from (10.27). However, if the noise terms in the state X and observation Y are independent we have

$$SX_k = C \operatorname{diag} X_k A'.$$

Taking $X_k = e_i$ and considering

$$\left\langle Se_i, f_r e_j' \right\rangle = \left\langle Ce_i, f_r \right\rangle \left\langle Ae_i, e_j \right\rangle$$

we see that if the noise terms are independent:

$$s_{rji} = c_{ri} a_{ji}$$

for $1 \le r \le M$, $1 \le i, j \le N$. If the noise terms are independent $\gamma_{k,k}\left(\mathscr{J}_k^{ij}\right)$, $\gamma_{k,k}\left(\mathscr{O}_k^i\right)$, and $\gamma_{k,k}\left(\mathscr{T}_k^{ij}\right)$ are given in Section 2.6. Taking inner products with $\underline{1} \in \mathbb{R}^N$ gives estimates for $\gamma_k\left(\mathscr{J}_k^{ij}\right)$, $\gamma_k\left(\mathscr{O}_k^i\right)$, and $\gamma_k\left(\mathscr{T}_k^{ij}\right)$, and substituting in (10.28) and (10.29) gives estimates for $\hat{a}_{ji}(k)$ and $\hat{c}_{ji}(k)$. Consequently, a test for independence is to check whether

$$\hat{s}_{rji}(k) = \hat{c}_{ri}(k) \cdot \hat{a}_{ji}(k).$$

Modification of our model and this test will give other tests for independence. For example, by enlarging the state space, so the state at time k is in fact (X_{k+1}, X_k) a test can be devised to check whether either the process X_k is Markov, or (X_{k+1}, X_k) is Markov, in a hidden Markov model situation. Alternatively, models can be considered where X_{k+1} and Y_{k+1} depend also on Y_k.

2.11 Problems and Notes

Problems

1. Show that $\overline{\Lambda}_k$ defined in Section 2.3 is a $(\overline{P}, \mathscr{G}_k)$-martingale, and Λ_k defined in Section 2.7 is a $(P, \mathscr{G}_k)$-martingale.
2. Fill in the details in the proof of Theorem 5.3.
3. Write $\rho_{m,k}(e_r) = \overline{E}\left[\langle X_m, e_r\rangle \overline{\Lambda}_k \mid \mathscr{Y}_k\right]$, $\overline{\Lambda}_{m,k} = \prod_{\ell=m}^{k} \overline{\gamma}_\ell$ and $\beta_{m,k}(e_r) = \overline{E}[\overline{\Lambda}_{m+2,k} \mid X_m = e_r, \mathscr{Y}_k]$. Show that $\beta_{m,k}$ satisfies the following backward recursive equation

$$\boxed{\beta_{m,k}(e_r) = M \sum_{\ell=1}^{N} \prod_{i=1}^{M} d_{i\ell}^{Y_{m+2}^i} \beta_{m,k}(e_\ell)\, p_{r\ell}}$$

and $\beta_{m,k}(\cdot) = \beta_{n-1,k}(\cdot) = 1$. Then verify that:

$$\boxed{\rho_{m,k}(e_r) = q_m(e_r)\, \beta_{m,k}(e_r) \prod_{i=1}^{M} d_{ir}^{Y_m^i},}$$

where $q_m(\cdot)$ is given recursively by (4.3).
4. Prove Theorem 7.4.
5. It is pointed out in Section 2.10 that alternatively, the transitions at time k of the processes X and Y could also depend on Y_{k-1}. Describe the dynamics of this model and define a new probability measure under which the observed process Y is a sequence of i.i.d. random variables uniformly distributed.
6. Using a "double change of measure" changing both processes X and Y into i.i.d. uniform random variables, rederive the recursions of Sections 2.4 to 2.6.

Notes

Hidden Markov models, HMMs, have found applications in many areas. The survey by Rabiner (1989) describes their role in speech processing. Stratonovich (1960) describes some similar models in Stratonovich (1960). The results of Aström (1965) are obtained using Bayes' rule, and the recursion he obtained is related to Theorem 4.3.

The expectation maximization, EM, algorithm was first introduced by Baum and Petrie (1966) and further developed by Dempster et al. (1977).

Our formulation, in terms of filters which estimate the number of jumps from one state to another $\mathcal{J}$, the occupation time $\mathcal{O}$, and the $\mathcal{T}$ process, avoids use of the forward-backward algorithm and does not require so much memory. However, it requires a larger number of calculations that can be done in parallel.

Related contributions can be found in Boel (1976) and Segall (1976b). The latter discusses only a single counting observation process. Boel has considered multidimensional point processes, but has not introduced Zakai equations or the change of measure.

The continuous-time versions of these results are presented in Chapters 7 and 8.

Chapter 3
Continuous-Range Observations

3.1 Introduction

This chapter first considers a discrete-time, finite-state Markov chain which is observed through a real- or vector-valued function whose values are corrupted by noise. For simplicity, we assume Gaussian noise. The main tool is again a version in discrete time of Girsanov's theorem. An explicit construction is given of a new measure $\overline{P}$ under which all components of the observation process are $N(0,1)$ i.i.d. random variables. Working under $\overline{P}$ we obtain unnormalized, recursive estimators and smoothers for the state of the Markov chain, given the observations. Furthermore, recursive estimators and smoothers are derived for the number of jumps of the chain from one state to another, for the occupation time in any state, and for processes related to the observation process. These estimators allow the parameters of our model, including the variance of the observation noise, to be reestimated using the EM algorithm (Baum, Petrie, Soules and Weiss 1970). Optimal recursive estimators for the state and parameters are obtained using techniques similar to those of Chapter 2.

In the later part of the chapter observations with colored noise are considered, that is, there is correlation between the noise terms in the signal at consecutive times. More generally, the case is discussed where a Markov chain influences a linear system which, in turn, is observed in noise. That is, mixed continuous-range and discrete-range state models are studied. All calculations take place under the reference probability measure $\overline{P}$ for which the observations are i.i.d.

3.2 State and Observation Processes

All processes will be defined on a complete probability space $(\Omega, \mathscr{F}, P)$. The discrete-time parameter k will take values in $\mathbb{N}$. Suppose $\{X_k, k \in \mathbb{N}\}$ is a finite-state chain representing the signal process. As in Chapter 2 the state space of X is the set of unit vectors

$$S_X = \{e_1, e_2, \ldots, e_N\}, \qquad e_i = (0, \ldots, 0, 1, 0, \ldots, 0)' \in \mathbb{R}^N.$$

We assume X_0 is given, or its distribution or mean $E[X_0]$ is known. Also, we assume as in Chapter 2 that X is a homogeneous Markov chain, so

$$P(X_{k+1} = e_j \mid \mathscr{F}_k) = P(X_{k+1} = e_j \mid X_k).$$

Suppose X is not observed directly, but rather there is an observation process $\{y_k, k \in \mathbb{N}\}$. For simplicity suppose y is scalar. The case of vector y is discussed in Section 3.8.

The signal model with real-valued y process has the form

$$\boxed{\begin{aligned} X_{k+1} &= AX_k + V_{k+1}, \\ y_k &= c(X_k) + \sigma(X_k) w_k. \end{aligned}} \tag{2.1}$$

Here $\{w_k\}$ is a sequence of zero mean, unit variance *normally distributed* $N(0,1)$ i.i.d. random variables. Because $X_k \in S_X$ the functions c and σ are determined by vectors $c = (c_1, c_2, \dots, c_N)'$ and $\sigma = (\sigma_1, \sigma_2, \dots, \sigma_N)'$ in $\mathbb{R}^N$; then $c(X_k) = \langle c, X_k \rangle$ and $\sigma(X_k) = \langle \sigma, X_k \rangle$ where $\langle\ ,\ \rangle$ denotes the inner product in $\mathbb{R}^N$.

We shall assume $\sigma_i \neq 0$ and thus without loss of generality that $\sigma_i > 0$, $1 \leq i \leq N$.

Notation 2.1 *$\{\mathscr{F}_k\}$, $k \in \mathbb{N}$, will denote the complete filtration generated by X; $\{\mathscr{Y}_k\}$, $k \in \mathbb{N}$, will denote the complete filtration generated by y; $\{\mathscr{G}_k\}$, $k \in \mathbb{N}$, will denote the complete filtration generated by X and y.*

Remark 2.2 The observation model $y_k = c(X_k) + \sigma(X_k) w_k$, $k \in \mathbb{N}$, will be discussed in Section 3.9.

■

3.3 Conditional Expectations

We first quickly derive the conditional distribution of X_k given $\mathscr{Y}_k$ using elementary considerations. Recall the w_k, $k \in \mathbb{N}$, are $N(0,1)$ i.i.d. random variables, so w_k is independent of $\mathscr{G}_k$ and, in particular, of $\mathscr{Y}_k \subset \mathscr{G}_k$.

For $t \in \mathbb{R}$ consider the conditional distribution

$$P(y_{k+1} \leq t \mid \mathscr{Y}_k) = \sum_{i=1}^{N} P(\sigma_i w_{k+1} \leq t - c_i) P(X_{k+1} = e_i \mid \mathscr{Y}_k).$$

Write $\hat{X}_k = E[X_k \mid \mathscr{Y}_k]$ and $\phi_i(x) = \left(2\pi\sigma_i^2\right)^{-1/2} \exp\left(-x^2/2\sigma_i^2\right)$ for the $N(0, \sigma_i)$ density. Then

$$P(y_{k+1} \leq t \mid \mathscr{Y}_k) = \sum_{i=1}^{N} \langle \hat{X}_{k+1}, e_i \rangle \int_{-\infty}^{t-c_i} \phi_i(x)\, dx.$$

The conditional density of y_{k+1} given $\mathscr{Y}_k$ is, thus,

$$\sum_{i=1}^{N}\left\langle \hat{X}_{k+1}, e_i\right\rangle \phi_i(t-c_i).$$

Now the joint distribution

$$\begin{aligned} P(X_{k+1}=e_i,\, y_{k+1}\le t \mid \mathscr{Y}_k) &= P(X_{k+1}=e_i \mid \mathscr{Y}_k)P(w_{k+1}\le t-c_i) \\ &= \left\langle \hat{X}_{k+1}, e_i\right\rangle \int_{-\infty}^{t-c_i}\phi_i(x)\,dx. \end{aligned}$$

Therefore, using Bayes' rule

$$\begin{aligned} E\left[\left\langle X_{k+1}, e_i\right\rangle \mid \mathscr{Y}_{k+1}\right] &= P(X_{k+1}=e_i \mid y_{k+1}, \mathscr{Y}_k) \\ &= \frac{\left\langle \hat{X}_{k+1}, e_i\right\rangle \phi_i(y_{k+1}-c_i)}{\sum_{j=1}^{N}\left\langle \hat{X}_{k+1}, e_j\right\rangle \phi_j(y_{k+1}-c_j)}. \end{aligned}$$

Consequently,

$$\begin{aligned} E[X_{k+1} \mid \mathscr{Y}_{k+1}] &= \sum_{i=1}^{N} E[\langle X_{k+1}, e_i\rangle \mid \mathscr{Y}_{k+1}]\, e_i \\ &= \frac{\sum_{i=1}^{N}\left\langle \hat{X}_{k+1}, e_i\right\rangle \phi_i(y_{k+1}-c_i)\, e_i}{\sum_{j=1}^{N}\left\langle \hat{X}_{k+1}, e_j\right\rangle \phi_j(y_{k+1}-c_j)}. \end{aligned} \tag{3.1}$$

The recursive filter for $\hat{X}_k$ follows:

Theorem 3.1

$$\hat{X}_{k+1} = E[X_{k+1} \mid \mathscr{Y}_{k+1}] = \frac{\sum_{i=1}^{N}\left\langle \hat{X}_k, e_i\right\rangle \phi_i(y_{k+1}-c_i)\, Ae_i}{\sum_{j=1}^{N}\left\langle \hat{X}_k, e_j\right\rangle \phi_j(y_{k+1}-c_j)}. \tag{3.2}$$

Proof V_{k+1} is an $\mathscr{F}_k$-martingale increment so $E[V_{k+1} \mid \mathscr{F}_k]=0$. However, the w_k are i.i.d. so

$$E[V_{k+1} \mid \mathscr{G}_k, w_{k+1}] = E[V_{k+1} \mid \mathscr{F}_k] = 0.$$

Consequently, $E[V_{k+1} \mid \mathscr{Y}_{k+1}] = E[E[V_{k+1} \mid \mathscr{G}_k, w_{k+1}]\mathscr{Y}_{k+1}] = 0$ and

$$\begin{aligned} \hat{X}_{k+1} &= E[X_{k+1} \mid \mathscr{Y}_{k+1}] = E[AX_k + V_{k+1} \mid \mathscr{Y}_{k+1}] \\ &= AE[X_k \mid \mathscr{Y}_{k+1}]. \end{aligned}$$

Substituting in (3.1) the result follows. □

Remark 3.2 A difficulty with the recursion (3.2) is that it is not linear in $\hat{X}_k$. ■

3.4 Change of Measure

Suppose $w(.)$ is a real random variable with density $\phi(w)$ and c and σ are known constants. Write $y(.) = c + \sigma w(.)$.

We wish to introduce a new probability measure $\overline{P}$, using a density λ, so that $d\overline{P}/dP = \lambda$, and under $\overline{P}$ the random variable y has density ϕ. That is,

$$\begin{aligned}\overline{P}(y \leq t) &= \int_{\Omega} I_{y \leq t}\, d\overline{P} \\ &= \int_{\Omega} I_{y \leq t} \lambda\, dP \\ &= \int_{-\infty}^{+\infty} I_{w \leq \frac{t-c}{\sigma}} \lambda(w)\phi(w)\, dw \\ &= \int_{-\infty}^{t} \lambda(w)\phi(w) \frac{dy}{\sigma}. \end{aligned} \tag{4.1}$$

The last equality holds since $y(.) = c + \sigma w(.)$. For

$$\overline{P}(y \leq t) = \int_{-\infty}^{t} \phi(y)\, dy \tag{4.2}$$

to hold we must have

$$\lambda(w) = \frac{\sigma\phi(y)}{\phi(w)}.$$

So far, on $(\Omega, \mathscr{F}, P)$, our observation process $\{y_k\}$, $k \in \mathbb{N}$, has the form $y_k = \langle c, X_k\rangle + \langle \sigma, X_k\rangle w_k$, where the w_k are $N(0,1)$ i.i.d. Write $\phi(\cdot)$ for the $N(0,1)$ density,

$$\begin{aligned}\lambda_\ell &= \frac{\langle \sigma, X_\ell\rangle \phi(y_\ell)}{\phi(w_\ell)}, \qquad \ell \in \mathbb{N}, \\ \Lambda_0 &= 1,\end{aligned}$$

and

$$\Lambda_k = \prod_{\ell=1}^{k} \lambda_\ell, \qquad k \geq 1.$$

Define a new probability measure $\overline{P}$ by setting the restriction of the Radon-Nikodym derivative to $\mathscr{G}_k$ equal to Λ_k: $\left(d\overline{P}/dP\right)|_{\mathscr{G}_k} = \Lambda_k$. The existence of $\overline{P}$ follows from Kolmogorov's Extension Theorem.

Lemma 4.1 *Under $\overline{P}$ the y_k are $N(0,1)$ i.i.d. random variables.*

Proof Now $\overline{P}(y_{k+1} \leq t \mid \mathscr{G}_k) = \overline{E}[I(y_{k+1} \leq t) \mid \mathscr{G}_k]$, and from a version of Bayes' Theorem this is

$$
\begin{aligned}
&= \frac{E\left[\Lambda_{k+1} I\left(y_{k+1} \leq t\right) \mid \mathscr{G}_k\right]}{E\left[\Lambda_{k+1} \mid \mathscr{G}_k\right]} \\
&= \frac{\Lambda_k}{\Lambda_k} \cdot \frac{E\left[\lambda_{k+1} I\left(y_{k+1} \leq t\right) \mid \mathscr{G}_k\right]}{E\left[\lambda_{k+1} \mid \mathscr{G}_k\right]}.
\end{aligned}
$$

Now under P, $w_{k+1} = \frac{y_{k+1} - \langle c, X_k \rangle}{\langle \sigma, X_k \rangle}$

$$
\begin{aligned}
E\left[\lambda_{k+1} \mid \mathscr{G}_k\right] &= E\left[E\left[\lambda_{k+1} \mid \mathscr{G}_k, X_{k+1}\right] \mid \mathscr{G}_k\right] \\
&= \int_{-\infty}^{\infty} \frac{\langle \sigma, X_{k+1} \rangle \phi\left(y_{k+1}\right)}{\phi\left(w_{k+1}\right)} \cdot \phi\left(w_{k+1}\right) \frac{1}{\langle \sigma, X_{k+1} \rangle} dy_{k+1} \\
&= \int_{-\infty}^{\infty} \phi\left(y_{k+1}\right) \cdot dy_{k+1} = 1,
\end{aligned}
$$

so

$$
\begin{aligned}
\overline{P}\left(y_{k+1} \leq t \mid \mathscr{G}_k\right) &= E\left[\lambda_{k+1} I\left(y_{k+1} \leq t\right) \mid \mathscr{G}_k\right] \\
&= \int_{-\infty}^{\infty} \frac{\langle \sigma, X_k \rangle \phi\left(y_{k+1}\right)}{\phi\left(w_{k+1}\right)} \cdot I\left(y_{k+1} \leq t\right) \phi\left(w_{k+1}\right) dw_{k+1} \\
&= \int_{-\infty}^{t} \phi\left(y_{k+1}\right) dy_{k+1} = \overline{P}\left(y_{k+1} \leq t\right).
\end{aligned}
$$

The result follows. □

Lemma 4.2 *Under the measure P, the process X remains a Markov process, with transition matrix A and initial distribution p_0.*

Proof To prove Lemma 4.2, we must show that under P, X has the Markov property and has both rate matrix A and initial distribution p_0. The equality of initial distributions for X under each measure is immediate from Bayes' rule, that is,

$$
\begin{aligned}
E\left[Z_0\right] &= \frac{\overline{E}\left[\Lambda_0^{-1} Z_0 \mid \{\Omega, \emptyset\}\right]}{\overline{E}\left[\Lambda_0^{-1} \mid \{\Omega, \emptyset\}\right]} \\
&= \frac{\overline{E}\left[Z_0\right]}{\overline{E}[1]} \\
&= \overline{E}\left[Z_0\right] = p_0.
\end{aligned} \tag{4.3}
$$

To complete the proof of Lemma 4.2, we examine the following expectation,

$$
\begin{aligned}
E\left[\langle X_k, e_j \rangle \mid \mathscr{G}_{k-1}\right] &= \frac{\overline{E}\left[\Lambda_k \langle X_k, e_j \rangle \mid \mathscr{G}_{k-1}\right]}{\overline{E}\left[\Lambda_k \mid \mathscr{G}_{k-1}\right]} \\
&= \frac{\Lambda_{k-1} \overline{E}\left[\lambda_k \langle X_k, e_j \rangle \mid \mathscr{G}_{k-1}\right]}{\Lambda_{k-1} \overline{E}\left[\lambda_k \mid \mathscr{G}_{k-1}\right]} \\
&= \frac{\overline{E}\left[\lambda_k \langle X_k, e_j \rangle \mid \mathscr{G}_{k-1}\right]}{\overline{E}\left[\lambda_k \mid \mathscr{G}_{k-1}\right]}.
\end{aligned} \tag{4.4}
$$

Using repeated conditioning, we rewrite the denominator in the last line of equation (4.4) as

$$\begin{aligned}\overline{E}\big[\lambda_k \mid \mathscr{G}_{k-1}\big] &= \overline{E}\left[\frac{\phi\Big(\big(y_k - \langle c, X_k\rangle\big)/\langle X_k, \sigma\rangle\Big)}{\langle X_k, \sigma\rangle \phi(y_k)} \mid \mathscr{G}_{k-1}\right] \\ &= \overline{E}\left[\overline{E}\left[\frac{\phi\Big(\big(y_k - \langle c, X_k\rangle\big)/\langle X_k, \sigma\rangle\Big)}{\langle X_k, \sigma\rangle \phi(y_k)} \mid X_k\,,\, \mathscr{G}_{k-1}\right] \mid \mathscr{G}_{k-1}\right]. \end{aligned} \tag{4.5}$$

The inner expectation of equation (4.5) can be evaluated by noting that,

$$\begin{aligned}&\overline{E}\left[\frac{\phi\Big(\big(y_k - \langle c, X_k\rangle\big)/\langle X_k, \sigma\rangle\Big)}{\langle X_k, \sigma\rangle \phi(y_k)} \mid X_k\,,\, \mathscr{G}_{k-1}\right] \\ &= \frac{1}{\langle X_k, \sigma\rangle}\int_{\mathbb{R}} \frac{\phi\Big(\big(\xi - \langle c, X_k\rangle\big)/\langle X_k, \sigma\rangle\Big)}{\phi(\xi)}\phi(\xi)d\xi \\ &= \frac{1}{\langle X_k, \sigma\rangle}\int_{\mathbb{R}} \phi\Big(\big(\xi - \langle c, X_k\rangle\big)/\langle X_k, \sigma\rangle\Big)d\xi \\ &= \frac{1}{\langle X_k, \sigma\rangle}\int_{\mathbb{R}} \phi(\varphi)\langle X_k, \sigma\rangle d\varphi \\ &= 1. \end{aligned} \tag{4.6}$$

It follows that

$$E\big[\langle X_k, e_i\rangle \mid \mathscr{G}_{k-1}\big] = \overline{E}\big[\lambda_k\langle X_k, e_i\rangle \mid \mathscr{G}_{k-1}\big]. \tag{4.7}$$

Again, using repeated conditioning, we see that

$$\begin{aligned}\overline{E}\big[\lambda_k\langle X_k, e_i\rangle \mid X_k \,\&\, \mathscr{G}_{k-1}\big] &= \frac{\langle X_k, e_i\rangle}{\langle X_k, \sigma\rangle}\int_{\mathbb{R}} \phi(\xi)\langle X_k, \sigma\rangle d\xi \\ &= \langle X_k, e_i\rangle. \end{aligned} \tag{4.8}$$

Consequently

$$\begin{aligned}E\big[\langle X_k, e_i\rangle \mid \mathscr{G}_{k-1}\big] = \overline{E}\big[\lambda_k\langle X_k, e_i\rangle \mid \mathscr{G}_{k-1}\big] &= \overline{E}\big[\langle X_k, e_i\rangle \mid \mathscr{G}_{k-1}\big] \\ &= \overline{E}\big[\langle X_k, e_i\rangle \mid X_{k-1}\big] \\ &= \langle AX_{k-1}, e_i\rangle. \end{aligned} \tag{4.9}$$

□

Conversely, we suppose we start with a probability measure $\overline{P}$ on $(\Omega, \mathscr{F})$ such that under $\overline{P}$

1. $\{X_k\}$, $k \in \mathbb{N}$, is a Markov chain with transition matrix A, so that $X_{k+1} = AX_k + V_{k+1}$, where $\overline{E}[V_{k+1} \mid \mathscr{F}_k] = 0$, and
2. $\{y_k\}$, $k \in \mathbb{N}$, is a sequence of $N(0,1)$ i.i.d. random variables, (which are, in particular, independent of the X_k).

We then wish to construct a probability measure P such that under P

$$w_{k+1} := \frac{y_{k+1} - \langle c, X_{k+1} \rangle}{\langle \sigma, X_{k+1} \rangle}, \qquad k \in \mathbb{N},$$

is a sequence of $N(0,1)$ i.i.d. random variables. That is, under P, $y_{k+1} = \langle c, X_{k+1} \rangle + \langle \sigma, X_{k+1} \rangle w_{k+1}$.

To construct P from $\overline{P}$ we introduce the inverses of λ_ℓ and Λ_k. Write

$$\overline{\lambda}_\ell = \lambda_\ell^{-1} = \frac{\phi(w_\ell)}{\langle \sigma, X_\ell \rangle \phi(y_\ell)},$$
$$\overline{\Lambda}_0 = 1$$

and

$$\overline{\Lambda}_k = \prod_{\ell=1}^{k} \overline{\lambda}_\ell, \qquad k \geq 1,$$

and define P by putting $(dP/d\overline{P})|_{\mathscr{G}_k} = \overline{\Lambda}_k$. Clearly the construction of P using the factors $\overline{\lambda}_k$ requires $\langle \sigma, X_k \rangle \neq 0$. The assumption that the observation process has nonsingular noise is standard in filtering theory. If the components of c are all different and $\langle \sigma, X_k \rangle = 0$ then observing $y_{k+1} = c_r$ implies $X_k = e_r$.

Lemma 4.3 *Under P the $\{w_k\}$, $k \in \mathbb{N}$, is a sequence of $N(0,1)$ i.i.d. random variables.*

Proof The proof is left as an exercise. □

Remark 4.4 We shall work under $\overline{P}$. However, it is under P that $y_k = \langle c, X_k \rangle + \langle \sigma, X_k \rangle w_k$ with the w_k, $N(0,1)$ and i.i.d.

■

3.5 Filter-Based State Estimation

The method we develop to estimate the parameters of the model, is based upon estimating the state process X and estimating a set of quantities derived from the process X. Two of these derived processes are related to the state process and one is related to both the state and observation process. These processes are listed below.

1. $N_k^{(j,i)}$, a discrete time counting process for the state transitions $e_i \to e_j$, where $i \neq j$,

$$N_k^{(j,i)} = \sum_{\ell=1}^{k} \langle X_{\ell-1}, e_i \rangle \langle X_\ell, e_j \rangle. \tag{5.1}$$

2. J_k^i, the cumulative sojourn time spent by the process X in state e_i,

$$J_k^i = \sum_{\ell=1}^{k} \langle X_{\ell-1}^k, e_i \rangle. \tag{5.2}$$

3. G_k^i, the level sum for the state e_i

$$G_k^i = \sum_{\ell=1}^{k} f(y_\ell) \langle X_{\ell-1}, e_i \rangle. \tag{5.3}$$

Here the function $f(\cdot)$ is any bounded mapping.

A Filter for the process X

To estimate the state process, given an observation set and an assumed model, we wish to evaluate the expectation $E\left[X_k \mid \mathscr{Y}_k\right]$. Using a version of Bayes' rule, this is,

$$E\left[X_k \mid \mathscr{Y}_k\right] = \frac{\overline{E}\left[\Lambda_k X_k \mid \mathscr{Y}_k\right]}{\overline{E}\left[\Lambda_k \mid \mathscr{Y}_k\right]}. \tag{5.4}$$

Write

$$q_k \triangleq \overline{E}[\Lambda_k X_k \mid \mathscr{Y}_k] \in \mathbb{R}^n. \tag{5.5}$$

Definition 5.1 We define the following quantities:

$$\gamma_j(y_{k+1}) \triangleq \frac{\phi\big((y_{k+1} - \langle c, e_j \rangle)/\langle \sigma, e_j \rangle\big)}{\langle \sigma, e_j \rangle \phi(y_{k+1})}, \tag{5.6}$$

$$B(y_{k+1}) \triangleq \begin{bmatrix} \gamma_1(y_{k+1}) & & \\ & \ddots & \\ & & \gamma_n(y_{k+1}) \end{bmatrix}, \tag{5.7}$$

$$\Gamma_{0,k} \triangleq B(y_k) A B(y_{k-1}) A B(y_{k-2}) \cdots A B(y_1). \tag{5.8}$$

Lemma 5.2 *The unnormalised probability vector q_k, is computed by the recursion*

$$\begin{aligned} q_k &= B(y_{k+1}) A q_k \\ &= \Gamma_{0,k} q_0. \end{aligned} \tag{5.9}$$

Proof

$$\begin{aligned}
q_{k+1} &= \overline{E}\left[\Lambda_{k+1}X_{k+1} \mid \mathscr{Y}_{k+1}\right] \\
&= \overline{E}\left[\Lambda_k \frac{\phi\left((y_{k+1}-\langle c, X_{k+1}\rangle)/\langle \sigma, X_{k+1}\rangle\right)}{\langle \sigma, X_{k+1}\rangle\, \phi(y_{k+1})} X_{k+1} | \mathscr{Y}_{k+1}\right] \\
&= \sum_{j=1}^{n} \overline{E}\left[\Lambda_k \langle X_{k+1}, e_j\rangle \mid \mathscr{Y}_k\right] \gamma_j(y_{k+1}) e_j \\
&= \sum_{j=1}^{n} \langle A q_k, e_j\rangle \gamma_j(y_{k+1}) e_j \\
&= B(y_{k+1}) A q_k.
\end{aligned}$$

□

To compute the normalised probability corresponding to $q_{0,k}$, one notes that $\overline{E}[\Lambda_k \mid \mathscr{Y}_k] = \langle q_{0,k}, \underline{1}\rangle$. Here $\underline{1} = (1,1,\dots,1)' \in \mathbb{R}^n$.

Filters for the processes $N^{(j,i)}$, J^i and G^i

To re-estimate parameters c and A we shall require estimates of the quantities defined by equations (5.1), (5.2) and (5.3). Rather than directly estimating the quantities, $N_k^{(j,i)}$, J_k^i, and G_k^i, recursive forms can be found to estimate of the related product-quantities, $N_k^{(j,i)} X_k \in \mathbb{R}^n$, $J_k^{(i)} X_k \in \mathbb{R}^n$ and $G_k^{(i)} X_k \in \mathbb{R}^n$. The outputs of these filters can then be manipulated to marginalise out the state process X, resulting in filtered estimates of the quantities of primary interest, namely $N^{(j,i)}$, J^i and G^i.

Write

$$q(N_k^{(j,i)} X_k) \triangleq \overline{E}\left[\Lambda_k N_k^{(j,i)} X_k \mid \mathscr{Y}_k\right]. \tag{5.10}$$

Lemma 5.3 *The process $q(N^{(j,i)}X)$ is computed recursively by the dynamics*

$$q(N_{k+1}^{(j,i)} X_{k+1}) = B(y_{k+1}) A q(N_k^{(j,i)} X_k) + a_{(j,i)} \langle q_k, e_i\rangle \gamma_j(y_{k+1}) e_j. \tag{5.11}$$

Proof

$$\begin{aligned}
q(N_{k+1}^{(j,i)} X_{k+1}) &= \overline{E}\left[\Lambda_{k+1} N_{k+1}^{(j,i)} X_{k+1} \mid \mathscr{Y}_{k+1}\right] \\
&= \overline{E}\left[\Lambda_k \frac{\phi\left((y_{k+1}-\langle c, X_{k+1}\rangle)/\langle \sigma, X_{k+1}\rangle\right)}{\langle \sigma, X_{k+1}\rangle \phi(y_{k+1})} \right. \\
&\qquad \left. \times \left(N_k^{(j,i)} + \langle X_k, e_i\rangle \langle X_{k+1}, e_j\rangle\right) X_{k+1} \mid \mathscr{Y}_{k+1}\right] \\
&= \sum_{\ell=1}^{n} \overline{E}\left[\Lambda_k \langle X_{k+1}, e_\ell\rangle (N_k^{(j,i)} + \langle X_k, e_i\rangle \langle X_{k+1}, e_j\rangle) \mid \mathscr{Y}_{0,k}\right] \gamma_\ell(y_{k+1}) e_\ell
\end{aligned}$$

$$
\begin{aligned}
&= \sum_{\ell=1}^{n} \Big\{ \overline{E}\big[\Lambda_k \langle AX_k, e_\ell \rangle N_k^{(j,i)} \mid \mathscr{Y}_{0,k}\big] \gamma_\ell(y_{k+1}) e_\ell \Big\} \\
&\qquad\qquad + \overline{E}\big[\Lambda_k \langle AX_k, e_j \rangle \langle X_k, e_i \rangle \mid \mathscr{Y}_{0,k}\big] \gamma_j(y_{k+1}) e_j \\
&= B(y_{k+1}) A q(N_k^{(j,i)} X_k) + a_{(j,i)} \langle q_k, e_i \rangle \gamma_j(y_{k+1}) e_j. \qquad (5.12)
\end{aligned}
$$

□

Write

$$
q(G_k^i X_k) \triangleq \overline{E}\big[\Lambda_k G_k^i X_k \mid \mathscr{Y}_k\big]. \qquad (5.13)
$$

Lemma 5.4 *The process $q(G^i X)$ is computed recursively by the dynamics*

$$
q(G_{k+1}^i X_{k+1}) = B(y_{k+1}) A q(G_k^i X_k) + f(y_{k+1}) \langle q_k, e_i \rangle B(y_{k+1}) A e_i. \qquad (5.14)
$$

Proof

$$
\begin{aligned}
\sigma(G_{k+1}^i X_{k+1}) &= \overline{E}\big[\Lambda_{k+1} G_{k+1}^i X_{k+1} \mid \mathscr{Y}_{k+1}\big] \\
&= \overline{E}\Bigg[\Lambda_k \frac{\phi\big((y_{k+1} - \langle c, X_{k+1}\rangle)/\langle \sigma, X_{k+1}\rangle\big)}{\langle \sigma, X_{k+1}\rangle \phi(y_{k+1})} \\
&\qquad\qquad \times \big(G_k^i + f(y_{k+1}) \langle X_k, e_i \rangle\big) X_{k+1} \mid \mathscr{Y}_{k+1}\Bigg] \\
&= \sum_{\ell=1}^{n} \overline{E}\big[\Lambda_k \langle X_{k+1}, e_\ell \rangle \gamma_\ell(y_{k+1}) G_k^i \mid \mathscr{Y}_{k+1}\big] e_\ell \\
&\qquad\qquad + \sum_{\ell=1}^{n} \overline{E}\big[\Lambda_k \langle X_{k+1}, e_\ell \rangle \langle X_k, e_i \rangle \mid \mathscr{Y}_k\big] f(y_{k+1}) \gamma_\ell(y_{k+1}) e_\ell \\
&= \sum_{\ell=1}^{n} \overline{E}\big[\Lambda_k \langle AX_k, e_\ell \rangle G_k^i \mid \mathscr{Y}_{k+1}\big] \gamma_\ell(y_{k+1}) e_\ell \\
&\qquad\qquad + \sum_{\ell=1}^{n} \overline{E}\big[\Lambda_k \langle X_k, e_i \rangle \mid \mathscr{Y}_{0,k}\big] a_{\ell i} f(y_{k+1}) \gamma_\ell(y_{k+1}) e_\ell \\
&= B(y_{k+1}) A q(G_k^i X_k) + f(y_{k+1}) \langle q_k, e_i \rangle B(y_{k+1}) A e_i. \qquad (5.15)
\end{aligned}
$$

□

Taking $f(y_k) \equiv 1$, we have the following recurrence for estimates of J:

Lemma 5.5

$$
q(J_{k+1}^i X_{k+1}) = B(y_{k+1}) A \sigma(J_k^i X_k) + \langle q_k, e_i \rangle B(y_{k+1}) A e_i. \qquad (5.16)
$$

Our primary objective in this article, is to estimate the parameters c, σ and A. The estimators for $\langle c, e_i \rangle$ and $a_{(i,j)}$ are given by, see Elliott (1993b), Elliott (1994a), Dembo and Zeitouni (1986), James, Krishnamurthy and Le Gland (1998),

$$\widehat{a}_{(i,j)} = E\left[a_{(i,j)} \mid \mathscr{Y}_{0,T}\right] = \frac{E\left[N_T^{(j,i)} \mid \mathscr{Y}_{0,T}\right]}{E\left[J_T^i \mid \mathscr{Y}_{0,T}\right]} = \frac{q(N_T^{(i,j)})}{q(J_T^i)},$$

$\forall$ pairs (i, j), $i \neq j$, (5.17)

$$\widehat{\langle c, e_i \rangle} = E\left[\langle c, e_i \rangle \mid \mathscr{Y}_{0,T}\right] = \frac{E\left[G_T^i \mid \mathscr{Y}_{0,T}\right]}{E\left[J_T^i \mid \mathscr{Y}_{0,T}\right]} = \frac{q(G_T^i)}{q(J_T^i)}. \tag{5.18}$$

Remark 5.6 Note that the estimator given by equation (5.17) is defined only for the off diagonal elements of the matrix A. However, one can compute the corresponding, (estimated), diagonal elements of A, by noting that $\forall i$, $\sum_{i=1}^{n} a_{(j,i)} = 1$. ■

The filter recursions given by Lemmata 5.3, 5.4 and 5.5, provide updates to estimate product processes, each involving the state process X. What we would like to do, is manipulate these filters so as to remove the dependence upon the process X. This manipulation is routine. Recall, for example, that

$$q(N_t^{(j,i)} X_t) \stackrel{\triangle}{=} \overline{E}\left[\Lambda_t N_t^{(j,i)} X_t \mid \mathscr{Y}_{0,t}\right]. \tag{5.19}$$

Since X takes values on a canonical basis of indicator functions, we see that

$$\begin{aligned}\langle q(N_k^{(j,i)} X_k), \mathbf{1}\rangle &= \langle \overline{E}\left[\Lambda_k N_k^{(j,i)} X_k \mid \mathscr{Y}_k\right], \mathbf{1}\rangle \\ &= \overline{E}\left[\Lambda_k N_k^{(j,i)} \langle X_k \rangle \mid \mathscr{Y}_k\right], \mathbf{1} \\ &= q(N_k^{(j,i)}).\end{aligned} \tag{5.20}$$

It follows that our quantities of interest are computed by

$$\sigma\left(N_k^{(j,i)}\right) = \langle \sigma\left(N_k^{(j,i)} X_k\right), \mathbf{1}\rangle, \tag{5.21}$$

$$\sigma\left(G_k^i\right) = \langle \sigma\left(G_k^i X_t\right), \mathbf{1}\rangle, \tag{5.22}$$

$$\sigma\left(J_k^i\right) = \langle \sigma\left(J_k^i X_k\right), \mathbf{1}\rangle. \tag{5.23}$$

3.6 Smoother-Based State Estimation

In this section we compute smoother update formulae for the process X and each of the quantities related to X given by equations (5.1), (5.2) and (5.3).

A Smoother for the process X

Suppose $0 \leq k \leq T$ and we are given the information $\mathscr{Y}_{0,T}$. We wish to estimate X_k given $\mathscr{Y}_{0,T}$. Again from Bayes' Theorem

$$E\big[X_k \mid \mathscr{Y}_{0,T}\big] = \frac{\overline{E}\big[\Lambda_{0,T}X_k \mid \mathscr{Y}_{0,T}\big]}{\overline{E}\big[\Lambda_{0,T} \mid \mathscr{Y}_{0,T}\big]}. \tag{6.1}$$

Now

$$\overline{E}\big[\Lambda_{0,T}X_k \mid \mathscr{Y}_{0,T}\big] = \overline{E}\big[\Lambda_k X_k \overline{E}\big[\Lambda_{k+1,T} \mid \mathscr{Y}_{0,T} \vee \mathscr{F}_k\big]\mathscr{Y}_{0,T}\big], \tag{6.2}$$

where $\Lambda_{k+1,T} = \prod_{\ell=k+1}^{T} \lambda_\ell$. Then by the Markov property

$$\overline{E}\big[\Lambda_{k+1,T} \mid \mathscr{Y}_{0,T} \vee \mathscr{F}_k\big] = \overline{E}\big[\Lambda_{k+1,T} \mid \mathscr{Y}_{0,T} \vee \sigma\{X_k\}\big]. \tag{6.3}$$

Write $v_{k,T} = \big(\langle v_{k,T}, e_1\rangle, \dots, \langle v_{k,T}, e_n\rangle\big)$, where

$$\langle v_{k,T}, e_i\rangle \triangleq \overline{E}\big[\Lambda_{k+1,T} \mid \mathscr{Y}_{0,T} \vee \{X_k = e_i\}\big]. \tag{6.4}$$

Lemma 6.1 *The process v, for $k \in \{0, 1, 2, \dots, T-1\}$, satisfies the backward dynamics, (dual to q), of the form:*

$$v_{k,T} = A'B(y_{k+1})v_{k+1,T}. \tag{6.5}$$

Here A' denotes the matrix transpose of A.

Proof

$$\begin{aligned}
\langle v_{k,T}, e_i\rangle &= \overline{E}\left[\Lambda_{k+2,T}\frac{\phi\big((y_{k+1} - \langle c, X_{k+1}\rangle)/\langle\sigma, X_{k+1}\rangle\big)}{\langle\sigma, X_{k+1}\rangle\phi(y_{k+1})}\middle|\mathscr{Y}_{0,T} \vee \{X_k = e_i\}\right] \\
&= \sum_{j=1}^{n} \overline{E}\big[\langle X_{k+1}, e_j\rangle\Lambda_{k+2,T} \mid \mathscr{Y}_{0,T} \vee \{X_k = e_i\}\big]\gamma_j(y_{k+1}) \\
&= \sum_{j=1}^{n} \overline{E}\big[\langle X_{k+1}, e_j\rangle\overline{E}\big[\Lambda_{k+2,T} \mid \mathscr{Y}_{0,T} \vee \{X_k = e_i\} \vee \\
&\qquad\qquad \{X_{k+1} = e_j\}\big] \mid \mathscr{Y}_{0,T} \vee \{X_k = e_i\}\big]\gamma_j(y_{k+1}) \\
&= \sum_{j=1}^{n} \overline{E}\big[\langle X_{k+1}, e_j\rangle\langle v_{k+1,T}, e_j\rangle \mid \mathscr{Y}_{0,T} \vee \{X_k = e_i\}\big]\gamma_j(y_{k+1}) \\
&= \sum_{j=1}^{n} a_{(j,i)}\langle v_{k+1,T}, e_j\rangle\gamma_j(y_{k+1}).
\end{aligned} \tag{6.6}$$

That is, $v_{k,T} = A'B(y_{k+1})v_{k+1,T}$. □

Write, for $k \in \{0, 1, \dots, T-1\}$

$$\Psi_{k+1,T} \triangleq A'B(y_{k+1})A'B(y_{k+2})\dots A'B(y_T) \in \mathbb{R}_{+}^{n\times n}, \tag{6.7}$$

$$\Psi_{T,T} \triangleq A'B(y_T). \tag{6.8}$$

Remark 6.2 Note that $v_{T,T} = \mathbf{1} \in \mathbb{R}^n$ so

$$v_{k,T} = \Psi_{k+1,T} \cdot \mathbf{1}. \tag{6.9}$$

■

Theorem 6.3 *The unnormalized smoothed estimate for the process X, at the time-index k, is computed by*

$$\overline{E}\big[\Lambda_{0,T} X_k \mid \mathcal{Y}_{0,T}\big] = \operatorname{diag}\Big\{ \langle q_k, e_i \rangle \Big\} v_{k,T}. \tag{6.10}$$

The corresponding normalised smoothed estimate is computed, for example, by

$$\begin{aligned} P(X_k = e_i \mid \mathcal{Y}_{0,T}) &= E\big[X_k = e_i \mid \mathcal{Y}_k\big] \\ &= \frac{1}{\langle \operatorname{diag}\{\langle q_k, e_i \rangle\} v_{k,T}, \underline{1} \rangle} \operatorname{diag}\Big\{ \langle q_k, e_i \rangle \Big\} v_{k,T} e_i. \end{aligned} \tag{6.11}$$

Proof Consider the event $\{X_k = e_i\}$, given the information $\mathcal{Y}_{0,T}$:

$$\begin{aligned} P(X_k = e_i \mid \mathcal{Y}_{0,T}) &= \overline{E}\big[\Lambda_{0,T} \langle X_k, e_i \rangle \mid \mathcal{Y}_{0,T}\big] \\ &= \overline{E}\big[\Lambda_k \langle X_k, e_i \rangle \overline{E}\big[\Lambda_{k+1,T} \mid \mathcal{Y}_{0,T} \vee \{X_k = e_i\}\big] \mathcal{Y}_{0,T}\big] \\ &= \langle q_k, e_i \rangle \langle v_{k,T}, e_i \rangle. \end{aligned} \tag{6.12}$$

Therefore,

$$\begin{aligned} \overline{E}\big[\Lambda_{0,T} X_k \mid \mathcal{Y}_{0,T}\big] &= \sum_{\ell=1}^{n} \langle q_k, e_\ell \rangle \langle v_{k,T}, e_\ell \rangle \\ &= \operatorname{diag}\Big\{ \langle q_k, e_\ell \rangle \Big\} v_{k,T}. \end{aligned} \tag{6.13}$$

□

Smoothers for the processes $N^{(j,i)}$, J^i and G^i

To compute smoother update formulae for the processes $N_k^{(j,i)}$, J_k^i, and G_k^i, we apply the same method used to establish Theorem 6.1. Consider the smoothed estimate

$$E\big[N_k^{(j,i)} X_k \mid \mathcal{Y}_{0,T}\big] = \frac{\overline{E}\big[\Lambda_{0,T} N_k^{(j,i)} X_k \mid \mathcal{Y}_{0,T}\big]}{\overline{E}\big[\Lambda_{0,T} \mid \mathcal{Y}_{0,T}\big]}. \tag{6.14}$$

The numerator of equation (6.14) is

$$\begin{aligned} &\overline{E}\big[\Lambda_k N_k^{(j,i)} X_k \overline{E}\big[\Lambda_{k+1,T} \mid \mathcal{Y}_{0,T} \vee \sigma\{X_k\}\big] \mathcal{Y}_{0,T}\big] \\ &\quad = \operatorname{diag}\Big\{ \langle q(N_k^{(j,i)} X_k), e_i \rangle \Big\} \cdot v_{k,T}. \end{aligned} \tag{6.15}$$

Therefore, the unnormalized smoothed estimate of $N_k^{(j,i)}$, given $\mathscr{Y}_{0,T}$, is computed by

$$\begin{aligned}\mathbf{1}'\operatorname{diag}\Big\{\langle q(N_k^{(j,i)}X_k),e_i\rangle\Big\}v_{k,T} &= \langle q(N_k^{(j,i)}X_k),v_{k,T}\rangle \\ &= \overline{E}\big[\Lambda_{0,T}N_k^{(j,i)} \mid \mathscr{Y}_{0,T}\big]. \end{aligned} \tag{6.16}$$

Given observations $\mathscr{Y}_{0,T}$, we are interested in $q(N_T^{(j,i)})$.

Theorem 6.4 *The smoothed estimate for the quantity $q(N_T^{(j,i)})$, is given by the sum*

$$q(N_T^{(j,i)}) = a_{(j,i)}\sum_{k=1}^{T}\langle q_{k-1},e_i\rangle\langle v_{k,T},e_j\rangle\gamma_j(y_k). \tag{6.17}$$

Proof Recalling Lemma 5.3, we see that

$$\begin{aligned}\langle q(N_{k+1}^{(j,i)}X_{k+1}),v_{k+1,T}\rangle &= \langle B(y_{k+1})Aq(N_k^{(j,i)}X_k),v_{k+1,T}\rangle \\ &\quad + a_{(j,i)}\gamma_j(y_{k+1})\langle q_k,e_i\rangle\langle v_{k+1,T},e_j\rangle \\ &= \langle q(N_k^{(j,i)}X_k),A'B(y_{k+1})v_{k+1,T}\rangle \\ &\quad + a_{(j,i)}\gamma_j(y_{k+1})\langle q_k,e_i\rangle\langle v_{k+1,T},e_j\rangle.\end{aligned}$$

That is,

$$\begin{aligned}\langle q(N_{k+1}^{(j,i)}X_{k+1}),v_{k+1,T}\rangle &= \langle q(N_k^{(j,i)}X_k),v_{k,T}\rangle \\ &\quad + a_{(j,i)}\gamma_j(y_{k+1})\langle q_k,e_i\rangle\langle v_{k+1,T},e_j\rangle \\ &= \langle q(N_{k-1}^{(j,i)}X_{k-1}),v_{k-1,T}\rangle \\ &\quad + a_{(j,i)}\gamma_j(y_k)\langle q_{k-1},e_i\rangle\langle v_{k,T},e_j\rangle \\ &\quad + a_{(j,i)}\gamma_j(y_{k+1})\langle q_k,e_i\rangle\langle v_{k+1,T},e_j\rangle \\ &= \langle q(N_{k-2}^{(j,i)}X_{k-2}),v_{k-2,T}\rangle \\ &\quad + a_{(j,i)}\gamma_j(y_{k-1})\langle q_{k-2},e_i\rangle\langle v_{k-1,T},e_j\rangle \\ &\quad + a_{(j,i)}\gamma_j(y_k)\langle q_{k-1},e_i\rangle\langle v_{k,T},e_j\rangle \\ &\quad + a_{(j,i)}\gamma_j(y_{k+1})\langle q_k,e_i\rangle\langle v_{k+1,T},e_j\rangle \\ &\;\;\vdots \\ &= a_{(j,i)}\sum_{k=1}^{T}\langle q_{k-1},e_i\rangle\langle v_{k,T},e_j\rangle\gamma_j(y_k).\end{aligned} \tag{6.18}$$

The last line in equation (6.18) follows by the boundary values, $N_0^{(j,i)} = 0$ and $v_{T,T} = \mathbf{1}$, consequently

$$\langle q(N_T^{(j,i)}X_T),v_{T,T}\rangle = q(N_T^{(j,i)}). \tag{6.19}$$

□

Similarly

$$E\left[G_k^i X_k \mid \mathscr{Y}_{0,T}\right] = \frac{\overline{E}\left[\Lambda_{0,T} G_k^i X_k \mid \mathscr{Y}_{0,T}\right]}{\overline{E}\left[\Lambda_{0,T} \mid \mathscr{Y}_{0,T}\right]}. \tag{6.20}$$

Here, the numerator is

$$\overline{E}\left[\Lambda_k G_k^i X_k \overline{E}\left[\Lambda_{k+1,T} \mid \mathscr{Y}_{0,T} \vee \sigma\{X_k\}\right] \mathscr{Y}_{0,T}\right] = \operatorname{diag}\left\{\langle q(G_k^i X_k), e_i\rangle\right\} v_{k,T}, \tag{6.21}$$

so again,

$$\begin{aligned}\mathbf{1}'\operatorname{diag}\left\{\langle q(G_k^i X_k), e_i\rangle\right\} v_{k,T} &= \overline{E}\left[\Lambda_{0,T} G_k^i \mid \mathscr{Y}_{0,T}\right] \\ &= \langle q(G_k^i X_k), v_{k,T}\rangle. \end{aligned} \tag{6.22}$$

Theorem 6.5 *The smoothed estimate for the quantity $q(G_T^i)$, is given by the sum*

$$q(G_T^i) = \sum_{k=1}^{T} f(y_k)\langle q_{k-1} e_i\rangle\langle v_{k-1}, e_i\rangle. \tag{6.23}$$

Proof Using Lemma 5.4, we note that,

$$\begin{aligned}\langle q(G_{k+1}^i X_{k+1}), v_{k+1,T}\rangle &= \langle B(y_{k+1}) A q(G_k^i X_k), v_{k+1,T}\rangle \\ &\quad + f(y_{k+1})\langle q_k, e_i\rangle\langle e_i, A' B(y_{k+1}) v_{k+1,T}\rangle \\ &= \langle q(G_k^i X_k), v_{k,T}\rangle \\ &\quad + f(y_{k+1})\langle q_k, e_i\rangle\langle v_{k,T}, e_i\rangle. \end{aligned} \tag{6.24}$$

That is,

$$\langle q(G_{k+1}^i X_{k+1}), v_{k+1,T}\rangle - \langle q(G_k^i X_k), v_{k,T}\rangle = f(y_{k+1})\langle q_k, e_i\rangle\langle v_{k,T}, e_i\rangle. \tag{6.25}$$

Since $G_0^i = 0$ and $v_{T,T} = \mathbf{1}$, it follows that

$$\begin{aligned}\langle q(G_T^i X_T), v_T\rangle &= q(G_T^i) \\ &= \sum_{k=1}^{T} f(y_k)\langle q_{k-1}, e_i\rangle\langle v_{k-1}, e_i\rangle. \end{aligned} \tag{6.26}$$

□

Corollary 6.6 *Taking $f(y_k) \equiv 1$,*

$$q(J_T^i) = \sum_{k=1}^{T} \langle q_{k-1}, e_i\rangle\langle v_{k-1}, e_i\rangle. \tag{6.27}$$

Remark 6.7 It is important to note that the smoothed estimates for the quantities, $N_k^{(j,i)}$, J_k^i, and G_k^i, depend only upon the process q and v.

■

Data-recursive smoothing formulae

To update smoothed estimates, we would like to compute expressions that do not require complete recalculation from the origin, in scenarios where one or more new observations are acquired. For example, suppose one has an observation record $\{y_0, y_1, \dots, y_T\}$, for which smoothed estimates have already been computed, that is, the process v has been computed for each time index $k = 0, 1, 2, \dots, T$. Further, suppose one acquires subsequent measurements $\{y_{T+1}, \dots, y_{T'}\}$. To utilize this new information, we would need the corresponding set $\{v_{T+1,T'}, v_{T+2,T'}, \dots, v_{T',T'} = \mathbf{1}\}$. Previous methods to update smoothed estimates have required, in this scenario, complete recalculation from the origin. The new formulae we present circumvent this problem and are afforded by the unnormalised reverse-time recursion for the process v and the matrix-valued process defined at equation (6.8). Recall that

$$\Psi_{k+1,T} = A'B(y_{k+1})A'B(y_{k+2}), \cdots, A'B(y_T). \tag{6.28}$$

Further, for the two epochs T and T', where $T' > T$, it follows that

$$\begin{aligned}\Psi_{k+1,T'} &= A'B(y_{k+1})A'B(y_{k+2}), \cdots, A'B(y_T)A'B(y_{T+1}), \cdots, A'B(y_{T'})\\ &= \Psi_{k+1,T}A'B(y_{T+1})A'B(y_{T+2}), \cdots, A'B(y_{T'})\\ &= \Psi_{k+1,T}\Psi_{T+1,T'}.\end{aligned} \tag{6.29}$$

Remark 6.8 The transitivity property for the matrix-valued process Ψ, shown by the calculation at (6.29), is critical in the development of data-recursive smoother update formulae.

■

Recalling Theorem (6.4), we now write

$$\begin{aligned}q(N_T^{(j,i)}) = a_{(j,i)}\Big\{\sum_k^{T-1} \gamma_j(y_k)\langle q_{k-1}, e_i\rangle\big(e_j'\Psi_{k+1,T}\big)\Big\}\mathbf{1}\\ + a_{(j,i)}\gamma_j(y_T)\langle q_{T-1}, e_i\rangle.\end{aligned} \tag{6.30}$$

Suppose now, that one receives a single new measurement y_{T+1} and one wishes to compute $q(N_{T+1}^{(j,i)})$. Using (6.30) and the transitivity of the matrix Ψ, we get

$$\begin{aligned}q(N_{T+1}^{(j,i)}) = a_{(j,i)}&\gamma_j(y_1)\langle q_0, e_i\rangle\big(e_j'\Psi_{2,T}\Psi_{T+1,T+1}\big)\mathbf{1}\\ &+ a_{(j,i)}\gamma_j(y_2)\langle q_1, e_i\rangle\big(e_j'\Psi_{3,T}\Psi_{T+1,T+1}\big)\mathbf{1}\\ &\vdots\\ &+ a_{(j,i)}\gamma_j(y_{T-1})\langle q_{T-2}, e_i\rangle\big(e_j'\Psi_{T,T}\Psi_{T+1,T+1}\big)\mathbf{1}\\ &+ a_{(j,i)}\gamma_j(y_T)\langle q_{T-1}, e_i\rangle\big(e_j'\Psi_{T+1,T+1}\big)\mathbf{1}\\ &+ a_{(j,i)}\gamma_j(y_{T+1})\langle q_T, e_j\rangle\end{aligned}$$

$$
\begin{aligned}
&= \Big\{a_{(j,i)} \sum_{k=1}^{T-1} \gamma_j(y_k)\langle q_{k-1}, e_i\rangle \big(e_j'\Psi_{k+1,T}\big)\Big\}\Psi_{T+1,T+1}\underline{1} \\
&\quad + a_{(j,i)}\gamma_j(y_T)\langle q_{T-1}, e_i\rangle \big(e_j' A' B(y_{T+1})\big)\underline{1} \\
&\quad + a_{(j,i)}\gamma_j(y_{T+1})\langle q_T, e_i\rangle.
\end{aligned} \tag{6.31}
$$

Suppose now, more generally, that a block of new measurements from $\{y_{T+1}, y_{T+2}, \ldots, y_{T'}\}$ is acquired. The update equation for this scenario is given in the following Lemma.

Lemma 6.9 *For the set* $\{y_0, y_1, \ldots, y_T, y_{T+1}, \ldots, y_{T'}\}$, *the information state estimate* $q(N_{T'}^{(j,i)})$ *is computed by the equation*

$$
\begin{aligned}
q(N_{T'}^{(j,i)}) &= \Big\{a_{(j,i)} \sum_{k=1}^{T-1} \gamma_j(y_k)\langle q_{k-1}, e_i\rangle \big(e_j'\Psi_{k+1,T}\big)\Big\}\Psi_{T+1,T'}\underline{1} \\
&\quad + a_{(j,i)}\Big\{\sum_{k=T}^{T'-1} \gamma_j(y_k)\langle q_k, e_i\rangle \big(e_j'\Psi_{k+1,T'}\big)\Big\}\mathbf{1} \\
&\quad + a_{(j,i)}\gamma_j(y_{T'})\langle q_{T'-1}, e_i\rangle.
\end{aligned} \tag{6.32}
$$

Remark 6.10 The update equation given by Lemma 6.9 shows how new information can be incorporated without complete recalculation from the origin. For example, to calculate $q(N_T^{(j,i)})$, it is necessary to compute the matrix

$$
\Gamma_{T-1}^{(j,i)} \triangleq a_{(j,i)}\Big\{\sum_{k}^{T-1} \gamma_j(y_k)\langle q_{k-1}, e_i\rangle \big(e_j'\Psi_{k+1,T}\big)\Big\} \in \mathbb{R}^{n\times n}. \tag{6.33}
$$

This matrix can easily be stored in memory and subsequently recalled upon the arrival of new measurements $\{y_{T+1}, \ldots, y_{T'}\}$ to compute the update given by in Lemma 6.9.

■

The smoother update equations corresponding to Theorem 6.5 and Corollary 6.6 are given in the following Lemma.

Lemma 6.11 *For the set* $\{y_0, y_1, \ldots, y_T, y_{T+1}, \ldots, y_{T'}\}$, *the information state estimate* $q(G_{T'}^{(i)})$ *is computed by the equation*

$$
\begin{aligned}
q(G_{T'}^{i}) &= \Big\{\sum_{k=1}^{T} f(y_k)\langle q_{k-1}, e_i\rangle \big(e_i'\Psi_{k,T}\big)\Big\}\Psi_{T+1,T'}\mathbf{1} \\
&\quad + \sum_{k=T+1}^{T'} f(y_k)\langle q_{k-1}, e_i\rangle \big(e_i'\Psi_{k,T}\big)\underline{1}.
\end{aligned} \tag{6.34}
$$

Lemma 6.12 *For the set* $\{y_0, y_1, \ldots, y_T, y_{T+1}, \ldots, y_{T'}\}$, *the information state estimate* $q(J_{T'}^{(i)})$ *is computed by setting, in Lemma* 6.11, $f(y_k) = 1, k \in \mathbb{N}$, *that is*

$$q(J^i_{T'}) = \Big\{ \sum_{k=1}^{T} \langle q_{k-1}, e_i \rangle \big(e_i' \Psi_{k,T} \big) \Big\} \Psi_{T+1,T'} \mathbf{1} \\ + \sum_{k=T+1}^{T'} \langle q_{k-1}, e_i \rangle \big(e_i' \Psi_{k,T} \big) \underline{1}. \tag{6.35}$$

Finally the EM algorithm parameters updates of our model are given by

$$\begin{aligned} \widehat{a}_{(j,i)} &= \frac{q(N_T^{(j,i)})}{q(J_T^i)} \\ \widehat{\langle c, e_i \rangle} &= \frac{q(G_T^i)}{q(J_T^i)} \\ \widehat{\langle \sigma, e_i \rangle} &= q(J_T^i)^{-1} \big[\sigma \big(G_T^i(y^2) \big) - 2 \langle c, e_i \rangle q(G_T^i) + \langle c, e_i \rangle^2 q(J_T^i) \big]. \end{aligned} \tag{6.36}$$

3.7 Vector Observations

Again, suppose the Markov chain has state space $S_X = \{e_1, \dots, e_N\}$ and $X_{k+1} = AX_k + V_{k+1}$, $k \in \mathbb{N}$.

Suppose now the observation process is d-dimensional with components:

$$\begin{aligned} y^1_{k+1} &= \langle c^1, X_k \rangle + \langle \sigma^1, X_k \rangle w^1_{k+1} \\ y^2_{k+1} &= \langle c^2, X_k \rangle + \langle \sigma^2, X_k \rangle w^2_{k+1} \\ &\vdots \\ y^d_{k+1} &= \langle c^d, X_k \rangle + \langle \sigma^d, X_k \rangle w^d_{k+1}, \qquad k \in \mathbb{N}. \end{aligned}$$

Here, for $1 \le j \le d$, $c^j = (c^j_1, c^j_2, \dots, c^j_N)' \in \mathbb{R}^N$, $\sigma^j = (\sigma^j_1, \sigma^j_2, \dots, \sigma^j_N)' \in \mathbb{R}^N$ and the w^j_ℓ, $1 \le j \le d$, $\ell \in \mathbb{N}$, are $N(0,1)$ i.i.d. random variables. We assume $\sigma^j_i > 0$ for $1 \le i \le N$, $1 \le j \le d$. Write $\phi^j_i(x) = (2\pi\sigma^j_i)^{-1/2} \exp\big(-x^2/2\sigma^j_i\big)$ for the $N(0, \sigma^j_i)$ density.

The analogs of the above results are easily derived. For example:

$$\begin{aligned} E[\langle X_k, e_i \rangle \mid \mathscr{Y}_{k+1}] &= \xi^i_k(y^1_{k+1}, y^2_{k+1}, \dots, y^d_{k+1}) \\ &= \frac{\big\langle \hat{X}_k, e_i \big\rangle \phi^1_i\,(y^1_{k+1} - c^1_i) \dots \phi^d_i\,(y^d_{k+1} - c^d_i)}{\sum_{j=1}^N \big\langle \hat{X}_k, e_j \big\rangle \phi^1_i\,(y^1_{k+1} - c^1_j) \dots \phi^d_i\,(y^d_{k+1} - c^k_j)}. \end{aligned}$$

For $1 \le i \le N$ write, with ϕ the $N(0,1)$ density,

$$\Gamma^i\big(\underline{y}_{k+1}\big) = \Gamma^i(y^1_{k+1}, \dots, y^d_{k+1}) = \prod_{j=1}^d \frac{\phi\left(\frac{y^j_{k+1} - c^j_i}{\sigma^j_i}\right)}{\sigma^j_i \, \phi\,(y^j_{k+1})} e_i.$$

Then we obtain

$$\gamma_{k+1}(X_{k+1}) = \sum_{i=1}^{N} \langle \gamma_k(X_k), \Gamma^i(\underline{y}_{k+1}) \rangle a_i,$$

$$\gamma_{m,k+1}(\langle X_m, e_r \rangle) = \sum_{i=1}^{N} \langle \gamma_{m,k}(\langle X_m, e_r \rangle), \Gamma^i(\underline{y}_{k+1}) \rangle a_i, \qquad m \le k,$$

$$\begin{aligned} \gamma_{k+1,k+1}(\mathscr{J}_{k+1}^{rs}) &= \sum_{i=1}^{N} \langle \gamma_{k,k}(\mathscr{J}_k^{rs}), \Gamma^i(\underline{y}_{k+1}) \rangle a_i \\ &\quad + \langle \gamma_k(X_k), \Gamma^r(\underline{y}_{k+1}) \rangle a_{sr} e_s, \\ \gamma_{m,k+1}(\mathscr{J}_m^{rs}) &= \sum_{i=1}^{N} \langle \gamma_{m,k}(\mathscr{J}_m^{rs}), \Gamma^i(\underline{y}_{k+1}) \rangle a_i, \qquad m \le k, \end{aligned}$$

$$\begin{aligned} \gamma_{k+1,k+1}(\mathscr{O}_{k+1}^{r}) &= \sum_{i=1}^{N} \langle \gamma_{k,k}(\mathscr{O}_k^{r}), \Gamma^i(\underline{y}_{k+1}) \rangle a_i \\ &\quad + \langle \gamma_k(X_k), \Gamma^r(\underline{y}_{k+1}) \rangle a_r, \\ \gamma_{m,k+1}(\mathscr{O}_m^{r}) &= \sum_{i=1}^{N} \langle \gamma_{m,k}(\mathscr{O}_m^{r}), \Gamma^i(\underline{y}_{k+1}) \rangle a_i, \qquad m \le k. \end{aligned}$$

Finally, with $\mathscr{T}_k^r f(y^j) = \sum_{\ell=1}^{k} \langle X_{\ell-1}, e_r \rangle f(y_\ell^j)$ we have

$$\begin{aligned} \gamma_{k+1,k+1}(\mathscr{T}_{k+1}^{r} f(y^j)) &= \sum_{i=1}^{N} \langle \gamma_{k,k}(\mathscr{T}_k^{r} f(y^j)), \Gamma^i(\underline{y}_{k+1}) \rangle a_i \\ &\quad + \langle \gamma_k(X_k), \Gamma^r(\underline{y}_{k+1}) \rangle f(y_{k+1}^i) y_{k+1}^j a_r, \\ \gamma_{m,k+1}(\mathscr{T}_m^{r} f(y^j)) &= \sum_{i=1}^{N} \langle \gamma_{m,k}(\mathscr{T}_m^{r} f(y^j)), e_i \rangle \Gamma^i(\underline{y}_{k+1}) a_i, \qquad m \le k. \end{aligned}$$

Thus in the vector case is

$$\hat{c}_r^j = \frac{\hat{G}_k^r(y^j)}{\hat{\mathscr{O}}_k^r} = \frac{\gamma_k\left(\mathscr{T}_k^r(y^j)\right)}{\gamma_k\left(\mathscr{O}_k^r\right)},$$

and

$$\begin{aligned} \hat{\sigma}_r^j &= \left(\hat{\mathscr{O}}_k^r\right)^{-1} \left(\hat{G}_k^r((y^j)^2) - 2c_r \hat{G}_k^r(y^j) + c_r^2 \hat{\mathscr{O}}_k^r\right) \\ &= \left(\gamma_k(\mathscr{O}_k^r)\right)^{-1} \left(\gamma_k\left(\mathscr{T}_k^r(y^j)^2\right) - 2c_r \gamma_k\left(\mathscr{T}_k^r(y^j)\right) + c_r^2 \gamma_k(\mathscr{O}_k^r)\right). \end{aligned}$$

3.8 Recursive Parameter Estimation

In this following three sections we discuss recursive methods for estimating the parameters in a manner similar to that described in Chapter 2. We shall suppose the signal model depends on a parameter θ which takes values in a measure space (Θ, β, λ). The value of θ is unknown and, in this section, we suppose it is constant. That is, for $1 \leq i, j \leq N$,

$$\begin{aligned} a_{ij}(\theta) &= P(X_{k+1} = e_i \mid X_k = e_j, \theta) \\ &= P(X_1 = e_i \mid X_0 = e_j, \theta). \end{aligned}$$

Write $A(\theta)$ for the $N \times N$ matrix $(a_{ij}(\theta))$, $1 \leq i, j \leq N$. Also, $\{\mathscr{F}_\ell\}, \ell \in \mathbb{N}$ will denote the complete filtration generated by X and θ, that is, for any $k \in \mathbb{N}, \mathscr{F}_k$ is the complete σ-field generated by X_ℓ, $\ell \leq k$, and θ.

We suppose the chain X is not observed directly; rather there is an observation process $\{y_\ell\}$, $\ell \in \mathbb{N}$, which, for simplicity, we suppose is real valued. The extension to vector observations is straightforward. The real observations process y has the form

$$y_k = c(\theta, X_k) + \sigma(\theta, X_k) w_k. \tag{8.1}$$

Here the w_ℓ, $\ell \in \mathbb{N}$, are real, i.i.d. random variables with a nonzero (positive) density ϕ. The extension to the situation where the w_ℓ, $\ell \in \mathbb{N}$, are independent but have possibly different nonzero density functions ϕ_ℓ, is immediate. Because X_k is always one of the unit vectors e_i, $1 \leq i \leq N$, for any $\theta \in \Theta$ and functions $c(\theta, \cdot)$ and $\sigma(\theta, \cdot)$ are determined by vectors

$$\begin{aligned} c(\theta) &= (c_1(\theta), c_2(\theta), \dots, c_N(\theta)) \in \mathbb{R}^N, \\ \sigma(\theta) &= (\sigma_1(\theta), \sigma_2(\theta), \dots, \sigma_N(\theta)) \in \mathbb{R}^N, \end{aligned}$$

so that

$$\begin{aligned} c(\theta, X_k) &= \langle c(\theta), X_k \rangle, \\ \sigma(\theta, X_k) &= \langle \sigma(\theta), X_k \rangle, \end{aligned}$$

where $\langle , \rangle$ denotes the inner product in $\mathbb{R}^N$.

Notation 8.1 *With* $\mathscr{G}_\ell^0 = \sigma\{X_0, X_1, \dots, X_\ell, y_1, \dots, y_{\ell-1}, \theta\}$*, then* $\{\mathscr{G}_\ell\}$*,* $\ell \in \mathbb{N}$*, will denote the complete filtration generated by* $\mathscr{G}_\ell^0$ *and* $\{\mathscr{Y}_\ell\}, \ell \in \mathbb{N}$*, will denote the complete filtration generated by* $\mathscr{Y}_\ell^0$ *where* $\mathscr{Y}_\ell^0 = \sigma\{y_0, y_1, \dots, y_\ell\}$.

The Recursive Densities

We shall work under probability measure $\overline{P}$, so that $\{y_\ell\}$, $\ell \in \mathbb{N}$, is a sequence of i.i.d. random variables with density ϕ, $\phi(y) > 0$, and $\{X_\ell\}$, $\ell \in \mathbb{N}$, is an independent Markov chain.

Notation 8.2 *Write $q_k^i(\theta)$, $k \in \mathbb{N}$, for the unnormalized conditional density such that*

$$\overline{E}\left[\overline{\Lambda}_k \langle X_k, e_i\rangle I(\theta \in d\theta) \mid \mathscr{Y}_k\right] = q_k^i(\theta)\, d\theta.$$

The existence of $q_k^i(\theta)$ will be discussed below.

Our main result follows. This provides a recursive, closed-form update for $q_k^i(\theta)$. The normalized conditional density

$$p_k^i(\theta)\, d\theta = E\left[\langle X_k, e_i\rangle I(\theta \in d\theta) \mid \mathscr{Y}_k\right]$$

is given in terms of q by:

$$p_k^i(\theta) = \frac{q_k^i(\theta)}{\sum_{j=1}^N \int_\Theta q_k^j(u)\, d\lambda(u)}.$$

Theorem 8.3

$$\boxed{q_{k+1}(u) = B(y_{k+1}, u)\, A\, q_k(u),} \tag{8.2}$$

where

$$B(y_{k+1}, u) = \operatorname{diag}\left(\sigma_i^{-1}(u)\, \phi^{-1}(y_{k+1})\, \phi\left(\frac{y_{k+1} - c_i(u)}{\sigma_i(u)}\right)\right), i = 1, \ldots, N.$$

Proof Suppose $f : \Theta \to \mathbb{R}$ is any measurable function. Then

$$\begin{aligned}
&\overline{E}\left[f(\theta)\langle X_{k+1}, e_i\rangle \overline{\Lambda}_{k+1} \mid \mathscr{Y}_{k+1}\right] \\
&= \sum_{j=1}^N \overline{E}\left[f(\theta)\langle X_k, e_j\rangle a_{ij}(\theta)\overline{\Lambda}_k \phi\left(\frac{y_{k+1} - c_i(\theta)}{\sigma_i(\theta)}\right) \mid \mathscr{Y}_k\right] \\
&\quad \times \phi(y_{k+1})^{-1} \sigma_i^{-1}(\theta) \\
&= \phi(y_{k+1})^{-1} \sum_{j=1}^N \int_\Theta f(u)\, a_{ij}(u)\, \phi\left(\frac{y_{k+1} - c_i(u)}{\sigma_i(u)}\right) \frac{q_k^j(u)}{\sigma_i(u)} d\lambda(u).
\end{aligned}$$

Therefore,

$$q_{k+1}^i(u) = \phi\left(\frac{y_{k+1} - c_i(u)}{\sigma_i(u)}\right) \sigma_i^{-1}(u)\, \phi^{-1}(y_{k+1}) \sum_{j=1}^N a_{ij}(u)\, q_k^j(u). \tag{8.3}$$

Using matrix notation the result follows. □

Remark 8.4 Suppose $\pi = (\pi_1, \ldots, \pi_N), \pi_i = P(X_0 = e_i)$, is the initial distribution for X_0 and $h(u)$, is the prior density for θ. Then

$$q_0^i(u) = \pi_i h(u),$$

and the updated estimates are obtained by substituting in (8.2) for $k \geq 1$.

■

If the prior estimates are delta functions (i.e., unit masses at particular values e_i and θ), then $q_1(u)$ and higher unnormalized conditional distributions can be calculated by formula (8.2). However, because no noise or dynamics enter into θ, if delta functions are taken as the prior distributions for θ no updating takes place (this is not the case with the distribution for X). This is to be expected because in the filtering procedure the prior does not represent an initial guess for θ given no information, but the best estimate for the distribution of θ given the initial information. Care must, therefore, be taken with the choice of the prior for θ and, unless there is reason to choose otherwise, priors should be taken so that they have support on the whole range of θ.

Vector Observations

Again, suppose the Markov chain has state space $\{e_1, \ldots, e_N\}$ and

$$X_{k+1} = A(\theta) X_k + V_{k+1}, \qquad k \in \mathbb{N},$$

for some (unknown) $\theta \in \Theta$.

Consider now the case where the observation process is d-dimensional with components:

$$\begin{aligned}
y^1_{k+1} &= \langle c^1(\theta), X_{k+1} \rangle + \langle \sigma^1(\theta), X_{k+1} \rangle w^1_{k+1} \\
y^2_{k+1} &= \langle c^2(\theta), X_{k+1} \rangle + \langle \sigma^2(\theta), X_{k+1} \rangle w^2_{k+1} \\
&\vdots \\
y^d_{k+1} &= \langle c^d(\theta), X_{k+1} \rangle + \langle \sigma^d(\theta), X_{k+1} \rangle w^d_{k+1}, \qquad k \in \mathbb{N}.
\end{aligned}$$

Here, for

$$1 \le j \le d, c^j(\theta) = \left(c^j_1(\theta), \ldots, c^j_N(\theta)\right)', \sigma^j(\theta) = \left(\sigma^j_1(\theta), \ldots, \sigma^j_N(\theta)\right)' \in \mathbb{R}^N.$$

Further, the w^j_ℓ, $1 \le j \le d, \ell \in \mathbb{N}$, are a family of independent random variables with nonzero densities $\phi_j(w)$.

The same techniques then establish the following result:

Theorem 8.5 *With*

$$\phi(y_{k+1}, u) = \sum_{j=1}^{d} \phi_j \left(\frac{y^j_{k+1} - c_i(u)}{\sigma_i(u)} \right) \left(\prod_{j=1}^{d} \phi_j(y_{k+1}) \right)^{-1}$$

$$\boxed{q^i_{k+1}(u) = \phi(y_{k+1}, u)\, \sigma_i^{-1}(u) \sum_{\ell=1}^{N} a_{i\ell}(u)\, q^\ell_k(u).} \tag{8.4}$$

3.9 HMMs with Colored Noise

In this section we extend the above results to observations with colored noise. We suppose the signal model parameters depend on some parameter θ which takes values in a measure space (Θ, β, λ). The value of θ is unknown, and we suppose it is constant. Then for $1 \leq i, j \leq N$, write $\mathscr{F}_k^0$ for the σ-field generated by $X_0, X_1, \ldots, X_k$ and θ and $\{\mathscr{F}_k\}, k \in \mathbb{N}$, for the complete filtration generated by $\mathscr{F}_k^0$.

$$\begin{aligned} a_{ij}(\theta) &= P(X_{k+1} = e_i \mid X_k = e_j, \theta) \\ &= P(X_1 = e_i \mid X_0 = e_j, \theta). \end{aligned}$$

Write $A(\theta)$ for the $N \times N$ matrix $(a_{ij}(\theta))$, $1 \leq i, j \leq N$. Then

$$X_{k+1} = A(\theta) X_k + V_{k+1}, \tag{9.1}$$

where $E[V_{k+1} \mid \mathscr{F}_k] = 0$.

The observation process $\{y_\ell\}$, $\ell \in \mathbb{N}$, which for simplicity is supposed to be real-valued, has the form

$$y_{k+1} = c(\theta, X_{k+1}) + d_1(\theta) w_k + \cdots + d_r(\theta) w_{k+1-r} + w_{k+1}. \tag{9.2}$$

Here $\{w_k\}$, $k \in \mathbb{N}$, is a sequence of i.i.d. random variables with nonzero density function ϕ. (The extension to time varying densities ϕ_k is immediate.) Suppose $d_r(\theta) \neq 0$.

Here $c(\theta, X_k)$ is a function, depending on a parameter θ, and the state X_k. Because X_k is always one of the unit vectors e_i the function $c(\theta, .)$ is determined by a vector

$$c(\theta) = (c_1(\theta), c_2(\theta), \ldots, c_N(\theta))$$

and

$$c(\theta, X_k) = \langle c(\theta), X_k \rangle,$$

where $\langle \, , \, \rangle$ denotes the inner product in Euclidean space.

Write $\bar{x}_{k+1} = (w_{k+1}, w_k, \ldots, w_{n-r+1})' \in \mathbb{R}^r$, $D = (1, 0, \ldots, 0)' \in \mathbb{R}^r$,

$$\Gamma(\theta) = \begin{pmatrix} -d_1(\theta) & -d_2(\theta) & \ldots & -d_{r-1}(\theta) & -d_r(\theta) \\ 1 & 0 & \ldots & 0 & 0 \\ 0 & 1 & \ldots & 0 & 0 \\ \vdots & \vdots & \ddots & \vdots & \vdots \\ 0 & 0 & \ldots & 1 & 0 \end{pmatrix}.$$

Then

$$\bar{x}_{k+1} = \Gamma(\theta) \bar{x}_k + D(y_{k+1} - \langle c(\theta), X_{k+1} \rangle)$$

and

$$y_{k+1} = \langle c(\theta), X_{k+1} \rangle + \langle d(\theta), \bar{x}_k \rangle + w_{k+1}.$$

The unobserved components are, therefore, X_{k+1}, $\bar{x}_k$, θ.

Again, because $d_r(\theta) \neq 0$

$$\Gamma(\theta)^{-1} = d_r^{-1}(\theta) \begin{pmatrix} 0 & d_r(\theta) & 0 & \dots & 0' \\ 0 & 0 & d_r(\theta) & \dots & 0 \\ \vdots & \vdots & \vdots & \ddots & \vdots \\ 0 & 0 & 0 & \dots & d_r(\theta) \\ -1 & -d_1(\theta) & -d_2(\theta) & \dots & -d_{r-1}(\theta) \end{pmatrix}.$$

Suppose f and h are arbitrary real-valued test functions. Then

$$\begin{aligned} &\overline{E}\left[f(\bar{x}_{k+1}) h(\theta) \langle X_{k+1}, e_i \rangle \overline{\Lambda}_{k+1} \mid \mathscr{Y}_{k+1}\right] \\ &= \iint f(\xi) h(u) q_{k+1}^i(\xi, u) \, d\xi \, d\lambda(u), \end{aligned} \tag{9.3}$$

where $q_{k+1}^i(\xi, u)$ is the unnormalized conditional density such that

$$\overline{E}\left[\langle X_{k+1}, e_i \rangle I(\theta \in d\theta) I(\bar{x}_{k+1} \in dz) \overline{\Lambda}_{k+1} \mid \mathscr{Y}_{k+1}\right] = q_{k+1}^i(z, \theta) \, dz \, d\theta.$$

Then (9.3) equals

$$\begin{aligned} &\overline{E}\Bigg[f(\Gamma(\theta)\bar{x}_k + D(y_{k+1} - c_i(\theta))) h(\theta) \\ &\quad \times \langle A(\theta) X_k + V_{k+1}, e_i \rangle \overline{\Lambda}_k \frac{\phi(y_{k+1} - c_i(\theta) - \langle d(\theta), \bar{x}_k \rangle)}{\phi(y_{k+1})} \mid \mathscr{Y}_{k+1}\Bigg] \\ &= \phi(y_{k+1})^{-1} \\ &\quad \times \iint \sum_{j=1}^{N} \{ f(\Gamma(u) z + D(y_{k+1} - c_i(u))) h(u) \\ &\qquad \times a_{ij}(u) \phi(y_{k+1} - c_i(u) - \langle d(u), z \rangle) q_k^i(z, u) \} dz \, d\lambda(u). \end{aligned} \tag{9.4}$$

Write

$$\xi = \Gamma(u) z + D(y_{k+1} - c_i(u))$$

so

$$z = \Gamma(u)^{-1} \{\xi - D(y_{k+1} - c_i(u))\}$$

and

$$dz \, d\lambda(u) = \Gamma(u)^{-1} d\xi \, d\lambda(u).$$

The functions f and h are arbitrary so from the equality of (9.3) and (9.4) we have the following result:

Theorem 9.1 *Write*

$$\Phi(y_{k+1},u,\xi) = \phi(y_{k+1})^{-1}\phi\Big(y_{k+1}-c_i(u) - \Big\langle d(u),\Gamma(u)^{-1}(\xi - D(y_{k+1}-c_i(u)))\Big\rangle\Big);$$

then for $1 \le i \le N$

$$\boxed{q^i_{k+1}(\xi,u) = \Phi(y_{k+1},u,\xi) \times \sum_{j=1}^{N} a_{ij}(u)\, q^j_k\Big(\Gamma(u)^{-1}(\xi - D(y_{k+1}-c_i(u)),u)\Big).} \tag{9.5}$$

3.10 Mixed-State HMM Estimation

In this section we consider the situation where a Markov chain influences a linear system which, in turn, is observed linearly in noise. The parameters of the model are supposedly unknown. Again a recursive expression is obtained for the unnormalized density of the state and parameters given the observations.

Again the state space of the Markov chain X is taken to be the set of unit vectors $\{e_1,\dots,e_N\}$ so that:

$$X_{k+1} = A(\theta)X_k + V_{k+1}, \qquad k \in \mathbb{N}.$$

The state of the linear system is given by a process x_k, $k \in \mathbb{N}$, taking values in $\mathbb{R}^d$, and its dynamics are described by the equation

$$x_{k+1} = F(\theta)x_k + G(\theta)X_k + v_{k+1}.$$

Here v_k, $k \in \mathbb{N}$, is a sequence of independent random variables with densities ψ_k.

The observation process has the form

$$y_{k+1} = C(\theta)x_k + w_{k+1}.$$

The w_k are independent random variables having strictly positive densities ϕ_k.

In summary, we have what we term a mixed-state HMM

$$\boxed{\begin{aligned} X_{k+1} &= A(\theta)X_k + V_{k+1},\\ x_{k+1} &= F(\theta)x_k + G(\theta)X_k + v_{k+1},\\ y_{k+1} &= C(\theta)x_k + w_{k+1}, \qquad k \in \mathbb{N}. \end{aligned}} \tag{10.1}$$

The parameter θ takes values in some measurable space (Θ,β,λ). Again write $q^i_k(z,\theta)$ for the unnormalized joint conditional density of x_k and θ, given that $X_k = e_i$

such that

$$q_k^i(z,\theta)\,dz\,d\theta = \overline{E}\left[\langle X_k, e_i\rangle I(x_k \in dz) I(\theta \in d\theta)\overline{\Lambda}_k \mid \mathscr{Y}_k\right].$$

For suitable test functions f and h consider

$$\begin{aligned}
&\overline{E}\left[\langle X_{k+1}, e_i\rangle f(x_{k+1}) h(\theta)\overline{\Lambda}_{k+1} \mid \mathscr{Y}_{k+1}\right] \\
&\quad = \iint f(\xi) h(u) q_{k+1}^i(\xi,u)\,d\xi\,d\lambda(u) \\
&\quad = \overline{E}\big[\langle A(\theta)X_k + V_{k+1}, e_i\rangle f(F(\theta)x_k + G(\theta)X_k + w_{k+1}) \\
&\qquad\quad \times h(\theta)\overline{\Lambda}_k \phi_{k+1}(y_{k+1} - C(\theta)x_k)\,\phi_{k+1}(y_{k+1})^{-1} \mid \mathscr{Y}_k\big] \\
&\quad = \phi_{k+1}(y_{k+1})^{-1} \\
&\qquad \times \sum_{j=1}^N \iiint \Big[a_{ij}(u) f(F(u)z + G(u)e_j + w) h(u) \\
&\qquad\qquad \times \phi_{k+1}(y_{k+1} - C(u)z)\,\psi_{k+1}(w)\,q_k^j(z,u)\Big]\,dz\,d\lambda(u)\,dw \\
&\quad = \phi_{k+1}(y_{k+1})^{-1} \\
&\qquad \times \sum_{j=1}^N \iiint \Big[a_{ij}(u) f(\xi) h(u)\,\phi_{k+1}(y_{k+1} - C(u)z) \\
&\qquad\qquad \times \psi_{k+1}(\xi - F(u)z + G(u)e_j)\,q_k^i(z,u)\Big]\,dz\,d\lambda(u)\,d\xi,
\end{aligned}$$

where the last equality follows by substituting $\xi = F(u)z + G(u)e_j + w$, $z = z$, $u = u$.

This identity holds for all test functions f and h, so we have the following result:

Theorem 10.1 *Write $\psi_{k+1}(\xi,u,z,e_j) = \psi_{k+1}(\xi - F(u)z + G(u)e_j)$; then*

$$\boxed{\begin{aligned} q_{k+1}^i(\xi,u) = \phi_{k+1}(y_{k+1})^{-1} \int \sum_{j=1}^N \big[a_{ij}(u)\,\phi_{k+1}(y_{k+1} - C(u)z) \\ \times \psi_{k+1}(\xi,u,z,e_j)\,q_k^j(z,u)\big]\,dz.\end{aligned}} \tag{10.2}$$

3.11 Problems and Notes

Problems

1. Prove Lemma 4.3.
2. Show that the sequences Λ_k and Λ_k^* defined in Section 3.7 are martingales.
3. For the model described in Section 3.8 obtain estimates for $\mathscr{J}_k^{rs}$, $\mathscr{O}_k^r$ and $\mathscr{T}_k^r$.
4. Derive the recursion for the unnormalized conditional density given in Theorem 8.5.

Notes

The basic method of the chapter is again the change of measure. The discrete-time version of Girsanov's theorem provides a new probability measure under which all the observations are i.i.d. random variables. Also, to obtain closed-form filters, estimates are obtained for processes $H_k X_k$. The result in the first part of the chapter were first reported in Elliott (1994b). Continuous-time versions of these results were obtained in Elliott (1993c) and can be found in Chapter 8.

Using filtering methods the forward-backward algorithm is not required in the implementation of the EM algorithm. For a description of the EM algorithm see Baum et al. (1970), and for recent applications see Krishnamurthy and Moore (1993) and Dembo and Zeitouni (1986). Earlier applications of the measure change methods can be found in Brémaud and van Schuppen (1976).

Chapter 4
Continuous-Range States and Observations

4.1 Introduction

The standard model for linear, discrete-time signals and observations is considered, in which the coefficient matrices depend on unknown, time-varying parameters. An explicit recursive expression is obtained for the unnormalized, conditional expectation, given the observations, of the state and the parameters. Results are developed for this special class of models, familiar in many signal-processing contexts, as a prelude to more general nonlinear models studied later in the chapter. Our construction of the equivalent measure is explicit and the recursions have simple forms.

As an interesting special case, we consider the parameter estimation problem for a general *autoregressive, moving average exogenous input* (ARMAX) model.

Also in this chapter, we explore the power of a double measure change technique for achieving both the measurements and states i.i.d.

4.2 Linear Dynamics and Parameters

All processes are defined initially on a probability space $(\Omega, \mathcal{F}, P)$. The discrete-time model we wish to discuss has the form

$$x_{k+1} = A(\theta_{k+1})x_k + B(\theta_{k+1})v_{k+1}, \tag{2.1}$$
$$y_k = C(\theta_k)x_k + D(\theta_k)w_k. \tag{2.2}$$

Here $k \in \mathbb{N}$ and x_0, or its distribution, are known. The signal process x_k takes values in some Euclidean space $\mathbb{R}^d$ while the observation process y_k takes values in $\mathbb{R}^q$. $\{v_\ell\}$, $\ell \in \mathbb{N}$, is a sequence of i.i.d., random variables, with density functions ψ, and v_ℓ has values in $\mathbb{R}^d$. Similarly, $\{w_\ell\}$, $\ell \in \mathbb{N}$, is a sequence of i.i.d. random variables with strictly positive density function ϕ, and w_ℓ also takes values in $\mathbb{R}^q$, that is, w_ℓ has the same dimensions as y_ℓ. The matrices $A(\theta_.)$, $B(\theta_.)$, $C(\theta_.)$, and $D(\theta_.)$ have appropriate dimensions and depend on the parameters θ.

For simplicity we suppose the parameters $\theta_k \in \mathbb{R}^p$ satisfy the dynamic equations

$$\theta_{k+1} = \alpha\theta_k + \nu_{k+1}. \tag{2.3}$$

Here either θ_0, or its distribution, is known, α is a real constant and $\{\nu_\ell\}$ is a sequence of i.i.d. random variables with density ρ. Finally, we suppose the matrices $B(r)$ and $D(r)$ are nonsingular for all $r \in \mathbb{R}$.

Notation 2.1 *Write* $\mathscr{G}^0_{k+1} = \sigma\{\theta_\ell, 1 \le \ell \le k+1, x_1, \dots, x_{k+1}, y_1, \dots, y_k\}$, $\mathscr{Y}^0_k = \sigma\{y_1, \dots, y_k\}$. $\{\mathscr{G}_k\}$ *and* $\{\mathscr{Y}_k\}$, $k \in \mathbb{N}$, *are the complete filtrations generated by the completions of* $\mathscr{G}^0_k$ *and* $\mathscr{Y}^0_k$, *respectively.*

Remarks 2.2 The above conditions can be modified. For example, the parameters θ can be vector valued.

■

Measure Change and Estimation

Write

$$\Lambda_k = \prod_{\ell=1}^{k} |\det D(\theta_k)| \frac{\phi(y_\ell)}{\phi(w_\ell)}.$$

A new probability measure $\overline{P}$ can be defined on $(\Omega, \bigvee_{\ell=1}^{\infty} \mathscr{G}_\ell)$ by setting the restriction to $\mathscr{G}_{k+1}$ of the Radon-Nikodym derivative $d\overline{P}/dP$ equal to Λ_k.

Lemma 2.3 *Under* $\overline{P}$ *the random variables* $\{y_\ell\}$, $\ell \in \mathbb{N}$, *are i.i.d. with density function* ϕ.

Proof For $t \in \mathbb{R}^q$ the event $\{y_k \le t\} = \{y_k^i \le t^i,\ i = 1, \dots, q\}$. Then

$$\begin{aligned}
\overline{P}(y_k \le t \mid \mathscr{G}_k) &= \overline{E}[I(y_k \le t) \mid \mathscr{G}_k] \\
&= \frac{E[\Lambda_k I(y_k \le t) \mid \mathscr{G}_k]}{E[\Lambda_k \mid \mathscr{G}_k]} \\
&= \frac{E\left[|\det D(\theta_k)| \frac{\phi(y_k)}{\phi(w_k)} I(y_k \le t) \mid \mathscr{G}_k\right]}{E\left[|\det D(\theta_k)| \frac{\phi(y_k)}{\phi(w_k)} \mid \mathscr{G}_k\right]}.
\end{aligned}$$

Now

$$E\left[|\det D(\theta_k)| \frac{\phi(y_k)}{\phi(w_k)} \mid \mathscr{G}_k\right] = \int_{\mathbb{R}^q} |\det D(\theta_k)| \phi(y_k)\, dw_k = 1$$

so

$$\overline{P}(y_k \le t \mid \mathscr{G}_k) = \int_{\mathbb{R}^q} I(y_k \le t) |\det D(\theta_k)| \phi(y_k)\, dw_k$$

$$= \int_{-\infty}^{t^1} \cdots \int_{-\infty}^{t^q} \phi(y_k)\, dy_k.$$

The result follows. □

Suppose we now start with a probability measure $\overline{P}$ on $(\Omega, \bigvee_{\ell=1}^{\infty} \mathscr{G}_\ell)$ such that under $\overline{P}$:

1. $\{y_k\}$, $k \in \mathbb{N}$, is a sequence of i.i.d. $\mathbb{R}^q$-valued random variables with positive density function ϕ;
2. $\{\theta_k\}$, $k \in \mathbb{N}$, are real variables satisfying (2.3);
3. $\{x_k\}$, $k \in \mathbb{N}$, is a sequence of $\mathbb{R}^d$-valued random variables satisfying (2.1).

Note in particular that under $\overline{P}$ the y_ℓ and x_ℓ are independent. We now construct, by an inverse procedure, a probability measure P such that under P, $\{w_\ell\}$, $\ell \in \mathbb{N}$, is a sequence of i.i.d. random variables with density ϕ, where $w_k := D(\theta_k)^{-1}(y_k - C(\theta_k)x_k)$.

To construct P from $\overline{P}$, write

$$\overline{\Lambda}_k = \prod_{\ell=1}^{k} |\det D(\theta_\ell)|^{-1} \frac{\phi(w_\ell)}{\phi(y_\ell)}.$$

P is defined by putting the restriction to $\mathscr{G}_k$ of the Radon-Nikodym derivative $dP/d\overline{P}$ equal to $\overline{\Lambda}_k$. The existence of $\overline{P}$ is a consequence of Kolmogorov's theorem.

Unnormalized Estimates

Write $q_k(z,\theta)$, $k \in \mathbb{N}$, for the unnormalized conditional density such that

$$\overline{E}\left[\overline{\Lambda}_k I(x_k \in dz) I(\theta_k \in d\theta) \mid \mathscr{Y}_k\right] = q_k(z,\theta)\, dz\, d\theta.$$

The existence of q_k will be discussed below.

We now derive a recursive update for q_k. The normalized conditional density

$$p_k(z,\theta)\, dz\, d\theta = E\left[I(x_k \in dz) I(\theta_k \in d\theta) \mid \mathscr{Y}_k\right]$$

is given by

$$p_k(z,\theta) = \frac{q_k(z,\theta)}{\int_{\mathbb{R}^d} \int_{\mathbb{R}^p} q_k(\xi,\lambda)\, d\xi\, d\lambda}.$$

Theorem 2.4 *For $k \in \mathbb{N}$,*

$$\boxed{\begin{aligned} q_{k+1}(z,\lambda) = \phi(y_{k+1})^{-1} \iint \Big[&\Delta_1(y_{k+1}, z, \lambda, \xi, \sigma) \\ &\times \psi\left(B(\sigma)^{-1}(z - A(\sigma)\xi)\right) q_k(\xi,\sigma)\Big] d\xi\, d\sigma, \end{aligned}} \tag{2.4}$$

where

$$\begin{aligned}&\Delta_1\left(y_{k+1},z,\lambda,\xi,\sigma\right)\\&=\left|\det D\left(\lambda\right)\right|^{-1}\phi\left(D\left(\lambda\right)^{-1}\left(y_{k+1}-C\left(\lambda\right)z\right)\right)\left|\det B\left(\sigma\right)\right|^{-1}\rho\left(\lambda-\alpha\sigma\right).\end{aligned}$$

Proof Suppose $f:\mathbb{R}^d\times\mathbb{R}\to\mathbb{R}$ is any Borel test function. Then

$$\begin{aligned}&\overline{E}\left[f\left(x_{k+1},\theta_{k+1}\right)\overline{\Lambda}_{k+1}\mid\mathscr{Y}_{k+1}\right]\\&\quad=\int_{\mathbb{R}^d}\int_{\mathbb{R}}f\left(z,\lambda\right)q_{k+1}\left(z,\lambda\right)dz\,d\lambda\\&\quad=\overline{E}\Big[f\left(A\left(\theta_k\right)x_k+B\left(\theta_k\right)v_{k+1},\alpha\theta_k+v_{k+1}\right)\overline{\Lambda}_k\left|\det D\left(\theta_{k+1}\right)\right|^{-1}\\&\qquad\times\phi\left(D\left(\theta_{k+1}\right)^{-1}\left(y_{k+1}-C\left(\theta_{k+1}\right)x_{k+1}\right)\right)\mid\mathscr{Y}_{k+1}\Big]\phi\left(y_{k+1}\right)^{-1}.\end{aligned}$$

Substituting for the remaining x_{k+1} and θ_{k+1} this is

$$\begin{aligned}&=\overline{E}\Big\{\int\!\!\int\Big[f\left(A\left(\theta_k\right)x_k+B\left(\theta_k\right)w,\alpha\theta_k+v\right)\overline{\Lambda}_k\left|\det D\left(\alpha\theta_k+v\right)\right|^{-1}\\&\qquad\times\phi\left(D\left(\alpha\theta_k+v\right)\right)^{-1}\\&\qquad\times\left(y_{k+1}-C\left(\alpha\theta_k+v\right)\left(A\left(\theta_k\right)x_k+B\left(\theta_k\right)w\right)\right)\\&\qquad\times\psi\left(w\right)\rho\left(v\right)\Big]dw\,dv\mid\mathscr{Y}_{k+1}\Big\}\phi\left(y_{k+1}\right)^{-1}.\\&=\int\!\!\int\!\!\int\!\!\int\Big[f\left(A\left(\lambda\right)z+B\left(\lambda\right)w,\alpha\lambda+v\right)\left|\det D\left(\alpha\lambda+v\right)\right|^{-1}\\&\qquad\times\phi\left(D\left(\alpha\lambda+v\right)\right)^{-1}\left(y_{k+1}-C(\alpha\lambda+v)\left(A(\lambda)z+B(\lambda)w\right)\right)\\&\qquad\times\psi\left(w\right)\rho\left(v\right)q_k\left(z,\lambda\right)\Big]dz\,dw\,d\lambda\,dv\,\phi\left(y_{k+1}\right)^{-1},\end{aligned}$$

where the last equality follows since the y_ℓ are independent.

Write $\xi=A\left(\lambda\right)z+B\left(\lambda\right)w$ and $\sigma=\alpha\lambda+v$. Then

$$dz\,dw\left(d\lambda\,dv\right)=\left|\det B\left(\lambda\right)\right|^{-1}dz\,d\xi\left(d\lambda\,d\sigma\right),$$

and the above integral equals

$$\begin{aligned}&\int\!\!\int\!\!\int\!\!\int\Big[f\left(\xi,\sigma\right)\left|\det D\left(\sigma\right)\right|^{-1}\phi\left(D\left(\sigma\right)^{-1}\left(y_{k+1}-C\left(\sigma\right)\xi\right)\right)\\&\qquad\times\psi\left(B\left(\lambda\right)^{-1}\left(\xi-A\left(\lambda\right)z\right)\right)\left|\det B\left(\lambda\right)\right|^{-1}\\&\qquad\times\rho\left(\sigma-\alpha\lambda\right)q_k\left(z,\lambda\right)\Big]dz\,d\xi\,d\lambda\,d\sigma.\end{aligned}$$

This identity holds for all Borel test functions f, so the result follows. □

Remark 2.5 Suppose $\pi(z)$ is the density of x_0, and $\rho_0(\lambda)$ is the density of θ_0. Then $q_0(z,\lambda) = \pi(z)\rho_0(\lambda)$ and updated estimates are obtained by substituting in (2.4). ■

Even if the *prior estimates* for x_0 or θ_0 are delta functions, the proof of Theorem 4.1 gives a function for $q_1(z,\lambda)$. In fact, if $\pi(z) = \delta(x_0)$ and $\rho_0(\lambda) = \delta(\theta_0)$ then we see

$$\begin{aligned} q_1(z,\lambda) = {} & |\det D(\lambda)|^{-1} \phi\left(D(\lambda)^{-1}(y_1 - C(\lambda)z)\right) \\ & \times \psi\left(B(\theta_0)^{-1}(z - A(\theta_0)x_0)\right) |\det B(\theta_0)|^{-1} \rho(\lambda - \alpha\theta_0), \end{aligned}$$

and further updates follow from (2.4).

If there are no dynamics in one of the parameters, so that $\alpha = 1$ and ρ is the delta mass at 0 giving $\theta_k = \theta_{k-1}$, $k \in \mathbb{N}$, then care must be taken with the choice of *prior distribution* for θ. In fact, if $\rho_0(\theta)$ is the prior distribution, the above procedure gives an unnormalized conditional density $q_k^\theta(z,\lambda)$ for each possible value of θ, and $q_k(z,\lambda) = q_k^\lambda(z,\lambda)\rho_0(\lambda)$.

4.3 The ARMAX Model

We now indicate how the general ARMAX model can be treated. Suppose $\{v_\ell\}$, $\ell \in \mathbb{N}$, is a sequence of (real) i.i.d. random variables with density ψ. Write

$$\begin{aligned} \theta^1 &= (a_1, \dots, a_{r_1}) \in \mathbb{R}^{r_1}, \\ \theta^2 &= (b_1, \dots, b_{r_2}) \in \mathbb{R}^{r_2}, \\ \theta^3 &= \left(c_1, \dots, c_{r_3}\right) \in \mathbb{R}^{r_3}, c_{r_3} \neq 0, \end{aligned} \tag{3.1}$$

for the unknown coefficient vectors, or parameters. An ARMAX system $\{y_\ell\}$ with *exogenous inputs* $\{u_\ell\}$, $\ell \in \mathbb{N}$, is then given by equations of the form

$$\begin{aligned} & y_{k+1} + a_1 y_k + \cdots + a_{r_1} y_{k+1-r_1} \\ & \quad = b_1 u_k + \cdots + b_{r_2} u_{k+1-r_2} + c_1 v_k + \cdots + c_{r_3} v_{k+1-r_3} + v_{k+1}. \end{aligned} \tag{3.2}$$

Write x_k for the column vector

$$\left(y_k, \dots, y_{k+1-r_1}, u_k, \dots, u_{k+1-r_2}, v_k, \dots, v_{k+1-r_3}\right)' \in \mathbb{R}^{r_1+r_2+r_3}.$$

Suppose $A(\theta)$ is the $(r_1+r_2+r_3) \times (r_1+r_2+r_3)$ matrix having $\left(-\theta^1, \theta^2, \theta^3\right)$ for its first row and 1 on the subdiagonal, with zeros elsewhere on other rows, except the (r_1+1) and (r_1+r_2+1) rows which are $0 \in \mathbb{R}^{r_1+r_2+r_3}$. B will denote the unit column vector in $\mathbb{R}^{r_1+r_2+r_3}$ having one in the (r_1) position and zeros elsewhere. C will denote the column vector in $\mathbb{R}^{r_1+r_2+r_3}$ having 1 in the first and (r_1+r_2+1) position and zeros elsewhere. The values of the u_ℓ are known exogenously; for example, if the variables u_ℓ are control variables u_k will depend on the values of $y_1, \dots, y_k$.

System (3.2) can then be written:

$$x_{k+1} = A(\theta)x_k + Bu_{k+1} + Cv_{k+1}, \tag{3.3}$$

$$y_{k+1} = \langle \theta, x_k \rangle + v_{k+1}. \tag{3.4}$$

Here $\theta = \left(-\theta^1, \theta^2, \theta^3\right)$ and $\langle\ ,\ \rangle$ denotes the scalar product in $\mathbb{R}^{r_1+r_2+r_3}$. Representation (3.3), (3.4) is not a *minimal representation*; see, for example, Anderson and Moore (1979) . However, it suffices for our discussion. Notice the same noise term v_{k+1} appears in (3.3) and (3.4). This is circumvented by substituting in (3.3) to obtain

$$x_{k+1} = \left(A(\theta) - C\theta'\right)x_k + Bu_{k+1} + Cy_{k+1} \tag{3.5}$$

together with

$$y_{k+1} = \langle \theta, x_k \rangle + v_{k+1}. \tag{3.6}$$

Write $\mathscr{Y}_k^0 = \sigma\{y_1, \dots, y_k\}$ and $\{\mathscr{Y}_\ell\}$, $\ell \in \mathbb{N}$, for the corresponding complete filtration. Write $\overline{x}_k$ for the column vector $\left(v_k, \dots, v_{k+1-r_3}\right)' \in \mathbb{R}^{r_3}$ so that $x_k' = (y_k, \dots, y_{k+1-r_1}, u_k, \dots, u_{k+1-r_2}, \overline{x}')$, and, given $\mathscr{Y}_k$, the $\overline{x}_k$ are the unknown components of x_k. Let $\alpha_{k+1} = y_{k+1} + \langle \theta^1, \left(y_k, \dots, y_{k+1-r_1}\right)' \rangle - \langle \theta^2, \left(u_k, \dots, u_{k+1-r_2}\right)' \rangle$ and write $\underline{\alpha}_k$ for the vector $\alpha_k \overline{C}$ where $\overline{C} = (1, 0, \dots, 0)' \in \mathbb{R}^{r_3}$. Then with $\Gamma\left(\theta^3\right)$ equal to the $r_3 \times r_3$ matrix

$$\Gamma\left(\theta^3\right) = \begin{pmatrix} -c_1 & -c_2 & \dots & -c_{r_3-1} & -c_{r_3} \\ 1 & 0 & \dots & 0 & 0 \\ 0 & 1 & \dots & 0 & 0 \\ \vdots & \vdots & \ddots & \vdots & \vdots \\ 0 & 0 & \dots & 1 & 0 \end{pmatrix}$$

we have $\overline{x}_{k+1} = \Gamma\left(\theta^3\right)\overline{x}_k + \underline{\alpha}_{k+1}$. The model is chosen so that $c_{r_3} \neq 0$; then

$$\Gamma\left(\theta^3\right)^{-1} = \begin{pmatrix} 0 & 1 & 0 & \dots & 0 & 0 \\ 0 & 0 & 1 & \dots & 0 & 0 \\ \vdots & \vdots & \vdots & \ddots & \vdots & \vdots \\ 0 & 0 & 0 & \dots & 1 & 0 \\ 0 & 0 & 0 & \dots & 0 & 1 \\ -\frac{1}{c_{r_3}} & -\frac{c_1}{c_{r_3}} & -\frac{c_2}{c_{r_3}} & \dots & -\frac{c_{r_3-2}}{c_{r_3}} & -\frac{c_{r_3-1}}{c_{r_3}} \end{pmatrix}.$$

Given $\mathscr{Y}_k$ we wish to determine the unnormalized conditional density of $\overline{x}_k$ and θ. Again, we suppose the processes are defined on $\left(\Omega, \mathscr{F}, \overline{P}\right)$ under which $\{y_\ell\}$, $\ell \in \mathbb{N}$, is a sequence of i.i.d. random variables with strictly positive densities ϕ. P is defined by putting the restriction of $dP/d\overline{P}$ to $\mathscr{G}_k$ equal to $\overline{\Lambda}_k$. Here $\overline{\Lambda}_k = \prod_{\ell=1}^k \phi\left(y_{\ell+1} - \langle \theta, x_\ell \rangle\right)/\phi\left(y_{\ell+1}\right)$. Write $q_k(\xi, \lambda)$ for the unnormalized conditional density such that

$$\overline{E}\left[I\left(\overline{x}_k \in d\xi\right) I\left(\theta \in d\lambda\right) \overline{\Lambda}_k \mid \mathscr{Y}_k\right] = q_k\left(\xi, \lambda\right) d\xi \, d\lambda.$$

Therefore we now consider any Borel test functions $f : \mathbb{R}^{r_3} \to \mathbb{R}$ and $g : \mathbb{R}^{r_1+r_2+r_3} \to \mathbb{R}$. Write $\bar{y}_k = \left(y_k, \ldots, y_{k+1-r_1}\right)'$ and $\overline{u}_k = \left(u_k, \ldots, u_{k+1-r_2}\right)'$. The same arguments to those used in Section 4.4 lead us to consider

$$\begin{aligned}
&\overline{E}\left[f\left(\overline{x}_{k+1}\right) g\left(\theta\right) \overline{\Lambda}_{k+1} \mid \mathscr{Y}_{k+1}\right] \\
&\quad = \iint f\left(\xi\right) g\left(\lambda\right) q_{k+1}\left(\xi, \lambda\right) d\xi \, d\lambda \\
&\quad = \overline{E}\Big[f\left(\Gamma\left(\theta^3\right) \overline{x}_k + \underline{\alpha}_{k+1}\right) g\left(\theta\right) \overline{\Lambda}_k \\
&\qquad\quad \times \phi\left(y_{k+1} + \langle \theta^1, \bar{y}_k \rangle - \langle \theta^2, \overline{u}_k \rangle - \langle \theta^3, \overline{x}_k \rangle\right) \mid \mathscr{Y}_{k+1}\Big] \phi\left(y_{k+1}\right)^{-1} \\
&\quad = \iint \Big[f\left(\Gamma\left(\lambda^3\right) z + \underline{\alpha}_{k+1}\right) g\left(\lambda\right) \\
&\qquad\quad \times \phi\left(y_{k+1} + \langle \lambda^1, \bar{y}_k \rangle - \langle \lambda^2, \overline{u}_k \rangle - \langle \lambda^3, z \rangle\right) \\
&\qquad\quad \times q_k\left(z, \lambda\right)\Big] dz \, d\lambda \, \phi\left(y_{k+1}\right)^{-1}.
\end{aligned} \tag{3.7}$$

Write

$$\xi = \Gamma\left(\lambda^3\right) z + \left(y_{k+1} + \langle \lambda^1, \bar{y}_k \rangle - \langle \lambda^2, \overline{u}_k \rangle\right) \overline{C}.$$

Then

$$z = \Gamma\left(\lambda^3\right)^{-1} \left(\xi - \left(y_{k+1} + \langle \lambda^1, \bar{y}_k \rangle - \langle \lambda^2, \overline{u}_k \rangle\right) \overline{C}\right)$$

and

$$dz \, d\lambda = dz \, d\lambda^1 \, d\lambda^2 \, d\lambda^3 = \Gamma\left(\lambda^3\right) d\xi \, d\lambda.$$

Substituting in (3.7) we have

$$\begin{aligned}
&\iint f\left(\xi\right) g\left(\lambda\right) q_{k+1}\left(\xi, \lambda\right) d\xi \, d\lambda \\
&\quad = \iint \Big[f\left(\xi\right) g\left(\lambda\right) \phi\Big(y_{k+1} + \langle \lambda^1, \bar{y}_k \rangle - \langle \lambda^2, \overline{u}_k \rangle \\
&\qquad - \Big\langle \lambda^3, \Gamma\left(\lambda^3\right)^{-1} \left(\xi - \left(y_{k+1} + \langle \lambda^1, \bar{y}_k \rangle - \langle \lambda^2, \overline{u}_k \rangle\right) \overline{C}\right) \Big\rangle \Big) \\
&\qquad \times \phi\left(y_{k+1}\right)^{-1} q_k\Big(\Gamma\left(\lambda^3\right)^{-1} \left(\xi - \left(y_{k+1} + \langle \lambda^1, \bar{y}_k \rangle - \langle \lambda^2, \overline{u}_k \rangle\right) \overline{C}\right), \lambda\Big) \\
&\qquad \times \Gamma\left(\lambda^3\right)^{-1}\Big] d\xi \, d\lambda.
\end{aligned}$$

We, therefore, have the following remarkable result for updating the unnormalized, conditional density of $\overline{x}_k$ and θ, given $\mathscr{Y}_k$:

Theorem 3.1

$$\boxed{\begin{aligned} q_{k+1}(\xi,\lambda) &= \Delta_2(y_{k+1},\bar{y}_k,\bar{u}_k,\xi,\lambda) \\ &\quad\times q_k\left(\Gamma(\lambda^3)^{-1}\left(\xi-\left(y_{k+1}+\langle\lambda^1,\bar{y}_k\rangle-\langle\lambda^2,u_k\rangle\right)\overline{C}\right),\lambda\right), \end{aligned}} \tag{3.8}$$

where $\Delta_2(y_{k+1},\bar{y}_k,\bar{u}_k,\xi,\lambda)=\frac{\phi(\xi_1)}{\phi(y_{k+1})}\Gamma(\lambda^3)^{-1}$ *and* $\xi_1=y_{k+1}+\langle\lambda^1,\bar{y}_k\rangle-\langle\lambda^2,\bar{u}_k\rangle-\langle\lambda^3,\Gamma(\lambda^3)^{-1}(\xi-(y_{k+1}+\langle\lambda^1,\bar{y}_k\rangle-\langle\lambda^2,\bar{u}_k\rangle)\overline{C})\rangle$.

Remark 3.2 This does not involve any integration.

■

If $\pi_0(\xi)$ is the prior density of x_0 and $\rho_0(\lambda)$ the prior density for λ, then $q_0(\xi,\lambda)=\pi_0(\xi)\rho_0(\lambda)$. The prior density must reflect information known about x_0 and θ, and not be just a guess. Because no dynamics or noise enter the parameters θ the estimation problem can be treated as though θ is fixed, followed by an averaging over θ using the density $\rho_0(\lambda)$.

4.4 Nonlinear Dynamics

In this section, nonlinear, vector-valued signal and observation dynamics are considered in discrete time, with additive (not necessarily Gaussian) noise. Here possibly *singular measures* describe the distribution of the state. The *forward recursion* for the *alpha* unnormalized, conditional density, and the *backward recursion* for the *beta* variable are derived. The unnormalized smoothed density is, as in the Baum-Welch (1966) situation, the product of alpha and beta.

The Baum-Welch algorithm usually discusses a Markov chain observed in Gaussian noise (Baum and Petrie 1966). The forward and backward Baum-Welch estimators are related to considering the observations under an equivalent probability measure; they provide unnormalized filtered and smoothed estimates of the state of the Markov chain, given the observations.

Suppose $\{x_k\}$, $k\in\mathbb{N}$, is a discrete-time stochastic state process taking values in some Euclidean space $\mathbb{R}^m$. We suppose that x_0 has a known distribution $\pi_0(x)$. The observation process $\{y_k\}$, $k\in\mathbb{N}$, takes value in some Euclidean space $\mathbb{R}^d$. The sets $\{v_k\}$ and $\{w_k\}$ $k\in\mathbb{N}$, will be sequences of independent, $\mathbb{R}^m$, $\mathbb{R}^d$-valued, random variables with probability distributions $d\Psi_k$ and densities ϕ_k, respectively. We assume the ϕ_k are *strictly positive*.

For $k\in\mathbb{N}$, $A_k:\mathbb{R}^m\to\mathbb{R}^m$, $C_k:\mathbb{R}^d\to\mathbb{R}^d$ are measurable functions, and we suppose for $k\geq 0$ that

$$\boxed{\begin{aligned} x_{k+1} &= A_{k+1}(x_k)+v_{k+1}, \\ y_k &= C_k(x_k)+w_k. \end{aligned}} \tag{4.1}$$

Measure Change for Observation Process

Define

$$\lambda_\ell = \frac{\phi_\ell(y_\ell)}{\phi_\ell(w_\ell)}, \quad \ell \in \mathbb{N}.$$

Write

$$\begin{aligned}\mathscr{G}_k^0 &= \sigma\{x_0, x_1, \ldots, x_k, y_1, \ldots, y_k\},\\ \mathscr{Y}_k^0 &= \sigma\{y_1, \ldots, y_k\},\end{aligned}$$

and $\{\mathscr{G}_k\}$, $\{\mathscr{Y}_k\}$, $k \in \mathbb{N}$, for the corresponding complete filtrations. With

$$\Lambda_k = \prod_{\ell=1}^{k} \lambda_\ell$$

a new probability measure $\overline{P}$ can be defined by setting the restriction of the Radon-Nikodym derivative $\left(d\overline{P}/dP\right)|_{\mathscr{G}_k}$ equal to Λ_k. The existence of $\overline{P}$ follows from Kolmogorov's theorem; under $\overline{P}$ the random variables y_ℓ, $\ell \in \mathbb{N}$, are independent and the density function of y_ℓ is ϕ_ℓ. Note that the process x is the same under both measures.

Note that under $\overline{P}$ the y_k are, in particular, independent of the x_k. To represent the situation where the state influences the observations we construct a probability measure P such that, under P, $w_{k+1} := y_{k+1} - C_{k+1}(x_k)$ is a sequence of independent random variables with positive density functions $\phi_{k+1}(\cdot)$. To construct P starting from $\overline{P}$ set

$$\begin{aligned}\overline{\lambda}_k &= \frac{\phi_k(w_k)}{\phi_k(y_k)},\\ \overline{\Lambda}_k &= \prod_{\ell=1}^{k} \overline{\lambda}_\ell,\end{aligned}$$

and

$$\left.\frac{dP}{d\overline{P}}\right|_{\mathscr{G}_k} = \overline{\Lambda}_k.$$

Under P the $\{v_\ell\}$ are independent random variables having densities ϕ_ℓ.

Recursive Estimates

We shall work under measure $\overline{P}$, so that the $\{y_k\}$, $k \in \mathbb{N}$, is a sequence of independent $\mathbb{R}^d$-valued random variables with densities ϕ_k and the $\{x_k\}$, $k \in \mathbb{N}$, satisfy the dynamics $x_{k+1} = A_{k+1}(x_k) + v_{k+1}$.

Notation 4.1 *Write* $d\alpha_k(x)$, $k \in \mathbb{N}$, *for the unnormalized conditional probability measure such that*

$$\overline{E}\left[\overline{\Lambda}_k I(x_k \in dx) \mid \mathscr{Y}_k\right] = d\alpha_k(x).$$

Theorem 4.2 *For* $k \in \mathbb{N}$, *a recursion for* $d\alpha_k(.)$ *is given by*

$$\boxed{d\alpha_k(z) = \frac{\phi_k(y_k - C_k(z))}{\phi_k(y_k)} \int_{\mathbb{R}^m} d\Psi_k(z - A_k(\xi))\, d\alpha_{k-1}(\xi).} \tag{4.2}$$

Proof Suppose $f : \mathbb{R}^m \to \mathbb{R}$ is any integrable Borel test function. Then

$$\overline{E}\left[\overline{\Lambda}_k f(x_k) \mid \mathscr{Y}_k\right] = \int_{\mathbb{R}^m} f(z)\, d\alpha_k(z).$$

However, denoting

$$\Phi_k(\xi, v_k) = \frac{\phi_k(y_k - C_k(A_k(\xi) + v_k))}{\phi_k(y_k)},$$

then

$$\begin{aligned}
&\overline{E}\left[\overline{\Lambda}_k f(x_k) \mid \mathscr{Y}_k\right] \\
&\quad = \overline{E}\left[\overline{\Lambda}_{k-1} \Phi_k(x_{k-1}, v_k) f(A_k(x_{k-1}) + v_k) \mid \mathscr{Y}_k\right] \\
&\quad = \overline{E}\left[\overline{\Lambda}_{k-1} \int_{\mathbb{R}^m} \Phi_k(x_{k-1}, v_k) f(A_k(x_{k-1}) + v_k)\, d\Psi_k(v_k) \,\Big|\, \mathscr{Y}_k\right] \\
&\quad = \int_{\mathbb{R}^m} \int_{\mathbb{R}^m} \Phi_k(\xi, v_k) f(A_k(\xi) + v_k)\, d\Psi_k(v_k)\, d\alpha_{k-1}(\xi).
\end{aligned}$$

Write

$$z = A_k(\xi) + v_k.$$

Consequently

$$\begin{aligned}
&\int_{\mathbb{R}^m} f(z)\, d\alpha_k(z) \\
&\quad = \int_{\mathbb{R}^m} \int_{\mathbb{R}^m} \frac{\phi_k(y_k - C_k(z))}{\phi_k(y_k)} f(z)\, d\Psi_k(z - A_k(\xi))\, d\alpha_{k-1}(\xi).
\end{aligned}$$

This identity holds for all Borel test functions f, so (4.2) follows. □

Notation 4.3 *In this section* $m, k \in \mathbb{N}$, *and* $m < k$. *Write* $\overline{\Lambda}_{m,k} = \prod_{\ell=m}^{k} \overline{\lambda}_\ell$ *and* $d\gamma_{m,k}(x)$ *for the unnormalized conditional probability measure such that*

$$\overline{E}\left[\overline{\Lambda}_k I(x_m \in dx) \mid \mathscr{Y}_k\right] = d\gamma_{m,k}(x).$$

Theorem 4.4 *For $m,k \in \mathbb{N}$, $m < k$*

$$\boxed{d\gamma_{m,k}(x) = \beta_{m,k}(x)\, d\alpha_m(x),} \tag{4.3}$$

where $d\alpha_m(x)$ is given recursively by Theorem 4.2 and

$$\beta_{m,k}(x) = \overline{E}\left[\overline{\Lambda}_{m+1,k} \mid x_m = x,\ \mathscr{Y}_k\right].$$

Proof For an arbitrary integrable function $f : \mathbb{R}^m \to \mathbb{R}$

$$\overline{E}\left[\overline{\Lambda}_k f(x_m) \mid \mathscr{Y}_k\right] = \int_{\mathbb{R}^m} f(x)\, d\gamma_{m,k}(x).$$

However,

$$\overline{E}\left[\overline{\Lambda}_k f(x_m) \mid \mathscr{Y}_k\right] = \overline{E}\left[\overline{\Lambda}_{1,m} f(x_m)\, \overline{E}\left[\overline{\Lambda}_{m+1,k} \mid x_0, \ldots, x_m, \mathscr{Y}_k\right] \mid \mathscr{Y}_k\right].$$

Now

$$\overline{E}\left[\overline{\Lambda}_{m+1,k} \mid x_m = x,\ \mathscr{Y}_k\right] := \beta_{m,k}(x).$$

Consequently,

$$\overline{E}\left[\overline{\Lambda}_k f(x_m) \mid \mathscr{Y}_k\right] = \overline{E}\left[\overline{\Lambda}_{1,m} f(x_m)\, \beta_{m,k}(x_m) \mid \mathscr{Y}_k\right]$$

and so, from Notation 4.1,

$$\int_{\mathbb{R}^m} f(x)\, d\gamma_{m,k}(x) = \int_{\mathbb{R}^m} f(x)\, \beta_{m,k}(x)\, d\alpha_m(x).$$

The function $f(x)$ is an arbitrary Borel test function; therefore, we see

$$d\gamma_{m,k}(x) = \beta_{m,k}(x)\, d\alpha_m(x).$$

□

Theorem 4.5 *$\beta_{m,k}(x)$ satisfies the backward recursive equation*

$$\boxed{\begin{aligned}\beta_{m,k}(x) &= \frac{1}{\phi_{m+1}(y_{m+1})} \\ &\quad \times \int_{\mathbb{R}^m} \left[\phi_{m+1}\left(y_{m+1} - C_{m+1}\left(A_{m+1}(x) + w\right)\right)\right. \\ &\quad\quad \left.\times\, \beta_{m+1,k}\left(A_{m+1}(x) + w\right)\right] d\Psi_{m+1}(w)\end{aligned}} \tag{4.4}$$

with $\beta_{n,n} = 1$.

Proof

$$
\begin{aligned}
\beta_{m,k}(x) &= \overline{E}\left[\overline{\Lambda}_{m+1,k} \mid x_m = x, \mathscr{Y}_k\right] \\
&= \overline{E}\left[\overline{\lambda}_{m+1}\overline{\Lambda}_{m+2,k} \mid x_m = x, \mathscr{Y}_k\right] \\
&= \overline{E}\left[\overline{\lambda}_{m+1}\overline{E}\left[\overline{\Lambda}_{m+2,k} \mid x_m = x, x_{m+1}, \mathscr{Y}_k\right] \mid x_m = x, \mathscr{Y}_k\right] \\
&= \overline{E}\left[\frac{\phi_{m+1}\left(y_{m+1} - C_{m+1}\left(A_{m+1}\left(x_m\right) + v_{m+1}\right)\right)}{\phi_{m+1}\left(y_{m+1}\right)}\right. \\
&\qquad \left. \times \beta_{m+1,k}\left(A_{m+1}\left(x_m\right) + v_{m+1}\right) \,\middle|\, x_m = x, \mathscr{Y}_k\right] \\
&= \frac{1}{\phi_{m+1}\left(y_{m+1}\right)} \int_{\mathbb{R}^m} \Big[\phi_{m+1}\left(y_{m+1} - C_{m+1}\left(A_{m+1}(x) + w\right)\right) \\
&\qquad \times \beta_{m+1,k}\left(A_{m+1}(x) + w\right)\Big] d\Psi_{m+1}(w).
\end{aligned}
$$

□

Change of Measure for the State Process

In this section we shall suppose that the noise in the state equation is not singular, that is, each v_k has a positive density function ψ_k. The observation process y is as described at the beginning of this section. Suppose $\overline{P}$ has been constructed. Define

$$\gamma_\ell = \frac{\psi_\ell\left(x_\ell\right)}{\psi_\ell\left(v_\ell\right)}$$

and

$$\Gamma_k = \prod_{\ell=1}^{k} \gamma_\ell;$$

then introduce a measure $\hat{P}$ by setting

$$\left.\frac{d\hat{P}}{d\overline{P}}\right|_{\mathscr{G}_k} = \Gamma_k.$$

Under $\hat{P}$ the random variables $\{x_\ell\}$, $\ell \in \mathbb{N}$, are independent with density function ψ_ℓ.

We now start with a probability measure $\hat{P}$ on $(\Omega, \bigvee_{n=1}^{\infty} \mathscr{G}_n)$ under which the process $\{x_\ell\}$ and $\{y_\ell\}$ are two sequences of independent random variables with respective densities ψ_ℓ and ϕ_ℓ. Note the x and y are independent of each other as well. To return to the real-world model described in Section 4.2 we must define a probability measure P by setting

$$\left.\frac{dP}{d\hat{P}}\right|_{\mathscr{G}_k} = \left.\frac{dP}{d\overline{P}}\right|_{\mathscr{G}_k} \left.\frac{d\overline{P}}{d\hat{P}}\right|_{\mathscr{G}_k} = \overline{\Lambda}_k\overline{\Gamma}_k.$$

Here $\overline{\Gamma}_k$ is the inverse of Γ_k, so that $\overline{\Gamma}_k = \prod_{\ell=1}^{k} \overline{\gamma}_\ell$, where $\overline{\gamma}_\ell = \psi_\ell(v_\ell) / \psi_\ell(x_\ell)$. Again the existence of P is guaranteed by Kolmogorov's Extension Theorem.

Recursive Estimates

We shall work under $\hat{P}$, so that $\{y_k\}$, $k \in \mathbb{N}$, and $\{x_k\}$, $k \in \mathbb{N}$, are two sequences of independent random variables with respective densities ϕ_k and ψ_k. Recall that a version of Bayes' theorem states that for a $\mathscr{G}$-adapted sequence $\{g_k\}$,

$$E\left[g_k \mid \mathscr{Y}_k\right] = \frac{\overline{E}\left[\overline{\Lambda}_k g_k \mid \mathscr{Y}_k\right]}{\overline{E}\left[\overline{\Lambda}_k \mid \mathscr{Y}_k\right]}.$$

Similarly,

$$\overline{E}\left[\overline{\Lambda} g_k \mid \mathscr{Y}_k\right] = \frac{\hat{E}\left[\overline{\Gamma}_k \overline{\Lambda}_k g_k \mid \mathscr{Y}_k\right]}{\hat{E}\left[\overline{\Gamma}_k \mid \mathscr{Y}_k\right]}.$$

Remark 4.6 The x_k sequence is independent of the y_k sequence under $\hat{P}$. Therefore conditioning on the x's it is easily seen that $\hat{E}\left[\overline{\Gamma}_k \mid \mathscr{Y}_k\right] = \hat{E}\left[\overline{\Gamma}_k\right] = \hat{E}\left[\overline{\Gamma}_{k-1}\right] = 1$.

■

Notation 4.7 *Suppose $\alpha_k(x)$, $k \in \mathbb{N}$, is the unnormalized conditional density such that*

$$\overline{E}\left[\overline{\Lambda}_k I(x_k \in dx) \mid \mathscr{Y}_k\right] = \alpha_k(x)\, dx.$$

We now rederive the recursive expression for α_k.

Theorem 4.8 *For $k \in \mathbb{N}$, a recursion for $\alpha_k(x)$ is given by*

$$\boxed{\alpha_k(x) = \frac{\phi_k(y_k - C_k(x))}{\phi_k(y_k)} \int_{\mathbb{R}^m} \psi_k(x - A_k(\xi))\, \alpha_{k-1}(\xi)\, d\xi.} \tag{4.5}$$

Proof Suppose $f : \mathbb{R}^m \to \mathbb{R}$ is any integrable Borel test function. Then

$$\overline{E}\left[f(x_k)\overline{\Lambda}_k \mid \mathscr{Y}_k\right] = \int_{\mathbb{R}^m} f(x)\, \alpha_k(x)\, dx. \tag{4.6}$$

However,

$$\overline{E}\left[f(x_k)\overline{\Lambda}_k \mid \mathscr{Y}_k\right] = \frac{\hat{E}\left[f(x_k)\overline{\Lambda}_k \overline{\Gamma}_k \mid \mathscr{Y}_k\right]}{\hat{E}\left[\overline{\Gamma}_k \mid \mathscr{Y}_k\right]}.$$

From Remark 4.6 the denominator equals 1. Using the independence of the x_k's and the y_k's under $\hat{P}$ we see

$$
\begin{aligned}
&\overline{E}\left[f(x_k)\overline{\Lambda}_k \mid \mathscr{Y}_k\right] \\
&= \hat{E}\left[f(x_k)\overline{\Lambda}_k\overline{\Gamma}_k \mid \mathscr{Y}_k\right] \\
&= \int_{\mathbb{R}^m}\int_{\mathbb{R}^m} \psi_k(x - A_k(\xi)) \frac{\phi_k(y_k - C_k(x))}{\phi_k(y_k)} \alpha_{k-1}(\xi)\, d\xi\, f(x)\, dx.
\end{aligned}
\tag{4.7}
$$

Since f is an arbitrary Borel test function, equations (4.6) and (4.7) yield at once (4.5). □

Notation 4.9 *For $m,k \in \mathbb{N}$, $m < k$, write $\overline{\Lambda}_{m,k} = \prod_{\ell=m}^{k} \overline{\lambda}_\ell$ and $\overline{\Gamma}_{m,k} = \prod_{\ell=m}^{k} \overline{\gamma}_\ell$. Write $\gamma_{m,k}(x)$ for the unnormalized conditional density such that*

$$\overline{E}\left[\overline{\Lambda}_k I(x_m \in dx) \mid \mathscr{Y}_k\right] = \gamma_{m,k}(x)\, dx.$$

It can be shown as in Theorems 4.2, 4.4, and 4.5 that

$$\boxed{\gamma_{m,k}(x) = \alpha_m(x)\, \beta_{m,k}(x),} \tag{4.8}$$

where $\alpha_m(x)$ is given recursively by Theorem 4.8 and $\beta_{m,k}(x)$ satisfies the backward recursive equation

$$
\boxed{
\begin{aligned}
\beta_{m,k}(x) = &\frac{1}{\phi_{m+1}(y_{m+1})} \\
&\times \int_{\mathbb{R}^m} \big[\psi_{m+1}(z - A_{m+1}(x)) \\
&\qquad \times \phi_{m+1}(y_{m+1} - C_{m+1}(z))\, \beta_{m+1,k}(z)\big]\, dz.
\end{aligned}
}
\tag{4.9}
$$

Notation 4.10 *For $m \in \mathbb{N}$, $m < k$, write $\xi_{m,m+1,k}(x^1, x^2)$ for the unnormalized conditional density such that*

$$\overline{E}\left[\overline{\Lambda}_k I(x_m \in dx^1)\, I(x_{m+1} \in dx^2) \mid \mathscr{Y}_k\right] = \xi_{m,m+1,k}(x^1, x^2)\, dx^1 dx^2.$$

Also write $\rho_{k+1,k}(x)$ for the unnormalized conditional density such that

$$\overline{E}\left[\overline{\Lambda}_{k+1} I(x_{k+1} \in dx) \mid \mathscr{Y}_k\right] = \rho_{k+1,k}(x)\, dx.$$

Theorem 4.11 *For $m,k \in \mathbb{N}$, $m < k$,*

$$
\begin{aligned}
&\xi_{m-1,m,k}(x^1, x^2) \\
&\quad = \alpha_{m-1}(x^1)\, \beta_{m,k}(x^2)\, \psi_m(x^2 - A_{m-1}(x^1)) \\
&\qquad \times \frac{\phi_m(y_m - C_m(x^2))}{\phi_m(y_m)}.
\end{aligned}
\tag{4.10}
$$

Proof Suppose $f, g : \mathbb{R}^m \to \mathbb{R}$ are arbitrary integrable Borel functions. Then

$$\begin{aligned}
&\hat{E}\left[f(x_{m-1})\,g(x_m)\,\overline{\Lambda}_k\overline{\Gamma}_k \mid \mathscr{Y}_k\right] \\
&= \int_{\mathbb{R}^{m-1}}\int_{\mathbb{R}^{m-1}} f(x^1)\,g(x^2)\,\xi_{m-1,m,k}(x^1,x^2)\,dx^1dx^2 \qquad (4.11)\\
&= \hat{E}\Big[\hat{E}\left[f(x_{m-1})\,g(x_m)\,\overline{\Lambda}_{0,m}\overline{\Lambda}_{m+1,k}\overline{\Gamma}_{0,m}\overline{\Gamma}_{m+1,k} \mid x_0,\dots,x_m,\mathscr{Y}_k\right] \mid \mathscr{Y}_k\Big] \\
&= \hat{E}\Big[f(x_{m-1})\,g(x_m)\,\overline{\Lambda}_{0,m}\overline{\Gamma}_{0,m} \\
&\qquad \times \hat{E}\left[\overline{\Lambda}_{m+1,k}\overline{\Gamma}_{m+1,k} \mid x_m,\mathscr{Y}_k\right] \mid \mathscr{Y}_k\Big] \\
&= \hat{E}\left[f(x_{m-1})\,g(x_m)\,\overline{\Lambda}_{0,m}\overline{\Gamma}_{0,m}\beta_{m,k}(x_m) \mid \mathscr{Y}_k\right] \\
&= \hat{E}\Big\{f(x_{m-1})\,\overline{\Lambda}_{0,m-1}\overline{\Gamma}_{0,m-1} \\
&\qquad \times \int_{\mathbb{R}^{m-1}}\Big[g(x^2)\,\psi_m(x^2 - A_m(x_{m-1})) \\
&\qquad\qquad \times \frac{\phi_m(y_m - C_m(x^2))}{\phi_m(y_m)}\beta_{m,k}(x^2)\Big]dx^2 \mid \mathscr{Y}_k\Big\} \\
&= \int_{\mathbb{R}^{m-1}}\int_{\mathbb{R}^{m-1}}\Big[f(x^1)\,g(x^2)\,\psi_m(x^2 - A_m(x^1)) \\
&\qquad\qquad \times \frac{\phi_m(y_m - C_m(x^2))}{\phi_m(y_m)}\beta_{m,k}(x^2)\,\alpha_{m-1}(x^1)\Big]dx^1dx^2.
\end{aligned} \qquad (4.12)$$

Since $f(x)$ and $g(x)$ are two arbitrary Borel test functions, comparing (4.11) with (4.12) gives equation (4.10). □

It is left an exercise to show that the one-step *predictor* satisfies the equation:

$$\boxed{\rho_{k,k-1}(x) = \int_{\mathbb{R}^d}\int_{\mathbb{R}^m} \phi_k(y_k - C_k(x))\,\psi_k(x - A_k(z))\,\alpha_{k-1}(z)\,dz\,dy.} \qquad (4.13)$$

Remark 4.12 The expression for unnormalized conditional density given by (4.13) can be easily generalized to the rth step predictor; the corresponding expression is

$$\begin{aligned}
&\rho_{k+r,k}(x) \\
&= \int_{\mathbb{R}^{mr}}\int_{\mathbb{R}^{dr}} H(y^1,\dots,y^r,x^1,\dots,x^{r-1},x) \\
&\quad \times \alpha_k(z)\,dy^1\dots dy^r dx^1\dots dx^{r-1}dz,
\end{aligned}$$

where

$$
\begin{aligned}
&H\left(y^1,\dots,y^r,x^1,\dots,x^{r-1},x\right)\\
&\quad=\phi_{k+1}\left(y^1-C_{k+1}\left(x^1\right)\right)\phi_{k+2}\left(y^2-C_{k+2}\left(x^1\right)\right)\cdots\phi_{k+r}\left(y^r-C_{k+r}\left(x\right)\right)\\
&\qquad\psi_{k+1}\left(x^1-A_{k+1}\left(z\right)\right)\psi_{k+2}\left(x^2-A_{k+2}\left(x^1\right)\right)\cdots\psi_{k+r}\left(x-A_{k+r}\left(x^{r-1}\right)\right).
\end{aligned}
$$

∎

4.5 Kalman Filter

Consider a system whose state at time $k=0,1,2,\dots,$ is $X_k\in\mathbb{R}^m$ and which can be observed only indirectly through another process $Y_k\in\mathbb{R}^d$.

Let $(\Omega,\mathscr{F},P)$ be a probability space upon which V_k and W_k are normally distributed with means 0 and respective covariance identity matrices $I_{m\times m}$ and $I_{d\times d}$. Assume that D_k is non singular, B_k is non singular and symmetric (for notational convenience). X_0 is a Gaussian random variable with zero mean and covariance matrix B_0^2 (of dimension $m\times m$). Let $\{\mathscr{F}_k\}$, $k\in\mathbb{N}$ be the complete filtration (that is $\mathscr{F}_0$ contains all the P-null events) generated by $\{X_0,X_1,\dots,X_k\}$.

The state and observations of the system satisfies the linear dynamics

$$X_{k+1}=A_kX_k+B_kV_{k+1}\in\mathbb{R}^m, \tag{5.1}$$

$$Y_k=C_kX_k+D_kW_k\in\mathbb{R}^d, \tag{5.2}$$

A_k, C_k are matrices of appropriate dimensions.

We shall also write $\{\mathscr{Y}_k\}$, $k\in\mathbb{N}$ for the complete filtration generated by $\{Y_0,Y_1,\dots Y_k\}$.

Using measure change techniques we shall derive a recursive expression for the conditional distribution of X_k given $\mathscr{Y}_k$.

We shall use the Matrix Inversion Lemma.

Lemma 5.1 *(Matrix Inversion Lemma). Assuming the required inverses exists, the matrix inversion lemma states that:*

$$(A_{11}-A_{12}A_{22}^{-1}A_{21})^{-1}=A_{11}^{-1}+A_{11}^{-1}A_{12}(A_{22}-A_{21}A_{11}^{-1}A_{12})^{-1}A_{21}A_{11}^{-1}.$$

Recursive Estimation

Initially we suppose all processes are defined on an "ideal" probability space $(\Omega,\mathscr{F},\overline{P})$; then under a new probability measure P, to be defined, the model dynamics (5.1) and (5.2) will hold.

Suppose that under $\overline{P}$:

1. $\{X_k\}$, $k \in \mathbb{N}$ is an i.i.d. $N(0, I_{m\times m})$ sequence with density function $\phi(x) = \dfrac{1}{(2\pi)^{m/2}} e^{-x'x/2}$,
2. $\{Y_k\}$, $k \in \mathbb{N}$ is an i.i.d. $N(0, I_{d\times d})$ sequence with density function $\psi(y) = \dfrac{1}{(2\pi)^{d/2}} e^{-y'y/2}$.

For any square matrix B write $|B|$ for the absolute value of its determinant.

For $l = 0$, $\overline{\lambda}_0 = \dfrac{\psi(D_0^{-1}(Y_0 - C_0 X_0))}{|D_0|\psi(Y_0)}$ and for $l = 1, 2, \dots$ define

$$\overline{\lambda}_l = \frac{\phi(B_{l-1}^{-1}(X_l - A_{l-1}X_{l-1}))\psi(D_l^{-1}(Y_l - C_l X_l))}{|B_{l-1}||D_l|\phi(X_l)\psi(Y_l)}, \tag{5.3}$$

$$\overline{\Lambda}_k = \prod_{l=0}^{k} \overline{\lambda}_l. \tag{5.4}$$

The process $\{\overline{\Lambda}_k\}$, $k \in \mathbb{N}$ is a $\overline{P}$-martingale with respect to the filtration $\{\mathscr{G}_k\}$, that is the complete σ-field generated by $\{X_0, X_1, \dots, X_k, Y_0, Y_1, \dots, Y_k\}$ for $k \in \mathbb{N}$.

Define P on $\{\Omega, \mathscr{F}\}$ by setting the restriction of the Radon-Nykodim derivative $\dfrac{dP}{d\overline{P}}$ to $\mathscr{G}_k$ equal to $\overline{\Lambda}_k$. It can be shown that on $\{\Omega, \mathscr{F}\}$ and under P, V_k and W_k are normally distributed with means 0 and respective covariance identity matrices $I_{m\times m}$ and $I_{d\times d}$, where

$$V_{k+1} \triangleq B_k^{-1}(X_{k+1} - A_k X_k),$$
$$W_k \triangleq D_k^{-1}(Y_k - C_k X_k).$$

Let $g : \mathbb{R}^m \to \mathbb{R}$ be a "test" function. Write $\alpha_k(.)$, $k \in \mathbb{N}$ for the density

$$\overline{E}[\overline{\Lambda}_k g(x_k) \mid \mathscr{Y}_k] = \int_{\mathbb{R}^m} g(x)\alpha_k(x)dx. \tag{5.5}$$

Then we have the following result:

Theorem 5.2

$$\alpha_{k+1}(x) = \frac{\psi(D_{k+1}^{-1}(Y_{k+1} - C_{k+1}x))}{|D_{k+1}||B_k|\psi(Y_{k+1})} \int \phi(B_k^{-1}(x - A_k z))\alpha_k(z)dz. \tag{5.6}$$

Proof The proof is as in Theorem 4.2. □

Remark 5.3 The linearity of (5.1) and (5.2) implies that (5.6) is proportional to a normal distribution with mean $\hat{X}_{k|k} = E[X_k \mid \mathscr{Y}_k]$ and error covariance matrix $\Sigma_{k|k} = E[(X_k - \hat{X}_{k|k})(X_k - \hat{X}_{k|k})' \mid \mathscr{Y}_k]$. Our purpose now is to give recursive estimates of $\hat{X}_{k|k}$ and $\Sigma_{k|k}$ using the recursion for $\alpha_k(x)$. ■

Suppose then that $\alpha_k(z) = \phi(\Sigma_{k|k}^{-1}(z - \hat{X}_{k|k}))$.
Write

$$\Psi(x, Y_{k+1}) = \frac{\psi(D_{k+1}^{-1}(Y_{k+1} - C_{k+1}x))}{|D_{k+1}||B_k|\psi(Y_{k+1})}(2\pi)^{-m/2}|\Sigma_{k|k}|^{-1/2}.$$

Then

$$\begin{aligned}
&\alpha_{k+1}(x) \\
&= \Psi(x, Y_{k+1}) \int_{\mathbb{R}^m} \exp(-1/2)\Big\{(x - A_k z))'(B_k^{-1}(x - A_k z)) \\
&\quad + (z - \hat{X}_{k|k})'\Sigma_{k|k}^{-1}(z - \hat{X}_{k|k})\Big\} dz \qquad (5.7) \\
&= K(x) \int_{\mathbb{R}^m} \exp(-1/2)\Big\{z'\sigma_k z - 2\delta_k' z\Big\} dz, \qquad (5.8)
\end{aligned}$$

where

$$K(x) = \frac{(2\pi)^{m/2}}{|\sigma_k|^{1/2}} \exp\Big\{(-1/2)(x'B_k^{-2}x - \delta_k'\sigma_k^{-1}\delta_k + \hat{X}_{k|k}'\Sigma_{k|k}^{-1}\hat{X}_{k|k})\Big\}$$

is independent of the variable z and

$$\begin{aligned}
\sigma_k &= A_k' B_k^{-2} A_k + \Sigma_{k|k}^{-1}, \\
\delta_k' &= x' B_k^{-2} A_k + \hat{X}_{k|k}' \Sigma_{k|k}^{-1}.
\end{aligned}$$

The next step is to complete the square in the argument of the exponential in (5.8) in order to rewrite the integrand as a normal density which, apart from some constant term, integrates out to 1.

Now

$$z'\sigma_k^{-1}z - 2\delta_k' z = (z - \sigma_k\delta_k)'\sigma_k^{-1}(z - \sigma_k\delta_k) - \delta_k'\sigma_k\delta_k, \qquad (5.9)$$

After substitution of (5.9) into (5.8) and integration we are left with only the x variable. Completing the square with respect to x

$$\alpha_{k+1}(x) = K_1 \exp\left\{(-\frac{1}{2})(x - \Sigma_{k+1|k+1}\Delta_k)'\Sigma_{k+1|k+1}^{-1}(x - \Sigma_{k+1|k+1}\Delta_k)\right\},$$

where K_1 is a constant independent of x and

$$\begin{aligned}
\Delta_k &= B_k^{-2} A_k \sigma_k \Sigma_{k|k}^{-1}\hat{X}_{k|k} + C_{k+1}' D_{k+1}^{-2} Y_{k+1} \\
\Sigma_{k+1|k+1}^{-1} &= B_k^{-2} - B_k^{-2} A_k \sigma_k A_k' B_k^{-2} \\
&\quad + C_{k+1}' D_{k+1}^{-2} C_{k+1}.
\end{aligned}$$

Using the Matrix Inversion Lemma 5.1 gives

$$\Sigma_{k+1|k+1}^{-1} = (B_k^{-2} + A_k \Sigma_{k|k} A_k')^{-1} + C_{k+1}' D_{k+1}^{-2} C_{k+1}.$$

The one-step ahead prediction version is

$$\begin{aligned}
\hat{X}_{k+1|k} &= E[X_{k+1} \mid \mathscr{Y}_k] \\
&= E[A_k X_k + B_k V_{k+1} \mid \mathscr{Y}_k] = A_k \hat{X}_{k|k}, \\
\Sigma_{k+1|k} &= E[(X_{k+1} - \hat{X}_{k+1|k})(X_{k+1} - \hat{X}_{k+1|k})' \mid \mathscr{Y}_k] \\
&= E[(X_{k+1} - A_k \hat{X}_{k|k})(X_{k+1} - A_k \hat{X}_{k|k})' \mid \mathscr{Y}_k] \\
&= A_k \Sigma_{k|k} A_k' + B_k^2.
\end{aligned}$$

Therefore

$$\Sigma_{k+1|k+1}^{-1} = C_{k+1}' D_{k+1}^{-2} C_{k+1} + \Sigma_{k+1|k}^{-1}.$$

Use of the Matrix Inversion Lemma 5.1 gives

$$\begin{aligned}
\Sigma_{k+1|k+1} &= \Sigma_{k+1|k} \\
&\quad - \Sigma_{k+1|k} C_{k+1}' (C_{k+1} \Sigma_{k+1|k} C_{k+1}' + D_{k+1}^2)^{-1} C_{k+1} \Sigma_{k+1|k}.
\end{aligned}$$

We have shown that

$$\alpha_{k+1}(x) = K_1 \exp\left\{ (-\frac{1}{2})(x - \Sigma_{k+1|k+1}\Delta_{k+1})' \Sigma_{k+1|k+1}^{-1} (x - \Sigma_{k+1|k+1}\Delta_{k+1}) \right\},$$

where K_1 is a constant independent of x. Therefore, when normalized $\alpha_{k+1}(x)$ is a normal density with mean $\mu_{k+1} = \Sigma_{k+1|k+1}\Delta_{k+1}$ and variance $\Sigma_{k+1|k+1}$.

We now compute $\mu_{k+1} = \Sigma_{k+1|k+1}\Delta_{k+1}$.

Write $(C_{k+1}\Sigma_{k+1|k}C_{k+1}' + D_{k+1}^2)^{-1} \stackrel{\triangle}{=} \mathfrak{C}_{k+1}$.

$$\begin{aligned}
\Sigma_{k+1|k+1}\Delta_{k+1} &= (\Sigma_{k+1|k} - \Sigma_{k+1|k} C_{k+1}' \mathfrak{C}_{k+1} C_{k+1} \Sigma_{k+1|k}) \\
&\quad \times (B_k^{-2} A_k \sigma_k \Sigma_{k|k}^{-1} \hat{X}_{k|k} + C_{k+1}' D_{k+1}^{-2} Y_{k+1}) \\
&= (\Sigma_{k+1|k} - \Sigma_{k+1|k} C_{k+1}' \mathfrak{C}_{k+1} C_{k+1} \Sigma_{k+1|k}) C_{k+1}' D_{k+1}^{-2} Y_{k+1} \\
&\quad + (\Sigma_{k+1|k} - \Sigma_{k+1|k} C_{k+1}' \mathfrak{C}_{k+1} C_{k+1} \Sigma_{k+1|k}) B_k^{-2} A_k \sigma_k \Sigma_{k|k}^{-1} \hat{X}_{k|k} \\
&= (\Sigma_{k+1|k} C_{k+1}' - \Sigma_{k+1|k} C_{k+1}' \mathfrak{C}_{k+1} (C_{k+1} \Sigma_{k+1|k} C_{k+1}' + D_{k+1}^2 - D_{k+1}^2)) \\
&\quad \times D_{k+1}^{-2} Y_{k+1} + \Sigma_{k+1|k+1} B_k^{-2} A_k \sigma_k \Sigma_{k|k}^{-1} \hat{X}_{k|k} \\
&= \Sigma_{k+1|k} C_{k+1}' \mathfrak{C}_{k+1} Y_{k+1} \\
&\quad + \Sigma_{k+1|k+1} B_k^{-2} A_k (\Sigma_{k|k}^{-1} + A_k' B_k^{-2} A_k)^{-1} \Sigma_{k|k}^{-1} \hat{X}_{k|k}
\end{aligned}$$

$$
\begin{aligned}
&= \Sigma_{k+1|k} C'_{k+1} \mathfrak{C}_{k+1} Y_{k+1} \\
&\quad + \Sigma_{k+1|k+1} B_k^{-2} A_k (\Sigma_{k|k} - \Sigma_{k|k} A'_k (B_k^2 + A_k \Sigma_{k|k} A'_k)^{-1} A_k \Sigma_{k|k}) \Sigma_{k|k}^{-1} \hat{X}_{k|k} \\
&= \Sigma_{k+1|k} C'_{k+1} \mathfrak{C}_{k+1} Y_{k+1} + \Sigma_{k+1|k+1} B_k^{-2} \\
&\quad \times (A_k \Sigma_{k|k} - (A_k \Sigma_{k|k} A'_k + B_k^2 - B_k^2)(B_k^2 + A_k \Sigma_{k|k} A'_k)^{-1} A_k \Sigma_{k|k}) \Sigma_{k|k}^{-1} \hat{X}_{k|k} \\
&= \Sigma_{k+1|k} C'_{k+1} \mathfrak{C}_{k+1} Y_{k+1} + \Sigma_{k+1|k+1} (B_k^2 + A_k \Sigma_{k|k} A'_k)^{-1} A_k \hat{X}_{k|k} \\
&= \Sigma_{k+1|k} C'_{k+1} \mathfrak{C}_{k+1} Y_{k+1} + (\Sigma_{k+1|k}^{-1} + C'_{k+1} D_{k+1}^{-2} C_{k+1})^{-1} \\
&\quad + (\Sigma_{k+1|k}^{-1} + C'_{k+1} D_{k+1}^{-2} C_{k+1} - C'_{k+1} D_{k+1}^{-2} C_{k+1}) A'_k)^{-1} A_k \hat{X}_{k|k} \\
&= \Sigma_{k+1|k} C'_{k+1} \mathfrak{C}_{k+1} Y_{k+1} + A_k \hat{X}_{k|k} - \Sigma_{k+1|k+1} C'_{k+1} D_{k+1}^{-2} C_{k+1}) A_k \hat{X}_{k|k} \\
&= \Sigma_{k+1|k} C'_{k+1} \mathfrak{C}_{k+1} Y_{k+1} + A_k \hat{X}_{k|k} \\
&\quad - (\Sigma_{k+1|k} - \Sigma_{k+1|k} C'_{k+1} \mathfrak{C}_{k+1} C_{k+1} \Sigma_{k+1|k}) C'_{k+1} D_{k+1}^{-2} C_{k+1} A_k \hat{X}_{k|k} \\
&= \Sigma_{k+1|k} C'_{k+1} \mathfrak{C}_{k+1} Y_{k+1} + A_k \hat{X}_{k|k} - (\Sigma_{k+1|k} C'_{k+1} - \Sigma_{k+1|k} C'_{k+1} \mathfrak{C}_{k+1} \\
&\quad \times (C_{k+1} \Sigma_{k+1|k} C'_{k+1} + D_{k+1}^2 - D_{k+1}^2)) D_{k+1}^{-2} C_{k+1} A_k \hat{X}_{k|k} \\
&= \Sigma_{k+1|k} C'_{k+1} \mathfrak{C}_{k+1} Y_{k+1} + A_k \hat{X}_{k|k} - \Sigma_{k+1|k} C'_{k+1} \mathfrak{C}_{k+1} C_{k+1} A_k \hat{X}_{k|k} \\
&= A_k \hat{X}_{k|k} + \Sigma_{k+1|k} C'_{k+1} \left(C_{k+1} \Sigma_{k+1|k} C'_{k+1} + D_{k+1}^2 \right)^{-1} \\
&\quad \times (Y_{k+1} - C_{k+1} \hat{X}_{k+|k}).
\end{aligned}
$$

Therefore

$$
\boxed{
\begin{aligned}
&\mu_{k+1} = A_k \hat{X}_{k|k} \\
&+ \Sigma_{k+1|k} C'_{k+1} \left(C_{k+1} \Sigma_{k+1|k} C'_{k+1} + D_{k+1}^2 \right)^{-1} (Y_{k+1} - C_{k+1} \hat{X}_{k+|k}).
\end{aligned}
}
$$

In Theorem 4.5 we established a backward recursion for $m < k$:

$$
\begin{aligned}
\beta_{m,k}(x) = \frac{1}{\phi_{m+1}(Y_{m+1})} \int_{\mathbb{R}^m} & \exp(-1/2) \Big\{ (B_m^{-1}(x - A_m z))' (B_m^{-1}(x - A_m z)) \\
& \times (D_{m+1}^{-1}(Y_{m+1} - C_{m+1} z))' (D_{m+1}^{-1}(Y_{m+1} - C_{m+1} z)) \Big\} \beta_{m+1,k}(z) dz,
\end{aligned}
$$

with $\beta_{k,k} = 1$.

Suppose

$$
\begin{aligned}
\beta_{m+1,k}(z) = & |\mathfrak{S}_{m+1|k}|^{-1/2} \\
& \times \exp(-1/2) \left\{ (z - \gamma_{m+1|k})' \mathfrak{S}_{m+1|k}^{-1} (z - \gamma_{m+1|k}) \right\}.
\end{aligned}
$$

Then the integral is

$$\beta_{m,k}(x) = \int_{\mathbb{R}^m} \exp(-1/2)\left\{(B_m^{-1}(x - A_m z))'(B_m^{-1}(x - A_m z))\right.$$
$$(D_{m+1}^{-1}(Y_{m+1} - C_{m+1}z))'(D_{m+1}^{-1}(Y_{m+1} - C_{m+1}z))$$
$$\left. + (z - \gamma_{m+1|k})'\mathfrak{S}_{m+1|k}^{-1}(z - \gamma_{m+1|k})\right\} dz$$
$$= K(x) \int_{\mathbb{R}^m} \exp(-1/2)\left\{z'\sigma_m^{-1}z - 2\delta_m' z\right\} dz,$$

where

$$K(x) = \exp\left\{(-1/2)(x'B_k^{-2}x - Y_{m+1}'D_{m+1}^{-2}Y_{m+1} + \gamma_{m+1|k}'\mathfrak{S}_{m+1|k}^{-1}\gamma_{m+1|k})\right\}$$

is independent of the variable z and

$$\sigma_m^{-1} = A_m'B_m^{-2}A_m + C_{m+1}'D_{m+1}^{-2}C_{m+1} + \mathfrak{S}_{m+1|k}^{-1},$$
$$\delta_m' = x'B_m^{-2}A_m + Y_{m+1}'D_{m+1}^{-2}C_{m+1} + \gamma_{m+1|k}\mathfrak{S}_{m+1|k}^{-1}.$$

Completing the square

$$z'\sigma_m^{-1}z - 2\delta_m' z = (z - \sigma_m\delta_m)'\sigma_m^{-1}(z - \sigma_m\delta_m) - \delta_m'\sigma_m\delta_m, \tag{5.10}$$

and integrating with respect to z leaves us with only the x variable. Completing the square with respect to x gives

$$\beta_{m,k}(x) = K_1 \exp\left\{(-\frac{1}{2})(x - \mathfrak{S}_{m|k}\Delta_m)'\mathfrak{S}_{m|k}^{-1}(x - \mathfrak{S}_{m|k}\Delta_m)\right\},$$

where K_1 is a constant independent of x and

$$\Delta_m = B_m^{-2}A_m\sigma_m C_{m+1}'D_{m+1}^{-2}Y_{m+1} + B_m^{-2}A_m\sigma_m\mathfrak{S}_{m+1|k}^{-1}\gamma_{m+1|k}$$
$$\mathfrak{S}_{m|k}^{-1} = B_m^{-2} - B_m^{-2}A_m\sigma_m A_m'B_m^{-2}.$$

Using the Matrix Inversion Lemma 5.1 and the above expression for σ_m^{-1} give

$$\mathfrak{S}_{m|k} = B_m^2 + A_m[C_{m+1}'D_{m+1}^{-2}C_{m+1} + \mathfrak{S}_{m+1|k}^{-1}]^{-1}A_m'.$$

We have shown that

$$\beta_{m,k} = K_1 \exp\left\{(-\frac{1}{2})(x - \mathfrak{S}_{m|k}\Delta_m)'\mathfrak{S}_{m|k}^{-1}(x - \mathfrak{S}_{m|k}\Delta_m)\right\},$$

where K_1 is a constant independent of x. Therefore, when normalized $\beta_{m,k}$ is a normal density with mean $\gamma_{m|k} = \mathfrak{S}_{m|k}\Delta_m$ and variance $\mathfrak{S}_{m|k}$.

We now compute

$$\gamma_{m|k} = \mathfrak{S}_{m|k}\Delta_m, \ \mathfrak{C}_{m+1} \triangleq \Big[C'_{m+1}D^{-2}_{m+1}C_{m+1} + \mathfrak{S}^{-1}_{m+1|k}\Big]^{-1},$$

then

$$\begin{aligned}\mathfrak{S}_{m|k}\Delta_m &= (B^2_m + A_m\mathfrak{C}_{m+1}A'_m) \\ &\quad \times (B^{-2}_m A_m\sigma_m C'_{m+1}D^{-2}_{m+1}Y_{m+1} + B^{-2}_m A_m\sigma_m\mathfrak{S}^{-1}_{m+1|k}\gamma_{m+1|k}) \\ &= \Big[A_m\sigma_m C'_{m+1}D^{-2}_{m+1} + A_m\mathfrak{C}_{m+1}A'_m B^{-2}_m A_m\sigma_m C'_{m+1}D^{-2}_{m+1}\Big]Y_{m+1} \\ &\quad + \Big[A_m\sigma_m + A_m\mathfrak{C}_{m+1}A'_m B^{-2}_m A_m\sigma_m\Big]\mathfrak{S}^{-1}_{m+1|k}\gamma_{m+1|k}.\end{aligned}$$

Now we apply the Matrix Inversion Lemma 5.1

$$\begin{aligned}\mathfrak{C}_{m+1} &= \Big[C'_{m+1}D^{-2}_{m+1}C_{m+1} + \mathfrak{S}^{-1}_{m+1|k}\Big]^{-1} \\ &= \mathfrak{S}_{m+1|k} - \mathfrak{S}_{m+1|k}C'_{m+1}\mathfrak{D}_{m+1}C_{m+1}\mathfrak{S}_{m+1|k},\end{aligned}$$

where

$$\mathfrak{D}_{m+1} = (D^2_{m+1} + C_{m+1}\mathfrak{S}_{m+1|k}C'_{m+1})^{-1}.$$

We first concentrate on the expression multiplying Y_{m+1}.

$$\begin{aligned}&A_m\sigma_m C'_{m+1}D^{-2}_{m+1} + A_m\mathfrak{C}_{m+1}A'_m B^{-2}_m A_m\sigma_m C'_{m+1}D^{-2}_{m+1} \\ &= A_m\Big[I + \mathfrak{S}_{m+1|k}A'_m B^{-2}_m A_m \\ &\qquad - \mathfrak{S}_{m+1|k}C'_{m+1}\mathfrak{D}_{m+1}C_{m+1}\mathfrak{S}_{m+1|k}A'_m B^{-2}_m A_m\Big]\sigma_m C'_{m+1}D^{-2}_{m+1}.\end{aligned}$$

Now recall that $\sigma^{-1}_m = A'_m B^{-2}_m A_m + C'_{m+1}D^{-2}_{m+1}C_{m+1} + \mathfrak{S}^{-1}_{m+1|k}$. Therefore $A'_m B^{-2}_m A_m = \sigma^{-1}_m - C'_{m+1}D^{-2}_{m+1}C_{m+1} - \mathfrak{S}^{-1}_{m+1|k}$, and

$$\begin{aligned}&A_m\Big[I + \mathfrak{S}_{m+1|k}A'_m B^{-2}_m A_m \\ &\qquad - \mathfrak{S}_{m+1|k}C'_{m+1}\mathfrak{D}_{m+1}C_{m+1}\mathfrak{S}_{m+1|k}A'_m B^{-2}_m A_m\Big]\sigma_m C'_{m+1}D^{-2}_{m+1} \\ &= A_m\Big[I + \mathfrak{S}_{m+1|k}\sigma^{-1}_m - \mathfrak{S}_{m+1|k}C'_{m+1}D^{-2}_{m+1}C_{m+1} - I \\ &\qquad - \mathfrak{S}_{m+1|k}C'_{m+1}\mathfrak{D}_{m+1}C_{m+1}\mathfrak{S}_{m+1|k}(\sigma^{-1}_m - C'_{m+1}D^{-2}_{m+1}C_{m+1} \\ &\qquad - \mathfrak{S}^{-1}_{m+1|k})\Big]\sigma_m C'_{m+1}D^{-2}_{m+1} \\ &= A_m\Big[\mathfrak{S}_{m+1|k}C'_{m+1} - \mathfrak{S}_{m+1|k}C'_{m+1}D^{-2}_{m+1}C_{m+1}\sigma_m C'_{m+1} \\ &\qquad - \mathfrak{S}_{m+1|k}C'_{m+1}\mathfrak{D}_{m+1}(C_{m+1}\mathfrak{S}_{m+1|k}C'_{m+1}\end{aligned}$$

$$
\begin{aligned}
&\quad - C_{m+1}\mathfrak{S}_{m+1|k}C'_{m+1}D_{m+1}^{-2}C_{m+1}\sigma_m C'_{m+1} \\
&\quad - C_{m+1}\sigma_m C'_{m+1})\Big]D_{m+1}^{-2} \\
&= A_m\mathfrak{S}_{m+1|k}C'_{m+1}\Big[I - D_{m+1}^{-2}C_{m+1}\sigma_m C'_{m+1} \\
&\quad - \mathfrak{D}_{m+1}(C_{m+1}\mathfrak{S}_{m+1|k}C'_{m+1}\{I - D_{m+1}^{-2}C_{m+1}\sigma_m C'_{m+1}\} \\
&\quad - C_{m+1}\sigma_m C'_{m+1})\Big]D_{m+1}^{-2} \\
&= A_m\mathfrak{S}_{m+1|k}C'_{m+1}\Big[I - D_{m+1}^{-2}C_{m+1}\sigma_m C'_{m+1} \\
&\quad - \Big(\mathfrak{D}_{m+1}\{C_{m+1}\mathfrak{S}_{m+1|k}C'_{m+1} + D_{m+1}^{2} - D_{m+1}^{2}\} \\
&\quad \times \{I - D_{m+1}^{-2}C_{m+1}\sigma_m C'_{m+1}\} - \mathfrak{D}_{m+1}C_{m+1}\sigma_m C'_{m+1}\Big)\Big]D_{m+1}^{-2} \\
&= A_m\mathfrak{S}_{m+1|k}C'_{m+1}\Big[I - D_{m+1}^{-2}C_{m+1}\sigma_m C'_{m+1} \\
&\quad - \Big(\{I - \mathfrak{D}_{m+1}D_{m+1}^{2}\}\{I - D_{m+1}^{-2}C_{m+1}\sigma_m C'_{m+1}\} \\
&\quad - \mathfrak{D}_{m+1}C_{m+1}\sigma_m C'_{m+1}\Big)\Big]D_{m+1}^{-2} \\
&= A_m\mathfrak{S}_{m+1|k}C'_{m+1}\mathfrak{D}_{m+1}.
\end{aligned}
$$

We now compute the coefficient of $\gamma_{m+1|k}$.

$$
\begin{aligned}
&\Big[A_m\sigma_m + A_m\mathfrak{C}_{m+1}A'_m B_m^{-2}A_m\sigma_m\Big]\mathfrak{S}_{m+1|k}^{-1} \\
&\quad = A_m\Big[I + \mathfrak{C}_{m+1}(\sigma_m^{-1} - C'_{m+1}D_{m+1}^{-2}C_{m+1} - \mathfrak{S}_{m+1|k}^{-1})\Big]\sigma_m\mathfrak{S}_{m+1|k}^{-1} \\
&\quad = A_m\Big[I + \mathfrak{C}_{m+1}\sigma_m^{-1} - \mathfrak{C}_{m+1}\{C'_{m+1}D_{m+1}^{-2}C_{m+1} + \mathfrak{S}_{m+1|k}^{-1} - \mathfrak{S}_{m+1|k}^{-1}\} \\
&\qquad - \mathfrak{C}_{m+1}\mathfrak{S}_{m+1|k}^{-1}\Big]\sigma_m\mathfrak{S}_{m+1|k}^{-1} \\
&\quad = A_m\mathfrak{C}_{m+1}\mathfrak{S}_{m+1|k}^{-1}
\end{aligned}
$$

where we used the relation $A'_m B_m^{-2}A_m = \sigma_m^{-1} - C'_{m+1}D_{m+1}^{-2}C_{m+1} - \mathfrak{S}_{m+1|k}^{-1}$.

Therefore

$$
\boxed{
\begin{aligned}
\gamma_{m|k} &= A_m\mathfrak{S}_{m+1|k}C'_{m+1}\Big(D_{m+1}^{2} + C_{m+1}\mathfrak{S}_{m+1|k}C'_{m+1}\Big)^{-1}Y_{m+1} \\
&\quad + A_m\Big(C'_{m+1}D_{m+1}^{-2}C_{m+1} + \mathfrak{S}_{m+1|k}^{-1}\Big)^{-1}\mathfrak{S}_{m+1|k}^{-1}\gamma_{m+1|k}.
\end{aligned}
}
$$

4.6 State and Mode Estimation for Discrete-Time Jump Markov Systems

We use the reference probability method to compute state and mode estimation scheme for a discrete-time stochastic hybrid dynamical system.

Gauss-Markov jump linear systems arise quite naturally in many practical settings, for example, tracking a manoeuvering object observed through RADAR. Here, no single set of dynamics will encapsulate all classes of motion, so one is led naturally to a hybrid collection of dynamics as a basic model. The estimation task for such models is significantly complicated by the need to jointly estimate a hidden state variable and the current model in effect. Currently, many of the standard techniques to solve this problem are ad hoc and not based upon the exact filter dynamics computed in Elliott, Dufour and Sworder (1996). In contrast to this situation, our new filters and smoothers for Gauss-Markov jump linear systems are developed from the exact hybrid dynamics. Using a general result, (see Lemma 6.3), we propose a new suboptimal algorithm which provides an exact hypothesis management scheme, circumventing geometric growth in algorithmic complexity. Our approach is based, in part, upon approximating probability densities by finite Gaussian mixtures and is justified by the basic results given in Sorenson and Alspach (1971).

In a simulation study, a comparison is given between the single Extended Kalman filter (EKF), the IMM, and our algorithm. We also compute a general smoothing algorithm for discrete-time hybrid dynamical system. To compute this algorithm, we exploit a duality between forwards and backwards (dual) dynamics. An interesting feature of our smoother, is it provides a new degree of freedom, that is, the product decomposition of the smoother density is approximated by mutually independent Gaussian mixtures, where the chosen accuracy of 'the past', (influencing the smoother density), is independent of the chosen accuracy of 'the future', influencing the smoother density.

Dynamics and Reference Probability

Initially we suppose all processes are defined on a fixed probability space $(\Omega, \mathscr{F}, P)$. For the class of Jump Markov systems considered, we will require three sets of dynamics. These are: the Markov chain dynamics, for the process whose value determines the model parameters, the indirectly observed state process, and the observation process.

Markov Chain Dynamics

To model parameter switching we consider a time homogeneous discrete time discrete state Markov chain Z. If the cardinality of the state space is m it is convenient, without loss of generality, to identify the state space of Z with the set

$S_X = \{e_1, \dots, e_m\}$, where e_i are unit vectors in $\mathbb{R}^m$ with unity as the ith element and zeros elsewhere. The dynamics for the process Z may be written as

$$Z_k = \Pi Z_{k-1} + L_k \in \mathbb{R}^m. \tag{6.1}$$

Here, $\Pi = \left[\pi_{(j,i)}\right]_{\substack{1\le j\le m \\ 1\le i\le m}}$ is the transition matrix of Z, with elements

$$\pi_{ji} \triangleq P(Z_k = e_j \mid Z_{k-1} = e_i) \tag{6.2}$$

for all $k \in \mathbb{N}$. The process L is a $(P, \sigma\{Z\})$-martingale increment, and we suppose $E\left[Z_0\right] = p_0$.

State Process Dynamics

We suppose the indirectly observed state vector $x \in \mathbb{R}^n$, has dynamics

$$x_k = \sum_{j=1}^{m} \langle Z_k, e_j \rangle A_j x_{k-1} + \sum_{j=1}^{m} \langle Z_k, e_j \rangle B_j w_k. \tag{6.3}$$

Here w is a vector-valued Gaussian process with $w \sim N(0, I_n)$. A_j and B_j are $n \times n$ matrices and for each $j \in \{1, 2, \dots, m\}$, are nonsingular. This condition can be relaxed (Elliott and Krishnamurthy 1999).

Observation Process Dynamics

Consider a vector-valued observation process with values in $\mathbb{R}^d$ and dynamics

$$y_k = \sum_{j=1}^{m} \langle Z_k, e_j \rangle C_j x_k + \sum_{j=1}^{m} \langle Z_k, e_j \rangle D_j v_k. \tag{6.4}$$

Here v is a vector-valued Gaussian process with $v \sim N(0, I_d)$. We suppose the matrices $D_j \in \mathbb{R}^{d \times d}$, for each $j \in \{1, 2, \dots, m\}$, are nonsingular. The systems we shall consider here are described by the dynamics (6.1), (6.3) and (6.4). The three stochastic processes Z, x and y are mutually statistically independent. Taken together, these dynamics form a triply stochastic system, with random inputs due to the processes Z, x and y. For example, if the Markov chain Z is in the state e_j, then the dynamical model with state x and observation y, is defined by the parameters set $\{A_j, B_j, C_j, D_j\}$.

Remark 6.1 At a cost of more complicated notation, we could consider observations of the form:

$$y_k = \sum_{j=1}^{m} \langle Z_k, e_j \rangle \left(C_j x_k + H_j\right) + \sum_{j=1}^{m} \langle Z_k, e_j \rangle D_j v_k. \tag{6.5}$$

This formulation includes scenarios where the Markov chain Z is observed directly. However, here we restrict our attention to cases where $H_j = 0$ for all $j \in \{1,2,\ldots,m\}$.

■

The following filtrations are complete.
$\mathscr{F}_k = \sigma\{x_\ell, 0 \le \ell \le k\}$, $\mathscr{Z}_k = \sigma\{Z_\ell, 0 \le \ell \le k\}$, $\mathscr{Y}_k = \sigma\{y_\ell, 0 \le \ell \le k\}$, $\mathscr{G}_k = \sigma\{Z_\ell, x_\ell, y_\ell, 0 \le \ell \le k\}$.

Reference Probability

The dynamics given at (6.1), (6.3) and (6.4), are each defined on a measurable space $(\Omega, \mathscr{F})$, under a measure P. However, consider a new measure $\overline{P}$, under which the dynamics for the processes Z, x and y, are, respectively

$$\overline{P}\begin{cases} Z_k & = \Pi Z_{k-1} + L_k, \\ x_k & \text{are iid and } N(0, I_n), \\ y_k & \text{are iid and } N(0, I_d). \end{cases}$$

The symbol $\Phi(\cdot)$ will be used to denote the zero mean normal density on $\mathbb{R}^d$: $\Phi(\xi) = (2\pi)^{-d/2}\exp\big(-\frac{1}{2}\xi'\xi\big)$. Similarly we shall also use the symbol $\Psi(\cdot)$ to denote a standardized Gaussian density. The space dimension on which these densities is defined will be clear by context. To avoid cumbersome notation with matrices, we sometimes denote the inverse of a matrix A by A^{-1}. Further, we denote the space of all $m \times n$ matrices by $\mathbb{M}^{m\times n}$. To compute the filter dynamics we now define the measure P by setting the restriction of its Radon-Nikodym derivative to $\mathscr{G}_k$ to

$$\Lambda_k \triangleq \frac{dP}{d\overline{P}}\Big|_{\mathscr{G}_k} = \prod_{\ell=0}^{k} \lambda_\ell,$$

where

$$\lambda_\ell = \sum_{j=1}^{m} \langle Z_\ell, e_j\rangle \frac{\Phi\big(D_j^{-1}(y_\ell - C_j x_\ell)\big)}{|D_j|\Phi(y_\ell)} \times \frac{\Psi\big(B_j^{-1}(x_\ell - A_j x_{\ell-1})\big)}{|B_j|\Psi(x_\ell)}.$$

The existence of P follows from the Kolmogorov extension Theorem.

For $k \ge 0$ define

$$v_k = \sum_{j=1}^{m} \langle Z_k, e_j\rangle D_j^{-1}(y_k - C_j x_k).$$

For $k \ge 1$ define

$$w_k = \sum_{j=1}^{m} \langle Z_k, e_j\rangle B_j^{-1}(x_k - A_j x_{k-1}).$$

Lemma 6.2 *Under the measure P, the dynamics for the Markov process Z are unchanged and x and y have dynamics given by* (6.3) *and* (6.4) *respectively.*

Proof Suppose $f : \mathbb{R}^n \to \mathbb{R}$ and $g : \mathbb{R}^d \to \mathbb{R}$ are any integrable Borel test functions. Then

$$\begin{aligned} E[f(w_k)g(v_k) \mid \mathscr{G}_{k-1}] &= \frac{\overline{E}[\Lambda_k f(w_k)g(v_k) \mid \mathscr{G}_{k-1}]}{\overline{E}[\Lambda_k \mid \mathscr{G}_{k-1}]} \\ &= \frac{\overline{E}[\lambda_k f(w_k)g(v_k) \mid \mathscr{G}_{k-1}]}{\overline{E}[\lambda_k \mid \mathscr{G}_{k-1}]}, \end{aligned} \tag{6.6}$$

since $\Lambda_k = \lambda_k \Lambda_{k-1}$ and Λ_{k-1} is known in $\mathscr{G}_{k-1}$.
Recalling the definition of λ_k, we see that

$$\begin{aligned} \overline{E}[\lambda_k \mid \mathscr{G}_{k-1}] &= \sum_{\ell=1}^{m} \overline{E}\left[\langle Z_k, e_l\rangle \frac{\Phi\big(D_\ell^{-1}(y_k - C_\ell x_k)\big)}{|D_\ell|\Phi(y_k)} \frac{\Psi\big(B_\ell^{-1}(x_k - A_\ell x_{k-1})\big)}{|B_\ell|\Psi(x_k)} \mid \mathscr{G}_{k-1}\right] \\ &= \sum_{\ell=1}^{m} \overline{E}\left[\overline{E}\left[\langle Z_k, e_l\rangle \frac{\Phi\big(D_\ell^{-1}(y_k - C_\ell x_k)\big)}{|D_\ell|\Phi(y_k)} \frac{\Psi\big(B_\ell^{-1}(x_k - A_\ell x_{k-1})\big)}{|B_\ell|\Psi(x_k)} \right.\right. \\ &\qquad \left.\left. \mid Z_k, x_k, \mathscr{G}_{k-1}\right] \mid \mathscr{G}_{k-1}\right] \\ &= \sum_{\ell=1}^{m} \overline{E}\left[\langle Z_k, e_l\rangle \frac{\Psi\big(B_\ell^{-1}(x_k - A_\ell x_{k-1})\big)}{|B_\ell|\Psi(x_k)} \overline{E}\left[\frac{\Phi\big(D_\ell^{-1}(y_k - C_\ell x_k)\big)}{|D_\ell|\Phi(y_k)} \right.\right. \\ &\qquad \left.\left. \mid Z_k, x_k, \mathscr{G}_{k-1}\right] \mid \mathscr{G}_{k-1}\right]. \end{aligned}$$

In the last line above, the inner expectation is equal to unity. To see this, we note that

$$\begin{aligned} \overline{E}\left[\frac{\Phi\big(D_\ell^{-1}(y_k - C_\ell x_k)\big)}{|D_\ell|\Phi(y_k)} \mid Z_k, x_k, \mathscr{G}_{k-1}\right] &= \int_{\mathbb{R}^d} \left[\frac{\Phi\big(D_\ell^{-1}(\xi - C_\ell x_k)\big)}{|D_\ell|\Phi(\xi)}\right] \Phi(\xi) d\xi \\ &= \int_{\mathbb{R}^d} \frac{\Phi\big(D_\ell^{-1}(\xi - C_\ell x_k)\big)}{|D_\ell|} d\xi. \end{aligned}$$

Write $z \triangleq D_\ell^{-1}(y_k - C_\ell x)$, then

$$\int_{\mathbb{R}^d} \frac{\Phi\big(D_\ell^{-1}(\xi - C_\ell x_k)\big)}{|D_\ell|} d\xi = \int_{\mathbb{R}^d} \frac{\Phi(z)}{|D_\ell|} |D_\ell| dz = 1.$$

So, the denominator at (6.7) reduces to,

$$
\begin{aligned}
\overline{E}[\lambda_k \mid \mathscr{G}_{k-1}] &= \sum_{\ell=1}^{m} \overline{E}\left[\langle Z_k, e_l \rangle \frac{\Psi\big(B_\ell^{-1}(x_k - A_\ell x_{k-1})\big)}{|B_\ell| \Psi(x_k)} \mid \mathscr{G}_{k-1} \right] \\
&= \sum_{\ell=1}^{m} \overline{E}\left[\overline{E}\Big[\langle Z_k, e_l \rangle \frac{\Psi\big(B_\ell^{-1}(x_k - A_\ell x_{k-1})\big)}{|B_\ell| \Psi(x_k)} \mid Z_k, \mathscr{G}_{k-1} \Big] \mid \mathscr{G}_{k-1} \right] \\
&= \sum_{\ell=1}^{m} \overline{E}\left[\langle Z_k, e_l \rangle \overline{E}\Big[\frac{\Psi\big(B_\ell^{-1}(x_k - A_\ell x_{k-1})\big)}{|B_\ell| \Psi(x_k)} \mid Z_k, \mathscr{G}_{k-1} \Big] \mid \mathscr{G}_{k-1} \right] \\
&= \sum_{\ell=1}^{m} \overline{E}\left[\langle Z_k, e_l \rangle \mid \mathscr{G}_{k-1} \right] \\
&= \overline{E}\left[\sum_{\ell=1}^{m} \langle Z_k, e_l \rangle \mid \mathscr{G}_{k-1} \right] \\
&= 1.
\end{aligned}
$$

Consequently,

$$
\begin{aligned}
&\overline{E}[\Lambda_k f(w_k) g(v_k) \mid \mathscr{G}_{k-1}] \\
&= \sum_{\ell=1}^{m} \overline{E}\Big[\langle Z_k, e_l \rangle f(B_\ell^{-1}(x_k - A_\ell x_{k-1})) g(D_\ell^{-1}(y_k - C_\ell x_k)) \\
&\quad \times \frac{\Phi\big(D_\ell^{-1}(y_k - C_\ell x_k)\big)}{|D_\ell| \Phi(y_k)} \frac{\Psi\big(B_\ell^{-1}(x_k - A_\ell x_{k-1})\big)}{|B_\ell| \Psi(x_k)} \mid \mathscr{G}_{k-1} \Big] \\
&= \sum_{\ell=1}^{m} \overline{E}\Big[\overline{E}\Big[\langle Z_k, e_l \rangle f(B_\ell^{-1}(x_k - A_\ell x_{k-1})) g(D_\ell^{-1}(y_k - C_\ell x_k)) \\
&\quad \times \frac{\Phi\big(D_\ell^{-1}(y_k - C_\ell x_k)\big)}{|D_\ell| \Phi(y_k)} \frac{\Psi\big(B_\ell^{-1}(x_k - A_\ell x_{k-1})\big)}{|B_\ell| \Psi(x_k)} \mid Z_k, x_k, \mathscr{G}_{k-1} \Big] \mid \mathscr{G}_{k-1} \Big] \\
&= \sum_{\ell=1}^{m} \overline{E}\Big[\langle Z_k, e_l \rangle f(B_\ell^{-1}(x_k - A_\ell x_{k-1})) \frac{\Psi\big(B_\ell^{-1}(x_k - A_\ell x_{k-1})\big)}{|B_\ell| \Psi(x_k)} \\
&\quad \times \overline{E}\Big[g(D_\ell^{-1}(y_k - C_\ell x_k)) \frac{\Phi\big(D_\ell^{-1}(y_k - C_\ell x_k)\big)}{|D_\ell| \Phi(y_k)} \mid Z_k, x_k, \mathscr{G}_{k-1} \Big] \mid \mathscr{G}_{k-1} \Big].
\end{aligned}
$$

The inner expectation is

$$
\int g(D_\ell^{-1}(y - C_\ell x_k)) \frac{\Phi\big(D_\ell^{-1}(y - C_\ell x_k)\big)}{|D_\ell|} dy = \int g(v) \Phi(v) dv.
$$

Here $v = D_\ell^{-1}(y - C_\ell x_k))$. Similarly:

$$\sum_{\ell=1}^{m} \overline{E}\left[\langle Z_k, e_l\rangle \overline{E}\left[f(B_\ell^{-1}(x_k - A_\ell x_{k-1}))\frac{\Psi\big(B_\ell^{-1}(x_k - A_\ell x_{k-1})\big)}{|B_\ell|\Psi(x_k)} \mid Z_k, \mathscr{G}_{k-1}\right] \mid \mathscr{G}_{k-1}\right]$$

$$= \sum_{\ell=1}^{m} \overline{E}\left[\langle Z_k, e_l\rangle \int f(w)\Psi(w)dw \mid \mathscr{G}_{k-1}\right] = \int f(w)\Psi(w)dw.$$

Hence

$$E[f(w_k)g(v_k) \mid \mathscr{G}_{k-1}] = \int f(w)\Psi(w)dw \int g(v)\Phi(v)dv.$$

To prove that the Markov chain Z is unchanged under the measure P requires two steps. We must show that the initial distribution for Z, that is p_0, is the same under P, as it is under $\overline{P}$. To complete the proof, we must show that under P, the process Z has the Markov property and also has a transition matrix Π.

Using Bayes' rule we see that,

$$E\left[Z_0\right] = \frac{\overline{E}\left[\Lambda_0^{-1} Z_0 \mid \{\Omega, \emptyset\}\right]}{\overline{E}\left[\Lambda_0^{-1} \mid \{\Omega, \emptyset\}\right]} = \frac{\overline{E}\left[Z_0\right]}{\overline{E}\left[1\right]} = \overline{E}\left[Z_0\right] = p_0.$$

To show that the process Z is Markov under P, with transition matrix Π, we consider the expectation

$$\begin{aligned} E[\langle Z_k, e_j\rangle \mid \mathscr{G}_{k-1}] &= \frac{\overline{E}[\Lambda_k \langle Z_k, e_j\rangle \mid \mathscr{G}_{k-1}]}{\overline{E}[\Lambda_k \mid \mathscr{G}_{k-1}]} \\ &= \frac{\overline{E}[\lambda_k \langle Z_k, e_j\rangle \mid \mathscr{G}_{k-1}]}{\overline{E}[\lambda_k \mid \mathscr{G}_{k-1}]} = \overline{E}[\lambda_k \langle Z_k, e_j\rangle \mid \mathscr{G}_{k-1}]. \end{aligned} \tag{6.7}$$

One can show that

$$\overline{E}[\lambda_k \langle Z_k, e_j\rangle \mid \mathscr{G}_{k-1}] = \langle \Pi Z_{k-1}, e_j\rangle.$$

Therefore, by (6.7)

$$E[\langle Z_k, e_j\rangle \mid \mathscr{G}_{k-1}] = \langle \Pi Z_{k-1}, e_j\rangle.$$

Consequently

$$E[\langle Z_k, e_j\rangle \mid \mathscr{Z}_{k-1}] = \langle \Pi Z_{k-1}, e_j\rangle.$$

□

Hybrid Filter Dynamics

Exact Filter Dynamics

The following Lemma is critical in what follows.

Lemma 6.3 *Suppose the random vector $\xi \in \mathbb{R}^n$, is normally distributed with $\xi \sim N(\mu, \Sigma)$. Further, suppose A is any matrix in $\mathbb{M}^{m\times n}$, y is a vector in $\mathbb{R}^n$ and the* *matrix $B \in \mathbb{M}^{m\times m}$, is nonsingular. With $p(\xi)$ denoting the Gaussian density function for ξ, the following identity holds,*

$$\begin{aligned}
&\int_{\mathbb{R}^n} \Psi\big(B^{-1}(y-A\xi)\big)p(\xi)d\xi \\
&= (2\pi)^{-n/2}E\Big[\exp\Big\{-\frac{1}{2}(y-A\xi)'(BB')^{-1}(y-A\xi)\Big\}\Big] \\
&= (2\pi)^{-n/2}|B||BB'+A\Sigma A'|^{-\frac{1}{2}} \\
&\quad\times \exp\Big\{-\frac{1}{2}\Big(y-A\mu\Big)'(BB'+A\Sigma A')^{-1}\Big(y-A\mu\Big)\Big\}. \qquad (6.8)
\end{aligned}$$

Proof (Elliott and Malcolm 2000) □

To compute a filter jointly estimating the density of the state vector x, and the state of the chain Z, consider the expectation:

$$E\big[\langle Z_k, e_j\rangle f(x_k) \mid \mathscr{Y}_k\big] = \frac{\overline{E}\big[\Lambda_k \langle Z_k, e_j\rangle f(x_k) \mid \mathscr{Y}_k\big]}{\overline{E}\big[\Lambda_k \mid \mathscr{Y}_k\big]}. \qquad (6.9)$$

Here the function $f(\cdot)$ is an arbitrary bounded measurable real-valued test function. Write the numerator as:

$$\sigma\big(\langle Z_k, e_j\rangle f(x_k)\big) \triangleq \overline{E}\big[\Lambda_k \langle Z_k, e_j\rangle f(x_k) \mid \mathscr{Y}_k\big]. \qquad (6.10)$$

Suppose we choose the test function $f(x) = 1$, then

$$\sum_{j=1}^{m} \sigma\big(\langle Z_k, e_j\rangle\big) = \sigma\Big(\sum_{j=1}^{m} \langle Z_k, e_j\rangle\Big) = \sigma(1) = \overline{E}\big[\Lambda_k \mid \mathscr{Y}_k\big].$$

Therefore, if the numerator (6.10) can be evaluated for any such f and all j, the denominator of (6.9) can be found. For each $j \in \{1,2,\dots,m\}$, the quantity $\sigma\big(\langle Z_k, e_j\rangle f(x_k)\big)$, defined at (6.10), is a continuous linear functional on the space of continuous functions and so defines a unique measure (see Theorem 2.14 p. 40, Rudin (1966)). Further, suppose that there exists a corresponding unique, unnormalised density function $q_k^j(x)$, such that

$$\sigma\big(\langle Z_k, e_j\rangle f(x_k)\big) \triangleq \int_{\mathbb{R}^n} f(\eta) q_k^j(\eta) d\eta. \qquad (6.11)$$

Then the normalised conditional density is:

$$P(x \in dx, Z_k = e_j \mid \mathscr{Y}_k) = \frac{q_k^j(x)dx}{\int_{\mathbb{R}^n} q_k^j(\xi)d\xi}.$$

Theorem 6.4 *The un-normalised probability density $q_k^j(x)$, satisfies the following integral-equation recursion,*

$$q_k^j(x) = \frac{\Phi\big(D_j^{-1}(y_k - C_j x)\big)}{\Phi(y_k)|D_j||B_j|} \sum_{r=1}^{m} \pi_{(j,r)} \int_{\mathbb{R}^n} \Psi\big(B_j^{-1}(x - A_j\xi)\big) q_{k-1}^r(\xi)\,d\xi. \tag{6.12}$$

Proof Recalling the definition at (6.11), note that

$$\begin{aligned}
&\int_{\mathbb{R}^n} f(\eta) q_k^j(\eta)\,d\eta \\
&\quad = \sigma\big(\langle Z_k, e_j\rangle f(x_k)\big) \\
&\quad = \overline{E}\big[\Lambda_k \langle Z_k, e_j\rangle f(x_k) \mid \mathscr{Y}_k\big] \\
&\quad = \overline{E}\big[\lambda_k \Lambda_{0,k-1} \langle Z_k, e_j\rangle f(x_k) \mid \mathscr{Y}_k\big] \\
&\quad = \overline{E}\Bigg[\Lambda_{0,k-1}\langle Z_k, e_j\rangle \frac{\Phi\big(D_j^{-1}(y_k - C_j x_k)\big)}{|D_j|\Phi(y_k)} \\
&\qquad \times \frac{\Psi\big(B_j^{-1}(x_k - A_j x_{k-1})\big)}{|B_j|\Psi(x_k)} f(x_k) \mid \mathscr{Y}_k\Bigg] \\
&\quad = \frac{1}{\Phi(y_k)|D_j||B_j|}\overline{E}\Bigg[\Lambda_{0,k-1}\langle \Pi Z_{k-1} + L_k, e_j\rangle \\
&\qquad \times \Phi\big(D_j^{-1}(y_k - C_j x_k)\big) \frac{\Psi\big(B_j^{-1}(x_k - A_j x_{k-1})\big)}{\Psi(x_k)} f(x_k) \mid \mathscr{Y}_k\Bigg] \\
&\quad = \frac{1}{\Phi(y_k)|D_j||B_j|} \sum_{r=1}^{m} \pi_{(j,r)} \overline{E}\Bigg[\Lambda_{k-1}\langle Z_{k-1}, e_r\rangle \\
&\qquad \times \Phi\big(D_j^{-1}(y_k - C_j x_k)\big) \frac{\Psi\big(B_j^{-1}(x_k - A_j x_{k-1})\big)}{\Psi(x_k)} f(x_k) \mid \mathscr{Y}_{k-1}\Bigg].
\end{aligned}$$

The final equality follows because under $\overline{P}$, the y_k are independent and $N(0, I_d)$. Further, since the expectations here are with respect to the measure $\overline{P}$, under which the random variables Z_k, x_k, y_k and x_{k-1}, are (mutually) independent, the expected value with respect to x_k is obtained by multiplying by the density $\Psi(x_k)$ and integrating. That is with

$$\begin{aligned}
H(x_{k-1};\Theta) &\triangleq \int_{\mathbb{R}^n} \left\{\Phi\big(D_j^{-1}(y_k - C_j\eta)\big) \frac{\Psi\big(B_j^{-1}(\eta - A_j x_{k-1})\big)}{\Psi(\eta)} f(\eta)\right\} \Psi(\eta)\,d\eta \\
&= \int_{\mathbb{R}^n} \Phi\big(D_j^{-1}(y_k - C_j\eta)\big)\Psi\big(B_j^{-1}(\eta - A_j x_{k-1})\big) f(\eta)\,d\eta,
\end{aligned}$$

then

$$
\begin{aligned}
\int_{\mathbb{R}^n} f(\eta) q_k^j(\eta) d\eta &= \frac{1}{\Phi(y_k)|D_j||B_j|} \sum_{r=1}^{m} \pi_{(j,r)} \overline{E}\big[\Lambda_{k-1}\langle Z_{k-1}, e_r\rangle H(x_{k-1};\Theta) \mid \mathscr{Y}_{k-1}\big] \\
&= \frac{1}{\Phi(y_k)|D_j||B_j|} \sum_{r=1}^{m} \pi_{(j,r)} \int_{\mathbb{R}^n} H(\xi;\Theta) q_{k-1}^r(\xi) d\xi \\
&= \frac{1}{\Phi(y_k)|D_j||B_j|} \sum_{r=1}^{m} \pi_{(j,r)} \int_{\mathbb{R}^n}\int_{\mathbb{R}^n} \Phi\big(D_j^{-1}(y_k - C_j\eta)\big) \\
&\quad \times \Psi\big(B_j^{-1}(\eta - A_j\xi)\big) q_{k-1}^r(\xi) f(\eta) d\eta d\xi. \qquad (6.13)
\end{aligned}
$$

Finally, as equation (6.13) holds for any suitable test function f, it follows, that

$$
q_k^j(x) = \frac{\Phi\big(D_j^{-1}(y_k - C_j x)\big)}{\Phi(y_k)|D_j||B_j|} \sum_{r=1}^{m} \pi_{(j,r)} \int_{\mathbb{R}^n} \Psi\big(B_j^{-1}(x - A_j\xi)\big) q_{k-1}^r(\xi) d\xi. \qquad (6.14)
$$

□

The recursion given at (6.14) is an exact filter; it is expressed as a density which is in general infinite dimensional. However, Gaussian densities are determined by their mean and variance. In what follows we will consider a finite Gaussian mixture representation of $q_{k-1}^r(x)$. The use of Gaussian sums is justified because any density can be approximated by the sum of Gaussian densities (see Sorenson and Alspach (1971)). Further, the noise in the state and the observation processes is assumed Gaussian. In turn, this implies that sums of Gaussian densities are introduced. For example, if the intial state is known exactly, the conditional state at the next time is described by a sum of Gaussian densities, and the number of terms in the sum increases at each time step by a factor equal to the number of states in the chain.

We can then compute the integrals in (6.14) by using Lemma 6.3.
Suppose $\Sigma_\alpha, \Sigma_\mu \in \mathbb{M}^{n\times n}$ are covariance matrices and $x, \alpha, \mu \in \mathbb{R}^n$. The following identity will be useful in the sequel:

$$
\begin{aligned}
&\exp\Big\{-\frac{1}{2}(x-\alpha)'\Sigma_\alpha(x-\alpha)\Big\}\exp\Big\{-\frac{1}{2}(x-\mu)'\Sigma_\mu(x-\mu)\Big\} \\
&= \exp\Big\{-\frac{1}{2}\Big(\alpha'\Sigma_\alpha\alpha + \mu'\Sigma_\mu\mu\Big)\Big\}\exp\Big\{\frac{1}{2}\Big(\Sigma_\alpha\alpha + \Sigma_\mu\mu\Big)'\big(\Sigma_\alpha + \Sigma_\mu\big)^{-1} \\
&\quad \times \Big(\Sigma_\alpha\alpha + \Sigma_\mu\mu\Big)\Big\}\exp\Big\{-\frac{1}{2}\Big(x - (\Sigma_\alpha + \Sigma_\mu)^{-1} \\
&\quad \times (\Sigma_\alpha\alpha + \Sigma_\mu\mu)\Big)'(\Sigma_\alpha + \Sigma_\mu)\Big(x - (\Sigma_\alpha + \Sigma_\mu)^{-1}(\Sigma_\alpha\alpha + \Sigma_\mu\mu)\Big)\Big\}. \qquad (6.15)
\end{aligned}
$$

Theorem 6.5 *Suppose the un-normalised probability density $q_{k-1}^r(\xi)$, (as it appears under the integral in equation (6.14)), can be written as a finite weighted Gaussian mixture with $M^q \in \mathbb{N}$ components. That is, for $k \in \{1, 2, \dots\}$, we suppose*

$$
\begin{aligned}
q^r_{k-1}(\xi) = \sum_{s=1}^{M^q} p^{r,s}_{k-1} \frac{1}{(2\pi)^{n/2}|\Sigma^{r,s}_{k-1|k-1}|^{\frac{1}{2}}} \\
\times \exp\Big\{-\frac{1}{2}(\xi - \alpha^{r,s}_{k-1|k-1})'(\Sigma^{r,s}_{k-1|k-1})^{-1}(\xi - \alpha^{r,s}_{k-1|k-1})\Big\}.
\end{aligned} \tag{6.16}
$$

Here $\Sigma^{r,s}_{k-1|k-1} \in \mathbb{M}^{n\times n}$, and $\alpha^{r,s}_{k-1|k-1} \in \mathbb{R}^n$, are both $\mathscr{Y}_{k-1}$-measurable functions for all pairs $(r,s) \in \{1,2,\dots,m\} \times \{1,2,\dots,M^q\}$. Using this Gaussian mixture (6.16)*, the recursion for the optimal un-normalised density process has the form*

$$
\begin{aligned}
q^j_k(x) \triangleq \frac{1}{(2\pi)^{(d+n)/2}\Phi(y_k)} \sum_{r=1}^{m}\sum_{s=1}^{M^q} K^q_{k,k-1}(j,r,s) \\
\times \exp\Big\{-\frac{1}{2}\Big(x - (\sigma^{j,r,s}_{k-1})^{-1}\delta^{j,r,s}_{k,k-1}\Big)'\sigma^{j,r,s}_{k-1}\Big(x - (\sigma^{j,r,s}_{k-1})^{-1}\delta^{j,r,s}_{k,k-1}\Big)\Big\}.
\end{aligned} \tag{6.17}
$$

The means of the exponential densities in (6.17) *are computed by the update equations*

$$
\begin{aligned}
(\sigma^{j,r,s}_{k-1})^{-1}\delta^{j,r,s}_{k,k-1} = A_r\alpha^{r,s}_{k-1|k-1} + \overline{\Sigma}^{j,r,s}_{k-1|k-1}C'_r\Big(C_r\overline{\Sigma}^{j,r,s}_{k-1|k-1}C'_r \\
+ D_rD'_r\Big)^{-1}\big(y_k - C_rA_r\alpha^{r,s}_{k-1|k-1}\big).
\end{aligned} \tag{6.18}
$$

Here

$$
\begin{aligned}
\overline{\Sigma}^{j,r,s}_{k-1|k-1} &\triangleq B_jB'_j + A_j\Sigma^{r,s}_{k-1|k-1}A'_j \in \mathbb{R}^{n\times n}, \\
\widetilde{u}^{j,r,s}_{k-1|k-1} &\triangleq A_j\alpha^{r,s}_{k-1|k-1} \in \mathbb{R}^n \\
\sigma^{j,r,s}_{k-1} &\triangleq C'_r(D_rD'_r)^{-1}C_r + (\overline{\Sigma}^{j,r,s}_{k-1|k-1})^{-1} \\
\delta^{j,r,s}_{k,k-1} &\triangleq (\overline{\Sigma}^{j,r,s}_{k-1|k-1})^{-1}\widetilde{u}^{j,r,s}_{k-1|k-1} + C'_r \in (D_rD'_r)^{-1}y_k \\
K^q_{k,k-1}(j,r,s) &\triangleq \frac{\pi_{(j,r)}p^{r,s}_{k-1}}{|\overline{\Sigma}^{j,r,s}_{k-1|k-1}|^{\frac{1}{2}}|D_j|}\exp\Big\{\frac{1}{2}(\delta^{j,r,s}_{k,k-1})'(\sigma^{j,r,s}_{k-1})^{-1}\delta^{j,r,s}_{k,k-1}\Big\} \\
&\quad\times \exp\Big\{-\frac{1}{2}\Big[y'_k(D_rD'_r)^{-1}y_k + (\widetilde{u}^{j,r,s}_{k-1|k-1})'(\overline{\Sigma}^{j,r,s}_{k-1|k-1})^{-1}\widetilde{u}^{j,r,s}_{k-1|k-1}\Big]\Big\}
\end{aligned}
$$

(Note that the square matrices $\overline{\Sigma}^{j,r,s}_{k-1|k-1}$ and $\sigma^{j,r,s}_{k-1}$ are symmetric.)

Proof To prove Theorem 6.5, we apply the identity given in Lemma 6.3 to the recursion at (6.12). In the first application, we eliminate the integral in equation (6.12). Using the finite-mixture representation (6.16), we write the recursion (6.12) for the function $q^j(x)$ as

$$q_k^j(x) = \frac{\Phi\big(D_j^{-1}(y_k - C_j x)\big)}{\Phi(y_k)|D_j||B_j|} \times \sum_{r=1}^{m} \pi_{(j,r)} \Bigg\{ \sum_{s=1}^{M^q} p_{k-1}^{r,s} \Bigg[\int_{\mathbb{R}^n} \Psi\big(B_j^{-1}(x - A_j\xi)\big) \frac{1}{(2\pi)^{n/2}|\Sigma_{k-1|k-1}^{r,s}|^{\frac{1}{2}}} \times \exp\Big\{-\frac{1}{2}\Big(\xi - \alpha_{k-1|k-1}^{r,s}\Big)'(\Sigma_{k-1|k-1}^{r,s})^{-1}\Big(\xi - \alpha_{k-1|k-1}^{r,s}\Big)\Big\} d\xi \Bigg]\Bigg\}.$$

Recalling Lemma 6.3, we see that

$$\begin{aligned}
&\int_{\mathbb{R}^n} \Psi\big(B_j^{-1}(x - A_j\xi)\big) \frac{1}{(2\pi)^{n/2}|\Sigma_{k-1|k-1}^{r,s}|^{\frac{1}{2}}} \\
&\quad \times \exp\Big\{-\frac{1}{2}\Big(\xi - \alpha_{k-1|k-1}^{r,s}\Big)'(\Sigma_{k-1|k-1}^{r,s})^{-1}\Big(\xi - \alpha_{k-1|k-1}^{r,s}\Big)\Big\} d\xi \\
&= \frac{1}{(2\pi)^{n/2}} |B_j||B_jB_j' + A_j\Sigma_{k-1|k-1}^{r,s}A_j'|^{-\frac{1}{2}} \exp\Big\{-\frac{1}{2}\Big(x - A_j\alpha_{k-1|k-1}^{r,s}\Big)' \\
&\quad \times (B_jB_j' + A_j\Sigma_{k-1|k-1}^{r,s}A_j')^{-1}\Big(x - A_j\alpha_{k-1|k-1}^{r,s}\Big)\Big\}.
\end{aligned}$$

Using this calculation and the definitions for the quantities $\overline{\Sigma}_{k-1|k-1}^{j,r,s}$ and $\widetilde{u}_{k-1|k-1}^{j,r,s}$, we can write,

$$\begin{aligned}
q_k^j(x) = \sum_{r=1}^{m}\sum_{s=1}^{M^q} &\frac{\pi_{(j,r)} p_{k-1}^{r,s}}{(2\pi)^{n/2}|\overline{\Sigma}_{k-1|k-1}^{j,r,s}|^{\frac{1}{2}}\Phi(y_k)|D_j|} \Phi\big(D_j^{-1}(y_k - C_j x)\big) \\
&\times \exp\Big\{-\frac{1}{2}\Big(x - \widetilde{u}_{k-1|k-1}^{j,r,s}\Big)' \in (\overline{\Sigma}_{k-1|k-1}^{j,r,s})^{-1}\Big(x - \widetilde{u}_{k-1|k-1}^{j,r,s}\Big)\Big\}.
\end{aligned}$$

The products of exponential functions in this equation can be simplified with the identity (6.15), resulting in

$$\begin{aligned}
q_k^j(x) = &\frac{1}{(2\pi)^{(d+n)/2}\Phi(y_k)} \sum_{r=1}^{m}\sum_{s=1}^{M^q} \frac{\pi_{(j,r)} p_{k-1}^{r,s}}{|\overline{\Sigma}_{k-1|k-1}^{j,r,s}|^{\frac{1}{2}}|D_j|} \\
&\times \exp\Big\{\frac{1}{2}(\delta_{k,k-1}^{j,r,s})'(\sigma_{k-1}^{j,r,s})^{-1}\delta_{k,k-1}^{j,r,s}\Big\} \\
&\times \exp\Big\{-\frac{1}{2}\Big[y_k'(D_rD_r')^{-1}y_k + (\widetilde{u}_{k-1|k-1}^{j,r,s})' \in (\overline{\Sigma}_{k-1|k-1}^{j,r,s})^{-1}\widetilde{u}_{k-1|k-1}^{j,r,s}\Big]\Big\} \\
&\times \exp\Big\{-\frac{1}{2}\Big(x - (\sigma_{k-1}^{j,r,s})^{-1}\delta_k^{j,r,s}\Big)'\sigma_{k-1}^{j,r,s}\Big(x - \in (\sigma_{k-1}^{j,r,s})^{-1}\delta_k^{j,r,s}\Big)\Big\}.
\end{aligned}$$

Finally, equation (6.18) is computed by the matrix inversion Lemma. □

Remark 6.6 It is interesting to note the similarity of Theorem 6.5 to the Kalman filter. The mean update equation given by (6.18), is precisely the Kalman filter state update equation, with Kalman gain $\overline{\Sigma}^{j,r,s}_{k-1|k-1}C'_r\left(C_r\overline{\Sigma}^{j,r,s}_{k-1|k-1}C'_r+D_rD'_r\right)^{-1}$. ■

Corollary 6.7 *The estimated conditional mode probability, for model j is given by*

$$p^j_k \triangleq P(Z_k=e_j \mid \mathscr{Y}_k)=\frac{\displaystyle\int_{\mathbb{R}^n} q^j_k(\xi)d\xi}{\displaystyle\sum_{\ell=1}^m\int_{\mathbb{R}^n} q^\ell_k(\xi)d\xi}=\frac{q^j_k}{\sum_{\ell=1}^m q^\ell_k}.$$

Here the estimated un-normalised mode probability q^j_k, is computed by the double summation

$$q^j_k=\frac{1}{(2\pi)^{d/2}\Phi(y_k)}\sum_{r=1}^m\sum_{s=1}^{M^q}\zeta^q_{k,k-1}(j,r,s), \tag{6.19}$$

where

$$\zeta^q_{k,k-1}(j,r,s)\triangleq K^q_{k,k-1}(j,r,s)|\sigma^{j,r,s}_{k-1}|^{-\frac{1}{2}}. \tag{6.20}$$

Proof The proof of Corollary 6.7 is immediate.

$$\begin{aligned}
q^j_k &\triangleq \int_{\mathbb{R}^n} q^j_k(\xi)d\xi\\
&=\frac{1}{(2\pi)^{(d+n)/2}\Phi(y_k)}\sum_{r=1}^m\sum_{s=1}^{M^q}K^q_{k,k-1}(j,r,s)\\
&\quad\times\int_{\mathbb{R}^n}\exp\Big\{-\frac{1}{2}\Big(\xi-(\sigma^{j,r,s}_{k-1})^{-1}\delta^{j,r,s}_{k,k-1}\Big)'\\
&\quad\times\sigma^{j,r,s}_{k-1}\Big(\xi-(\sigma^{j,r,s}_{k-1})^{-1}\delta^{j,r,s}_{k,k-1}\Big)\Big\}d\xi\\
&=\frac{1}{(2\pi)^{d/2}\Phi(y_k)}\sum_{r=1}^m\sum_{s=1}^{M^q}K^q_{k,k-1}(j,r,s)|\sigma^{j,r,s}_{k-1}|^{-\frac{1}{2}}.
\end{aligned}$$

□

Remark 6.8 The scalar-valued quantities $\zeta^q_{k,k-1}(j,r,s)$, are all non-negative weights. These quantities form the double sum at equation (6.19), whose value is the estimated mode probability q^j_k. In what follows, we shall exploit the form of equation (6.20) to fix the memory requirements of our suboptimal filters. Note that if $\sigma^{j,r,s}_{k-1}$ is relatively large, then $|\sigma^{j,r,s}_{k-1}|^{-\frac{1}{2}}$ makes a small contribution to (6.19). That is, larger (error) covariances give smaller weights in the sum (6.19). Accordingly, a subset of the terms $\zeta^q_{k,k-1}(j,r,s)$ will be identified from the the set of all $m\times M^q$ such quantities. The elements of this subset will be identified through their relative magnitudes, as the most substantial contributors to the sum at (6.19) and therefore the quantity q_k^j. Ultimately, this subset of quantities will be used to identify, at every time k, the most significant components of the Gaussian mixture for the density $q_k(x)$. ■

Sub-Optimal Filter Dynamics

In this section we develop a sub-optimal recursive filter by extending an idea due to Viterbi (Viterbi 1969). The motivation to develop a sub-optimal filter is immediate from the dynamics of the un-normalised density (6.17). Suppose at time $k = 1$, these dynamics involve $m \times M^q$ densities. Then the next time, $k = 2$, these dynamics require $m \times m \times M^q$ densities. It is clear, that the demand on memory requirements is exponential in time, with the number of densities required at time k being $m^k \times M^q$. What we wish to do is to circumvent this growth by identifying a subset of candidate mixture densities, from which we construct a sub-optimal density with fixed, (in time), memory requirements.

Write

$$\Gamma^q \triangleq \{1,2,\ldots,m\} \times \{1,2,\ldots,M^q\},$$
$$\widetilde{S}^q_{k,k-1}(j,r,s) \triangleq \{\zeta^q_{k,k-1}(j,r,s)\}_{(r,s)\in\Gamma^q}.$$

To remove the growth in memory requirements, we propose to identify, at each time k, the M^q-best candidate densities for each suboptimal density $q_k^j(x)$, using the corresponding estimated mode probabilities. The key to this idea is to identify M^q optimal densities, that is, the M^q components in the Gaussian mixture, through their corresponding set of estimated mode probabilities, q^j, $j \in \{1,2,\ldots,m\}$.

Since the estimated mode probabilities, given by equation (6.19), are formed by a summation over non-negative quantities, we can identify the M^q largest contributors to this sum and then use the corresponding indexes to identify the M^q-best Gaussian densities. This maximisation procedure is as follows:

$$\zeta^q_{k,k-1}(j,r^*_{k,1},s^*_{k,1}) \triangleq \max_{(r,s)\in\Gamma^q} \widetilde{S}^q_{k,k-1}(j,r,s),$$
$$\zeta^q_{k,k-1}(j,r^*_{k,2},s^*_{k,2}) \triangleq \max_{(r,s)\in\Gamma^q\setminus(r^*_{k,1},s^*_{k,1})} \widetilde{S}^q_{k,k-1}(j,r,s),$$
$$\vdots$$
$$\zeta^q_{k,k-1}(j,r^*_{k,M^q},s^*_{k,M^q}) \triangleq \max_{(r,s)\in\Gamma^q\setminus\{(r^*_{k,1},s^*_{k,1}),\ldots,(r^*_{k,M^q-1},s^*_{k,M^q-1})\}} \widetilde{S}^q_{k,k-1}(j,r,s).$$

Note that we are not directly interested in the quantities $\zeta^q_{k,k-1}(j,r^*_{k,\ell},s^*_{k,\ell})$, rather, the indexes that locate these quantities. The optimal index set, for the density of model j, is:

$$\mathscr{I}_k^j \triangleq \{(r^*_{k,1},s^*_{k,1}),(r^*_{k,2},s^*_{k,2}),\ldots,(r^*_{k,M^q},s^*_{k,M^q})\}.$$

Using these indexes, we approximate the suboptimal un-normalised density, of order M^q, corresponding to the density $q_k^j(x)$ as

$$q_k^j(x) \triangleq \frac{1}{(2\pi)^{(d+n)/2}\Phi(y_k)} \sum_{\ell=1}^{M^q} K_{k,k-1}^q(j, r_{k,\ell}^*, s_{k,\ell}^*) \times \exp\left\{-\frac{1}{2}\left(x - \mathfrak{Z}_k^{-1}\mathfrak{R}_k\right)' \mathfrak{Z}\left(x - \mathfrak{Z}_k^{-1}\mathfrak{R}_k\right)\right\}, \tag{6.21}$$

where $\mathfrak{Z}_k^{-1} = (\sigma_{k-1}^{j,r_{k,\ell}^*,s_{k,\ell}^*})^{-1}$, and $\mathfrak{R}_k = \delta_{k,k-1}^{j,r_{k,\ell}^*,s_{k,\ell}^*}$. Further, by Corollary 6.7, the un-normalised mode probability corresponding to the expectation $E\left[Z_k = e_j \mid \mathscr{Y}_k\right]$, is approximated by

$$q_k^j \triangleq \frac{1}{(2\pi)^{d/2}\Phi(y_k)} \sum_{\ell=1}^{M^q} \zeta_{k,k-1}^q(j, r_{k,\ell}^*, s_{k,\ell}^*).$$

To normalise the function $q_k^j(x)$, we need the sum of all terms $\left\{q_k^1, \dots, q_k^m\right\}$ and so write $\varphi_{k|k} \triangleq \sum_{j=1}^{m} q_k^j$.

Densities such as $q_k^j(x)$, provide all the information available. However, what is often required in state estimation (filtering), is an expression for the state estimate of x_k at time k given $\mathscr{Y}_k$, that is: $\widehat{x}_{k|k} \triangleq E\left[x_k \mid \mathscr{Y}_k\right]$.

Lemma 6.9 *The sub-optimal state estimate $\widehat{x}_{k|k}$, for a Gaussian mixture of order M^q, has the representation*

$$\begin{aligned}\widehat{x}_{k|k} = {} & \frac{1}{(2\pi)^{d/2}\Phi(y_k)\varphi_{k|k}} \sum_{j=1}^{m}\sum_{\ell=1}^{M^q} \frac{K_{k,k-1}^q(j, r_{k,\ell}^*, s_{k,\ell}^*)}{|\mathfrak{Z}|^{\frac{1}{2}}} \\ & \times \left[A_{r_{k,\ell}^*}\alpha_{k-1|k-1}^{r_{k,\ell}^*,s_{k,\ell}^*} + \overline{\Sigma}_{k-1|k-1}^{j,r_{k,\ell}^*,s_{k,\ell}^*} C'_{r_{k,\ell}^*}\left(D_{r_{k,\ell}^*}D'_{r_{k,\ell}^*} + C_{r_{k,\ell}^*}\overline{\Sigma}_{k-1|k-1}^{j,r,s}C'_{r_{k,\ell}^*}\right)^{-1}\right. \\ & \left.\times \left(y_k - C_{r_{k,\ell}^*}A_{r_{k,\ell}^*}\alpha_{k-1|k-1}^{r_{k,\ell}^*,s_{k,\ell}^*}\right)\right].\end{aligned} \tag{6.22}$$

Here $\varphi_{k|k} = \frac{1}{(2\pi)^{d/2}\Phi(y_k)} \sum_{j=1}^{m}\sum_{\ell=1}^{M^q} K_{k,k-1}^q(j, r_{k,\ell}^*, s_{k,\ell}^*)|\sigma_{k-1}^{j,r_{k,\ell}^*,s_{k,\ell}^*}|^{-\frac{1}{2}}$.

Proof

$$\begin{aligned}E\left[x_k \mid \mathscr{Y}_k\right] &= \frac{1}{\varphi_{k|k}} \int_{\mathbb{R}^n} \xi \left[\sum_{j=1}^{m} q_k^j(\xi)\right] d\xi \\ &= \frac{1}{\varphi_{k|k}} \sum_{j=1}^{m} \left[\int_{\mathbb{R}^n} \xi\, q_k^j(\xi) d\xi\right] \\ &= \frac{1}{\varphi_{k|k}} \sum_{j=1}^{m} \int_{\mathbb{R}^n} \xi \left[\frac{1}{(2\pi)^{(d+n)/2}\Phi(y_k)} \sum_{\ell=1}^{M^q} K_{k,k-1}^q(j, r_{k,\ell}^*, s_{k,\ell}^*)\right.\end{aligned}$$

$$\times \exp\Big\{-\frac{1}{2}\Big(\xi - \mathfrak{Z}^{-1}\mathfrak{R}_k\Big)'\mathfrak{Z}\Big(\xi - \mathfrak{Z}_k^{-1}\mathfrak{R}_k\Big)\Big\}\Big] d\xi$$

$$= \frac{1}{(2\pi)^{(d+n)/2}\Phi(y_k)\varphi_{k|k}} \sum_{j=1}^{m}\sum_{\ell=1}^{M^q} K^q_{k,k-1}(j, r^*_{k,\ell}, s^*_{k,\ell}) \int_{\mathbb{R}^n} \xi$$

$$\times \Big[\exp\Big\{-\frac{1}{2}\Big(\xi - \mathfrak{Z}^{-1}\mathfrak{R}_k\Big)'\mathfrak{Z}\Big(\xi - \mathfrak{Z}_k^{-1}\mathfrak{R}_k\Big)\Big\}\Big] d\xi$$

$$= \frac{1}{(2\pi)^{d/2}\Phi(y_k)\varphi_{k|k}} \sum_{j=1}^{m}\sum_{\ell=1}^{M^q} \frac{K^q_{k,k-1}(j, r^*_{k,\ell}, s^*_{k,\ell})}{|\mathfrak{Z}|^{\frac{1}{2}}} \mathfrak{Z}^{-1}\mathfrak{R}_k.$$

The proof is completed by applying the matrix inversion Lemma to the term $\mathfrak{Z}^{-1}\mathfrak{R}_k$. □

It is also of practical interest to have a relation for the corresponding error covariance, that is:

$$\Sigma_{k|k} \triangleq E\big[(x_k - \widehat{x}_{k|k})(x - \widehat{x}_{k|k})' \mid \mathscr{Y}_k\big]$$
$$= \frac{1}{\varphi_{k|k}} \int_{\mathbb{R}^n} (\xi - \widehat{x}_{k|k})(\xi - \widehat{x}_{k|k})' \Big(\sum_{j=1}^{m} q_k^j(\xi)\Big) d\xi = \widehat{\Sigma}_{k|k}, \text{ say.}$$

Lemma 6.10 *The filter state error covariance $\widehat{\Sigma}_{k|k}$, for a Gaussian mixture of order M^q, has the following representation;*

$$\widehat{\Sigma}_{k|k} = \frac{1}{(2\pi)^{d/2}\Phi(y_k)\varphi_{k|k}} \sum_{j=1}^{m}\sum_{\ell=1}^{M^q} \frac{K^q_{k,k-1}(j, r^*_{k,\ell}, s^*_{k,\ell})}{|\sigma_{k-1}^{j, r^*_{k,\ell}, s^*_{k,\ell}}|^{\frac{1}{2}}} \Big[\mathfrak{Z}_k^{-1}$$
$$+ \Big(\mathfrak{Z}^{-1}\mathfrak{R}_k - \widehat{x}_{k|k}\Big)\Big(\mathfrak{Z}^{-1}\delta_{k,k-1}^{j, r^*_{k,\ell}, s^*_{k,\ell}} - \widehat{x}_{k|k}\Big)'\Big]. \tag{6.23}$$

Proof Recalling the approximate un-normalised density (6.21), we see that

$$\widehat{\Sigma}_{k|k} = \frac{1}{(2\pi)^{(d+n)/2}\Phi(y_k)\varphi_{k|k}} \sum_{j=1}^{m}\sum_{\ell=1}^{M^q} K^q_{k,k-1}(j, r^*_{k,\ell}, s^*_{k,\ell})$$

$$\times \int_{\mathbb{R}^n} (\xi - \widehat{x}_{k|k})(\xi - \widehat{x}_{k|k})' \Big[\exp\Big\{-\frac{1}{2}\Big(\xi - \mathfrak{Z}^{-1}\mathfrak{R}_k\Big)'\mathfrak{Z}\Big(\xi - \mathfrak{Z}^{-1}\mathfrak{R}_k\Big)\Big\}\Big] d\xi$$

$$= \frac{1}{(2\pi)^{d/2}\Phi(y_k)\varphi_{k|k}} \sum_{j=1}^{m}\sum_{\ell=1}^{M^q} \frac{K^q_{k,k-1}(j, r^*_{k,\ell}, s^*_{k,\ell})}{|\mathfrak{Z}|^{\frac{1}{2}}} \int_{\mathbb{R}^n} (\xi - \widehat{x}_{k|k})(\xi - \widehat{x}_{k|k})'$$

$$\times \Big[\frac{|\mathfrak{Z}|^{\frac{1}{2}}}{(2\pi)^{n/2}} \exp\Big\{-\frac{1}{2}\Big(\xi - \mathfrak{Z}^{-1}\mathfrak{R}_k\Big)'\mathfrak{Z}\Big(\xi - \mathfrak{Z}^{-1}\mathfrak{R}_k\Big)\Big\}\Big] d\xi$$

Writing

$$(\xi - \widehat{x}_{k|k}) = \left(\xi - \mathfrak{Z}^{-1}\mathfrak{R}_k + \mathfrak{Z}^{-1}\mathfrak{R}_k - \widehat{x}_{k|k}\right),$$

(6.32) is the sum of the integrals:

$$\frac{|\mathfrak{Z}|^{\frac{1}{2}}}{(2\pi)^{n/2}} \int_{\mathbb{R}^n} \left(\xi - \mathfrak{Z}^{-1}\mathfrak{R}_k\right)\left(\xi - \mathfrak{Z}^{-1}\mathfrak{R}_k\right)' \times \exp\left\{-\frac{1}{2}\left(\xi - \mathfrak{Z}^{-1}\mathfrak{R}_k\right)' \mathfrak{Z}\left(\xi - \mathfrak{Z}^{-1}\mathfrak{R}_k\right)\right\} d\xi = (\mathfrak{Z})^{-1},$$

$$\frac{|\mathfrak{Z}|^{\frac{1}{2}}}{(2\pi)^{n/2}} \int_{\mathbb{R}^n} \left(\xi - \mathfrak{Z}^{-1}\mathfrak{R}_k\right) \times \exp\left\{-\frac{1}{2}\left(\xi - \mathfrak{Z}^{-1}\mathfrak{R}_k\right)' \mathfrak{Z}\left(\xi - \mathfrak{Z}^{-1}\mathfrak{R}_k\right)\right\} d\xi \times (\mathfrak{Z}^{-1}\mathfrak{R}_k - \widehat{x}_{k|k})' = 0,$$

$$\frac{|\mathfrak{Z}|^{\frac{1}{2}}}{(2\pi)^{n/2}} \left(\mathfrak{Z}^{-1}\mathfrak{R}_k - \widehat{x}_{k|k}\right)' \int_{\mathbb{R}^n} (\xi - \mathfrak{Z}^{-1}\mathfrak{R}_k) \times \exp\left\{-\frac{1}{2}\left(\xi - \mathfrak{Z}^{-1}\mathfrak{R}_k\right)' \mathfrak{Z}\left(\xi - \mathfrak{Z}^{-1}\mathfrak{R}_k\right)\right\} d\xi = 0,$$

and

$$\frac{|\mathfrak{Z}|^{\frac{1}{2}}}{(2\pi)^{n/2}} \left(\mathfrak{Z}^{-1}\mathfrak{R}_k - \widehat{x}_{k|k}\right)\left(\mathfrak{Z}^{-1}\mathfrak{R}_k - \widehat{x}_{k|k}\right)' \times \int_{\mathbb{R}^n} \exp\left\{-\frac{1}{2}\left(\xi - \mathfrak{Z}^{-1}\mathfrak{R}_k\right)' \mathfrak{Z}\left(\xi - \mathfrak{Z}^{-1}\mathfrak{R}_k\right)\right\} d\xi = \left(\mathfrak{Z}^{-1}\mathfrak{R}_k - \widehat{x}_{k|k}\right)\left(\mathfrak{Z}^{-1}\mathfrak{R}_k - \widehat{x}_{k|k}\right)'.$$

Finally, combining the calculations above the result follows. □

Remark 6.11 It is interesting to note the equation (6.32) is essentially in the form of an additive decomposition of two components, that is, a weighted sum of the two terms $(\mathfrak{Z})^{-1}$ and $(\mathfrak{Z}^{-1}\mathfrak{R}_k - \widehat{x}_{k|k})(\mathfrak{Z}^{-1}\mathfrak{R}_k - \widehat{x}_{k|k})'$. This suggests that the estimation error arises from two sources: the approximation by a finite mixture, and the standard error in the estimate of x. The matrix $(\mathfrak{Z})^{-1}$ is related, for example, to the covariance $\Sigma^{r,s}_{k-1|k-1}$, which contributes to component 's', in the finite mixture representation of the r-th un-normalised density (recall equation (6.16)). The matrix $(\mathfrak{Z}^{-1}\mathfrak{R}_k - \widehat{x}_{k|k})(\mathfrak{Z}^{-1}\mathfrak{R}_k - \widehat{x}_{k|k})'$ however, computes the error covariance between the state estimate $\widehat{x}_{k|k}$ and the approximate mean of the state x. This decomposition could provide another method to determine the appropriate number of components for the Gaussian mixtures.

■

Hybrid Smoother Dynamics

Smoothing is a term used for a form of offline estimation where one has information (from the observation process), beyond the current time. Suppose one has a set of observations generated by the dynamics (6.4), for example, $\{y_1, y_2, \dots, y_T\}$ and we wish to estimate

$$E\big[x_\ell \mid \mathscr{Y}_{0,T}\big], 0 \le \ell \le T.$$

Currently, all smoothing schemes for Gauss-Markov jump linear systems are ad hoc and are not based upon the exact hybrid filter dynamics, for example, see Chen and Tugnait (1986), Chen and Tugnait (2000).

To compute smoothers, we exploit a duality between the forward function $q_k^j(x)$ which evolves forwards in time and a related function $v_k^j(x)$ which evolves backwards in time (see Malcolm and Elliott (1999) and Elliott and Malcolm (2001)). To construct our smoothing algorithm, we compute a finite memory time-reversed recursion for the functions $v_k^j(x)$. A sub-optimal v is then defined using an extension of the idea of Viterbi, by replacing a summation with a maximisation.

Exact Smoother Dynamics

Recalling the form of Bayes' rule, we note that

$$E\big[\langle Z_k, e_j\rangle f(x_k) \mid \mathscr{Y}_{0,T}\big] = \frac{\overline{E}\big[\Lambda_{0,T}\langle Z_k, e_j\rangle f(x_k) \mid \mathscr{Y}_{0,T}\big]}{\overline{E}\big[\Lambda_{0,T} \mid \mathscr{Y}_T\big]}. \tag{6.24}$$

Write

$$\widetilde{\mathscr{F}}_k \triangleq \sigma\{x_\ell, Z_\ell, 0 \le \ell \le k\}.$$

Consider the numerator in the quotient of equation (6.24)

$$\begin{aligned}
&\overline{E}\big[\Lambda_{0,T}\langle Z_k, e_j\rangle f(x_k) \mid \mathscr{Y}_{0,T}\big] = \overline{E}\big[\Lambda_k \Lambda_{k+1,T}\langle Z_k, e_j\rangle f(x_k) \mid \mathscr{Y}_{0,T}\big] \\
&= \overline{E}\Big[\overline{E}\big[\Lambda_k \Lambda_{k+1,T}\langle Z_k, e_j\rangle f(x_k) \mid \widetilde{\mathscr{F}}_k \vee \mathscr{Y}_{0,T}\big] \mid \mathscr{Y}_{0,T}\Big] \\
&= \overline{E}\Big[\Lambda_k\langle Z_k, e_j\rangle f(x_k)\overline{E}\big[\Lambda_{k+1,T} \mid \widetilde{\mathscr{F}}_k \vee \mathscr{Y}_{0,T}\big] \mid \mathscr{Y}_{0,T}\Big].
\end{aligned}$$

As our processes are Markov

$$\overline{E}\big[\Lambda_{k+1|T} \mid \widetilde{\mathscr{F}}_k \vee \mathscr{Y}_{0,T}\big] = \overline{E}\big[\Lambda_{k+1|T} \mid Z_k, x_k, \mathscr{Y}_{0,T}\big].$$

Write

$$v_k^j(x) \triangleq \overline{E}\left[\Lambda_{k+1,T} \mid Z_k = e_j, x_k = x, \mathscr{Y}_{0,T}\right], \tag{6.25}$$

and, $\overline{E}\left[\Lambda_k \langle Z_k, e_j\rangle f(x_k) v_k^j(x_k) \mid \mathscr{Y}_{0,T}\right] = \int_{\mathbb{R}^n} \left\{ q_k^j(\xi) v_k^j(\xi) \right\} f(\xi) d\xi.$

We now compute a backwards recursion for the function $v_k^j(x)$, similar in form to the forward recursion for the function $q_k^j(x)$, given in equation (6.12).

Theorem 6.12 *The un-normalised function $v_k^j(x)$, satisfies the backward recursion:*

$$v_k^j(x) = \sum_{r=1}^m \frac{\pi_{(r,j)}}{|D_r||B_r|\Phi(y_{k+1})} \int_{\mathbb{R}^n} \Phi\big(D_r^{-1}(y_{k+1} - C_r\xi)\big)\Psi(B_r^{-1}(\xi - A_r x)) v_{k+1}^r(\xi) d\xi.$$

Proof From definition (6.25), we again use repeated conditioning to write

$$v_k^j(x) = \overline{E}\left[\Lambda_{k+1,T} \mid Z_k = e_j, x_k = x, \mathscr{Y}_{0,T}\right] = \overline{E}\left[\lambda_{k+1}\Lambda_{k+2,T} \mid Z_k = e_j, x_k = x, \mathscr{Y}_{0,T}\right]$$

$$Z_k = e_j, x_k = x, \mathscr{Y}_{0,T}\Big]$$

$$= \overline{E}\Big[\lambda_{k+1}\overline{E}\left[\Lambda_{k+2,T} \mid Z_{k+1}, x_{k+1}, Z_k = e_j, x_k = x, \mathscr{Y}_{0,T}\right] \mid$$

$$Z_k = e_j, x_k = x, \mathscr{Y}_{0,T}\Big]$$

$$= \sum_{r=1}^m \overline{E}\Big[\lambda_{k+1}\langle Z_{k+1}, e_r\rangle \overline{E}\left[\Lambda_{k+2,T} \mid Z_{k+1} = e_r, x_{k+1}, Z_k = e_j, x_k = x, \mathscr{Y}_{0,T}\right] \mid$$

$$Z_k = e_j, x_k = x, \mathscr{Y}_{0,T}\Big]$$

$$= \sum_{r=1}^m \overline{E}\Big[\lambda_{k+1}\langle Z_{k+1}, e_r\rangle \overline{E}\left[\Lambda_{k+2,T} \mid Z_{k+1} = e_r, x_{k+1}, \mathscr{Y}_{0,T}\right] \mid$$

$$Z_k = e_j, x_k = x, \mathscr{Y}_{0,T}\Big]$$

(since Z is a Markov chain and under $\overline{P}$, the process x is iid)

$$= \sum_{r=1}^m \overline{E}\Big[\langle Z_{k+1}, e_r\rangle \frac{\Phi\big(D_r^{-1}(y_{k+1} - C_r x_{k+1})\big)}{|D_r|\Phi(y_{k+1})} \frac{\Psi\big(B_r^{-1}(x_{k+1} - A_r x_k)\big)}{|B_r|\Psi(x_{k+1})}$$

$$\times\, v_{k+1}^r(x_{k+1}) \mid Z_k = e_j, x_k = x, \mathscr{Y}_{0,T}\Big]$$

$$= \sum_{r=1}^{m} \frac{\pi_{(r,j)}}{|D_r||B_r|\Phi(y_{k+1})} \int_{\mathbb{R}^n} \Phi\big(D_r^{-1}(y_{k+1} - C_r\xi)\big)\Psi(B_r^{-1}(\xi - A_r x))v^r_{k+1}(\xi)d\xi.$$

□

Theorem 6.13 *Suppose the un-normalised function $v^r_{k+1}(\xi)$, (as it appears under the integral in Theorem 6.12), may be approximated as a finite weighted Gaussian mixture with $M^v \in \mathbb{N}$ components. That is, suppose*

$$v^r_{k+1}(\xi) = \sum_{s=1}^{M^v} \rho^{r,s}_{k+1} \frac{1}{(2\pi)^{n/2}|\Sigma^{r,s}_{k+1|T}|^{\frac{1}{2}}} \times \exp\Big\{-\frac{1}{2}(\xi - \alpha^{r,s}_{k+1|T})'(\Sigma^{r,s}_{k+1|T})^{-1}(\xi - \alpha^{r,s}_{k+1|T})\Big\}.$$

Here $\Sigma^{r,s}_{k+1|T} \in \mathbb{M}^{n\times n}$, and $\alpha^{r,s}_{k+1|T} \in \mathbb{R}^n$, are both $\mathscr{Y}_{k+1,T}$-measurable functions for all pairs $(r,s) \in \{1,2,\dots,m\} \times \{1,2,\dots,M^v\}$. Using the Gaussian mixture in Theorem 6.13, the recursion for $v^j_k(x)$, at times $k \in \{1,2,\dots,T-1\}$, has the form

$$v^j_k(x) \triangleq \frac{1}{(2\pi)^{(n+d)/2}\,\Phi(y_{k+1})} \sum_{r=1}^{m}\sum_{s=1}^{M^v} K^v_{k+1,T}(j,r,s) \times \exp\Big\{-\frac{1}{2}\Big(x - (S^{r,s}_{k+1|T})^{-1}\tau^{r,s}_{k+1|T}\Big)' S^{r,s}_{k+1,T}\Big(x - (S^{r,s}_{k+1|T})^{-1}\tau^{r,s}_{k+1|T}\Big)\Big\}.$$

At the final time T, $\forall\, j \in \{1,2,\dots,m\}$, $v^j_T(x) \triangleq 1$. Here

$$\begin{aligned}
K^v_{k+1,T}(j,r,s) \triangleq{}& \frac{\pi_{(r,j)}\rho^{r,s}_{k+1}}{|D_r||B_r||\Sigma^{r,s}_{k+1|T}|^{\frac{1}{2}}\,\big|\gamma^r + \big(\Sigma^{r,s}_{k+1|T}\big)^{-1}\big|^{\frac{1}{2}}}\\
&\times \exp\Big\{-\frac{1}{2}(\alpha^{r,s}_{k+1|T})'(\Sigma^{r,s}_{k+1|T})^{-1}\Big[I - \widetilde{\Sigma}^{r,s}_{k+1|T}\big(\Sigma^{r,s}_{k+1|T}\big)^{-1}\Big]\alpha^{r,s}_{k+1|T}\Big\}\\
&\times \exp\Big\{-\frac{1}{2}y'_{k+1}(D_rD'_r)^{-1}y_{k+1}\Big\}\\
&\times \exp\Big\{-\frac{1}{2}\Big(\widetilde{\Sigma}^{r,s}_{k+1|T}\big(\Sigma^{r,s}_{k+1|T}\big)^{-1}\alpha^{r,s}_{k+1|T}\Big)'\\
&\times \big(\widetilde{\Sigma}^{r,s}_{k+1|T}\big)^{-1}\Big(\widetilde{\Sigma}^{r,s}_{k+1|T}\big(\Sigma^{r,s}_{k+1|T}\big)^{-1}\alpha^{r,s}_{k+1|T}\Big)\Big\}\\
&\times \exp\Big\{\frac{1}{2}\Big(C_r(D_rD'_r)^{-1}y_{k+1} + \big(\Sigma^{r,s}_{k+1|T}\big)^{-1}\alpha^{r,s}_{k+1|T}\Big)'\\
&\times \big(\widetilde{\Sigma}^{r,s}_{k+1|T}\big)^{-1}\Big(C_r(D_rD'_r)^{-1}y_{k+1} + \big(\Sigma^{r,s}_{k+1|T}\big)^{-1}\alpha^{r,s}_{k+1|T}\Big)\Big\}\\
&\times \exp\Big\{\frac{1}{2}(\tau^{r,s}_{k+1|T})'(S^{r,s}_{k+1|T})^{-1}\tau^{r,s}_{k+1|T}\Big\} \in \mathbb{R},
\end{aligned}$$

$$S^{r,s}_{k+1|T} \triangleq A'_r(B_rB'_r)^{-1}A_r - A'_r(B_rB'_r)^{-1}\widetilde{\Sigma}^{r,s}_{k+1|T}(B_rB'_r)^{-1}A_r \in \mathbb{R}^{n\times n},$$

$$\tau^{r,s}_{k+1|T} \triangleq A'_r(B_rB'_r)^{-1}\widetilde{\Sigma}^{r,s}_{k+1|T}\left(C'_r(D_rD'_r)^{-1}y_{k+1} + (\Sigma^{r,s}_{k+1|T})^{-1}\alpha^{r,s}_{k+1|T}\right) \in \mathbb{R}^n,$$

$$\gamma_r \triangleq (B_rB'_r)^{-1} + C'_r(D_rD'_r)^{-1}C_r \in \mathbb{R}^{n\times n},$$

$$\mu^r_{k+1} \triangleq C'_r(D_rD'_r)^{-1}y_{k+1} + (B_rB'_r)^{-1}A_rx \in \mathbb{R}^n,$$

$$\widetilde{\Sigma}^{r,s}_{k+1|T} \triangleq \left(\gamma^r + \Sigma^{r,s}_{k+1|T}\right)^{-1} \in \mathbb{R}^{n\times n}.$$

(Note that the square matrices γ_r and $\widetilde{\Sigma}^{r,s}_{k+1|T}$ are symmetric.)

Proof To prove Theorem 6.13, we first evaluate the integral in the recursion in Theorem 6.12. To complete the proof, the result of this calculation is written as a weighted sum of Gaussian densities.
Write

$$v^j_k(x) = \sum_{r=1}^{m} \frac{\pi_{(r,j)}}{|D_r||B_r|\Phi(y_{k+1})} I_a, \tag{6.26}$$

where $I_a \triangleq \int_{\mathbb{R}^n} \Phi\big(D_r^{-1}(y_{k+1} - C_r\xi)\big)\Psi(B_r^{-1}(\xi - A_rx))v^r_{k+1}(\xi)d\xi$.

Completing the square of the exponentials in I_a, we see that

$$I_a = \frac{1}{(2\pi)^{(n+d)/2}} \exp\Big\{\frac{1}{2}(\mu^r_{k+1})'(\gamma^r)^{-1}\mu^r_{k+1}\Big\}$$
$$\times \exp\Big\{-\frac{1}{2}y'_{k+1}(D_rD'_r)^{-1}y_{k+1}\Big\}\exp\Big\{-\frac{1}{2}x'A'_r(B_rB'_r)^{-1}A_rx\Big\}I_b.$$

Here

$$I_b = \int_{\mathbb{R}^n} \exp\Big\{-\frac{1}{2}\Big(\xi - (\gamma^r)^{-1}\mu^r_{k+1}\Big)'\gamma_r\Big(\xi - (\gamma^r)^{-1}\mu^r_{k+1}\Big)\Big\}v^r_{k+1}(\xi)d\xi$$
$$\gamma^r \triangleq C'_r(D_rD'_r)^{-1}C_r + (B_rB'_r)^{-1}$$
$$\mu^r_{k+1} \triangleq C'_r(D_rD'_r)^{-1}y_{k+1} + (B_rB'_r)^{-1}A_rx.$$

Since $v^r_{k+1}(\xi)$ is a Gaussian mixture, the integrand of I_b is a product of Gaussian densities and can be evaluated directly, that is,

$$I_b = \sum_{s=1}^{M^v} \rho^{r,s}_{k+1} \int_{\mathbb{R}^n} \exp\Big\{-\frac{1}{2}\Big(\xi - (\gamma^r)^{-1}\mu^r_{k+1}\Big)'\gamma_r\Big(\xi - (\gamma^r)^{-1}\mu^r_{k+1}\Big)\Big\}$$
$$\times \frac{1}{(2\pi)^{n/2}|\Sigma^{r,s}_{k+1|T}|^{\frac{1}{2}}} \exp\Big\{-\frac{1}{2}\Big(\xi - \alpha^{r,s}_{k+1|T}\Big)'(\Sigma^{r,s}_{k+1|T})^{-1}\Big(\xi - \alpha^{r,s}_{k+1|T}\Big)\Big\}d\xi$$

$$= \sum_{s=1}^{M^v} \rho_{k+1}^{r,s} \left[\frac{\left|\left(\gamma^r + (\Sigma_{k+1|T}^{r,s})^{-1}\right)^{-1}\right|^{\frac{1}{2}}}{|\Sigma_{k+1|T}^{r,s}|^{\frac{1}{2}}} \exp\left\{-\frac{1}{2}(\mu_{k+1}^r)'(\gamma^r)^{-1}\mu_{k+1}^r\right\} \right.$$
$$\times \exp\left\{-\frac{1}{2}(\alpha_{k+1|T}^{r,s})'(\Sigma_{k+1|T}^{r,s})^{-1}\alpha_{k+1|T}^{r,s}\right\}$$
$$\times \exp\left\{\frac{1}{2}\left(\mu_{k+1}^r + (\Sigma_{k+1|T}^{r,s})^{-1}\alpha_{k+1|T}^{r,s}\right)'\left(\gamma^r + (\Sigma_{k+1|T}^{r,s})^{-1}\right)^{-1}\left(\mu_{k+1}^r\right.\right.$$
$$\left.\left.\left. + (\Sigma_{k+1|T}^{r,s})^{-1}\alpha_{k+1|T}^{r,s}\right)\right\}\right].$$

Recalling equation (6.26) and the terms I_a and I_b, we see that

$$v_k^j(x) = \frac{1}{(2\pi)^{(d+n)/2}\,\Phi(y_{k+1})} \sum_{r=1}^{m}\sum_{s=1}^{M^v} \frac{\pi_{(r,j)}\rho_{k+1}^{r,s}\left|\left(\gamma^r + (\Sigma_{k+1|T}^{r,s})^{-1}\right)^{-1}\right|^{\frac{1}{2}}}{|D_r||B_r||\Sigma_{k+1|T}^{r,s}|^{\frac{1}{2}}}$$
$$\times \exp\left\{-\frac{1}{2}x'A_r'(B_rB_r')^{-1}A_rx\right\}$$
$$\times \exp\left\{-\frac{1}{2}y_{k+1}'(D_rD_r')^{-1}y_{k+1}\right\}\exp\left\{-\frac{1}{2}(\alpha_{k+1|T}^{r,s})'(\Sigma_{k+1|T}^{r,s})^{-1}\alpha_{k+1|T}^{r,s}\right\}$$
$$\times \exp\left\{\frac{1}{2}(\alpha_{k+1|T}^{r,s})'(\Sigma_{k+1|T}^{r,s})^{-1}\widetilde{\Sigma}_{k+1|T}^{r,s}\Sigma_{k+1|T}^{r,s}\alpha_{k+1|T}^{r,s}\right\}$$
$$\times \exp\left\{\frac{1}{2}\left[(\mu_{k+1}^r)'\widetilde{\Sigma}_{k+1|T}^{r,s}\mu_{k+1}^r + (\mu_{k+1}^r)'\widetilde{\Sigma}_{k+1|T}^{r,s}(\Sigma_{k+1|T}^{r,s})^{-1}\alpha_{k+1|T}^{r,s}\right.\right.$$
$$\left.\left. + (\alpha_{k+1|T}^{r,s})'(\Sigma_{k+1|T}^{r,s})^{-1}\widetilde{\Sigma}_{k+1|T}^{r,s}\mu_{k+1}^r\right]\right\}.$$

Consequently

$$v_k^j(x) = \frac{1}{(2\pi)^{(d+n)/2}\,\Phi(y_{k+1})} \sum_{r=1}^{m}\sum_{s=1}^{M^v} \frac{\pi_{(r,j)}\rho_{k+1}^{r,s}\left|\left(\gamma^r + (\Sigma_{k+1|T}^{r,s})^{-1}\right)^{-1}\right|^{\frac{1}{2}}}{|D_r||B_r||\Sigma_{k+1|T}^{r,s}|^{\frac{1}{2}}}$$
$$\times \exp\left\{-\frac{1}{2}(\alpha_{k+1|T}^{r,s})'(\Sigma_{k+1|T}^{r,s})^{-1}\left[I - \widetilde{\Sigma}_{k+1|T}^{r,s}(\Sigma_{k+1|T}^{r,s})^{-1}\right]\alpha_{k+1|T}^{r,s}\right\}$$
$$\times \exp\left\{-\frac{1}{2}y_{k+1}'(D_rD_r')^{-1}y_{k+1}\right\}$$
$$\times \exp\left\{-\frac{1}{2}\left(\widetilde{\Sigma}_{k+1|T}^{r,s}(\Sigma_{k+1|T}^{r,s})^{-1}\alpha_{k+1|T}^{r,s}\right)'\right.$$
$$\left.\times (\widetilde{\Sigma}_{k+1|T}^{r,s})^{-1}\left(\widetilde{\Sigma}_{k+1|T}^{r,s}(\Sigma_{k+1|T}^{r,s})^{-1}\alpha_{k+1|T}^{r,s}\right)\right\}$$
$$\times \exp\left\{\frac{1}{2}\left(C_r(D_rD_r')^{-1}y_{k+1} + (\Sigma_{k+1|T}^{r,s})^{-1}\alpha_{k+1|T}^{r,s}\right)'\right.$$
$$\left.\times (\widetilde{\Sigma}_{k+1|T}^{r,s})^{-1}\left(C_r(D_rD_r')^{-1}y_{k+1} + (\Sigma_{k+1|T}^{r,s})^{-1}\alpha_{k+1|T}^{r,s}\right)\right\}$$

$$\times \exp\left\{\frac{1}{2}(\tau^{r,s}_{k+1|T})'(S^{r,s}_{k+1|T})^{-1}\tau^{r,s}_{k+1|T}\right\}$$
$$\times \exp\left\{-\frac{1}{2}\left(x-(S^{r,s}_{k+1|T})^{-1}\tau^{r,s}_{k+1|T}\right)' S^{r,s}_{k+1,T}\left(x-(S^{r,s}_{k+1|T})^{-1}\tau^{r,s}_{k+1|T}\right)\right\}.$$

□

Corollary 6.14 *Write* $v^j_k \triangleq \int_{\mathbb{R}^n} v^j_k(\xi)d\xi$.

The scalar-valued quantity v^j_k, *is computed by the double sum*

$$v^j_k = \frac{1}{(2\pi)^{d/2}\,\Phi(y_{k+1})}\sum_{r=1}^{m}\sum_{s=1}^{M^v}\vartheta^v_{k+1,T}(j,r,s).$$

Here $\vartheta^v_{k+1,T}(j,r,s) \triangleq K^v_{k+1,T}(j,r,s)|S^{r,s}_{k+1|T}|^{-\frac{1}{2}}$.

The proof of Corollary 6.14 is immediate.

Sub-Optimal Smoother Dynamics

Write

$$\Gamma^v \triangleq \{1,2,\ldots,m\}\times\{1,2,\ldots,M^v\},$$
$$\widetilde{S}^v_{k+1,T}(j,r,s) \triangleq \{\vartheta^v_{k+1,T}(j,r,s)\}_{(r,s)\in\Gamma^v}.$$

As before, we propose, at each time k, to identify the M^v-best candidate functions, (components in the Gaussian mixture), for each function $v^j_k(x)$. This maximisation procedure is as follows:

$$\vartheta^v_{k+1,T}(j,r^*_1,s^*_1) \triangleq \max_{(r,s)\in\Gamma^v}\widetilde{S}^v_{k+1,T}(j,r,s),$$
$$\vartheta^v_{k+1,T}(j,r^*_2,s^*_2) \triangleq \max_{(r,s)\in\Gamma^v\setminus(r^*_1,s^*_1)}\widetilde{S}^v_{k+1,T}(j,r,s),$$
$$\vdots$$
$$\vartheta^v_{k+1,T}(j,r^*_{M^v},s^*_{M^v}) \triangleq \max_{(r,s)\in\Gamma^v\setminus\{(r^*_1,s^*_1),\ldots,(r^*_{M^v-1},s^*_{M^v-1})\}}\widetilde{S}^v_{k+1,T}(j,r,s).$$

The optimal index set, for function $v^j_k(x)$, is:

$$\mathscr{I}_k(j) \triangleq \{(r^*_{k,1},s^*_{k,1}),(r^*_{k,2},s^*_{k,2}),\ldots,(r^*_{k,M^v},s^*_{k,M^v})\}.$$

Using these indexes, the order M^v equation for $v^j_k(x)$, whose memory requirements are fixed in time, has the form;

$$v_k^j(x) \triangleq \frac{1}{(2\pi)^{(d+n)/2}\Phi(y_{k+1})} \sum_{\ell=1}^{M^v} K^v_{k+1,T}(j, r^*_{k+1,\ell}, s^*_{k+1,\ell})$$
$$\times \exp\Big\{-\frac{1}{2}\Big(x - (S^{r^*_{k+1,\ell},s^*_{k+1,\ell}}_{k+1|T})^{-1}\tau^{r^*_{k+1,\ell},s^*_{k+1,\ell}}_{k+1|T}\Big)' S^{r^*_{k+1,\ell},s^*_{k+1,\ell}}_{k+1,T}$$
$$\times \Big(x - (S^{r^*_{k+1,\ell},s^*_{k+1,\ell}}_{k+1|T})^{-1}\tau^{r^*_{k+1,\ell},s^*_{k+1,\ell}}_{k+1|T}\Big)\Big\}.$$

Write, for the smoother density normalisation constant,

$$\varphi_{k|T} \triangleq \sum_{j=1}^{m}\Big(\int_{\mathbb{R}^n}\Big[q_k^j(\xi)v_k^j(\xi)\Big]d\xi\Big).$$

Combining the functions $q_k^j(x)$ and $v_k^j(x)$, we see that the fixed interval smoothed estimate of state may be written as,

$$\begin{aligned}\widehat{x}_{k|T} \triangleq E\big[x_t \mid \mathscr{Y}_{0,T}\big] &= \frac{\int_{\mathbb{R}^n}\xi\Big[\sum_{j=1}^m q_k^j(\xi)v_k^j(\xi)\Big]d\xi}{\int_{\mathbb{R}^n}\Big[\sum_{j=1}^m q_k^j(\xi)v_k^j(\xi)\Big]d\xi}\\ &= \frac{1}{\varphi_{k|T}}\int_{\mathbb{R}^n}\xi\Big[\sum_{j=1}^m q_k^j(\xi)v_k^j(\xi)\Big]d\xi \qquad (6.27)\\ &= \frac{1}{\varphi_{k|T}}\sum_{j=1}^m\Big[\int_{\mathbb{R}^n}\xi\Big[q_k^j(\xi)v_k^j(\xi)\Big]d\xi\Big].\end{aligned}$$

Since the functions $q_k(x)$ and $v_k(x)$ are each weighted Gaussian mixtures, the integrals in (6.27) can be evaluated exactly.

Lemma 6.15 *The smoothed state estimate $\widehat{x}_{k|T}$, for Gaussian mixtures with orders M^q and M^v, has the following representation*

$$\widehat{x}_{k|T} = \frac{1}{\varphi_{k|T}}\sum_{j=1}^m\sum_{\ell=1}^{M^q}\sum_{i=1}^{M^v}\Big[K^q_{k,k-1}(j, r^*_{k,\ell}, s^*_{k,\ell})K^v_{k+1,T}(j, r^*_{k,i}, s^*_{k,i})\Big|$$
$$\times\Big(\sigma^{r^*_{k,\ell},s^*_{k,\ell}}_{k-1} + S^{r^*_{k+1,\ell},s^*_{k+1,\ell}}_{k+1|T}\Big)^{-1}\Big|^{\frac{1}{2}} \times \exp\Big\{\frac{1}{2}\Big(\delta^{r^*_{k,\ell},s^*_{k,\ell}}_k + \tau^{r^*_{k+1,i},s^*_{k+1,i}}_{k+1|T}\Big)'$$
$$\times\Big(\sigma^{r^*_{k,\ell},s^*_{k,\ell}}_{k-1} + S^{r^*_{k+1,\ell},s^*_{k+1,\ell}}_{k+1|T}\Big)^{-1}\Big(\delta^{r^*_{k,\ell},s^*_{k,\ell}}_k + \tau^{r^*_{k+1,i},s^*_{k+1,i}}_{k+1|T}\Big)\Big\}$$
$$\times\exp\Big\{-\frac{1}{2}\Big[(\tau^{r^*_{k+1,i},s^*_{k+1,i}}_{k+1|T})'(S^{r^*_{k+1,\ell},s^*_{k+1,\ell}}_{k+1|T})^{-1}\tau^{r^*_{k+1,i},s^*_{k+1,i}}_{k+1|T} + (\delta^{r^*_{k,\ell},s^*_{k,\ell}}_k)'$$
$$\times(\sigma^{r^*_{k,\ell},s^*_{k,\ell}}_{k-1})^{-1}\delta^{r^*_{k,\ell},s^*_{k,\ell}}_k\Big]\Big\}$$
$$\times\Big(\sigma^{r^*_{k,\ell},s^*_{k,\ell}}_{k-1} + S^{r^*_{k+1,\ell},s^*_{k+1,\ell}}_{k+1|T}\Big)^{-1}\Big(\delta^{r^*_{k,\ell},s^*_{k,\ell}}_k + \tau^{r^*_{k+1,i},s^*_{k+1,i}}_{k+1|T}\Big)\Big].$$

Here

$$\begin{aligned}
\varphi_{k|T} = \sum_{j=1}^{m}\sum_{\ell=1}^{M^q}\sum_{i=1}^{M^v}\Bigg[& K^q_{k,k-1}(j, r^*_{k,\ell}, s^*_{k,\ell}) \\
& \times K^v_{k+1,T}(j, r^*_{k,i}, s^*_{k,i}) \exp\Big\{\frac{1}{2}\Big(\delta_k^{r^*_{k,\ell},s^*_{k,\ell}} + \tau_{k+1|T}^{r^*_{k+1,i},s^*_{k+1,i}}\Big)' \\
& \times \Big(\sigma_{k-1}^{r^*_{k,\ell},s^*_{k,\ell}} + S_{k+1|T}^{r^*_{k+1,\ell},s^*_{k+1,\ell}}\Big)^{-1}\Big(\delta_k^{r^*_{k,\ell},s^*_{k,\ell}} + \tau_{k+1|T}^{r^*_{k+1,i},s^*_{k+1,i}}\Big)\Big\} \\
& \times \exp\Big\{-\frac{1}{2}\Big[(\tau_{k+1|T}^{r^*_{k+1,i},s^*_{k+1,i}})'(S_{k+1|T}^{r^*_{k+1,\ell},s^*_{k+1,\ell}})^{-1}\tau_{k+1|T}^{r^*_{k+1,i},s^*_{k+1,i}} + (\delta_k^{r^*_{k,\ell},s^*_{k,\ell}})' \\
& \times (\sigma_{k-1}^{r^*_{k,\ell},s^*_{k,\ell}})^{-1}\delta_k^{r^*_{k,\ell},s^*_{k,\ell}}\Big]\Big\}\Big|\Big(\sigma_{k-1}^{r^*_{k,\ell},s^*_{k,\ell}} + S_{k+1|T}^{r^*_{k+1,\ell},s^*_{k+1,\ell}}\Big)^{-1}\Big|^{\frac{1}{2}}\Bigg] \in \mathbb{R}.
\end{aligned} \tag{6.28}$$

Proof

$$\begin{aligned}
\widehat{x}_{k|T} &= \frac{1}{\varphi_{k|T}}\sum_{j=1}^{m}\Bigg[\int_{\mathbb{R}^n}\xi\Big[q_k^j(\xi)v_k^j(\xi)\Big]d\xi\Bigg] \\
&= \frac{1}{\varphi_{k|T}}\sum_{j=1}^{m}\Bigg[\int_{\mathbb{R}^n}\xi\Bigg(\frac{1}{(2\pi)^{(d+n)/2}\Phi(y_k)}\sum_{\ell=1}^{M^q}K^q_{k,k-1}(j, r^*_{k,\ell}, s^*_{k,\ell}) \\
&\quad \times \exp\Big\{-\frac{1}{2}\Big(\xi - (\sigma_{k-1}^{r^*_{k,\ell},s^*_{k,\ell}})^{-1}\delta_k^{r^*_{k,\ell},s^*_{k,\ell}}\Big)'\sigma_{k-1}^{r^*_{k,\ell},s^*_{k,\ell}} \\
&\quad \times \Big(\xi - (\sigma_{k-1}^{r^*_{k,\ell},s^*_{k,\ell}})^{-1}\delta_k^{r^*_{k,\ell},s^*_{k,\ell}}\Big)\Big\}\Bigg) \\
&\quad \times \Bigg(\frac{1}{(2\pi)^{(d+n)/2}\Phi(y_{k+1})}\sum_{i=1}^{M^v}K^v_{k+1,T}(j, r^*_{k,i}, s^*_{k,i}) \\
&\quad \times \exp\Big\{-\frac{1}{2}\Big(\xi - (S_{k+1|T}^{r^*_{k+1,\ell},s^*_{k+1,\ell}})^{-1}\tau_{k+1|T}^{r^*_{k+1,\ell},s^*_{k+1,\ell}}\Big)' \\
&\quad \times S_{k+1,T}^{r^*_{k+1,\ell},s^*_{k+1,\ell}}\Big(\xi - (S_{k+1|T}^{r^*_{k+1,\ell},s^*_{k+1,\ell}})^{-1}\tau_{k+1|T}^{r^*_{k+1,\ell},s^*_{k+1,\ell}}\Big)\Big\}\Bigg)d\xi\Bigg] \\
&= \frac{1}{(2\pi)^{(d+n)}\Phi(y_k)\Phi(y_{k+1})\varphi_{k|T}}\sum_{j=1}^{m}\sum_{\ell=1}^{M^q}\sum_{i=1}^{M^v}\Bigg[K^q_{k,k-1}(j, r^*_{k,\ell}, s^*_{k,\ell}) \\
&\quad \times K^v_{k+1,T}(j, r^*_{k,i}, s^*_{k,i})\int_{\mathbb{R}^n}\xi\Big(\exp\Big\{-\frac{1}{2}\Big(\xi - (\sigma_{k-1}^{r^*_{k,\ell},s^*_{k,\ell}})^{-1}\delta_k^{r^*_{k,\ell},s^*_{k,\ell}}\Big)'\sigma_{k-1}^{r^*_{k,\ell},s^*_{k,\ell}} \\
&\quad \times \Big(\xi - (\sigma_{k-1}^{r^*_{k,\ell},s^*_{k,\ell}})^{-1}\delta_k^{r^*_{k,\ell},s^*_{k,\ell}}\Big)\Big\} \\
&\quad \times \exp\Big\{-\frac{1}{2}\Big(\xi - (S_{k+1|T}^{r^*_{k+1,\ell},s^*_{k+1,\ell}})^{-1}\tau_{k+1|T}^{r^*_{k+1,\ell},s^*_{k+1,\ell}}\Big)'S_{k+1,T}^{r^*_{k+1,\ell},s^*_{k+1,\ell}} \\
&\quad \times \Big(\xi - (S_{k+1|T}^{r^*_{k+1,i},s^*_{k+1,i}})^{-1}\tau_{k+1|T}^{r^*_{k+1,i},s^*_{k+1,i}}\Big)\Big\}\Big)d\xi\Bigg]
\end{aligned}$$

$$
\begin{aligned}
&= \frac{1}{(2\pi)^{(d+n)}\Phi(y_k)\Phi(y_{k+1})\varphi_{k|T}} \sum_{j=1}^{m}\sum_{\ell=1}^{M^q}\sum_{i=1}^{M^v}\Bigg[K^q_{k,k-1}(j, r^*_{k,\ell}, s^*_{k,\ell}) \\
&\times K^v_{k+1,T}(j, r^*_{k,i}, s^*_{k,i}) \exp\Big\{\frac{1}{2}\Big(\delta_k^{r^*_{k,\ell},s^*_{k,\ell}} + \tau_{k+1|T}^{r^*_{k+1,i},s^*_{k+1,i}}\Big)' \\
&\times \Big(\sigma_{k-1}^{r^*_{k,\ell},s^*_{k,\ell}} + S_{k+1|T}^{r^*_{k+1,\ell},s^*_{k+1,\ell}}\Big)^{-1}\Big(\delta_k^{r^*_{k,\ell},s^*_{k,\ell}} + \tau_{k+1|T}^{r^*_{k+1,i},s^*_{k+1,i}}\Big)\Big\} \\
&\times \exp\Big\{-\frac{1}{2}\Big[\big(\tau_{k+1|T}^{r^*_{k+1,i},s^*_{k+1,i}}\big)'\big(S_{k+1|T}^{r^*_{k+1,\ell},s^*_{k+1,\ell}}\big)^{-1}\tau_{k+1|T}^{r^*_{k+1,i},s^*_{k+1,i}} + \big(\delta_k^{r^*_{k,\ell},s^*_{k,\ell}}\big)' \\
&\times \big(\sigma_{k-1}^{r^*_{k,\ell},s^*_{k,\ell}}\big)^{-1}\delta_k^{r^*_{k,\ell},s^*_{k,\ell}}\Big]\Big\}(2\pi)^{n/2}\Big|\Big(\sigma_{k-1}^{r^*_{k,\ell},s^*_{k,\ell}} + S_{k+1|T}^{r^*_{k+1,\ell},s^*_{k+1,\ell}}\Big)^{-1}\Big|^{\frac{1}{2}} \\
&\times \Big(\sigma_{k-1}^{r^*_{k,\ell},s^*_{k,\ell}} + S_{k+1|T}^{r^*_{k+1,\ell},s^*_{k+1,\ell}}\Big)^{-1}\Big(\delta_k^{r^*_{k,\ell},s^*_{k,\ell}} + \tau_{k+1|T}^{r^*_{k+1,i},s^*_{k+1,i}}\Big)\Bigg].
\end{aligned}
$$

□

Remark 6.16 It is interesting to note, that in the formulation of the smoother defined above, the integers M^q and M^v are mutually independent. This feature suggests one is afforded an extra degree of freedom, that is, the practitioner can explicitly impose a desired accuracy in the contribution of the past, (choice of M^q), and the contribution of the future, (choice of M^v).

■

The smoother error covariance is defined as:

$$
\begin{aligned}
\Sigma_{k|T} &\triangleq E\big[(x_k - \widehat{x}_{k|T})(x_k - \widehat{x}_{k|T})' \mid \mathscr{Y}_T\big] \\
&= \frac{1}{\varphi_{k|T}}\sum_{j=1}^{m}\int_{\mathbb{R}^n}(\xi - \widehat{x}_{k|T})(\xi - \widehat{x}_{k|T})' \\
&\times \Bigg[\Bigg(\frac{1}{(2\pi)^{(d+n)/2}\Phi(y_k)}\sum_{\ell=1}^{M^q}K^q_{k,k-1}(j, r^*_{k,\ell}, s^*_{k,\ell}) \\
&\times \exp\Big\{-\frac{1}{2}\Big(\xi - \big(\sigma_{k-1}^{r^*_{k,\ell},s^*_{k,\ell}}\big)^{-1}\delta_k^{r^*_{k,\ell},s^*_{k,\ell}}\Big)'\sigma_{k-1}^{r^*_{k,\ell},s^*_{k,\ell}} \\
&\times \Big(\xi - \big(\sigma_{k-1}^{r^*_{k,\ell},s^*_{k,\ell}}\big)^{-1}\delta_k^{r^*_{k,\ell},s^*_{k,\ell}}\Big)\Big\}\Bigg) \\
&\times \Bigg(\frac{1}{(2\pi)^{(d+n)/2}\Phi(y_{k+1})}\sum_{i=1}^{M^v}K^v_{k+1,T}(j, r^*_{k,i}, s^*_{k,i}) \\
&\times \exp\Big\{-\frac{1}{2}\Big(\xi - \big(S_{k+1|T}^{r^*_{k+1,\ell},s^*_{k+1,\ell}}\big)^{-1}\tau_{k+1|T}^{r^*_{k+1,\ell},s^*_{k+1,\ell}}\Big)' \\
&\times S_{k+1,T}^{r^*_{k+1,\ell},s^*_{k+1,\ell}}\Big(\xi - \big(S_{k+1|T}^{r^*_{k+1,\ell},s^*_{k+1,\ell}}\big)^{-1}\tau_{k+1|T}^{r^*_{k+1,\ell},s^*_{k+1,\ell}}\Big)\Big\}\Bigg)\Bigg]d\xi = \widehat{\Sigma}_{k|T}.
\end{aligned}
$$

Lemma 6.17 *The smoother state error covariance $\widehat{\Sigma}_{k|T}$, for Gaussian mixtures with orders M^q and M^v, has the following representation*

$$
\begin{aligned}
\widehat{\Sigma}_{k|T} = & \frac{1}{\varphi_{k|T}} \sum_{j=1}^{m} \sum_{\ell=1}^{M^q} \sum_{i=1}^{M^v} \\
& \times \Bigg[K^q_{k,k-1}(j, r^*_{k,\ell}, s^*_{k,\ell}) K^v_{k+1,T}(j, r^*_{k,i}, s^*_{k,i}) \Big| (\sigma_{k-1}^{r^*_{k,\ell}, s^*_{k,\ell}} + S_{k+1|T}^{r^*_{k+1,\ell}, s^*_{k+1,\ell}})^{-1} \Big|^{\frac{1}{2}} \\
& \times \exp\Big\{ \frac{1}{2} \Big((\tau_{k+1|T}^{r^*_{k+1,i}, s^*_{k+1,i}})' (S_{k+1|T}^{r^*_{k+1,\ell}, s^*_{k+1,\ell}})^{-1} \tau_{k+1|T}^{r^*_{k+1,i}, s^*_{k+1,i}} \\
& + (\delta_k^{r^*_{k,\ell}, s^*_{k,\ell}})' (\sigma_{k-1}^{r^*_{k,\ell}, s^*_{k,\ell}})^{-1} \delta_k^{r^*_{k,\ell}, s^*_{k,\ell}} \Big) \Big\} \\
& \times \Big[\Big(\sigma_{k-1}^{r^*_{k,\ell}, s^*_{k,\ell}} + S_{k+1|T}^{r^*_{k+1,\ell}, s^*_{k+1,\ell}} \Big)^{-1} \\
& + \Big((\sigma_{k-1}^{r^*_{k,\ell}, s^*_{k,\ell}} + S_{k+1|T}^{r^*_{k+1,\ell}, s^*_{k+1,\ell}})^{-1} (\delta_k^{r^*_{k,\ell}, s^*_{k,\ell}} + \tau_{k+1|T}^{r^*_{k+1,i}, s^*_{k+1,i}}) - \widehat{x}_{k|T} \Big) \\
& \times \Big((\sigma_{k-1}^{r^*_{k,\ell}, s^*_{k,\ell}} + S_{k+1|T}^{r^*_{k+1,\ell}, s^*_{k+1,\ell}})^{-1} (\delta_k^{r^*_{k,\ell}, s^*_{k,\ell}} + \tau_{k+1|T}^{r^*_{k+1,i}, s^*_{k+1,i}}) - \widehat{x}_{k|T} \Big)' \Big] \Bigg].
\end{aligned}
$$

Here the scalar quantity $\varphi_{k|T}$ is as defined by equation (6.28). The proof of the lemma is similar to the proof of Lemma 6.10 and so it omitted.

Example

In this section we illustrate the performance of of the new hybrid filter state estimator. Two scenarios are considered, in the first example, the Markov chain transition matrix is fixed and simulations are computed for five different values of Signal to Noise Ratio (SNR). In the second example, we consider two different transition matrices. For each of our examples, comparisons are provided against the IMM algorithm (with standard Kalman filters). All the examples consider the same scalar-valued state process x and we set $m = 3$. Further, the order of the Gaussian mixtures is fixed at $M^q = 5$. For our first example, we set

$$
\begin{aligned}
H_1 &\triangleq \{A_{H_1} = 1, B_{H_1} = 0.05, C_{H_1} = 1\}, \\
H_2 &\triangleq \{A_{H_2} = 0.9, B_{H_2} = 0.01, C_{H_2} = 2\}, \\
H_3 &\triangleq \{A_{H_3} = 1.1, B_{H_3} = 0.1, C_{H_3} = 1.5\}
\end{aligned}
$$

and

$$
\Pi \triangleq \begin{bmatrix} 0.8 & 0.2 & 0.2 \\ 0.1 & 0.6 & 0.2 \\ 0.1 & 0.2 & 0.6 \end{bmatrix}. \tag{6.29}
$$

The initial state x_0 is a Gaussian random variable, with mean 10 and covariance 10. Each of the five Figures 6.1(a)–6.1(e), show simulations at different levels of SNR, which was varied only through the noise gain on the state dynamics, that is: SNR-Case 1 $\{D_{H_1} = 0.5, D_{H_2} = 0.25, D_{H_3} = 0.5\}$, SNR-Case 2 $\{D_{H_1} = 1, D_{H_2} = 0.5, D_{H_3} = 1\}$, SNR-Case 3 $\{D_{H_1} = 2, D_{H_2} = 1, D_{H_3} = 2\}$, SNR-Case 4 $\{D_{H_1} = 4, D_{H_2} = 2, D_{H_3} = 4\}$, SNR-Case 5 $\{D_{H_1} = 8, D_{H_2} = 4, D_{H_3} = 8\}$. For these five scenarios, Monte Carlo simulation were computed, consisting of 100 trials and discrete time indexed from $k = 1$ to 100. The corresponding mean square errors for the IMM and the new hybrid filter are given in Figures 6.1(a)–6.1(e).

In our second example, two different values of the parameter Π were considered, with

$$\Pi\text{-Case } 6 \Pi \triangleq \begin{bmatrix} 0.8 & 0.1 & 0.1 \\ 0.1 & 0.8 & 0.1 \\ 0.1 & 0.1 & 0.8 \end{bmatrix}, \Pi\text{-Case } 7 \Pi \triangleq \begin{bmatrix} 0.4 & 0.3 & 0.3 \\ 0.3 & 0.4 & 0.3 \\ 0.3 & 0.3 & 0.4 \end{bmatrix}.$$

In this example, the same model parameters from the first example were used, that is, H_1, H_2 and H_3, with the state noise gains set at $\{D_{H_1} = 1, D_{H_2} = 0.5, D_{H_3} = 1\}$. In all these cases, the new hybrid filter significantly outperforms the IMM. In cases 4 and 5 (see Figures 6.1(d) and 6.1(e)), when the signal to noise ratio is very low the performances of these two algorithms are comparable. For the cases 1 to 3, when the signal to noise ratio is standard or high, the suboptimal filter is more efficient than the IMM (see Figures 6.1(a)–6.1(c)). The same conclusion can be made when the Markov chain Z jumps slowly or rapidly, (see cases 6 and 7 and Figures 6.2(a), 6.2(b)).

Pseudo Code Algorithm

In this section we define a three-step pseudo code form of our estimation algorithm. Following the completion of an initialisation step, our three algorithm steps are:

Step 1 Compute Gaussian-mixture Densities,
Step 2 Hypothesis management,
Step 3 Updating.

In details these steps are given below.

Initialisation

- Choose initial Gaussian mixtures statistics:

$$\mathscr{A}_0 \triangleq \begin{bmatrix} \alpha_{0|0}^{1,1} & \alpha_{0|0}^{1,2} & \cdots & \alpha_{0|0}^{1,M^q} \\ \alpha_{0|0}^{2,1} & \alpha_{0|0}^{2,2} & \cdots & \alpha_{0|0}^{2,M^q} \\ \vdots & \vdots & \vdots & \vdots \\ \alpha_{0|0}^{m,1} & \alpha_{0|0}^{m,2} & \cdots & \alpha_{0|0}^{m,M^q} \end{bmatrix}$$

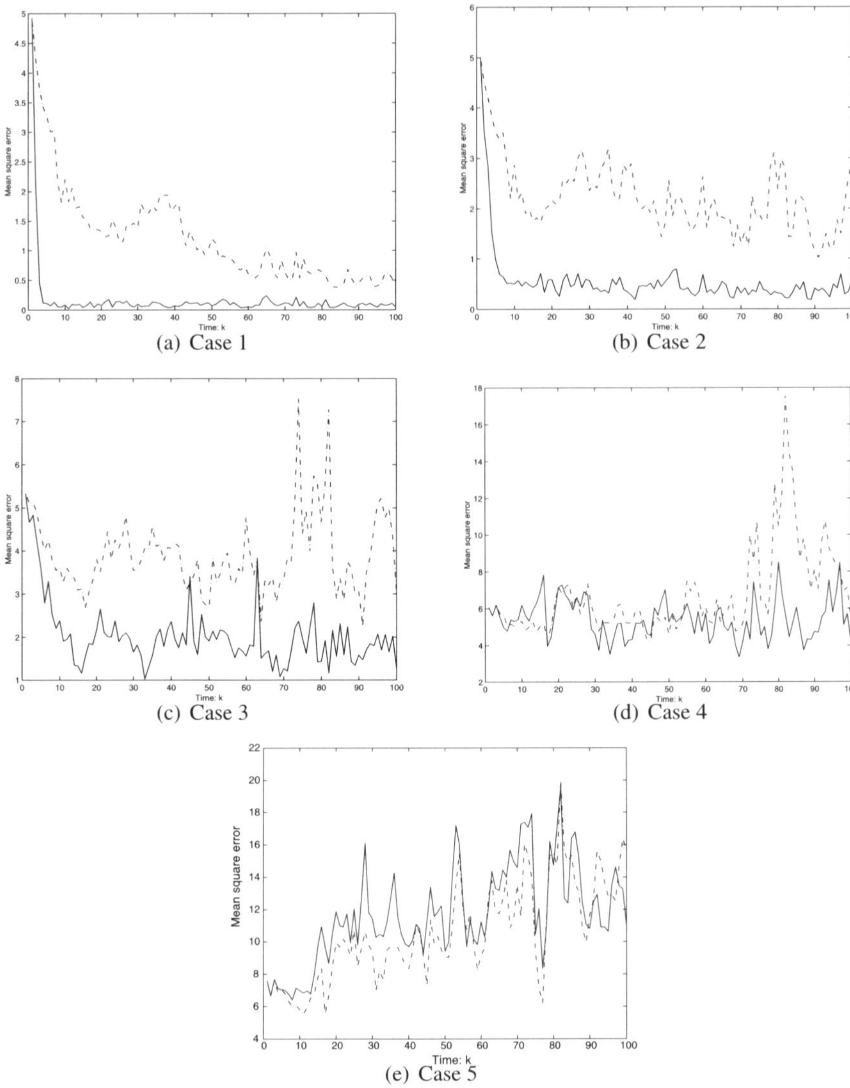

Fig. 6.1 SNR performance of the IMM and the new hybrid filter, with $M^q = 5$

and

$$\mathscr{B}_0 \triangleq \begin{bmatrix} \Sigma_{0|0}^{1,1} & \Sigma_{0|0}^{1,2} & \cdots & \Sigma_{0|0}^{1,M^q} \\ \Sigma_{0|0}^{2,1} & \Sigma_{0|0}^{2,2} & \cdots & \Sigma_{0|0}^{2,M^q} \\ \vdots & \vdots & \vdots & \vdots \\ \Sigma_{0|0}^{m,1} & \Sigma_{0|0}^{m,2} & \cdots & \Sigma_{0|0}^{m,M^q} \end{bmatrix}.$$

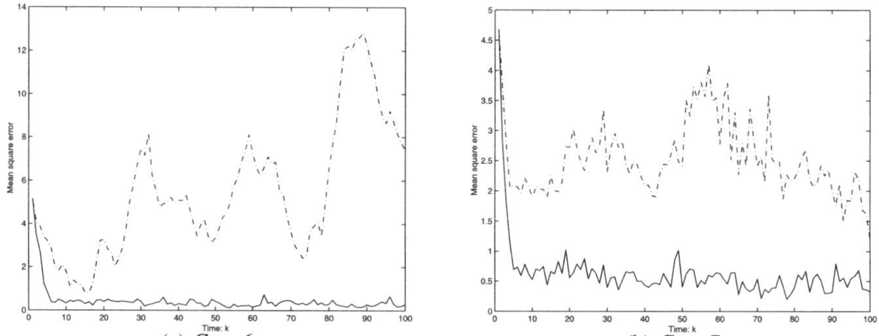

Fig. 6.2 Fast and slow switching Markov chain comparison of the IMM and the new hybrid filter, with $M^q = 5$

- For each $j \in \{1, 2, \dots, m\}$, choose, by some means, the initial sets of Gaussian-mixture weights $\{W_0^{(j,1)}, W_0^{(j,2)}, \dots, W_0^{(j,M^q)}\}$.

For $k = 1, 2, \dots N$, repeat:

Step 1, Compute Gaussian Mixture Densities:

- Define an index set, $\Gamma \triangleq \{1, 2, \dots, m\} \times \{1, 2, \dots, M^q\}$.
- For each $j \in \{1, 2, \dots, m\}$, at each time k, compute of the following $m \times M^q$ quantities, each indexed by the pairs $(r, s) \in \Gamma$:

$$
\begin{aligned}
\overline{\Sigma}^{j,(r,s)}_{k-1|k-1} &\triangleq B_j B_j' + A_j \Sigma^{(r,s)}_{k-1|k-1} A_j' \\
\widetilde{u}^{j,(r,s)}_{k-1|k-1} &\triangleq A_j \alpha^{(r,s)}_{k-1|k-1} \\
\sigma^{j,(r,s)} &\triangleq C_r' \in (D_r D_r') C_r + \in \left(\overline{\Sigma}^{j,(r,s)}_{k-1|k-1}\right)^{-1} \\
\left(\sigma^{j,(r,s)}\right)^{-1} &\triangleq \overline{\Sigma}^{j,(r,s)}_{k-1|k-1} - \\
&\quad \overline{\Sigma}^{j,(r,s)}_{k-1|k-1} C_r' \left(C_r \overline{\Sigma}^{j,(r,s)}_{k-1|k-1} C_r' + D_r D_r'\right)^{-1} C_r \overline{\Sigma}^{j,(r,s)}_{k-1|k-1} \\
\delta^{j,(r,s)} &\triangleq (\overline{\Sigma}^{j,(r,s)}_{k-1|k-1})^{-1} \widetilde{u}^{j,(r,s)}_{k-1|k-1} + C_r' (D_r D_r')^{-1} y_k \\
K^{j,(r,s)} &\triangleq \frac{\pi_{(j,r)} W^{(r,s)}_{k-1}}{|\overline{\Sigma}^{j,(r,s)}_{k-1|k-1}|^{\frac{1}{2}} |D_j|} \\
&\quad \times \exp\left\{\frac{1}{2} (\delta^{j,(r,s)})' (\sigma^{j,(r,s)})^{-1} \delta^{j,(r,s)}\right\} \exp\left\{-\frac{1}{2}\left[y_k' (D_r D_r')^{-1} y_k \right.\right. \\
&\quad \left.\left. + (\widetilde{u}^{j,(r,s)}_{k-1|k-1})' \in (\overline{\Sigma}^{j,(r,s)}_{k-1|k-1})^{-1} \widetilde{u}^{j,(r,s)}_{k-1|k-1}\right]\right\}.
\end{aligned}
\tag{6.30}
$$

Step 2, Hypothesis Management:

- For each $j \in \{1, 2, \ldots, m\}$, compute the quantities

$$\mathscr{C}^j \triangleq \begin{bmatrix} \zeta^{j,(1,1)} & \zeta^{j,(1,2)} & \ldots & \zeta^{j,(1,M^q)} \\ \zeta^{j,(2,1)} & \zeta^{j,(2,2)} & \ldots & \zeta^{j,(2,M^q)} \\ \vdots & \vdots & \vdots & \vdots \\ \zeta^{j,(m,1)} & \zeta^{j,(m,2)} & \ldots & \zeta^{j,(m,M^q)} \end{bmatrix}.$$

  Here, for example,

$$\zeta^{j,(r,s)} = K^{j,(r,s)} |\sigma^{j,(r,s)}|^{-\frac{1}{2}}.$$

- Using each matrix $\{\mathscr{C}^1, \mathscr{C}^2, \ldots, \mathscr{C}^m\}$, compute the optimal index sets

$$\begin{aligned} \mathscr{I}^1 &= \{(r^*_{1,1}, s^*_{1,1}), \ldots, (r^*_{1,M^q}, s^*_{1,M^q})\} \\ \mathscr{I}^2 &= \{(r^*_{2,1}, s^*_{2,1}), \ldots, (r^*_{2,M^q}, s^*_{2,M^q})\} \\ &\vdots \\ \mathscr{I}^m &= \{(r^*_{m,1}, s^*_{m,1}), \ldots, (r^*_{m,M^q}, s^*_{m,M^q})\}, \end{aligned}$$

  where, for example, the index set $\mathscr{I}^j$ is computed via the successive maximisations:

$$\begin{aligned} \zeta^{j,(r^*_{j,1}, s^*_{j,1})} &\triangleq \max_{(r,s) \in \Gamma} \{\mathscr{C}^j\}, \\ \zeta^{j,(r^*_{j,2}, s^*_{j,2})} &\triangleq \max_{(r,s) \in \Gamma \backslash (r^*_{j,1}, s^*_{j,1})} \{\mathscr{C}^j\}, \\ &\vdots \\ \zeta^{j,(r^*_{j,M^q}, s^*_{j,M^q})} &\triangleq \max_{(r,s) \in \Gamma \backslash \{(r^*_{j,1}, s^*_{j,1}), \ldots\}} \{\mathscr{C}^j\}. \end{aligned}$$

Step 3, Updating:

- The two estimates $\widehat{x}_{k|k} = E[x_k \mid \mathscr{Y}_k]$ and $\Sigma_{k|k} = E[(x_k - \widehat{x}_{k|k})(x_k - \widehat{x}_{k|k})' \mid \mathscr{Y}_k]$, are computed, respectively, by the formulae:

$$\begin{aligned} \widehat{x}_{k|k} &= \frac{\sum_{j=1}^m \sum_{\ell=1}^{M^q} \frac{K^{j,(r^*_{j,\ell}, s^*_{j,\ell})}}{|\sigma^{j,r^*_{j,\ell}, s^*_{j,\ell}}|^{\frac{1}{2}}} \in \left(\sigma^{j,r^*_{j,\ell}, s^*_{j,\ell}}\right)^{-1} \delta^{j,(r^*_{j,\ell}, s^*_{j,\ell})}}{\sum_{j=1}^m \sum_{\ell=1}^{M^q} K^{j,(r^*_{j,\ell}, s^*_{j,\ell})} |\sigma^{j,(r^*_{j,\ell}, s^*_{j,\ell})}|^{-\frac{1}{2}}} \\ &= \left\{ \frac{1}{\sum_{j=1}^m \sum_{\ell=1}^{M^q} K^{j,(r^*_{j,\ell}, s^*_{j,\ell})} |\sigma^{j,(r^*_{j,\ell}, s^*_{j,\ell})}|^{-\frac{1}{2}}} \right\} \end{aligned}$$

$$\times \sum_{j=1}^{m} \sum_{\ell=1}^{M^q} \left\{ \frac{K^{j,(r^*_{j,\ell},s^*_{j,\ell})}}{|\sigma^{j,r^*_{j,\ell},s^*_{j,\ell}}|^{\frac{1}{2}}} \left[A_{r^*_{j,\ell}} \alpha^{r^*_{j,\ell},s^*_{j,\ell}}_{k-1|k-1} + \overline{\Sigma}^{j,r^*_{j,\ell},s^*_{j,\ell}}_{k-1|k-1} C'_{r^*_{j,\ell}} \right.\right.$$
$$\times \left(D_{r^*_{j,\ell}} D'_{r^*_{j,\ell}} + C_{r^*_{j,\ell}} \overline{\Sigma}^{j,(r^*_{j,\ell},s^*_{j,\ell})}_{k-1|k-1} C^{-1}_{r^*_{j,\ell}} \right)^{-1}$$
$$\left.\left. \times \left(y_k - C_{r^*_{j,\ell}} A_{r^*_{j,\ell}} \alpha^{r^*_{j,\ell},s^*_{j,\ell}}_{k-1|k-1} \right) \right] \right\} \tag{6.31}$$

and

$$\widehat{\Sigma}_{k|k} = \left\{ \frac{1}{\sum_{j=1}^{m} \sum_{\ell=1}^{M^q} K^{j,(r^*_{j,\ell},s^*_{j,\ell})} |\sigma^{j,(r^*_{j,\ell},s^*_{j,\ell})}|^{-\frac{1}{2}}} \right\}$$
$$\times \sum_{j=1}^{m} \sum_{\ell=1}^{M^q} \left\{ \frac{K^{j,(r^*_{j,\ell},s^*_{j,\ell})}}{|\sigma^{j,(r^*_{j,\ell},s^*_{j,\ell})}|^{\frac{1}{2}}} \left[\in \left(\sigma^{j,(r^*_{j,\ell},s^*_{j,\ell})} \right)^{-1} \right.\right.$$
$$+ \left(\in \left(\sigma^{j,(r^*_{j,\ell},s^*_{j,\ell})} \right)^{-1} \delta^{j,(r^*_{j,\ell},s^*_{j,\ell})} - \widehat{x}_{k|k} \right)$$
$$\left.\left. \times \left(\left(\sigma^{j,(r^*_{j,\ell},s^*_{j,\ell})} \right)^{-1} \delta^{j,(r^*_{j,\ell},s^*_{j,\ell})} - \widehat{x}_{k|k} \right)' \right] \right\}. \tag{6.32}$$

- Update the matrices $\mathscr{A}_k$ and $\mathscr{B}_k$.

$$\mathscr{A}_k = \begin{bmatrix} \alpha^{1,1}_{k|k} & \alpha^{1,2}_{k|k} & \dots & \alpha^{1,M^q}_{k|k} \\ \alpha^{2,1}_{k|k} & \alpha^{2,2}_{k|k} & \dots & \alpha^{2,M^q}_{k|k} \\ \vdots & \vdots & \vdots & \vdots \\ \alpha^{m,1}_{k|k} & \alpha^{m,2}_{k|k} & \dots & \alpha^{m,M^q}_{k|k} \end{bmatrix}$$
$$\triangleq \left[\in \left(\sigma^{\gamma,r^*_{\gamma,\eta},s^*_{\gamma,\eta}} \right)^{-1} \delta^{\gamma,(r^*_{\gamma,\eta},s^*_{\gamma,\eta})} \right]_{\substack{1 \le \gamma \le m \\ 1 \le \eta \le M^q}}$$

and

$$\mathscr{B}_k = \begin{bmatrix} \Sigma^{1,1}_{k|k} & \Sigma^{1,2}_{k|k} & \dots & \Sigma^{1,M^q}_{k|k} \\ \Sigma^{2,1}_{k|k} & \Sigma^{2,2}_{k|k} & \dots & \Sigma^{2,M^q}_{k|k} \\ \vdots & \vdots & \vdots & \vdots \\ \Sigma^{m,1}_{k|k} & \Sigma^{m,2}_{k|k} & \dots & \Sigma^{m,M^q}_{k|k} \end{bmatrix}$$
$$\triangleq \left[\in \left(\sigma^{\gamma,r^*_{\gamma,\eta},s^*_{\gamma,\eta}} \right)^{-1} \right]_{\substack{1 \le \gamma \le m \\ 1 \le \eta \le M^q}}.$$

- The un-normalised mode probability corresponding to the expectation $E\big[Z_k = e_j \mid \mathscr{Y}_k\big]$, is approximated by

$$q_k^j \triangleq \frac{1}{(2\pi)^{d/2}\Phi(y_k)} \sum_{\ell=1}^{M^q} \zeta^{j,(r^*_{j,\ell},s^*_{j,\ell})}. \tag{6.33}$$

Normalising the function $q_k^j(x)$, we have

$$P(Z_k = e_j \mid \mathscr{Y}_k) = \frac{\langle \underline{q}_k, e_j\rangle}{\langle \underline{q}_k, \underline{1}\rangle}.$$

- For each $j \in \{1,2,\dots,m\}$ update the normalised weights $\{W_k^{(j,1)}, W_k^{(j,2)}, \dots, W_k^{(j,M^q)}\}$, with the equation

$$W_k^{(j,\ell)} \triangleq \frac{\zeta^{j,(r^*_{j,\ell},s^*_{j,\ell})}}{\sum_{j=1}^m \sum_{\ell=1}^{M^q} \zeta^{j,(r^*_{j,\ell},s^*_{j,\ell})}}.$$

Return to Step 1

Remark 6.18 It is worth noting that the recursions given at (6.31) and (6.30) bear a striking resemblance to the form of the well know Kalman filter. This is not surprising. Suppose one knew the state of the Markov chain Z at each time k, consequently one would equivalently know the parameters of the dynamical system at this time. In such a scenario the Kalman filter in fact the optimal filter.

■

Remark 6.19 The alogrithm presented in this article can be formulated with no hypothesis management, at the cost of accuracy, by setting $M^q = 1$.

■

Nonlinear Systems

In many practical examples, inherently nonlinear stochastic hybrid systems can be successfully linearised and thereafter techniques such as those presented in this article may be routinely applied. Typical scenarios, are those with a hybrid collection of state dynamics, each observed through a common nonlinear mapping, or, alternatively, a hybrid collection of models, some linear, some nonlinear. One well known example of hybrid state dynamics observed through a nonlinear mapping is the bearings only tracking problem. In this case the hidden state process is measured through an inverse tan function and then corrupted by additive noise. The standard approach to such a problem is to first linearise the dynamics. However, to apply such techniques, a one-step-ahead prediction of the hidden state process is required. If the hybrid system being considered has either deterministic or constant parameters, this calculation is routine. By contrast, the same calculation for a stochastic hybrid system requires the joint prediction the future state and the future model. Extending the state estimator given at (6.31), we give the one-step-ahead predictor in the next Lemma.

Lemma 6.20 *Write*

$$\widehat{x}_{k|k}(j) \triangleq E\big[\langle Z_k, e_j\rangle x_k \mid \mathcal{Y}_k\big]. \tag{6.34}$$

The quantity defined at (6.34), *is a conditional mean estimate of x_k computed on those ω-sets which realise the event $Z_k = e_j$. For the hybrid stochastic system defined by the dynamics at* (6.3) *and* (6.4), *the one-step prediction is computed by the following equation*

$$\widehat{x}_{k+1|k} \triangleq E[x_{k+1} \mid \mathcal{Y}_k] = \sum_{i=1}^{m}\sum_{j=1}^{m} \pi_{(j,i)} A_i \widehat{x}_{k|k}(j).$$

Proof

$$\begin{aligned}
\widehat{x}_{k+1|k} &= \sum_{i=1}^{m} E\big[\langle Z_{k+1}, e_j\rangle A_i x_k \mid \mathcal{Y}_k\big] \\
&\quad + \sum_{i=1}^{m} E\big[\langle Z_{k+1}, e_j\rangle B_i w_{k+1} \mid \mathcal{Y}_k\big] \\
&= \sum_{i=1}^{m} E\big[\langle \Pi Z_k, e_j\rangle A_i x_k \mid \mathcal{Y}_k\big] \\
&\quad + \sum_{i=1}^{m} E\big[\langle \Pi Z_k, e_j\rangle B_i w_{k+1} \mid \mathcal{Y}_k\big] \\
&= \sum_{i=1}^{m} E\Big[\sum_{j=1}^{m} \langle Z_k, e_j\rangle \langle \Pi Z_k, e_j\rangle A_i x_k \mid \mathcal{Y}_k\Big] \\
&= \sum_{i=1}^{m}\sum_{j=1}^{m} \pi_{(j,i)} A_i E\big[\langle Z_k, e_j\rangle x_k \mid \mathcal{Y}_k\big] \\
&= \sum_{i=1}^{m}\sum_{j=1}^{m} \pi_{(j,i)} A_i \widehat{x}_{k|k}(j)
\end{aligned}$$

□

The $j = 1, 2, \ldots, m$ quantities $\hat{x}_{k|k}(j)$ may be computed as follows:

$$\widehat{x}_{k|k}(j) = \frac{\bigoplus \Sigma_{\ell=1}^{M^q}\left\{ \dfrac{K^{j,(r^*_{j,\ell}, s^*_{j,\ell})}}{|\sigma^{j, r^*_{j,\ell}, s^*_{j,\ell}}|^{\frac{1}{2}}} \in \{\sigma^{j, r^*_{j,\ell}, s^*_{j,\ell}}\}^{-1} \delta^{j,(r^*_{j,\ell}, s^*_{j,\ell})} \right\}}{\bigoplus \Sigma_{\ell=1}^{M^q}\left\{ K_{k,k-1}^{j,(r^*_{j,\ell}, s^*_{j,\ell})} |\sigma^{j,(r^*_{j,\ell}, s^*_{j,\ell})}|^{-\frac{1}{2}} \right\}}.$$

4.7 Example

To demonstrate the performance of the algorithm described above, we consider a vector-valued state and observation process and a two-state Markov chain. Our Markov chain has a transition matrix

$$\Pi = \begin{bmatrix} 0.98 & 0.02 \\ 0.01 & 0.99 \end{bmatrix}$$

and an initial distribution $\left[0.4, 0.6\right]'$. The stochastic system model parameters $\{A, B, C, D\}$, for the two models considered, are each listed below.

$$A_1 = \begin{bmatrix} -0.8 & 0 \\ 0 & 0.2 \end{bmatrix}, \qquad A_2 = \begin{bmatrix} 0.8 & 0 \\ 0 & -0.2 \end{bmatrix},$$

$$B_1 = B_2 = \begin{bmatrix} 1 & 0 \\ 0 & 1 \end{bmatrix},$$

$$C_1 = \begin{bmatrix} 2 & 0.5 \\ 0.2 & 1 \end{bmatrix}, \qquad C_2 = \begin{bmatrix} 1 & 0.5 \\ 2 & 0.2 \end{bmatrix},$$

$$D_1 = D_2 = \begin{bmatrix} 2 & 0 \\ 0 & 2 \end{bmatrix}.$$

A single realisation of this hybrid state and observation process was generated for 100 samples. Typical realisations of the estimated mode probability and the estimated state process are given, respectively, in Figures 7.1 and 7.2. The estimation of the hidden state process was compared against the exact filter, that is, the Kalman filter supplied with the exact parameter values A_k, B_k, C_k, D_k. This comparison is somewhat artificial as, one never has knowledge of the the hidden Markov chain in real problem settings, nonetheless, this example does serve to show the encouraging performance of the Gaussian mixture estimator.

4.8 Problems and Notes

Problems

1. Show that the unnormalized conditional density of the one-step predictor $\rho_{k+1,k}(x) = \overline{E}\left[\Lambda_{k+1} I(x_{k+1} \in dx) \mid \mathscr{Y}_k\right]$ is given recursively by (4.13).
2. Of interest in applications is the linear model with a singular matrix coefficient in the noise term of the state dynamics:

$$\begin{aligned} x_{k+1} &= Ax_k + Bv^*_{k+1} \in \mathbb{R}^m, \\ y_k &= Cx_k + w_k \in \mathbb{R}^d. \end{aligned}$$

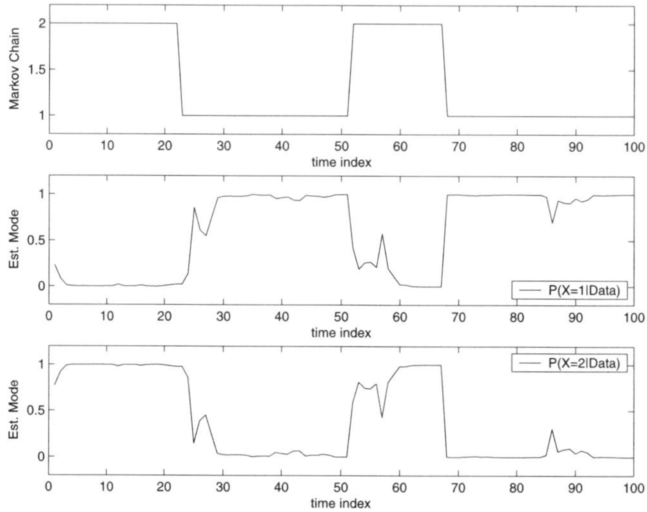

Fig. 7.1 Exact Markov chain and the estimated mode probability generated by the Gaussian mixture scheme

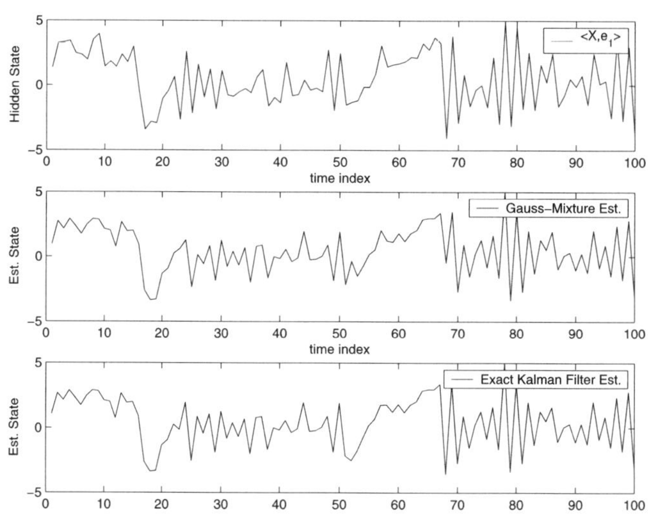

Fig. 7.2 True hidden state process and two estimates of this process. Here we plot only the first component of the state process. *Subplot 2* is the state estimate process generated from the exact Kalman filter. *Subplot 3* (the lowest subplot) is the state estimate process generated by the Gaussian mixture filter

Obtain recursive estimators (*Hint*: Set $v_k = Bv_k^*$. Even if B is *singular*, v_k always has probability distribution. The support of this distribution is the set of values of Bv_k^*. See also Section 4.4.)

3. Consider the model whose dynamics are given by (2.1), (2.2), and (2.3). Suppose now that the dynamics of the parameters θ^i are described by finite-state Markov chains
$$\theta^i_{k+1} = A^i \theta^i_k + V^i_{k+1},$$
where the V^i are martingale increments (see Chapter 2). Find recursive estimates of the conditional distribution of the state of the model.

Notes

The results of this chapter are familiar in many signal processing contexts. The ARMAX model is widely studied in the adaptive estimation and control literature as well as in time-series analysis and economics. In this ARMAX model case it is remarkable that the recursive formulae for the unnormalized densities do not involve any integration.

Chapter 5
A General Recursive Filter

5.1 Introduction

In this chapter more general models are considered. We use again the same reference measure methods. Both nonlinear with nonadditive noise and linear dynamics are considered. In Section 5.7 the results are extended to a parameter estimation problem. In this case the same noise enters the signal and observations. In Section 5.8, an abstract formulation is given in terms of *transition densities*. Finally, in Section 5.9 we discuss a correlated noise case, where the noise in the observations appears in the state dynamics as well.

5.2 Signal and Observations

All processes are defined initially on a probability space $(\Omega, \mathscr{F}, P)$.

Suppose $\{x_\ell\}$, $\ell \in \mathbb{N}$, is a discrete-time stochastic process taking values in some Euclidean space $\mathbb{R}^d$. Then $\{v_\ell\}$, $\ell \in \mathbb{N}$, will denote a sequence of independent $\mathbb{R}^n$-valued random variables. The density function of v_ℓ is ψ_ℓ.

$$a : \mathbb{R}^d \times \mathbb{R}^n \to \mathbb{R}^d$$

is a measurable function, and we suppose for $k \geq 0$ that

$$x_{k+1} = a(x_k, v_{k+1}). \tag{2.1}$$

We suppose that x_0, or its density $\pi_0(x)$, is known. Finally, we assume there is an inverse map $d : \mathbb{R}^d \times \mathbb{R}^d \to \mathbb{R}^n$ such that if $x_{k+1} = a(x_k, v_{k+1})$ then

$$v_{k+1} = d(x_{k+1}, x_k). \tag{2.2}$$

Note this condition is fulfilled if the noise is additive, so $x_{k+1} = a(x_k) + v_{k+1}$. We require d to be *differentiable* in the first variable.

The observation process $\{y_k\}$, $k \in \mathbb{N}$, for simplicity, we suppose is real valued; the extension to vector observations is immediate.

The observation noise $\{w_k\}$ will be a sequence of independent real-valued random variables. We suppose each w_k has a strictly positive density function ϕ_k. Also $c : \mathbb{R}^d \times \mathbb{R} \to \mathbb{R}$ is a measurable function and we suppose for $k \geq 0$ that

$$y_k = c(x_k, w_k). \tag{2.3}$$

Again we assume there is an inverse map $g : \mathbb{R} \times \mathbb{R}^d \to \mathbb{R}$ such that if $y_k = c(x_k, w_k)$ then

$$w_k = g(y_k, x_k). \tag{2.4}$$

Again, this condition is satisfied for additive noise, when $y_k = c(x_k) + w_k$. Finally we require the derivatives

$$C(x_k, w_k) := \left.\frac{\partial c(x_k, w)}{\partial w}\right|_{w=w_k}, \qquad G(y_k, x_k) := \left.\frac{\partial g(y, x_k)}{\partial y}\right|_{y=y_k} \tag{2.5}$$

to be nonsingular.

Remarks 2.1 The above conditions can be weakened or modified. For example, appropriate changes would allow x_k to be of the form $x_{k+1} = a(x_k, x_{k-1}, v_{k+1})$. ■

In summary, the dynamics of our model are given by:

$$\boxed{\begin{aligned} x_{k+1} &= a(x_k, v_{k+1}), \\ y_k &= c(x_k, w_k), \end{aligned}} \tag{2.6}$$

v_k and w_k are sequences of independent $\mathbb{R}^n$- and $\mathbb{R}$-valued random variables with density functions ψ_k and ϕ_k, respectively, and the functions $a(.,.)$, $c(.,.)$ satisfy the above conditions.

5.3 Change of Measure

Define

$$\Lambda_k = \prod_{\ell=1}^{k} \frac{\phi_\ell(y_\ell)}{\phi_\ell(w_\ell)} G(y_\ell, x_\ell)^{-1}.$$

Write

$$\begin{aligned} \mathscr{G}_{k+1}^0 &= \sigma\{x_0, x_1, \ldots, x_{k+1}, y_1, \ldots, y_k\}, \\ \mathscr{Y}_k^0 &= \sigma\{y_1, \ldots, y_k\}, \end{aligned}$$

and $\{\mathscr{G}_k\}$, $\{\mathscr{Y}_k\}$, $k \in \mathbb{N}$, for the corresponding complete filtrations. A new probability measure P can be defined by setting the restriction of the Radon-Nikodym derivative $(d\overline{P}/dP)|_{\mathscr{G}_{k+1}}$ equal to Λ_k. Then simple manipulations as in earlier studies, and paralleling those below, show that under $\overline{P}$ the random variables $\{y_\ell\}$, $\ell \in \mathbb{N}$, are independent with density functions ϕ_ℓ and the dynamics of x are as under P.

Starting with the probability measure $\overline{P}$ we can recover the probability measure P such that under P

$$w_k := g(y_k, x_k)$$

is a sequence of independent random variables with positive densities $\phi_k(b)$. To construct P starting from $\overline{P}$ write

$$\overline{\Lambda}_k = \prod_{\ell=1}^{k} \overline{\lambda}_\ell = \prod_{\ell=1}^{k} \frac{\phi_\ell(w_\ell)}{\phi_\ell(y_\ell)} C(x_\ell, w_\ell)^{-1}.$$

P is defined by putting the restriction of the Radon-Nikodym derivative $(dP/d\overline{P})|_{\mathscr{G}_k}$ equal to $\overline{\Lambda}_k$. The existence of P follows from Kolmogorov's extension theorem.

Lemma 3.1 *Under P the $\{w_\ell\}, \ell \in \mathbb{N}$, are independent random variables having densities ϕ_ℓ.*

Proof First

$$\overline{E}\left[\overline{\lambda}_k \mid \mathscr{G}_k\right] = \int_{-\infty}^{\infty} \frac{\phi_k(w_k)}{\phi_k(y_k)} C(x_k, w_k)^{-1} \phi_k(y_k)\, dy_k = \int_{-\infty}^{\infty} \phi_k(w_k)\, dw_k$$

(given $\mathscr{G}_k$ so that $dx_k = 0$)

$$= 1.$$

Next

$$\begin{aligned} P(w_k \le t \mid \mathscr{G}_k) &= E\left[I(w_k \le t) \mid \mathscr{G}_k\right] \\ &= \frac{\overline{E}\left[\overline{\Lambda}_k I(w_k \le t) \mid \mathscr{G}_k\right]}{\overline{E}\left[\overline{\Lambda}_k \mid \mathscr{G}_k\right]} \\ &= \overline{E}\left[\overline{\lambda}_k I(w_k \le t) \mid \mathscr{G}_k\right] \\ &= \int_{-\infty}^{\infty} I(w_k \le t) \frac{\phi_k(w_k)}{\phi_k(y_k)} \cdot C(x_k, w_k)^{-1} \phi_k(y_k)\, dy_k \\ &= \int_{-\infty}^{t} \phi_k(w_k)\, dw_k = P(w_k \le t). \end{aligned}$$

That is under P, the w_k are independent with densities ϕ_k. □

5.4 Recursive Estimates

We shall work under measure $\overline{P}$, so that $\{y_k\}$, $k \in \mathbb{N}$, is a sequence of independent real random variables with densities ϕ_k; furthermore, $x_{k+1} = a(x_k, v_{k+1})$ where the v_k are independent random variables with densities ψ_k. With $\xi = a(z, v)$, consider the inverse mapping $v = d(\xi, z)$ and derivative

$$D(\xi_k, z_k) = \left.\frac{\partial d(\xi, z_k)}{\partial \xi}\right|_{\xi=\xi_k}. \tag{4.1}$$

Notation 4.1 *Write $q_k(z)$, $k \in \mathbb{N}$, for the unnormalized conditional density such that*

$$\overline{E}\left[\overline{\Lambda}_k I(x_k \in dz) \mid \mathscr{Y}_k\right] = q_k(z)\, dz.$$

The existence of q_k will be discussed below. We denote $|D| = |\det D|$.

Theorem 4.2 *For $k \in \mathbb{N}$ we have the recursive estimates:*

$$\boxed{q_{k+1}(\xi) = \Delta_1(y_{k+1}, \xi) \int_{\mathbb{R}^d} \psi_{k+1}(d(\xi, z)) \mid D(\xi, z) \mid q_k(z)\, dz,} \tag{4.2}$$

where $\Delta_1(y_{k+1}, \xi) = [\phi_{k+1}(g(y_{k+1}, \xi)) / \phi_{k+1}(y_{k+1})] C(\xi, g(y_{k+1}, \xi))^{-1}$, *and* $d(\xi, z) = v$.

Proof Suppose $f : \mathbb{R}^d \to \mathbb{R}$ is any Borel test function. Then

$$\begin{aligned}
&\overline{E}\left[f(x_{k+1}) \overline{\Lambda}_{k+1} \mid \mathscr{Y}_{k+1}\right] \\
&\quad = \int_{\mathbb{R}^d} f(z) q_{k+1}(z)\, dz \\
&\quad = \overline{E}\Big[f(a(x_k, v_{k+1})) \overline{\Lambda}_k \phi_{k+1}(g(y_{k+1}, x_{k+1})) \\
&\qquad\qquad \times C(x_{k+1}, g(y_{k+1}, x_{k+1}))^{-1} \mid \mathscr{Y}_{k+1}\Big] \phi_{k+1}(y_{k+1})^{-1} \\
&\quad = \overline{E}\Big\{\int_{\mathbb{R}^n} \Big[f(a(x_k, v)) \overline{\Lambda}_k \phi_{k+1}(g(y_{k+1}, a(x_k, v))) \\
&\qquad\qquad \times C(a(x_k, v), g(y_{k+1}, a(x_k, v)))^{-1} \\
&\qquad\qquad \times \psi_{k+1}(v)\Big] dv \mid \mathscr{Y}_{k+1}\Big\} \phi_{k+1}(y_{k+1})^{-1} \\
&\quad = \int_{\mathbb{R}^d} \int_{\mathbb{R}^n} \Big[f(a(z, v)) \phi_{k+1}(g(y_{k+1}, a(z, v))) \\
&\qquad\qquad \times C(a(z, v), g(y_{k+1}, a(z, v)))^{-1} \\
&\qquad\qquad \times \psi_{k+1}(v) q_k(z)\Big] dv\, dz\, \phi_{k+1}(y_{k+1})^{-1}.
\end{aligned}$$

Now

$$dv\,dz = \left| \det \begin{vmatrix} \frac{\partial v}{\partial \xi} & \frac{\partial v}{\partial z} \\ \frac{\partial z}{\partial \xi} & \frac{\partial z}{\partial z} \end{vmatrix} \right| d\xi\,dz = |D(\xi,z)|\,d\xi\,dz,$$

Consequently,

$$\begin{aligned}&\int_{\mathbb{R}^d} f(\xi)\,q_{k+1}(\xi)\,d\xi \\ &\quad = \int_{\mathbb{R}^d}\int_{\mathbb{R}^d}\Big[f(\xi)\,\phi_{k+1}(g(y_{k+1},\xi))\,C(\xi,g(y_{k+1},\xi))^{-1} \\ &\qquad\qquad \times \psi_{k+1}(d(\xi,z))\,q_k(z)\,|D(\xi,z)|\Big]\,d\xi\,dz\,\phi_{k+1}(y_{k+1})^{-1}.\end{aligned}$$

This identity holds for all Borel test functions f, so the result follows. □

Remarks 4.3 Suppose $\pi(z)$ is the density of x_0, so for any Borel set $A \subset \mathbb{R}^d$

$$P(x_o \in A) = \overline{P}(x_0 \in A) = \int_A \pi(z)\,dz.$$

Then $q_0(z) = \pi(z)$ and updated estimates are obtained by substituting in (4.2). Even if the prior estimates of x_0 are delta functions the proof of Theorem 4.2 gives a function for $q_1(z)$. For example, suppose $\pi(z) = \delta(x_0)$, the unit mass at x_0. Then the argument of Theorem 4.2 gives

$$q_1(\xi) = \phi_1(y_1)^{-1}\,\phi_1(g(y_1,\xi))\,C(\xi,g(y_1,\xi))^{-1}\,\psi_1(d(\xi,x_0))\,|D(\xi,x_0)|.$$

Subsequent recursions follow from (4.2).

■

A Second Information Pattern

Suppose as in Section 5.2 that $x_{k+1} \in \mathbb{R}^d$ and

$$x_{k+1} = a(x_k, v_{k+1}), \tag{4.3}$$

where the v_k are independent random variables with density functions ψ_k. We now suppose the scalar observation process y_k depends on the value of x at the previous time, that is,

$$y_{k+1} = c(x_k, w_{k+1}). \tag{4.4}$$

Again, the w_k are independent random variables with positive density functions ϕ_k. If (4.3) holds we assume $v_{k+1} = d(x_{k+1}, x_k)$; if (4.4) holds we assume $w_{k+1} = g(y_{k+1}, x_k)$. Write $\mathcal{G}_k^o = \sigma\{x_0, x_1, \dots, x_k, y_1, \dots, y_k\}$ and $\mathcal{Y}_k^o = \sigma\{y_1, \dots, y_k\}$ and

$\{\mathcal{G}_k\}$ and $\{\mathcal{Y}_k\}$ for the corresponding complete filtrations. Again, $q_k(z)$ will be the unnormalized density defined by

$$\overline{E}\left[\overline{\Lambda}_k I(x_k \in dz) \mid \mathcal{Y}_k\right] = q_k(z)\,dz.$$

The same method then gives the following result:

Theorem 4.4 *With the dynamics and information pattern given by* (4.3) *and* (4.4)

$$\boxed{q_{k+1}(\xi) = \int_{\mathbb{R}^d} \Delta_2(y_{k+1},\xi,z)\,\psi_{k+1}(d(\xi,z))\,q_k(z)\,dz,} \tag{4.5}$$

where

$$\Delta_2(y_{k+1},\xi,z) = \frac{\phi_{k+1}(g(y_{k+1},z))}{\phi_{k+1}(y_{k+1})} C(z,g(y_{k+1},z))^{-1}\,|D(\xi,z)|.$$

This again gives the recursive update of the unnormalized density. Suppose the dynamics are linear, so that

$$\begin{aligned} x_{k+1} &= Ax_k + v_{k+1}, \\ y_{k+1} &= \langle c, x_k\rangle + w_{k+1}. \end{aligned}$$

Here c is an $\mathbb{R}^d$-valued row vector. In this case (4.5) specializes to give the following:

Corollary 5.4.1

$$q_{k+1}(\xi) = \phi_{k+1}(y_{k+1})^{-1} \int_{\mathbb{R}^d} \phi_{k+1}(y_{k+1} - \langle c,z\rangle)\,\psi_{k+1}(\xi - Az)\,q_k(z)\,dz.$$

5.5 Extended Kalman Filter

The Kalman filter can be recovered for linear models as in Chapter 4, Section 5.7. Indeed, the Kalman filter can reasonably be applied to linearizations of certain non-linear models. These well-known results are included for completeness; see also Anderson and Moore (1979).

Consider the restricted class of nonlinear models, in obvious notation

$$x_{k+1} = a(x_k) + v_{k+1}, \qquad v_k \sim N[0,Q_k]. \tag{5.1}$$
$$y_k = c(x_k) + w_k, \qquad w_k \sim N[0,R_k]. \tag{5.2}$$

Let us here continue to use the notation $\hat{x}_{k|k}$ to denote some estimate of x_k given measurements up until time k, even though this may not be a conditional mean estimate. Let us also denote

$$A_{k+1} = \left.\frac{\partial a(x)}{\partial x}\right|_{x=\hat{x}_{k|k}}, \qquad C_k = \left.\frac{\partial c(x)}{\partial x}\right|_{x=\hat{x}_{k|k}}$$

assuming the derivatives exist. If $a(x), c(x)$ are sufficiently smooth, and $\hat{x}_{k|k}$ is close to x_k, the following linearized model could be a reasonable approximation to (5.1) and (5.2)

$$\begin{aligned} x_{k+1} &= A_k x_k + v_{k+1} + \bar{u}_k, \\ y_k &= C_k x_k + w_k + \bar{y}_k, \end{aligned}$$

where $\bar{u}_k = a\left(\hat{x}_{k|k}\right) - A_k \hat{x}_{k|k}, \bar{y}_k = c\left(\hat{x}_{k|k-1}\right) - C_k \hat{x}_{k|k-1}$ are known if $\hat{x}_{k|k}, \hat{x}_{k|k-1}$ are known. The Kalman filter for this approximate model is a slight variation of that of Chapter 4, Section 5.5, and, suitably initialized, is

$$\boxed{\begin{aligned} \hat{x}_{k|k} &= \hat{x}_{k|k-1} + K_k\left(y_k - c\left(\hat{x}_{k|k-1}\right)\right), \\ \hat{x}_{k+1|k} &= a\left(\hat{x}_{k|k}\right), \\ K_k &= \Sigma_{k|k-1} C_k'\left(C_k \Sigma_{k|k-1} C_k' + R_k\right)^{-1}, \\ \Sigma_{k|k} &= \Sigma_{k|k-1} - \Sigma_{k|k-1} C_k'\left(C_k \Sigma_{k|k-1} C_k' + R_k\right)^{-1} C_k \Sigma_{k|k-1}, \\ \Sigma_{k+1|k} &= A_{k+1} \Sigma_{k|k} A_{k+1}' + Q_{k+1}. \end{aligned}} \tag{5.3}$$

These are as in Chapter 4. We stress that this is not an optimal filter unless the linearization is exact. Indeed, the extended Kalman filter (EKF) can be far from optimal, particularly in high noise environments, for poor initial conditions or, indeed, when the nonlinearities are not suitably smooth. Even so, it is widely used in applications (see also Chapter 6).

5.6 Parameter Identification and Tracking

Again suppose the state of a system is described by a discrete-time stochastic process $\{x_k\}$, $k \in \mathbb{N}$, taking values in $\mathbb{R}^d$. The noise sequence $\{v_k\}$, $k \in \mathbb{N}$, is a family of independent $\mathbb{R}^n$-valued random variables. The density function of v_k is ψ_k, and we suppose $\psi_k(v) > 0$, $\forall v \in \mathbb{R}^n$. Further, we now suppose there is an unknown parameter θ which takes a value in a measure space (Ω, β, λ). For $k \geq 0$ we suppose

$$x_{k+1} = a\left(x_k, \theta, v_{k+1}\right). \tag{6.1}$$

It is assumed that x_0 or its distribution, is known. The observation process is again of the form

$$y_{k+1} = c\left(x_k, v_{k+1}\right). \tag{6.2}$$

Note the same noise v appears in both signal and observation. We suppose the observation process y is the same dimension n as the noise process v, and there is an inverse map $g : \mathbb{R}^n \times \mathbb{R}^d \to \mathbb{R}^n$ such that if $y_{k+1} = c\left(x_k, v_{k+1}\right)$ then

$$v_{k+1} = g\left(y_{k+1}, x_k\right). \tag{6.3}$$

Again, this is the case if the noise is additive.

Substituting (6.3) in (6.1) we see

$$x_{k+1} = a\left(x_k, \theta, g\left(y_{k+1}, x_k\right)\right).$$

We suppose there is an inverse map d such that $\xi = a\left(z, \theta, g\left(y, z\right)\right)$ with

$$D\left(\xi_k, \theta_k, y_k\right) := \left. \frac{\partial d\left(\xi, \theta_k, y_k\right)}{\partial \xi} \right|_{\xi = \xi_k} \tag{6.4}$$

then

$$z = d\left(\xi, \theta, y\right). \tag{6.5}$$

Writing

$$\Lambda_k = \prod_{\ell=1}^{k} \frac{\psi_\ell\left(y_\ell\right)}{\psi_\ell\left(v_\ell\right)} G\left(y_\ell, x_{\ell-1}\right)^{-1}$$

a new probability measure $\overline{P}$ can be defined by putting $\left(d\overline{P}/dP\right)|_{\mathscr{G}_k} = \Lambda_k$.

Calculations as before show that, for any Borel set $A \subset \mathbb{R}^n$,

$$\overline{P}\left(y_k \in A \mid \mathscr{G}_k\right) = \int_A \psi_k\left(y\right) dy,$$

so that the y_k are independent under $\overline{P}$ with densities ψ_k. Therefore, we start with a probability space $\left(\Omega, \mathscr{F}, \overline{P}\right)$, such that under $\overline{P}$

$$x_{k+1} = a\left(x_k, \theta, v_{k+1}\right)$$

and the $\{y_k\}$, $k \in \mathbb{N}$, are a sequence of independent random variables with strictly positive density functions ψ_ℓ.

Writing

$$\overline{\Lambda}_k = \prod_{\ell=1}^{k} \frac{\psi_\ell\left(g\left(y_\ell, x_{\ell-1}\right)\right)}{\psi_\ell\left(y_\ell\right)} C\left(x_{\ell-1}, v_\ell\right)^{-1},$$

a new probability measure $\overline{P}$ can be defined by putting $\left(dP/d\overline{P}\right)|_{\mathscr{G}_k} = \overline{\Lambda}_k$. As before, under P the random variables $\{v_k\}$, $k \in \mathbb{N}$, are independent with strictly positive densities ψ_ℓ.

Write $q_k\left(\xi, \theta\right)$ for the unnormalized conditional density such that

$$q_k\left(z, \theta\right) dz\, d\theta = \overline{E}\left[I\left(x_k \in dz\right) I\left(\theta \in d\theta\right) \overline{\Lambda}_k \mid \mathscr{Y}_k\right].$$

Let $f : \mathbb{R}^d \to \mathbb{R}$ and $h : \Theta \to \mathbb{R}$ be suitable test functions. Then

$$\begin{aligned}
&\overline{E}\left[f\left(x_{k+1}\right) h\left(\theta\right) \overline{\Lambda}_{k+1} \mid \mathscr{Y}_{k+1}\right] \\
&\quad = \iint f\left(\xi\right) h\left(u\right) q_{k+1}\left(\xi, u\right) d\xi\, d\lambda\left(u\right)
\end{aligned}$$

$$= \overline{E}\left[f(a(x_k,\theta,g(y_{k+1},x_k)))\,h(\theta)\,\overline{\Lambda}_k \frac{\psi_{k+1}(g(y_{k+1},x_k))}{\psi_{k+1}(y_{k+1})} \right.$$
$$\left. \times C(x_k,g(y_{k+1},x_k))^{-1} \;\middle|\; \mathscr{Y}_{k+1} \right]$$
$$= \psi_{k+1}(y_{k+1})^{-1} \iint \Big[f(a(z,u,g(y_{k+1},z)))\,h(u)\,\psi_{k+1}(g(y_{k+1},z))$$
$$\times C(z,g(y_{k+1},z))^{-1} q_k(z,u) \Big] d\xi\, d\lambda(u).$$

Writing $\xi = a(z,u,g(y_{k+1},z))$, $u = u$, the above is

$$= \psi_{k+1}(y_{k+1})^{-1} \iint \Big[f(\xi)\,h(u)\,\psi_{k+1}(g(y_{k+1},d(\xi,u,y_{k+1})))$$
$$\times C(d(\xi,u,y_{k+1}),g(y_{k+1},d(\xi,u,y_{k+1})))^{-1}$$
$$\times q_k(d(\xi,u,y_{k+1}),u)\,|D(\xi,u,y_{k+1})| \Big] d\xi\, d\lambda(u).$$

This identity is true for all test functions f and h, so we deduce the following result:

Theorem 6.1 *The following algebraic recursion updates the estimates* q_{k+1}

$$\boxed{\begin{aligned} q_{k+1}(\xi,u) &= \Delta_3(\xi,u,y_{k+1})\,\psi_{k+1}(g(y_{k+1},d(\xi,u,y_{k+1}))) \\ &\quad \times (\psi_{k+1}(y_{k+1}))^{-1} q_k(d(\xi,u,y_{k+1}),u), \end{aligned}} \tag{6.6}$$

where,

$$\Delta_3(\xi,u,y_{k+1}) = C(d(\xi,u,y_{k+1}),g(y_{k+1},d(\xi,u,y_{k+1})))^{-1}\,|D(\xi,u,y_{k+1})|.$$

Remark 6.2 It is of interest that (6.6) does not involve any integration.

■

If $x_{k+1} = A(\theta)x_k + v_{k+1}$ and $y_{k+1} = cx_k + v_{k+1}$ for suitable matrices $A(\theta), c$ then $v_{k+1} = g(y_{k+1},x_k) = y_{k+1} - c_k$ so $x_{k+1} = (A(\theta)-c)x_k + y_{k+1}$.

In this case x_{k+1}, y_{k+1} and v_{k+1} have the same dimension. Therefore, the inverse function d of (6.5) exists if $(A(\theta)-c)$ is nonsingular, and then

$$x_k = (A(\theta)-c)^{-1}(x_{k+1} - y_{k+1}) = d(x_{k+1},\theta,y_{k+1}).$$

In the case of parameter tracking the parameter θ varies with time. Suppose θ_{k+1} takes values in $\mathbb{R}^p$ and

$$\theta_{k+1} = A_\theta \theta_k + \nu_{k+1} \tag{6.7}$$

for some $p \times p$ matrix A_θ. Here $\{\nu_\ell\}$ is a sequence of independent random variables; ν_ℓ has density ϕ_ℓ. The dynamics, together with (6.7), are as before

$$x_{k+1} = a(x_k, \theta_k, v_{k+1}), \tag{6.8}$$
$$y_{k+1} = c(x_k, v_{k+1}). \tag{6.9}$$

If (6.9) holds then, as before, we suppose

$$v_{k+1} = g(y_{k+1}, x_k). \tag{6.10}$$

Consequently,

$$x_{k+1} = a(x_k, \theta_k, g(y_{k+1}, x_k)). \tag{6.11}$$

Arguing as above we are led to consider

$$\begin{aligned}
&\overline{E}\left[f(x_{k+1})h(\theta_{k+1})\overline{\Lambda}_{k+1} \mid \mathscr{Y}_{k+1}\right] \\
&\quad = \iint f(\xi)h(u)q_{k+1}(\xi,u)\,d\xi\,d\lambda(u) \\
&\quad = \overline{E}\Bigg[f(a(x_k,\theta_k,g(y_{k+1},x_k)))h(A_\theta\theta_k + v_{k+1})\overline{\Lambda}_k \frac{\psi_{k+1}(g(y_{k+1},x_k))}{\psi_{k+1}(y_{k+1})} \\
&\qquad\quad \times C(x_k, g(y_{k+1},x_k))^{-1} \,\Big|\, \mathscr{Y}_{k+1}\Bigg] \\
&\quad = \psi_{k+1}(y_{k+1})^{-1} \\
&\qquad \times \iint \Big[f(a(z,\theta,g(y,z)))h(A_\theta\theta + v)\phi(v)\psi_{k+1}(g(y_{k+1},z)) \\
&\qquad\qquad \times C(z, g(y_{k+1},z))^{-1} q_k(z,\theta)\Big]\,dz\,d\theta\,dv.
\end{aligned}$$

This time substitute

$$\begin{aligned}
\xi &= a(z,\theta,g(y,z)), \\
u &= A_\theta\theta + v, \\
\theta &= \theta.
\end{aligned}$$

The above conditional expectation then is equal to:

$$\begin{aligned}
&\psi_{k+1}(y_{k+1})^{-1} \\
&\quad \times \iint \Big[f(\xi)h(u)\phi(u - A_\theta\theta)\psi_{k+1}(g(y_{k+1}, d(\xi,u,y_{k+1}))) \\
&\qquad\quad \times q_k(d(\xi,u,y_{k+1}),\theta)\,C(d(\xi,u,y_{k+1}), g(y_{k+1}, d(\xi,u,y_{k+1})))^{-1} \\
&\qquad\quad \times |D(\xi,u,y_{k+1})|\Big]\,d\theta\,d\xi\,d\lambda(u).
\end{aligned}$$

Again, the above identity is true for all test functions f and h and gives the following results.

Theorem 6.3 *The following recursion updates the unconditional density $q_k(\xi,u)$ of the state x_k and parameter θ_k:*

$$\boxed{q_{k+1}(\xi,u) = \Delta_4(\xi,u,y_{k+1}) \int \phi(u - A_\theta \theta)\, q_k(d(\xi,u,y_{k+1}),\theta)\, d\theta,} \tag{6.12}$$

where,

$$\begin{aligned}\Delta_4(\xi,u,y_{k+1}) &\\ = \psi_{k+1}(y_{k+1})^{-1}\, &\psi_{k+1}(g(y_{k+1},d(\xi,u,y_{k+1})))\\ &\times C(d(\xi,u,y_{k+1}),g(y_{k+1},d(\xi,u,y_{k+1})))^{-1}\,|D(\xi,u,y_{k+1})|.\end{aligned}$$

5.7 Formulation in Terms of Transition Densities

In this section we give a formulation of the above results when the dynamics of the state and observation processes are described by *transition densities*.

Again, all processes are supposed defined initially on a complete probability space $(\Omega,\mathscr{F},P)$.

Consider a signal process $\{x_k\}$, $k \in \mathbb{N}$, which takes values in $\mathbb{R}^d$. We suppose x is a Markov process with transition densities $p_k(x,z)$. That is,

$$P(x_{k+1} \in dx \mid x_k = z) = p_{k+1}(x,z)\, dx$$

and

$$E[f(x_{k+1}) \mid x_k = z] = \int_{\mathbb{R}^d} f(x)\, p_{k+1}(x,z)\, dx.$$

We suppose x_0, or its distribution, is known. The process x is observed through a $\mathbb{R}^n$-valued process y whose transitions are a function of x. That is, we suppose there is a strictly positive density $\rho_k(y,x)$ such that

$$P(y_{k+1} \in dy \mid x_k = x) = \rho_{k+1}(y,x)\, dy. \tag{7.1}$$

Note, we suppose that under P the eevent $[y_{k+1} \in dy]$ is independent of values of x at all times other than k and all y except y_{k+1}.

Again

$$E[f(y_{k+1}) \mid x_k = x] = \int_{\mathbb{R}^n} f(y)\, \rho_{k+1}(y,x)\, dy.$$

Suppose $\rho_k(y)$ is the unconditional probability density of y_k, and that $\rho_k(y) > 0$ for all $y \in \mathbb{R}^n$. That is,

$$E[f(y_k)] = \int_{\mathbb{R}^n} f(y)\, \rho_k(y)\, dy.$$

Again, $\mathscr{G}_k^0 = \sigma\{x_0, x_1, \ldots, x_k, y_1, \ldots, y_k\}$ and $\mathscr{Y}_k^0 = \sigma\{y_1, \ldots, y_k\}$; $\{\mathscr{G}_k\}$ and $\{\mathscr{Y}_k\}$ are, respectively, the complete filtrations generated by $\mathscr{G}_k^0$ and $\mathscr{Y}_k^0$.

Write

$$\Lambda_k = \prod_{\ell=1}^{k} \frac{\rho_\ell(y_\ell)}{\rho_\ell(y_\ell, x_{\ell-1})}$$

and define a new probability measure by putting $\left(d\overline{P}/dP\right)|_{\mathscr{G}_k} = \Lambda_k$. Then under $\overline{P}$ the y_k are independent random variables with density $\rho_k(y) > 0$. Suppose, therefore, we start with a probability space $\left(\Omega, \mathscr{F}, \overline{P}\right)$ such that, under $\overline{P}$, $\{x_k\}$ is a Markov process with transition densities $p_k(x,z)$ and the y_k are independent random variables with positive densities $\rho_k(y)$. Now write

$$\overline{\Lambda}_k = \prod_{\ell=1}^{k} \frac{\rho_\ell(y_\ell, x_{\ell-1})}{\rho_\ell(y_\ell)}.$$

Define a probability measure P by putting $\left(dP/d\overline{P}\right)|_{\mathscr{G}_k} = \overline{\Lambda}_k$. Consider any Borel function $f : \mathbb{R}^d \to \mathbb{R}$ with compact support and write $q_k(\xi)$ for the unnormalized conditional density such that

$$\overline{E}\left[I(x_k \in d\xi)\overline{\Lambda}_k \mid \mathscr{Y}_k\right] = q_k(\xi)\,d\xi;$$

then

$$\begin{aligned}
&\overline{E}\left[f(x_{k+1})\overline{\Lambda}_{k+1} \mid \mathscr{Y}_{k+1}\right] \\
&\quad = \int_{\mathbb{R}^d} f(\xi)\, q_{k+1}(\xi)\, d\xi \\
&\quad = \overline{E}\left[\int_{\mathbb{R}^d} f(\xi)\, p_{k+1}(\xi, x_k)\, d\xi\, \overline{\Lambda}_k \frac{\rho_{k+1}(y_{k+1}, x_k)}{\rho_{k+1}(y_{k+1})} \;\Big|\; \mathscr{Y}_{k+1}\right].
\end{aligned}$$

The y_{k+1} are independent (of x) under $\overline{P}$, so this is equal to

$$\rho_{k+1}(y_{k+1})^{-1} \int_{\mathbb{R}^d}\int_{\mathbb{R}^d} f(\xi)\, p_{k+1}(\xi, z)\, \rho_{k+1}(y_{k+1}, z)\, q_k(z)\, dz\, d\xi.$$

This identity holds for all such functions f; therefore, we have:

Theorem 7.1 *The following recursive expression updates q_k:*

$$\boxed{q_{k+1}(\xi) = \rho_{k+1}(y_{k+1})^{-1} \int_{\mathbb{R}^d} p_{k+1}(\xi, z)\, \rho_{k+1}(y_{k+1}, z)\, q_k(z)\, dz.} \tag{7.2}$$

5.8 Dependent Case

Dynamics

Suppose $\{x_k\}$, $k \in \mathbb{N}$, is a discrete-time stochastic process taking values in some Euclidean space $\mathbb{R}^m$. We suppose that x_0 has a known distribution $\pi_0(x)$. $\{v_k\}$, $k \in \mathbb{N}$, will be a sequence of independent, $\mathbb{R}^n$-valued, random variables with probability distributions $d\psi_k$ and $\{w_k\}$ a sequence of independent $\mathbb{R}^m$-valued, random variables with positive densities ϕ_k. For $k \in \mathbb{N}$, $a_{k+1} : \mathbb{R}^m \times \mathbb{R}^n \to \mathbb{R}^m$ are measurable functions, and we suppose for $k \geq 0$ that

$$x_{k+1} = a_{k+1}(x_k, v_{k+1}) + w_{k+1}. \tag{8.1}$$

The observation process $\{y_k\}$, $k \in \mathbb{N}$, takes values in some Euclidean space $\mathbb{R}^m$. For $k \in \mathbb{N}$, $c_k : \mathbb{R}^m \to \mathbb{R}^m$ are measurable functions and we suppose for $k \geq 1$ that

$$y_k = c_k(x_k) + w_k. \tag{8.2}$$

We assume that there is an inverse map $d_k : \mathbb{R}^m \times \mathbb{R}^n \to \mathbb{R}^m$ such that if (8.1) holds then

$$v_{k+1} = d_{k+1}(x_{k+1} - w_{k+1}, x_k).$$

We also assume that there is an inverse map $V_k : \mathbb{R}^m \to \mathbb{R}^m$ such that if $c_k(x_k) + x_k := W_k(x_k)$ then $V_k(W_k(x_k)) = x_k$. We now suppose we start with a probability measure $\overline{P}$ on $(\Omega, \bigvee_{n=1}^{\infty} \mathcal{G}_n)$ such that under $\overline{P}$:

1. $\{y_k\}$, $k \in \mathbb{N}$, is a sequence of independent random variables having density functions $\phi_k > 0$.
2. $x_{k+1} = a_{k+1}(x_k, v_{k+1}) + w_{k+1} = a_{k+1}(x_k, v_{k+1}) + y_{k+1} - c_{k+1}(x_{k+1})$ or, using our assumptions,

$$x_{k+1} = V_{k+1}(a_{k+1}(x_k, v_{k+1}) + y_{k+1}).$$

Working under the probability measure $\overline{P}$, and denoting $d\alpha_k(x)$, $k \in \mathbb{N}$, for the unnormalized conditional probability measure such that

$$\overline{E}\left[\overline{\Lambda}_k I(x_k \in dx) \mid \mathcal{Y}_k\right] := d\alpha_k(x),$$

Write

$$\overline{\Lambda}_k = \prod_{\ell=1}^{k} \frac{\phi_\ell(y_\ell - c_\ell(x_\ell))}{\phi_\ell(y_\ell)}.$$

With $D(.,\xi) := \frac{\partial d(.,\xi)}{\partial \xi}$, we have

Theorem 8.1 *For $k \in \mathbb{N}$, a recursion for $d\alpha_k(\cdot)$ is given by*

$$d\alpha_{k+1}(z) = \frac{\phi_{k+1}(y_{k+1} - c_{k+1}(z))}{\phi_{k+1}(y_{k+1})} \times \int_{\mathbb{R}^m} |D(c_{k+1}(z) + z - y_{k+1}, \xi)| \times d\alpha_k(\xi)\, d\psi_{k+1}(d_{k+1}(c_{k+1}(z) + z - y_{k+1}, \xi)).$$

Notation 8.2 *Here $m, k \in \mathbb{N}$ and $m < k$. Write $\overline{\Lambda}_{m,k} = \prod_{\ell=m}^{k} \overline{\lambda}_\ell$ and $d\gamma_{m,k}(x)$ for the unnormalized conditional probability measure such that*

$$\overline{E}\left[\overline{\Lambda}_k I(x_m \in dx) \mid \mathscr{Y}_k\right] := d\gamma_{m,k}(x).$$

$d\gamma_{m,k}(x)$ describes the smoothed estimate of x_m given $\mathscr{Y}_k$, $m < n$.

Lemma 8.3 *For $m, k \in \mathbb{N}$, $m < k$ the smoothed estimate of x_m given $\mathscr{Y}_k$ is given by:*

$$d\gamma_{m,k}(x) = \beta_{m,k}(x)\, d\alpha_m(x)$$

where $d\alpha_m(x)$ is given recursively by Theorem 8.1 *and*

$$\beta_{m,k}(x) = \overline{E}\left[\overline{\Lambda}_{m+1,k} \mid x_m = x, \mathscr{Y}_k\right].$$

Proof Let $f : \mathbb{R}^m \to \mathbb{R}$ be an arbitrary integrable function

$$\overline{E}\left[\overline{\Lambda}_k f(x_m) \mid \mathscr{Y}_k\right] = \int_{\mathbb{R}^m} f(x)\, d\gamma_{m,k}(x).$$

However,

$$\overline{E}\left[\overline{\Lambda}_k f(x_m) \mid \mathscr{Y}_k\right] = \overline{E}\left[\overline{\Lambda}_{1,m} f(x_m) \overline{E}\left[\overline{\Lambda}_{m+1,k} \mid x_0, \dots, x_m, \mathscr{Y}_k\right] \mid \mathscr{Y}_k\right].$$

Now

$$\overline{E}\left[\overline{\Lambda}_{m+1,k} \mid x_m = x, \mathscr{Y}_k\right] := \beta_{m,k}(x).$$

Consequently,

$$\overline{E}\left[\overline{\Lambda}_k f(x_m) \mid \mathscr{Y}_k\right] = \overline{E}\left[\overline{\Lambda}_{1,m} f(x_m)\, \beta_{m,k}(x_m) \mid \mathscr{Y}_k\right]$$

and so, from our notation,

$$\int_{\mathbb{R}^m} f(x)\, d\gamma_{m,k}(n) = \int_{\mathbb{R}^m} f(x)\, \beta_{m,k}(x)\, d\alpha_m(x),$$

which yields at once the result. □

Lemma 8.4 *$\beta_{m,k}(x)$ satisfies the backward recursive equation*

$$\begin{aligned}\beta_{m,k}(x) = & \left(\phi_{m+1}(y_{m+1})\right)^{-1} \\ & \times \int_{\mathbb{R}^m} \phi_{m+1}\left(y_{m+1} - c_{m+1}\left(V_{m+1}\left(a_{m+1}(x,w) + y_{m+1}\right)\right)\right) \\ & \times \beta_{m+1,k}\left(V_{m+1}\left(a_{m+1}(x,w) + y_{m+1}\right)\right) d\psi_{m+1}(w).\end{aligned}$$

Proof

$$\begin{aligned}\beta_{m,k}(x) &= \overline{E}\left[\overline{\Lambda}_{m+1,k} \mid x_m = x, \mathscr{Y}_k\right] \\ &= \overline{E}\left[\overline{\lambda}_{m+1}\overline{\Lambda}_{m+2,n} \mid x_m = x, \mathscr{Y}_k\right] \\ &= \overline{E}\left[\frac{\phi_{m+1}\left(y_{m+1} - c_{m+1}\left(V_{m+1}\left(a_{m+1}(x_m, v_{m+1}) + y_{m+1}\right)\right)\right)}{\phi_{m+1}(y_{m+1})}\right. \\ &\qquad \left. \times \overline{E}\left[\overline{\Lambda}_{m+2,n} \mid x_m = x, x_{m+1}, \mathscr{Y}_k\right] \,\middle|\, x_m = x, \mathscr{Y}_k\right] \\ &= \overline{E}\left[\frac{\phi_{m+1}\left(y_{m+1} - c_{m+1}\left(V_{m+1}\left(a_{m+1}(x_m, v_{m+1}) + y_{m+1}\right)\right)\right)}{\phi_{m+1}(y_{m+1})}\right. \\ &\qquad \left. \times \beta_{m+1,k}\left(V_{m+1}\left(a_{m+1}(x_m, v_{m+1}) + y_{m+1}\right)\right) \,\middle|\, x_m = x, \mathscr{Y}_k\right] \\ &= \frac{1}{\phi_{m+1}(y_{m+1})} \\ &\qquad \times \int_{\mathbb{R}^m} \left[\phi_{m+1}\left(y_{m+1} - c_{m+1}\left(V_{m+1}\left(a_{m+1}(x,w) + y_{m+1}\right)\right)\right)\right. \\ &\qquad\qquad \left. \times \beta_{m+1,k}\left(V_{m+1}\left(a_{m+1}(x,w) + y_{m+1}\right)\right)\right] d\psi_{m+1}(w).\end{aligned}$$

□

We can also obtain the one-step predictor.

Notation 8.5 *Write $d\rho_{k+1,k}(x)$ for the unnormalized conditional probability measure such that*

$$\overline{E}\left[\overline{\Lambda}_{k+1} I\left(x_{k+1} \in dx\right) \mid \mathscr{Y}_k\right] := d\rho_{k+1,k}(x).$$

Lemma 8.6 *The one-step predictor is given by the following equation:*

$$\begin{aligned}d\rho_{k+1,k}(x) = & \int_{\mathbb{R}^m} \int_{\mathbb{R}^m} \phi_{k+1}\left(y - c_{k+1}(x)\right) \left| \frac{\partial d_{k+1}\left(c_{k+1}(x) + x - y, z\right)}{\partial x} \right| \\ & \times d\psi_{k+1}\left(d_{k+1}\left(c_{k+1}(x) + x - y, z\right)\right) d\rho_k(z)\, dy.\end{aligned}$$

Proof Suppose f is an arbitrary integrable Borel function. Then

$$\begin{aligned}
&\overline{E}\left[f(x_{k+1})\overline{\Lambda}_{k+1} \mid \mathscr{Y}_k\right]\\
&= \int_{\mathbb{R}^m} f(x)\, d\rho_{k+1,k}(x)\\
&= \overline{E}\Bigg[\overline{\Lambda}_k \overline{E}\Bigg[\frac{\phi_{k+1}(y_{k+1} - c_{k+1}(V_{k+1}(a_{k+1}(x_k, v_{k+1}) + y_{k+1})))}{\phi_{k+1}(y_{k+1})}\\
&\qquad \times f(V_{k+1}(a_{k+1}(x_k, v_{k+1}) + y_{k+1})) \,\Big|\, x_0, \dots, x_k, \mathscr{Y}_k\Bigg] \,\Big|\, \mathscr{Y}_k\Bigg]\\
&= \overline{E}\Bigg[\overline{\Lambda}_k \int_{\mathbb{R}^m}\int_{\mathbb{R}^m} \phi_{k+1}(y - c_{k+1}(V_{k+1}(a_{k+1}(x_k, w) + y)))\\
&\qquad \times f(V_{k+1}(a_{k+1}(x_k, w) + y))\, dy\, d\psi_{k+1}(w) \,\Big|\, \mathscr{Y}_k\Bigg]\\
&= \int_{\mathbb{R}^m}\int_{\mathbb{R}^m}\int_{\mathbb{R}^m} \Big[f(V_{k+1}(a_{k+1}(z, w) + y))\\
&\qquad \times \phi_{k+1}(y - c_{k+1}(V_{k+1}(a_{k+1}(z, w) + y)))\Big]\\
&\qquad \times d\psi_{k+1}(w)\, d\rho_k(z)\, dy.
\end{aligned}$$

Let $x = V_{k+1}(a_{k+1}(z, w) + y)$. Then $w = d_{k+1}(c_{k+1}(x) + x - y, z)$. Hence

$$\begin{aligned}
&\int_{\mathbb{R}^m} f(x)\, d\rho_{k+1,k}(x)\\
&= \int_{\mathbb{R}^m}\int_{\mathbb{R}^m}\int_{\mathbb{R}^m} f(x)\, \phi_{k+1}(y - c_{k+1}(x)) \left|\frac{\partial d_{k+1}(c_{k+1}(x) + x - y, z)}{\partial x}\right|\\
&\qquad \times d\psi_{k+1}(d_{k+1}(c_{k+1}(x) + x - y, z))\, d\rho_k(z)\, dy,
\end{aligned}$$

and the result follows. □

A Second Model

Suppose $\{x_k\}$, $k \in \mathbb{N}$, is a discrete time stochastic process taking values in some Euclidean space $\mathbb{R}^d$. We suppose that x_0 has a known distribution $\pi_0(x)$. The set $\{v_k\}$, $k \in \mathbb{N}$, is a sequence of independent, $\mathbb{R}^n$-valued, random variables with probability distributions $d\psi_k$ and $\{w_k\}$ a sequence of $\mathbb{R}^m$-valued, random variables with positive densities ϕ_k. For $k \in \mathbb{N}$, $a_k : \mathbb{R}^d \times \mathbb{R}^n \times \mathbb{R}^m \to \mathbb{R}^d$ are measurable functions, and we suppose for $k \geq 0$ that

$$x_{k+1} = a_{k+1}(x_k, v_{k+1}, w_{k+1}), \tag{8.3}$$

and x_k is not observed directly. There is an observation process $\{y_k\}$, $k \in \mathbb{N}$ taking values in some Euclidean space $\mathbb{R}^p$ and for $k \in \mathbb{N}$, $c_k : \mathbb{R}^p \to \mathbb{R}^q$ are measurable functions such that

$$y_{k+1} = c_{k+1}(x_k, w_{k+1}). \tag{8.4}$$

We assume that for each $k \in \mathbb{N}$ there is an inverse map $d_k : \mathbb{R}^d \times \mathbb{R}^d \times \mathbb{R}^m \to \mathbb{R}^n$ such that

$$v_{k+1} = d_{k+1}(x_{k+1}, x_k, w_{k+1}). \tag{8.5}$$

We require d_k to be differentiable in the first variable for each k. We also assume that for each k, there is an inverse map $g_k : \mathbb{R}^p \times \mathbb{R}^d \to \mathbb{R}^m$ such that if (4.4) holds

$$w_{k+1} = g_{k+1}(y_{k+1}, x_k).$$

Finally, we require

$$C_{k+1}(x_k, w_{k+1}) = \left.\frac{\partial c_{k+1}(x, w_{k+1})}{\partial x}\right|_{x=x_k}, \tag{8.6}$$

$$G_{k+1}(y_{k+1}, x_k) = \left.\frac{\partial g_{k+1}(y, x_k)}{\partial y}\right|_{y=y_{k+1}},$$

to be nonsingular for each $k \in \mathbb{N}$. Here the σ-field generated by the signal and observation processes is $\mathcal{G}_k^0 = \sigma(x_0, x_1, \dots, x_k, y_1, \dots, y_k)$ with complete version $\{\mathcal{G}_k\}$ for $k \in \mathbb{N}$. The new probability measure $\overline{P}$, under which the $\{y_\ell\}$ are independent with densities ϕ_ℓ, is obtained if we define

$$\left.\frac{d\overline{P}}{dP}\right|_{\mathcal{F}_k} = \Lambda_k := \prod_{\ell=1}^{n} \frac{\phi_\ell(y_\ell)}{\phi_\ell(w_\ell)} G_\ell(y_\ell, x_{\ell-1})^{-1}.$$

We immediately have, for $k \in \mathbb{N}$, a recursion for the unnormalized conditional probability measure $\overline{E}\left[\overline{\Lambda}_k I(x_k \in dx) \mid \mathcal{Y}_k\right] := d\alpha_k(x)$

$$d\alpha_{k+1}(x) = \int_{\mathbb{R}^d} \Phi(x, z, y_{k+1}) \, d\psi_{k+1}(d_{k+1}(x, z, g_{k+1}(y_{k+1}, z))) \, d\alpha_k(z).$$

Here

$$\Phi(x, z, y_{k+1}) = \frac{\phi_{k+1}(g_{k+1}(y_{k+1}, z))}{\phi_{k+1}(y_{k+1})} c_{k+1}(z, g_{k+1}(y_{k+1}, z))^{-1} \times \left| \frac{\partial d_{k+1}(x, z, g_{k+1}(y_{k+1}, z))}{\partial x} \right|.$$

5.9 Recursive Prediction Error Estimation

The case of optimal estimation for linear models with unknown parameters θ, as studied in Chapter 4, Section 5.2 is a specialization of the work of the previous sections. In this case, the unknown constant parameters θ can be viewed also as states, denoted θ_k with $\theta_{k+1} = \theta_k = \theta$, and the state space model with states x_k, viz.

$$\begin{aligned} x_{k+1} &= A(\theta)x_k + B(\theta)v_{k+1}, \\ y_k &= C(\theta)x_k + w_k, \end{aligned} \tag{9.1}$$

can be rewritten as a nonlinear system with states x_k, θ_k as follows

$$\begin{aligned} \begin{bmatrix} \theta_{k+1} \\ x_{k+1} \end{bmatrix} &= \begin{bmatrix} I & 0 \\ 0 & A(\theta_k) \end{bmatrix} \begin{bmatrix} \theta_k \\ x_k \end{bmatrix} + \begin{bmatrix} 0 \\ B(\theta_k) \end{bmatrix} v_{k+1} \\ y_k &= C(\theta_k)x_k + w_k. \end{aligned} \tag{9.2}$$

Readers with a background in system identification will be aware that such models are not usually the best ones for the purposes of identification; see, for example, Ljung and Söderström (1983). The parametrization is not unique in general. However, if one is uncertain of the signal model, why not work with the unique signal model with a known stable universe, termed the *innovations model*, or the *canonical model*. This model generates the same statistics for the measurements y_k, and is simply derived from the optimal conditional state estimator yielding conditional state estimates $E[x_k \mid y_{k-1}, \theta]$, denoted $\hat{x}_{k|k-1,\theta}$. Indeed, the innovations model is the right inverse of this system, and is given as, with $\theta_0 = \theta$

$$\begin{bmatrix} \theta_{k+1} \\ \hat{x}_{k+1|k,\theta} \end{bmatrix} = \begin{bmatrix} I & 0 \\ 0 & A(\theta_k) \end{bmatrix} \begin{bmatrix} \theta_k \\ \hat{x}_{k|k-1,\theta} \end{bmatrix} + \begin{bmatrix} 0 \\ K(\theta_k) \end{bmatrix} v_k, \tag{9.3}$$

$$y_k = C(\theta_k)\hat{x}_{k|k-1,\theta} + v_k, \tag{9.4}$$

where $K(\theta_k) = K(\theta)$ is the so-called conditional Kalman gain, and v_k is the innovations white noise process. It usually makes sense in the parametrization process to have $K(\theta)$ as a subset of θ. This avoids the need to calculate the nonlinearities $K(.)$ via the Riccati equation of the Kalman filter.

To achieve practical (but inevitably suboptimal) recursive filters for this model, a first suggestion might be to apply the extended Kalman filter (EKF) of Section 5.5, which is the Kalman filter applied to a linearized version of (9.4). This leads to parameter estimates denoted $\hat{\theta}_k$, and to estimates of the optimal conditional estimate $\hat{x}_{k|k-1,\theta}$ which, in turn, one can conveniently denote $\hat{x}_{k|k-1,\hat{\Theta}_{k-1}}$, where $\hat{\Theta}_k$ denotes the set $\{\hat{\theta}_0, \dots, \hat{\theta}_k\}$. There is a global convergence theory providing reasonable conditions for $\hat{\theta}_k$ to approach θ as k approaches ∞. Indeed, results are stronger in that the EKF estimates approach the performance of the optimal filter asymptotically, and so this method is termed asymptotically optimal. Furthermore, the algorithms can be simplified by neglecting certain terms that asymptotically approach zero in the EKF, without compromising the convergence theory. The resulting algorithm is identical to a recursive prediction error (RPE) estimator, as formulated in Moore and Weiss (1979); see Ljung and Söderström (1983) for a more recent and complete treatment.

The RPE scheme seeks to select estimates $\hat{\theta}_k$ on-line so as to minimize a prediction error $\left(y_k - C(\hat{\theta}_k)\hat{x}_{k|k-1,\hat{\Theta}_{k-1}}\right)$ in a squared average sense. Such schemes, along with the EKF, apply to more general parametrized models than linearly parametrized models, (9.2) and (9.4). Indeed, the next chapter presents one such application.

Suffice it is to say here, the EKF and RPE schemes consist of three subalgorithms coupled together. The first is the conditional state estimator driven from both y_k and parameter estimates $\hat{\theta}_k$, in lieu of θ, which yields state estimates $\hat{x}_{k|k-1,\hat{\Theta}_k}$ in lieu of $\hat{x}_{k|k-1,\hat{\theta}}$ as desired. There is a sensitivity filter, yielding estimates of the sensitivity of the state estimates and prediction errors to parameter estimates as

$$\left.\frac{\partial \hat{x}_{k+1|k,\hat{\Theta}_{k-1},\theta}}{\partial \theta}\right|_{\theta=\hat{\theta}_k}, \qquad \psi_{k+1} = \left.\frac{\partial \left(y_{k+1} - C(\theta)\,\hat{x}_{k+1|k,\hat{\Theta}_{k-1},\theta}\right)}{\partial \theta}\right|_{\theta=\hat{\theta}_k}$$

These are the states and output, respectively, of a filter also driven by y_k and $\hat{\theta}_k$. Details for this filter are not given here. Finally, there is an update for the parameter estimates driven from y_k and ψ_k, as

$$\begin{aligned} \hat{\theta}_k &= \hat{\theta}_{k-1} + P_k \psi_k \left(y_k - C(\theta_k)\,\hat{x}_{k|k-1,\hat{\Theta}_{k-1}}\right), \\ P_k^{-1} &= P_{k-1}^{-1} + \psi_k \psi_k' \end{aligned}$$

suitably initialized. The rationale for this is not developed here.

These subfilters and estimators are coupled to form the RPE scheme. The full suite of the RPE components appears formidable for any application, but it is *finite-dimensional* and *asymptotically optimal* under reasonable conditions. Further development of this topic is omitted, save that in the next chapter an application within the context of information state filtering is developed.

5.10 Problems and Notes

Problems

1. Establish the recursion given in Theorem 8.1
2. Suppose that $x_k \in \mathbb{R}^m$ and

$$x_{k+1} = A_{k+1}x_k + v_{k+1} + w_{k+1}.$$

Here A_k are, for each $k \in \mathbb{N}$, $m \times m$ matrices, $\{v_\ell\}$ is a sequence of independent $\mathbb{R}^m$-valued random variables with probability distribution $d\psi_\ell$ and $\{w_\ell\}$ is a sequence of independent $\mathbb{R}^m$-valued random variables with positive densities $\phi_\ell(b)$. Further, suppose the $\mathbb{R}^m$-valued observation process has the form

$$y_k = C_k x_k + w_k.$$

Here C_k are, for each $k \in \mathbb{N}$, $m \times m$ matrices. Find recursions for the conditional probability measures $E\left[I\left(x_k \in dx\right) \mid \mathscr{Y}_k\right]$ and $E\left[I\left(x_m \in dx\right) \mid \mathscr{Y}_k\right]$, $m \neq k$.

3. Assume here that the signal and observation processes are given by the dynamics

$$\begin{aligned} x_{k+1} &= a_{k+1} x_k + v_{k+1} + w_{k+1} \in \mathbb{R}, \\ y_k &= c_k x_k + w_k \in \mathbb{R}. \end{aligned}$$

Here, a_k, c_k are real numbers, v_k and w_k are normally distributed with means 0 and respective variances σ_k^2 and γ_k^2. Derive recursive estimates of the conditional mean and variance of the process x_k given the observations up to time k.

4. Repeat Problem 3 for vector-valued x and y.

Notes

Section 5.8 is closer in spirit to earlier work of Ho and Lee (1968) and Anderson (1968). However, the role of the dynamics is not so apparent. Related ideas can be found in Jazwinski (1970) and McGarthy (1974), but the measure transformation technique is not used in these references. A related formula can be found in Campillo and le Gland (1989), which discusses discretizations of the Zakai equation and applications of the expectation maximization algorithm.

Chapter 6
Practical Recursive Filters

6.1 Introduction

Hidden Markov models with states in a finite discrete set and uncertain parameters have been widely applied in areas such as communication systems, speech processing, and biological signal processing (Rabiner 1989; Chung, Krishnamurthy and Moore 1991). A limitation of the techniques presented in the previous chapters for such applications is the curse of dimensionality which arises because the computational effort, speed, and memory requirements are at least in proportion to the square of the number of states of the Markov chain, even with known fixed parameters. With unknown parameters in a continuous range, the optimal estimators are infinite-dimensional. In more practical reestimation algorithms which involve multipasses through the data memory requirements are also proportional to the length of data being processed. There is an incentive to explore finite-dimensional on-line (sequential) practical algorithms, to seek improvements in terms of memory and computational speed, and also to cope with slowly varying unknown HMM parameters. This leads inevitably into suboptimal schemes, which are quite difficult to analyze, but the reward is practicality for engineering applications.

The key contribution of this chapter is to formulate HMMs, with states in a finite-discrete set and with unknown parameters, in such a way that HMM filters and the *extended Kalman filter* (EKF) or related *recursive prediction error* (RPE) techniques of previous chapters can be applied in tandem. The EKF and RPE methods are, in essence, Kalman filters (KF) designed for linearized signal models with states in a continuous range. The RPE methods are specializations of EKF algorithms for the case when the unknown constant parameters of the model are viewed, and estimated, as states. Certain EKF terms which go to zero asymptotically, in this EKF case, can be set to zero without loss of convergence properties. This simplification is, in fact, the RPE scheme. For HMM models, the parameters to be estimated are the HMM transition probabilities, the N state values of the Markov chain, and the measurement noise variance. The computational complexity for computing these estimates is of order at least N^2 per time instant. The particular model parametrization we consider here uses the square root of the transition probabilities constrained to the surface of a sphere $\mathscr{S}^{N-1}$ in $\mathbb{R}^N$. The advantage of working on the sphere is that estimates of transition probabilities are nonnegative and remain normalized, as required. Of course, in practice, the model parameters are often not constant but time varying, and then the RPE approach breaks down.

In order to illustrate how optimal filter theory gives insight for real applications, we next address the task of demodulating signals for communication systems with fading noisy transmission channels. Such channels can be the limiting factor in communications systems, particularly with *multipath* situations arising from mobile receivers or transmitters.

Signal models in this situation have slowly varying parameters so that RPE methods can not be used directly. We consider *quadrature amplitude modulation* (QAM), *frequency modulation* (FM) and *phase modulation* (PM). Related schemes are *frequency shift keying* (FSK) and *phase shift keying* (PSK). Of course, traditional *matched filters* (MF), *phase locked loops* (PLL), and *automatic gain controllers* (AGC) can be effective, but they are known to be far from optimal, particularly in high noise. Optimal schemes, on the other hand, are inherently infinite-dimensional and are thus impractical. Also they may not be robust to modeling errors. The challenge is to devise suboptimal robust demodulation schemes which can be implemented by means of a digital signal processing chip. Our approach here is to use KF techniques coupled with optimal (HMM) filtering, for demodulation of modulated signals in complex Rayleigh fading channels.

State-Space Signal Model

As in earlier chapters, let X_k be a discrete-time homogeneous, first-order Markov process belonging to a finite discrete set. The state space of X, without loss of generality, can be identified with the set of unit vectors $S_X = \{e_1, e_2, \dots, e_N\}$, $e_i = (0, \dots, 0, 1, 0, \dots, 0)' \in \mathbb{R}^N$ with 1 in the ith position. The transition probability matrix is

$$A = (a_{ji})\, 1 \leq i, j \leq N \quad \text{where } a_{ji} = P\left(X_{k+1} = e_j \mid X_k = e_i\right)$$

so that $E\left[X_{k+1} \mid X_k\right] = AX_k$ Of course $a_{ji} \geq 0, \sum_{j=1}^{N} a_{ji} = 1$, for each j. We also denote $\{\mathscr{F}_l\}$, $l \in \mathbb{N}$ the complete filtration generated by X, that is, for any $k \in \mathbb{N}$, $\mathscr{F}_k$ is the complete σ-field generated by $X_l, l \leq k$.

Recall that the dynamics of X_k are given by the state equation

$$X_{k+1} = AX_k + V_{k+1}, \tag{1.1}$$

where V_{k+1} is a $\mathscr{F}_k$ martingale increment, in that $E\left[V_{k+1} \mid \mathscr{F}_k\right] = 0$.

We assume that X_k is hidden, that is, indirectly observed by measurements y_k. The *observation process* y_k has the form

$$y_k = \langle c, X_k \rangle + w_k, \qquad w_k \text{ are i.i.d.} \sim N\left[0, \sigma_w^2\right] \tag{1.2}$$

with $\langle , \rangle$ denoting the inner product in $\mathbb{R}^N$, and where $c \in \mathbb{R}^N$ is the vector of state values of the Markov chain. Let $\mathscr{Y}_l$ be the σ-field generated by y_k, $k \le l$ and let $\mathscr{G}_k$ be the complete filtration generated by $X_k, \mathscr{Y}_k'$. We shall denote parametrized probability densities as $b_k(i) := b(y_k, c_i) = P[y_k \in dy \mid X_k = e_i, \theta]$, where

$$b(y_k, c_i) = \frac{1}{\sqrt{2\pi\sigma_w^2}} \exp\left[-\frac{(y_k - c_i)^2}{2\sigma_w^2} \right]. \tag{1.3}$$

Because w_k are i.i.d., the independence property

$$E(y_k \mid X_{k-1} = e_i, \mathscr{F}_{k-2}, \mathscr{Y}_{k-1}) = E(y_k \mid X_{k-1} = e_i)$$

holds and is essential for formulating the problem as an HMM. Also we assume that the initial state probability vector for the Markov chain $\underline{\pi} = (\pi_i)$ is defined from $\pi_i = P(X_0 = e_i)$. The HMM is denoted $\lambda = (A, c, \underline{\pi}, \sigma_w^2)$.

Model Parametrization

Suppose that λ is parametrized by an unknown vector θ so that $\lambda(\theta) = (A(\theta), c(\theta), \underline{\pi}, \sigma_w^2(\theta))$. We propose a parametrization, the dimension of which is $N_\theta = N + N^2 + 1$, representing N state values, N^2 transition probabilities, and the noise variance. We consider this parametrization

$$\theta = (c_1, \ldots, c_N, s_{11}, \ldots, s_{1N}, s_{21}, \ldots, s_{NN}, \sigma_w^2)',$$

where $a_{ji} = s_{ji}^2$. The benefit of this parametrization is that the parameters s_{ji} belong to the *sphere* $\mathscr{S}^{N-1} := \{s_{ji} : \sum_{j=1}^N s_{ji}^2 = 1\}$ which is a smooth *manifold*. This is perhaps preferable to parametrizations of a_{ji} on a simplex $\Delta_N := \{a_{ji} \mid \sum_{j=1}^N a_{ji} = 1, a_{ij} \ge 0\}$, where the boundary constraint $a_{ij} \ge 0$ can be a problem in estimation. Actually, working with angle parametrizations on the sphere could be a further simplification to avoid the normalization constraint.

Conditional Information-State Model

Let $\hat{X}_k(\theta)$ denote the conditional filtered-state estimate of X_k at time k, being given by $\hat{X}_k(\theta) := E[X_k \mid \mathscr{Y}_k, \theta]$. It proves convenient to work with the unnormalized conditional estimates $q_k(\theta)$, termed here "forward" *information states*. Thus

$$\hat{X}_k(\theta) := E(X_k \mid \mathscr{Y}_k, \theta) = \langle q_k(\theta), \underline{1} \rangle^{-1} q_k(\theta), \tag{1.4}$$

where $\underline{1}$ is the column vector containing all ones. Here $q_k(\theta)$ is conveniently computed using the following "forward" recursion; see Chapter 3 or Rabiner (1989):

$$q_{k+1}(\theta) = B(y_{k+1}, \theta) A(\theta) q_k(\theta), \tag{1.5}$$

where $B(y_k,\theta) = \text{diag}(b(y_k,c_1),\ldots,b(y_k,c_N))$. Letting $\hat{y}_{k+1}(\theta)$ denote the prediction $E[y_{k+1} \mid \mathscr{Y}_k,\theta]$, then

$$\hat{y}_{k+1}(\theta) = \langle c, A(\theta) q_k(\theta) < q_k(\theta), \underline{1} >^{-1} \rangle$$

and the prediction error $n_{k+1}(\theta) := y_{k+1} - \hat{y}_{k+1}(\theta)$ is a martingale increment. Thus, the signal model (1.1) and (1.2) can now be written as (1.5) together with

$$y_{k+1} = \langle c, A(\theta) q_k(\theta) \langle q_k(\theta), \underline{1} \rangle^{-1} \rangle + n_{k+1}(\theta). \tag{1.6}$$

This can be referred to as a *conditional innovations model* or an *information state-model.*

6.2 Recursive Prediction Error HMM Algorithm

For the information-state model in (1.5) and (1.6), we present in this section an on-line prediction error algorithm for estimating the parameters θ, assumed to be constant. Here $q_k(\theta)$ is recursively estimated at each iteration, using obvious notation, as follows

$$\boxed{\hat{q}_{k+1}(\hat{\Theta}_k) = B(y_{k+1}, \hat{\theta}_k) A(\hat{\theta}_k) \hat{q}_k(\hat{\Theta}_{k-1}),} \tag{2.1}$$

where $\hat{\theta}_k$ is the recursive estimate of the parameter vector based on $\mathscr{Y}_k$, and $\hat{\Theta}_k := \{\hat{\theta}_1,\ldots,\hat{\theta}_k\}$. Let $\hat{y}_{k+1}(\hat{\Theta}_k)$ denote the predicted output at time $k+1$ based on measurements up to time k. Then

$$\boxed{\hat{y}_{k+1}(\hat{\Theta}_k) = \langle c(\hat{\theta}_k), \langle \hat{q}_k(\hat{\Theta}_{k-1}), \underline{1} \rangle^{-1} A(\hat{\theta}_k) \hat{q}_k(\hat{\Theta}_{k-1}) \rangle.} \tag{2.2}$$

The RPE parameter update equations are (Ljung and Söderström 1983)

$$\boxed{\hat{\theta}_{k+1} = \Gamma_{\text{proj}} \{ \hat{\theta}_k + \gamma_{k+1} R_{k+1}^{-1} \psi_{k+1} \hat{n}_{k+1}(\hat{\Theta}_k) \},} \tag{2.3}$$

where

$$\hat{n}_{k+1}(\hat{\Theta}_k) := y_{k+1} - \hat{y}_{k+1}(\hat{\Theta}_k) \tag{2.4}$$

$$R_{k+1}^{-1} = \frac{1}{1-\gamma_{k+1}} \left(R_k^{-1} - \frac{\gamma_{k+1} R_k^{-1} \psi_{k+1} \psi_{k+1}' R_k^{-1}}{(1-\gamma_{k+1}) + \gamma_{k+1} \psi_{k+1}' R_k^{-1} \psi_{k+1}} \right). \tag{2.5}$$

Here γ_k is a gain sequence (often referred to as step size) satisfying,

$$\gamma_k \geq 0, \sum_{k=1}^{\infty} \gamma_k = \infty, \sum_{k=1}^{\infty} \gamma_k^2 < \infty. \tag{2.6}$$

A selection $\gamma_k = k^{-1}$ is often used. Also ψ_k is the gradient

$$\psi_k' := \left(-d\hat{n}_k(\hat{\Theta}_{k-2}), \theta / d\theta \right) \big|_{\theta = \hat{\theta}_{k-1}}$$

and R_k is the Hessian , or covariance matrix, approximation. For general RPE schemes, the notation $\Gamma_{\text{proj}}\{.\}$ represents a projection into the stability domain. In our case, stability is guaranteed in the constraint domain, so that $\Gamma_{\text{proj}}\{.\}$ can be simply a projection into the constraint domain. This is discussed below.

We now present gradient and projection calculations for the RPE based algorithm (2.3)–(2.5), to estimate the HMM parameters θ. The details are not important for the reader unless an implementation is required.

The derivative vector, ψ_k, defined above is given, for $m,n \in \{1,\dots,N\}$, by

$$\begin{aligned}\psi_{k+1} &= \left.\frac{\partial \hat{y}_{k+1}(\hat{\Theta}_{k-1},\theta)}{\partial \theta}\right|_{\theta=\hat{\theta}_k}\\ &= \left.\left(\frac{\partial \hat{y}_{k+1}(\hat{\Theta}_{k-1},\theta)}{\partial c_m}, \frac{\partial \hat{y}_{k+1}(\hat{\Theta}_{k-1},\theta)}{\partial s_{mn}}, \frac{\partial \hat{y}_{k+1}(\hat{\Theta}_{k-1},\theta)}{\partial \sigma_w^2}\right)'\right|_{\theta=\hat{\theta}_k}. \end{aligned} \tag{2.7}$$

To evaluate ψ_k first note that (2.2) can be rewritten as

$$\hat{y}_{k+1}(\hat{\Theta}_{k-1},\theta) = \alpha_{k-1}\sum_{j=1}^{N}\left(c_j \sum_{i=1}^{N} a_{ij}\hat{q}_k^i\right), \quad \text{where } \alpha_k = \langle \hat{q}_k, \underline{1}\rangle^{-1}. \tag{2.8}$$

Here we omit the obvious dependence of $\hat{q}_k$ on $\hat{\Theta}_{k-1}$, and c_i, a_{ij} on θ. The derivatives with respect to the discrete-state values, c_i, are obtained by differentiating (2.8) to yield

$$\begin{aligned}\frac{\partial \hat{y}_{k+1}(\hat{\Theta}_{k-1},\theta)}{\partial c_m} &= \alpha_k \sum_{i=1}^{N} a_{im}\hat{q}_k^i + \alpha_k \sum_{j=1}^{N}\left(c_j \sum_{i=1}^{N} a_{ij}\eta_k^i(m)\right)\\ &\quad - \alpha_k^2 \left(\sum_{i=1}^{N}\eta_k^i(m)\right)\sum_{j=1}^{N}\left(c_i \sum_{i=1}^{N} a_{ij}\hat{q}_k^i\right), \end{aligned} \tag{2.9}$$

where $\eta_{k+1}^i(m) := \partial \hat{q}_{k+1}^i / \partial c_m$, from which we have

$$\eta_{k+1}^i(m) = \begin{cases} \displaystyle\sum_{i=1}^{N} \eta_k^i(m)\, a_{ij} b(y_{k+1}, c_j) & \text{if } j \neq m\\ \displaystyle\sum_{i=1}^{N} \eta_k^i(m)\, a_{ij} b(y_{k+1}, c_j) & \\ \displaystyle\quad + \frac{(y_{k+1}-c_j)}{\sigma_w^2}\left(\sum_{i=1}^{N}\hat{q}_k^i a_{ij}\right) b(y_{k+1}, c_j) & \text{if } j = m. \end{cases} \tag{2.10}$$

Recalling that $s_{ij} \in \mathscr{S}^{N-1}$, then derivatives with respect to the transition probabilities are those on the tangent space of $\mathscr{S}^{N-1}$. Thus

$$\begin{aligned}\frac{\partial \hat{y}_{k+1}(\hat{\Theta}_k,\theta)}{\partial s_{mn}} &= 2\alpha_k \hat{q}_k^m \left(c_n s_{mn} - \sum_{\ell=1}^{N} c_\ell s_{mn} s_{m\ell}^2 \right) \\ &\quad + \alpha_k \sum_{j=1}^{N} \left(c_j \sum_{i=1}^{N} s_{ij}^2 \xi_k^i(m,n) \right) \\ &\quad - \alpha_k^2 \left(\sum_{i=1}^{N} \xi_k^i(m,n) \right) \sum_{j=1}^{N} \left(c_j \sum_{i=1}^{N} s_{ij}^2 \hat{q}_k^i \right), \end{aligned} \tag{2.11}$$

where $\xi_{k+1}^j(m,n) := \partial \hat{q}_{k+1}^j / \partial s_{mn}$, and

$$\xi_{k+1}^j(m,n) = \begin{cases} \sum_{i=1}^{N} \xi_k^i(m,n)\, s_{ij}^2 b(y_{k+1},c_j) & \\ \quad - 2 s_{mj}^2 s_{mn} \hat{q}_k^m b(y_{k+1},c_j) & \text{if } j \neq n \\ \sum_{i=1}^{N} \xi_k^i(m,n)\, s_{ij}^2 b(y_{k+1},c_j) & \\ \quad + 2 s_{mj}\left(1 - s_{mj}^2\right) \hat{q}_k^m b(y_{k+1},c_j) & \text{if } j = n. \end{cases} \tag{2.12}$$

In achieving an update estimate of s_{ij} at time $k+1$ via (2.7), there is a required projection $\Gamma_{\text{proj}}\{.\}$ into the constraint domain (the surface of a unit sphere in $\mathbb{R}^N$). Thus, in updating $\hat{s}_{ij}$, first an unconstrained update, denoted $\hat{s}_{ij}^U$, is derived then projected onto the sphere by renormalization as follows,

$$\hat{s}_{ij}^2 = \frac{\left(\hat{s}_{ij}^U\right)^2}{\sum_{j=1}^{N} \left(\hat{s}_{ij}^U\right)^2} \tag{2.13}$$

to achieve $\sum_{j=1}^{N} \hat{s}_{ij}^2 = 1$ as required.

The derivative with respect to the measurement noise variance, σ_w^2, is given by

$$\begin{aligned}\frac{\partial \hat{y}_{k+1}(\hat{\Theta}_k,\theta)}{\partial \sigma_w^2} &= \alpha_k \sum_{j=1}^{N} \left(c_j \sum_{i=1}^{N} a_{ij} \rho_k^i \right) \\ &\quad - \alpha_k^2 \left(\sum_{i=1}^{N} \rho_k^i \right) \sum_{j=1}^{N} \left(c_j \sum_{i=1}^{N} a_{ij} \hat{q}_k^i \right), \end{aligned} \tag{2.14}$$

where $\rho_{k+1}^j := \partial \hat{q}_{k+1}^j / \partial \sigma_w^2$, and

$$\begin{aligned}\rho_{k+1}^j &= \sum_{i=1}^{N} \rho_k^i a_{ij} b(y_{k+1},c_j) \\ &\quad + \left(\frac{(y_{k+1}-c_j)^2}{2\sigma_w^4} - \frac{1}{2\sigma_w^2} \right) \left(\sum_{i=1}^{N} \hat{q}_k^i a_{ij} \right) b(y_{k+1},c_j). \end{aligned} \tag{2.15}$$

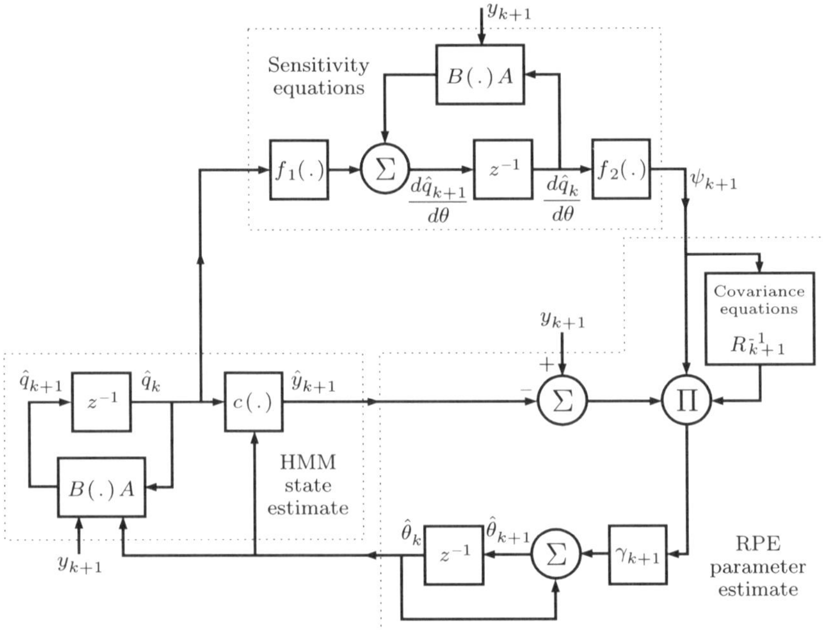

Where $f_1(.)$ is given implicitly in (2.10), (2.12), and (2.15), and $f_2(.)$ is given implicitly in (2.9), (2.11), and (2.14).

Fig. 2.1 RPE/HMM scheme for HMM identification

Remarks 2.1

1. Proofs of convergence of $\hat{\theta}_k$ to θ as $k \to \infty$ under reasonable identifiability and excitation conditions for this scheme are a little unconventional since V_k is a martingale increment rather than white noise, and is beyond the scope of this chapter.
2. The RPE technique is in fact a Newton-type off-line reestimation approach and so, when convergent, is quadratically convergent.
3. Figure 2.1 shows the recursive prediction error HMM scheme, where $\hat{q}_k$ denotes $\hat{q}_k(\Theta_{k-1})$.

■

Scaling From (1.5), (2.10), (2.12), and (2.15) it is noted that as k increases, q_k^i and derivatives of q_k^i, decrease exponentially and can quickly exceed the numerical range of the computer. In order to develop a consistent scaling strategy the scaling factor must cancel out in the final update equations, (2.3) and (2.5), to ensure it has no effect on the estimates. We suggest the following strategy for scaling based on techniques in Rabiner (1989). Let q_k be the actual unscaled forward variable defined in (1.5), $\bar{q}_k$ be the unscaled updated version based on previous scaled version, and

Table 2.1 Parameter estimation error for SNR = 0 dB

Iterations	ERR($\hat{c}_1$)	ERR($\hat{c}_2$)	ERR($\hat{a}_{11}$)	ERR($\hat{a}_{22}$)
25000	0.085	0.088	0.042	0.047
50000	0.058	0.041	0.017	0.013
75000	0.045	0.041	0.015	0.010
100000	0.036	0.033	0.011	0.012

$\widetilde{q}_k$ be the scaled updated version based on previous scaled version, that is,

$$\widetilde{q}_k^i = f_k \bar{q}_k^i \qquad \text{where } f_k := \frac{1}{\sum_{i=1}^N \bar{q}_k^i}. \tag{2.16}$$

It follows from (2.16) that $\widetilde{q}_k(i) = f_T q_k(i)$ with $f_T = (f_k f_{k-1} \cdots f_0)$. For the derivative terms (η, ξ, ρ) similar expressions can be derived using the same scaling factor, f_T. That is, $\widetilde{\eta}_k(i) = f_T \eta_k^i$, $\widetilde{\xi}_k^i = f_T \xi_k^i$ and $\widetilde{\rho}_k^i = f_T \rho_k^i$. It can be shown, by direct substitution into (2.10), (2.12), and (2.15), that derivatives of ψ_{k+1} evaluated with $\widetilde{q}_k$, $\widetilde{\eta}_k$, $\widetilde{\xi}_k$, and $\widetilde{\rho}_k$ are equivalent to the case where no scaling is to take place.

Increased Step Size and Averaging Equations (2.3) and (2.5) show how the gain sequence γ_k scales the update of both R_k and $\hat{\theta}_k$. Apart from satisfying the restrictions in (2.6) it can be any function. Generally, it has the form $\gamma_k = \gamma_0/k^n, n \in \mathbb{R}$. In the derivation of (2.5), $\gamma_k = \frac{1}{k}$ is assumed. In practice, for this case, γ_k tends to become too small too quickly, and does not allow fast convergence for initial estimates chosen far from the minimum error point. To overcome this problem, Polyak and Juditsky (1992) suggest a method for applying a larger step size, (i.e., $0 \le n \le 1$), and then averaging the estimate. Averaging is used to get a smoother estimate, as the larger step will mean higher sensitivity to noise, and also to ensure that the third requirement in (2.6) remains satisfied. In our simulations we chose $n = 0.5$.

Simulation Studies Presented in Tables 2.1 to 2.4 are results of simulations carried out using two-state Markov chains in white Gaussian noise. Each table is generated from 50 simulations, and the error function used is given by

$$\mathrm{ERR}(\hat{x}) = \sqrt{\frac{1}{50} \sum_{i=1}^{50} (x_i - \hat{x}_i)^2}.$$

The parameters of the Markov chain are $c = [0 \quad 1]'$ and $a_{ii} = 0.9$. The *signal-to-noise ratio* SNR is therefore given by $10 \log\left(1/\sigma_w^2\right)$. Initial parameter estimates used in generating Tables 2.1 and 2.2 are $c = [0.4 \quad 0.6]'$ and $a_{ii} = 0.5$. In some inadequately excited cases the state value estimates collapse to a single state.

The algorithms have been shown to work for Markov chains with up to six states, and no limit to the number of states is envisaged. For further examples of such simulation studies, the reader is directed to Collings, Krishnamurthy and Moore (1993).

Table 2.2 Parameter estimation error for SNR = −12.0 dB

Iterations	ERR($\hat{c}_1$)	ERR($\hat{c}_2$)	ERR($\hat{a}_{11}$)	ERR($\hat{a}_{22}$)
25000	0.271	0.238	0.131	0.146
50000	0.245	0.232	0.159	0.137
75000	0.226	0.190	0.127	0.110
100000	0.210	0.210	0.099	0.085

Table 2.3 Effect of variations in initial transition probability estimates

$\hat{a}_{ii}(0)$	ERR($\hat{c}_1$)	ERR($\hat{c}_2$)	ERR($\hat{a}_{11}$)	ERR($\hat{a}_{22}$)
0.7	0.107	0.115	0.038	0.023
0.5	0.160	0.095	0.063	0.031
0.3	0.182	0.113	0.067	0.045
0.1	0.143	0.077	0.057	0.033

Results after 25000 Iterations: $\hat{c}(0) = [0,1]'$, SNR = 0 dB

Table 2.4 Effect of variations in initial level estimates

$\hat{c}_1(0)$	$\hat{c}_2(0)$	ERR($\hat{c}_1$)	ERR($\hat{c}_2$)	ERR($\hat{a}_{11}$)	ERR($\hat{a}_{22}$)
0.1	0.9	0.081	0.094	0.078	0.099
0.3	0.7	0.078	0.070	0.022	0.025
0.5	0.5	0.120	0.135	0.073	0.083

Results after 25000 Iterations: $\hat{a}_{ii}(0) = 0.9$, SNR = 0 dB

6.3 Example: Quadrature Amplitude Modulation

Digital information grouped into fixed-length bit strings, is frequently represented by suitably spaced points in the complex plane. Quadrature amplitude modulation (QAM) transmission schemes are based on such a representation. In this section, we first present the usual (QAM) signal model and then propose a reformulation so as to apply hidden Markov model (HMM) and Kalman filtering (KF) methods.

The technical approach presented here is to work with the signals in a discrete set and associate with these a discrete state vector X_k which is an indicator function for the signal. Here, X_k belongs to a discrete set of unit vectors. The states X_k are assumed to be first-order Markov with known transition probability matrix $\mathbf{A}$ and state values $\mathbf{Z}$. Associated with the channel are time-varying parameters (gain, phase shift, and noise color), which are modeled as states x_k, in a continuous range $x_k \in \mathbb{R}^n$. The channel parameters arise from a known linear time-invariant stochastic system. State-space models are formulated involving a mixture of the states X_k and x_k, and are termed *mixed-state models*. These are reformulated using HMM filtering theory to achieve a nonlinear representation with a state vector consisting of q_k and x_k, where q_k is an unnormalized information state, representing a discrete-state conditional probability density. These reformulated models are termed *conditional information-state models*. Next, the EKF algorithm, or some derivative scheme, can be applied to this model for state estimation, thereby achieving both signal and channel estimation, also a coupled HMM/KF algorithm.

In this section we present the QAM signal model in the HMM framework. In Section 6.5 we present a coupled HMM/KF algorithm, which we apply to the model, in the simulation studies which follow.

Signal Model

Let m_k be a complex discrete-time signal ($k \in \mathbb{N}$) where for each k,

$$m_k \in \mathbf{Z} = \left\{z^{(1)}, \dots, z^{(2^N)}\right\}, \quad z^{(i)} \in \mathbb{C},\, N \in \mathbb{N}. \tag{3.1}$$

We also define the vector

$$z = z^R + \mathbf{j}z^I = \left(z^{(1)}, \dots, z^{(2^N)}\right)' \in \mathbb{C}^{2^N}. \tag{3.2}$$

For digital transmission, each element of $\mathbf{Z}$ is used to represent a string of N bits. In the case of QAM, each of these complex elements, $z^{(i)}$, is chosen so as to generate a rectangular grid of equally spaced points in the complex space $\mathbb{C}$. A 16-state ($N = 4$) QAM signal constellation is illustrated in Figure 3.1.

Now we note that at any time k, the message $m_k \in \mathbf{Z}$ is complex valued and can be represented in either polar or rectangular form, in obvious notation,

$$m_k = \rho_k \exp\left[\mathbf{j}\Upsilon_k\right] = m_k^R + \mathbf{j}m_k^I. \tag{3.3}$$

The real and imaginary components of m_k can then be used to generate piecewise constant-time signals, $m(t) = m_k$ for $t \in [t_k, t_{k+1})$, where t_k arises from regular sampling. The messages are then modulated and transmitted in quadrature as a QAM bandpass signal

$$s(t) = A_c\left[m^R(t)\cos(2\pi f t + \theta) + m^I(t)\sin(2\pi f t + \theta)\right], \tag{3.4}$$

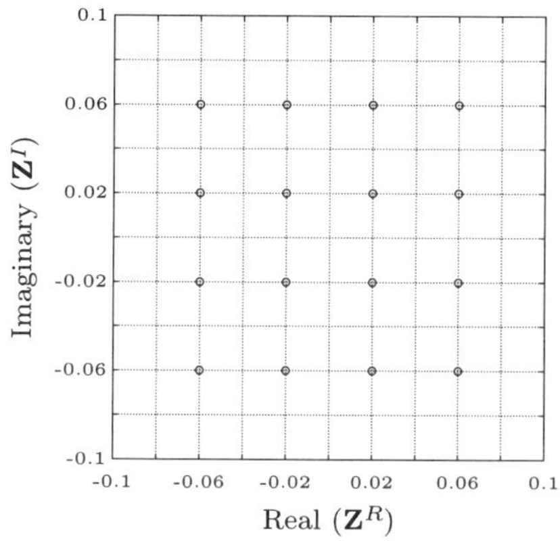

Fig. 3.1 16-state QAM signal constellation

where the carrier amplitude A_c, frequency f, and phase θ are constant. This transmission scheme is termed QAM because the signal is *quadrature* in nature, where the real and imaginary components of the message are transmitted as two *amplitudes* which *modulate* quadrature and in-phase carriers.

Channel Model

The QAM signal is passed through a channel which can cause amplitude and phase shifts, as, for example, in fading channels due to multiple transmission paths. The channel can be modeled by a multiplicative disturbance resulting in a discrete time base-band disturbance,

$$c_k = \kappa_k \exp[\mathbf{j}\phi_k] = c_k^R + \mathbf{j}c_k^I \in \mathbb{C} \tag{3.5}$$

which introduces time-varying gain and phase changes to the signal. The time variations in c_k are realistically assumed to be slow in comparison to the discrete-time message rate.

The baseband output of the channel, corrupted by additive noise w_k, is therefore given in discrete time, by

$$y_k = c_k s_k + w_k \in \mathbb{C}, \tag{3.6}$$

where c_k is given in (3.5). Assume that $w_k \in \mathbb{C}$ has i.i.d. real and imaginary parts, w_k^R and w_k^I, respectively, with zero mean and Gaussian, so that $w_k^R, w_k^I \sim N\left[0, \sigma_w^2\right]$.

Let us we work with the vector $x_k \in \mathbb{R}^2$ associated with the real and imaginary parts of c_k, as

$$x_k = \begin{pmatrix} \kappa_k \cos\phi_k \\ \kappa_k \sin\phi_k \end{pmatrix} = \begin{pmatrix} c_k^R \\ c_k^I \end{pmatrix}. \tag{3.7}$$

This is referred to as a Cartesian coordinate representation of the state.

Assumption on Channel Fading Characteristics Consider that the dynamics of x_k, from (3.7), are given by

$$x_{k+1} = F x_k + v_{k+1}, \qquad v_k \text{ are i.i.d.} \sim N[0, Q_k] \tag{3.8}$$

for some known F, (usually with $\lambda(F) < 1$, where λ indicates eigenvalues, to avoid unbounded x_k, and typically with $F = fI$ for some scalar $0 \ll f < 1$).

Another useful channel model can be considered using polar coordinates consisting of channel gain κ_k and phase ϕ_k as follows

$$\begin{aligned} \kappa_{k+1} &= f_\kappa \kappa_k + v_{k+1}^\kappa \qquad && \text{where } v_k^\kappa \text{ is Rayleigh distributed } \left[m_\kappa, \sigma_\kappa^2\right] \\ \phi_{k+1} &= f_\phi \phi_k + v_{k+1}^\phi \qquad && \text{where } v_k^\phi \text{ is uniform } [0, 2\pi) \end{aligned} \tag{3.9}$$

and typically, $0 \ll f_\kappa < 1$ and $0 \ll f_\phi < 1$.

It is usual to assume that the variations associated with the magnitude of the channel gain κ and the phase shift ϕ are independent, with variances given by σ_κ^2

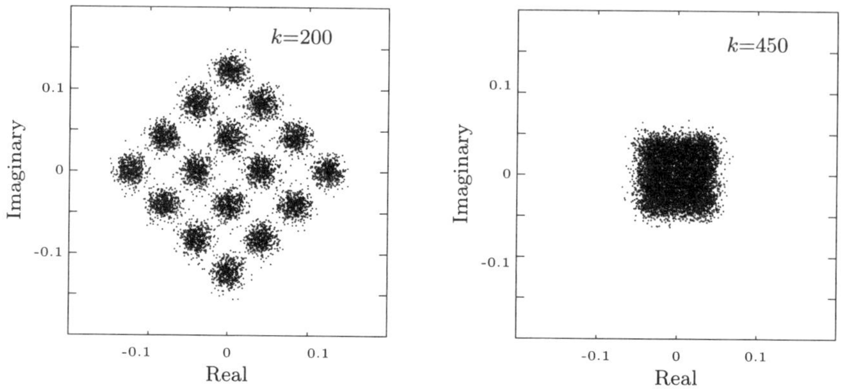

Fig. 3.2 16-state QAM signal constellation output from channel

and σ_ϕ^2, respectively. It follows, as in Anderson and Moore (1979), that an appropriate covariance matrix of the corresponding Cartesian channel model noise vector v_k, is given by

$$Q_k = E\left[v_k v_k'\right] \simeq \begin{bmatrix} \sigma_\kappa^2 \cos^2\phi_k + \kappa_k^2\sigma_\phi^2 \sin^2\phi_k & \left(\sigma_\kappa^2 - \kappa_k^2\sigma_\phi^2\right)\sin\phi_k\cos\phi_k \\ \left(\sigma_\kappa^2 - \kappa_k^2\sigma_\phi^2\right)\sin\phi_k\cos\phi_k & \sigma_\kappa^2\sin^2\phi_k + \kappa_k^2\sigma_\phi^2\cos^2\phi_k \end{bmatrix}. \tag{3.10}$$

Here we prefer to work with the Cartesian channel model, as it allows us to write the system in the familiar state-space form driven by Gaussian noise, thus facilitating the application of the EKF scheme presented later.

Remark 3.1 In Figure 3.2 the output constellation is presented, with signal-to-noise ratio SNR $= 6$ dB, from a channel with fading characteristics given in Example 1 of Section 6.5. The plots show 1000 data points at each of the constellation points for times $k = 200$ and $k = 450$, and give an indication of how the channel affects the QAM signal.

■

State-Space Signal Model

Consider the following assumption on the message sequence.

$$m_k \text{ is a first-order homogeneous Markov process.} \tag{3.11}$$

Remark 3.2 This assumption enables us to consider the signal in a Markov model framework, and thus allows Markov filtering techniques to be applied. It is a reasonable assumption on the signal, given that error correcting coding has been employed in transmission. Coding techniques, such as convolutional coding, produce signals which are not i.i.d. and yet to a first approximation display first-order Markov properties. Of course, i.i.d. signals can be considered in this framework too, since a Markov chain with a transition probability matrix which has all elements the same gives rise to an i.i.d. process.

■

Let us define the vector X_k to be an indicator function associated with m_k. Thus the state space of X, without loss of generality, can be identified with the set of unit vectors $\mathscr{S} = \{e_1, e_2, \dots, e_{2^N}\}$, where as earlier $e_i = (0,\dots,0,1,0,\dots,0)' \in \mathbb{R}^{2^N}$ with 1 in the ith position. Then

$$m_k = z' X_k, \tag{3.12}$$

where z is defined in (3.2). Under the assumption (3.11) the transition probability matrix associated with m_k, in terms of X_k, is

$$A = (a_{ji}), \ 1 \le i, j \le 2^N \qquad \text{where } a_{ij} = P(X_{k+1} = e_i \mid X_k = e_j)$$

so that

$$E[X_{k+1} \mid X_k] = AX_k.$$

Of course $a_{ij} \ge 0, \sum_{i=1}^{2^N} a_{ij} = 1$ for each j. We also denote $\{\mathscr{F}_l, l \in \mathbb{N}\}$ the complete filtration generated by X. The dynamics of X_k are given by the state equation

$$X_{k+1} = AX_k + V_{k+1}, \tag{3.13}$$

where V_{k+1} is a $\mathscr{F}_k$ martingale increment, in that $E[V_{k+1} \mid \mathscr{F}_k] = 0$.

The *observation process* from (3.6), for the Cartesian channel model, can be expressed in terms of the state X_k as

$$\begin{pmatrix} y_k^R \\ y_k^I \end{pmatrix} = \begin{pmatrix} (z^R)' X_k & -(z^I)' X_k \\ (z^I)' X_k & (z^R)' X_k \end{pmatrix} \begin{pmatrix} c_k^R \\ c_k^I \end{pmatrix} + \begin{pmatrix} w_k^R \\ w_k^I \end{pmatrix} \tag{3.14}$$

or equivalently, with the appropriate definition of $h(.)$, $\mathbf{y}_k$, $\mathbf{w}_k$,

$$\mathbf{y}_k = h(X_k) x_k + \mathbf{w}_k, \qquad \mathbf{w}_k \text{ are i.i.d.} \sim N[0, R_k]. \tag{3.15}$$

Note that, $E\left[w_{k+1}^R \mid \mathscr{G}_k\right] = 0$ and $E\left[w_{k+1}^I \mid \mathscr{G}_k\right] = 0$, where $\mathscr{Y}_l$ is the σ-field generated by $y_k, k \le l$, and $\mathscr{G}_k$ is the complete filtration generated by $X_k, \mathscr{Y}_k$. It is usual to assume that w^R and w^I are independent so that an appropriate covariance matrix associated with the measurement noise vector $\mathbf{w}_k$ has the form

$$R_k = \begin{bmatrix} \sigma_{w^R}^2 & 0 \\ 0 & \sigma_{w^I}^2 \end{bmatrix}. \tag{3.16}$$

For simplicity we take $\sigma^2_{w^R} = \sigma^2_{w^I} = \sigma^2_w$. It is now readily seen that

$$E\left[V_{k+1} \mid \mathscr{G}_k\right] = 0. \tag{3.17}$$

In order to demonstrate the attractiveness of the Cartesian channel model, we now use the properties of the indicator function X_k to express the observations (3.15) in a linear form with respect to X_k and x_k,

$$\begin{aligned} \mathbf{y}_k &= h(X_k)x_k + \mathbf{w}_k \\ &= [h(e_1)x_k h(e_2)x_k \cdots h(e_{2^N})x_k] X_k + \mathbf{w}_k \\ &= H[I_{2^N} \otimes x_k] X_k + \mathbf{w}_k, \end{aligned} \tag{3.18}$$

where $H = [h(e_1) \cdots h(e_{2^N})]$ and $\otimes$ denotes a *Kronecker product*. Recall that

$$A \otimes B = \begin{bmatrix} a_{11}B & a_{12}B & \dots \\ a_{21}B & a_{22}B & \dots \\ \vdots & \vdots & \ddots \end{bmatrix}.$$

The observations (3.18) are now in a form which is bilinear in X_k and x_k.

We shall define the vector of parametrized probability densities (which we will loosely call symbol probabilities), as $B_k = (b_k(i))$, for $b_k(i) := P[\mathbf{y_k} \mid X_k = e_i, x_k]$, where

$$\begin{aligned} b_k(i) = \frac{1}{2\pi\sigma_w^2} \exp\Bigg(&-\frac{\left\{y_k^R - \left[(z^R)' e_i c_k^R - (z^I)' e_i c_k^I\right]\right\}^2}{2\sigma_w^2} \\ &-\frac{\left\{y_k^I - \left[(z^I)' e_i c_k^R + (z^R)' e_i c_k^I\right]\right\}^2}{2\sigma_w^2}\Bigg). \end{aligned} \tag{3.19}$$

Now because w_k^R and w_k^I are i.i.d., $E(y_k \mid X_{k-1} = e_i, \mathscr{F}_{k-2}, \mathscr{Y}_{k-1}) = E(y_k \mid X_{k-1} = e_i)$ which is essential for formulating the signal model as an HMM, parametrized by the fading channel model parameters x_k.

To summarize, we now have the following lemma,

Lemma 3.3 *Under assumptions* (3.11) *and* (3.8), *the QAM signal model* (3.1)–(3.6) *has the following state-space representation in terms of the* 2^N*-dimension discrete-state message indicator function* X_k, *and* x_k, *the continuous-range state associated with the fading channel characteristics.*

$$\boxed{\begin{aligned} X_{k+1} &= AX_k + V_{k+1}, \\ x_{k+1} &= Fx_k + v_{k+1}, \\ \mathbf{y}_k &= H[I_{2^N} \otimes x_k] X_k + \mathbf{w}_k. \end{aligned}} \tag{3.20}$$

The signal model state equations are linear and the measurements are bilinear in X_k and x_k.

Remarks 3.4

1. If x_k is known, then the model specializes to an HMM denoted $\lambda = (A, \mathbf{Z}, \underline{\pi}, \sigma_\omega^2, x_k)$, where $\underline{\pi} = (\pi_i)$, defined from $\pi_i = P(X_1 = e_i)$, is the initial state probability vector for the Markov chain.
2. If X_k is known, then the model specializes to a linear state-space model.
3. By way of comparison, for a polar coordinate channel representation, the observation process can only be expressed in terms of a linear operator on the channel gain, with a nonlinear operator on the phase. Thus, if X_k and ϕ_k, or κ_k and ϕ_k, are known, then the model specializes to a linear state-space model, but not if X_k and κ_k are known and ϕ_k is unknown.

■

Conditional Information-State Signal Model

Let $\hat{X}_k(\mathscr{X})$ denote the conditional filtered state estimate of X_k at time k, given the channel parameters $\mathscr{X}_k = \{x_0, \ldots, x_k\}$, i.e.,

$$\hat{X}_k(\mathscr{X}) := E(X_k \mid \mathscr{Y}_k, \mathscr{X}_k, \lambda). \tag{3.21}$$

Let us again denote $\underline{1}$ to be the column vector containing all ones, and introduce "forward" variable (Rabiner 1989) $q_k(\mathscr{X})$ is such that the ith element $q_k^i(\mathscr{X}) := P(y_0, \ldots, y_k, X_k = e_i \mid \mathscr{X}_k)$. Observe that $\hat{X}_{k|\mathscr{X}}$ can be expressed in terms of $q_k(\mathscr{X})$ by

$$\hat{X}_k(\mathscr{X}) = \langle q_k(\mathscr{X}), \underline{1} \rangle^{-1} q_k(\mathscr{X}). \tag{3.22}$$

Here $q_k(\mathscr{X})$ is conveniently computed using the following "forward" recursion:

$$q_{k+1}(\mathscr{X}) = B(y_{k+1}, x_{k+1}) A q_k(\mathscr{X}), \tag{3.23}$$

where $B(y_{k+1}, x_{k+1}) = \operatorname{diag}\left(b_{k+1}(1), \ldots, b_{k+1}(2^N)\right)$.

We now seek to express the observations y_k in terms of the unnormalized conditional estimates $q_k(\mathscr{X})$.

Lemma 3.5 *The conditional measurements* $\mathbf{y}_k(\mathscr{X})$ *can be expressed as*

$$\mathbf{y}_k(\mathscr{X}) = H[I_{2^N} \otimes x_k] \langle q_{k-1}(\mathscr{X}), \underline{1} \rangle^{-1} A q_{k-1}(\mathscr{X}) + n_k(\mathscr{X}), \tag{3.24}$$

where $q_k(\mathscr{X})$ *given from* (3.23) *and* $n_k(\mathscr{X})$ *is a* $(\mathscr{X}_k, \mathscr{Y}_{k-1})$ *martingale increment with covariance matrix of the conditional noise term* $n_k(\mathscr{X})$ *is given by*

$$R_n = \sigma_w^2 I + H[I_{2^N} \otimes x_k] \left\{ \hat{X}_k^D(\mathscr{X}) - \hat{X}_k(\mathscr{X}) \hat{X}_k(\mathscr{X})' \right\} [I_{2^N} \otimes x_k]' H'. \tag{3.25}$$

Here $\hat{X}_k^D(\mathscr{X})$ is the matrix that has diagonal elements which are the elements of $\hat{X}_k(\mathscr{X})$.

Proof Following standard arguments, since $q_k(\mathscr{X})$ is measurable with respect to $\{\mathscr{X}_k, \mathscr{Y}_k\}$, then $E\left[w_{k+1}^R \mid \mathscr{Y}_k\right] = 0$, $E\left[w_{k+1}^I \mid \mathscr{Y}_k\right] = 0$ and $E\left[V_{k+1} \mid \mathscr{Y}_k\right] = 0$, so that

$$\begin{aligned}
&E\left[n_k(\mathscr{X}) \mid \mathscr{X}_k, \mathscr{Y}_{k-1}\right] \\
&\quad = E\Big[H\left[I_{2^N} \otimes x_k\right] X_k + \mathbf{w}_k \\
&\qquad\quad - H\left[I_{2^N} \otimes x_k\right] \langle q_{k-1}(\mathscr{X}), \underline{1}\rangle^{-1} A q_{k-1}(\mathscr{X}) \mid \mathscr{X}_k, \mathscr{Y}_{k-1}\Big] \\
&\quad = H\left[I_{2^N} \otimes x_k\right]\left(A\hat{X}_{k-1}(\mathscr{X}) - \langle q_{k-1}(\mathscr{X}), \underline{1}\rangle^{-1} A q_{k-1}(\mathscr{X})\right) \\
&\quad = 0.
\end{aligned}$$

Also

$$\begin{aligned}
R_n &= E\left[n_k(\mathscr{X}) n_k(\mathscr{X})' \mid \mathscr{X}_k, \mathscr{Y}_{k-1}\right] \\
&= E\left[\left(w_k + H\left[I_{2^N} \otimes x_k\right]\left(X_k - \frac{A q_{k-1}}{\langle q_{k-1}, \underline{1}\rangle}\right)\right)^2 \mid \mathscr{X}_k, \mathscr{Y}_{k-1}\right] \\
&= E\left[w_k^2 \mid \mathscr{X}_k, \mathscr{Y}_{k-1}\right] \\
&\quad + E\Big[H\left[I_{2^N} \otimes x_k\right]\left(X_k - \frac{A q_{k-1}}{\langle q_{k-1}, \underline{1}\rangle}\right)\left(X_k - \frac{A q_{k-1}}{\langle q_{k-1}, \underline{1}\rangle}\right)' \\
&\qquad\quad \times \left[I_{2^N} \otimes x_k\right]' H' \,\Big|\, \mathscr{X}_k, \mathscr{Y}_{k-1}\Big] \\
&= \sigma_w^2 I + H\left[I_{2^N} \otimes x_k\right] \\
&\quad \times E\left[\left(X_k - \hat{X}_{k|\mathscr{X}}\right)\left(X_k - \hat{X}_k(\mathscr{X})\right)' \mid \mathscr{X}_k, \mathscr{Y}_{k-1}\right]\left[I_{2^N} \otimes x_k\right]' H' \\
&= \sigma_w^2 I + H\left[I_{2^N} \otimes x_k\right]\left\{\hat{X}_k^D(\mathscr{X}) - \hat{X}_k(\mathscr{X})\hat{X}_k(\mathscr{X})'\right\}\left[I_{2^N} \otimes x_k\right]' H'.
\end{aligned}$$

□

We summarize the conditional information-state model in the following lemma.

Lemma 3.6 *The state-space representation* (3.20) *can be reformulated to give the following conditional information-state signal model with states $q_k(\mathscr{X})$,*

$$\boxed{\begin{aligned}
q_{k+1|\mathscr{X}} &= B(y_{k+1}, x_{k+1}) A q_k(\mathscr{X}), \\
x_{k+1} &= F x_k + v_k, \\
\mathbf{y}_k(\mathscr{X}) &= H\left[I_{2^N} \otimes x_k\right] \langle q_{k-1}(\mathscr{X}), \underline{1}\rangle^{-1} A q_{k-1}(\mathscr{X}) + n_k(\mathscr{X}).
\end{aligned}} \tag{3.26}$$

Remark 3.7 When $F \equiv I$ and $v \equiv 0$, then x_k is constant. Under these conditions, the problem of channel-state estimation reduces to one of parameter identification, and recursive prediction error techniques can be used, as in Collings et al. (1993). However, an EKF or some derivative scheme is required for parameter tracking when x_k is not constant, as in Section 6.5.

■

6.4 Example: Frequency Modulation

Frequency modulation (FM) is a common method for information transmission. Frequency-modulated signals carry the information message in the frequency component of the signal. In this section, we first present the usual FM signal model including a fading channel, and then propose a reformulation involving discrete-time sampling and quantization so as to apply hidden Markov model (HMM) and extended Kalman filtering (EKF) methods. The resulting mixed-continuous-discrete-state FM model is similar in form to the QAM model of Section 6.3. However, there is an added layer of complexity in that an integration (summation) of message states is included in the model. The observations are a function of the phase of the signal while the message is contained in the frequency (phase is an integration of frequency in this case), whereas for QAM, the observations are a function of the complex-valued amplitude of the signal, which directly represents the message.

In addition to this added layer of complexity, in order to implement a demodulation scheme taking advantage of an adaptive HMM approach, it is necessary to quantize the frequency space. The quantization and digital sampling rate are design parameters which introduce suboptimality to the FM receiver. However, if the quantization is fine enough, and the sampling rate fast enough, then the loss in performance due to digitization will be outweighed by the performance gain (over more standard schemes) from the demodulation scheme presented here. If the signal is a digital frequency shift-keyed (FSK) signal, then these quantization errors would not arise. Our schemes are equally applicable to analog FM and digital FSK. Also, it should be noted that the techniques presented in this paper are applicable to other frequency/phase-based modulation schemes. In particular, continuous phase modulated (CPM) signals can be seen as a derivative of these FM models. CPM transmission schemes have a reduced-order model due to the message being carried in the phase, as opposed to the frequency. In fact, the CPM model has the same form as that for quadrature amplitude modulation (QAM).

We now reformulate FM signal models with fading channels in a state-space form involving discrete states X_k and continuous range states x_k.

Signal Model

Let f_k be a real-valued discrete-time signal, where for each k,

$$f_k \in \mathbf{Z}_f = \left\{z_f^{(1)}, \dots, z_f^{(L_f)}\right\}, \qquad z_f^{(i)} = \left(i/L_f\right)\pi \in \mathbb{R}, \qquad L_f \in \mathbb{N}. \tag{4.1}$$

We also denote the vector

$$z_f = \left(z_f^{(1)}, \ldots, z_f^{(L_f)}\right)' \in \mathbb{R}^{L_f}. \tag{4.2}$$

Therefore, each of the $M \in \mathbb{N}$ elements of $\mathbf{Z}_f$ is an *equally spaced* frequency in the range $[0, \pi)$. Now we note that at any k, the message, $f_k \in \mathbf{Z}_f$, is real valued and can be used to generate piecewise constant signals by $f(t) = f_k$ for $t \in [t_k, t_{k+1})$. For transmission, the instantaneous frequency, $f_i(t)$ is varied linearly with the baseband signal $f(t)$, $f_i(t) = f_c + f(t)$. This gives the following transmitted signal.

$$s_{\text{trans}}(t) = A_c \cos\left[2\pi f_c t + 2\pi\theta(t)\right], \qquad \theta(t) = \int_0^t f(\tau)\, d\tau, \tag{4.3}$$

where the carrier amplitude A_c and frequency f_c are constant, and $\theta(t)$ is the phase of the signal. For the formulation which follows it is convenient to represent the FM signal, assumed sampled in a quadrature and in-phase manner, in a complex baseband notation, as

$$s(t) = A_c \exp\left[\mathbf{j}\theta(t)\right], \qquad s_k = A_c \exp\left[\mathbf{j}\theta_k\right], \tag{4.4}$$

where the amplitude A is a known constant and

$$\theta_k = \left(\theta_{k-1} + f_k\right)_{2\pi}. \tag{4.5}$$

Here $(.)_{2\pi}$ denotes modulo 2π addition.

Channel Model

The FM signal is passed through a fading noise channel as in the case of the QAM signals. Thus, the channel can be modeled by a multiplicative disturbance, $c(t)$, resulting in a discrete-time baseband disturbance c_k as in (3.5). The baseband output of the channel, corrupted by additive noise w_k, is therefore given by (3.6). The channel states are given by (3.7).

State-Space Signal Model

For the FM signal, a discrete-time state space signal model is now generated. Consider the following assumption on the message signal.

$$f_k \text{ is a first-order homogeneous Markov process.} \tag{4.6}$$

Remarks 4.1

1. This assumption enables us to consider the signal in a Markov model framework, and thus allows Markov filtering techniques to be applied. It is a reasonable as-

sumption on the signal if the transition probability matrix used is a diagonally dominated, Toeplitz, circulant matrix. In the case of digital frequency shift-keyed (FSK) signals, the assumption is still valid, given that error-correcting coding has been employed in transmission. Coding techniques, such as convolutional coding produce signals which are not i.i.d., yet display Markov properties. Of course, i.i.d. signals can be considered in this framework too, since a Markov chain with a transition probability matrix which has all elements the same gives rise to an i.i.d. process.
2. Higher-order message signal models are discussed below. It is known that HMM signal processing is robust to modeling errors. Therefore, in the absence of knowledge about the true message signal model, reasonable estimates will still be generated.

■

Let us define a vector X_k^f to be an indicator function associated with f_k. Thus, the state-space of X^f, without loss of generality, can be identified with the set of unit vectors $S_{X^f} = \{e_1, e_2, \ldots, e_{L_f}\}$, where $e_i = (0, \ldots, 0, 1, 0, \ldots, 0)' \in \mathbb{R}^{L_f}$ with 1 in the ith position, so that

$$f_k = z_f' X_k^f. \tag{4.7}$$

Under the assumption (4.6) the transition probability matrix associated with f_k, in terms of X_k^f, is

$$A^f = \left(a_{ji}^f\right) \qquad 1 \leq i, j \leq L_f \text{ where } a_{ji}^f = P\left(X_{k+1}^f = e_j \mid X_k^f = e_i\right)$$

so that

$$E\left[X_{k+1}^f \mid X_k^f\right] = A^f X_k^f.$$

Of course, $a_{ij}^f \geq 0, \sum_{i=1}^{L_f} a_{ij}^f = 1$, for each j. We also denote $\{\mathscr{F}_l, l \in \mathbb{N}\}$ the complete filtration generated by X^f, that is, for any $k \in \mathbb{N}, \mathscr{F}_k$ is the complete σ-field generated by $X_k^f, l \leq k$.

As before, the dynamics of X_k^f are given by the state equation

$$X_{k+1}^f = A^f X_k^f + V_{k+1}, \tag{4.8}$$

where V_{k+1} is a $\mathscr{F}_k$ martingale increment.

As noted previously, the states represented by X^f are each characterized by a real value $z_f^{(i)}$ corresponding to the unit vector $e_i \in S_{X^f}$. These are termed the state values of the Markov chain.

When considering the finite-discrete set of possible message signals $\mathbf{Z}_f$, it becomes necessary to quantize the time-sum of these message signals, θ_k, given from (4.5). We introduce the set

$$\mathbf{Z}_\theta = \left\{z_\theta^{(1)}, \ldots, z_\theta^{(L_\theta)}\right\} \qquad \text{where } z_\theta^{(i)} = \frac{2\pi i}{(L_\theta)} \in \mathbb{R} \tag{4.9}$$

and corresponding vector

$$z_\theta = \left(z_\theta^{(1)}, \ldots, z_\theta^{(L_\theta)}\right)' \in \mathbb{R}^{L_\theta}. \tag{4.10}$$

Lemma 4.2 *Given the discrete-state message $f_k \in \mathbf{Z}_f$ from* (4.1)*, the phase θ_k from* (4.5) *and the set $\mathbf{Z}_\theta$ of* (4.9)*, then $\theta_k \in \mathbf{Z}_\theta$ for any k, iff $L_\theta = 2nL_f$, for some $n \in \mathbb{N}$.*

Proof For any $a \in \{1,\ldots,L_\theta\}$ and $b \in \{1,\ldots,L_f\}$

$$\begin{aligned}\theta_{k+1} &= (\theta_k + f_k)_{2\pi} \\ &= \left(a\frac{2\pi}{L_\theta} + b\frac{2\pi n}{L_\theta}\right)_{2\pi} \in \mathbf{Z}_\theta.\end{aligned}$$

□

Let us now define a vector $X_k^\theta \in S_{X^f} = \{e_1,\ldots,e_{L_\theta}\}$ to be an indicator function associated with θ_k, so that when $\theta_k = z_\theta^{(i)}$, $X_k^\theta = e_i$. Now given (4.5), X_{k+1}^θ is a "rotation" on X_k^θ by an amount determined from X_{k+1}^f. In particular,

$$X_{k+1}^\theta = A^\theta\left(X_{k+1}^f\right).X_k^\theta, \tag{4.11}$$

where $A^\theta(.)$ is a transition probability matrix given by

$$A^\theta\left(X_{k+1}^f\right) = S^{r_{k+1}}, r_k = [1,2,\ldots,L_f]\,X_k^f \tag{4.12}$$

and S is the rotation operator

$$S = \begin{pmatrix} 0 & 0 & \ldots & 0 & 1 \\ 1 & 0 & \ldots & 0 & 0 \\ 0 & 1 & \ldots & 0 & 0 \\ \vdots & \vdots & \ddots & \vdots & \vdots \\ 0 & 0 & \ldots & 1 & 0 \end{pmatrix}. \tag{4.13}$$

The *observation process* (3.5) can now be expressed in terms of the state X_k^θ

$$\begin{pmatrix} y_k^R \\ y_k^I \end{pmatrix} = \begin{pmatrix} A_c\cos\left[z_\theta' X_k^\theta\right] & -A_c\sin\left[z_\theta' X_k^\theta\right] \\ A_c\sin\left[z_\theta' X_k^\theta\right] & A_c\cos\left[z_\theta' X_k^\theta\right] \end{pmatrix}\begin{pmatrix} c_k^R \\ c_k^I \end{pmatrix} + \begin{pmatrix} w_k^R \\ w_k^I \end{pmatrix} \tag{4.14}$$

or, equivalently, with the appropriate definition of $h_\theta(.)$,

$$\begin{aligned}\mathbf{y}_k &= h_\theta\left(X_k^\theta\right)x_k + \mathbf{w}_k, \qquad \mathbf{w}_k \text{ are i.i.d.} \sim N[0,R_k] \\ &= \left[h_\theta(e_1)x_k h_\theta(e_2)x_k \cdots h_\theta\left(e_{L_\theta}\right)x_k\right] X_k^\theta + \mathbf{w}_k \\ &= H_\theta\left[I_{L_\theta} \otimes x_k\right] X_k^\theta + \mathbf{w}_k, \end{aligned} \tag{4.15}$$

where the augmented matrix $H_\theta = \left[h_\theta(e_1) \dots h_\theta(e_{L_\theta})\right]$. Here we see that the Cartesian coordinates for the channel model allow the observations to be written in a form which is linear in both X_k^θ and x_k. Note that $E\left[w_{k+1}^R \mid \mathscr{G}_k\right] = 0$ and $E\left[w_{k+1}^I \mid \mathscr{G}_k\right] = 0$, where $\mathscr{Y}_l$ is the σ-field generated by $\mathscr{Y}_k, k \leq l$. It is usual to assume that w^R and w^I are independent so that the covariance matrix associated with the measurement noise vector $\mathbf{w}_k$ has the form

$$R_k = \begin{bmatrix} \sigma_{w^R}^2 & 0 \\ 0 & \sigma_{w^I}^2 \end{bmatrix}. \tag{4.16}$$

Here we take $\sigma_{w^R}^2 = \sigma_{w^I}^2 = \sigma_w^2$.

It is now readily seen that

$$E\left[V_{k+1} \mid \mathscr{G}_k\right] = 0. \tag{4.17}$$

We now define the vector of parametrized probability densities as $\mathbf{b}_k^\theta = \left(b_k^\theta(i)\right)$, for $b_k^\theta(i) := P\left[y_k \mid X_k^\theta = e_i, x_k\right]$,

$$\begin{aligned} b_k^\theta(i) = \frac{1}{2\pi\sigma_w^2} \exp\Bigg(&- \frac{\left[y_k^R - A_c\left(\cos\left[z_\theta' e_i\right] c_k^R - \sin\left[z_\theta' e_i\right] c_k^I\right)\right]^2}{2\sigma_w^2} \\ &- \frac{\left[y_k^I - A_c\left(\sin\left[z_\theta' e_i\right] c_k^R + \cos\left[z_\theta' e_i\right] c_k^I\right)\right]^2}{2\sigma_w^2}\Bigg). \end{aligned} \tag{4.18}$$

In summary, we have the following lemma:

Lemma 4.3 *Under assumptions* (4.6) *and* (3.8), *the FM signal models* (4.1)–(4.5) *have the following state-space representation in terms of the L_f and L_θ indicator functions X_k^θ and X_k^f, respectively,*

$$\boxed{\begin{aligned} X_{k+1}^f &= A^f X_k^f + V_{k+1}^f, \\ X_{k+1}^\theta &= A^\theta\left(X_{k+1}^f\right) . X_k^\theta, \\ x_{k+1} &= F x_k + v_{k+1}, \\ \mathbf{y}_k &= H_\theta\left[I_{L_\theta} \otimes x_k\right] X_k^\theta + \mathbf{w}_k. \end{aligned}} \tag{4.19}$$

Remarks 4.4

1. Observe that the model is in terms of the channel parameters (states) x_k in a continuous range, and in terms of indicator functions (states) which belong to a finite-discrete set, being the vertices of a simplex.
2. This model (4.19) has linear dynamics for the states x_k and X_k^f, but X_{k+1}^θ is bilinear in X_k^θ and X_{k+1}^f. The measurements are bilinear in x_k and X_k^θ.

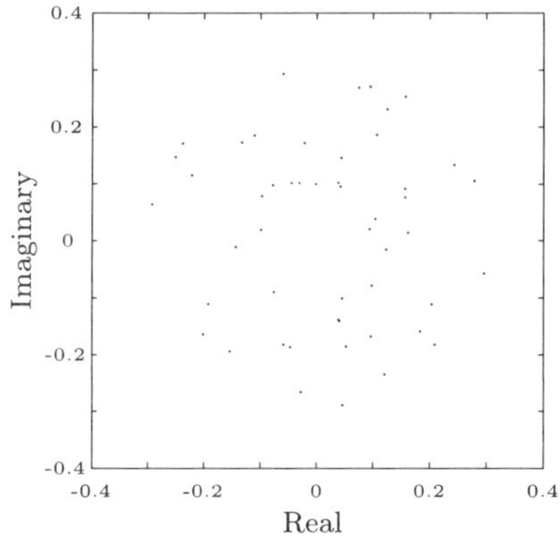

$\kappa(t) = 1 + 0.5\sin(2\pi t/5000)$, $\phi(t) = 0.04\pi\cos(2\pi t/5000)$, SNR = 2.4 dB

Fig. 4.1 FM signal constellation output from channel

3. In Figure 4.1 we present the channel output for 5000 data points, with channel noise variance $\sigma_w^2 = 0.00001$, and channel dynamics $\kappa(t) = 1 + 0.5\sin\left(2\pi t/5000\right)$, $\phi(t) = 0.04\pi\cos\left(2\pi t/5000\right)$. In our simulations we use much more rapidly changing channel dynamics, given in Example 1, over only 1000 points. Figure 4.1 is used here merely to illustrate the nature of the channel's effect on the signal constellation.

■

Conditional Information-State Signal Models

Let $\hat{X}_k^f\left(\mathscr{X}, \mathscr{X}^\theta\right)$ and $\hat{X}_k^\theta\left(\mathscr{X}, \mathscr{X}^f\right)$ denote the conditional filtered state estimates of X_k^f and X_k^θ, respectively, where $\mathscr{X}_k = \left\{x_0, \ldots, x_k\right\}$, $\mathscr{X}_k^\theta = \left\{X_0^\theta, \ldots, X_k^\theta\right\}$, and $\mathscr{X}_k^f = \left\{X_0^f, \ldots, X_k^f\right\}$. In particular,

$$\hat{X}_k^f\left(\mathscr{X}, \mathscr{X}^\theta\right) := E\left(X_k^f \mid \mathscr{Y}_k, \mathscr{X}_k, \mathscr{X}_{k-1}^\theta, \lambda\right), \tag{4.20}$$

$$\hat{X}_k^\theta\left(\mathscr{X}, \mathscr{X}^f\right) := E\left(X_k^\theta \mid \mathscr{Y}_k, \mathscr{X}_k, \mathscr{X}_k^f, \lambda\right). \tag{4.21}$$

Let us again denote the column vector containing all ones as $\underline{1}$, and introduce "forward" variables, $q_k^f\left(\mathscr{X}, \mathscr{X}^\theta\right)$ and $q_k^\theta\left(\mathscr{X}, \mathscr{X}^f\right)$, such that their ith elements are given, respectively, by $P\left(y_0, \ldots, y_k, X_k^f = e_i \mid \mathscr{X}_k, \mathscr{X}_{k-1}^\theta\right)$ and $P\left(y_0, \ldots, y_k, X_k^\theta = e_i \mid \mathscr{X}_k, \mathscr{X}_k^f\right)$.

Observe that $\hat{X}_k^f(\mathscr{X},\mathscr{X}^\theta)$ and $\hat{X}_k^\theta(\mathscr{X},\mathscr{X}^f)$ can be expressed in terms of $q_k^f(\mathscr{X},\mathscr{X}^\theta)$ and $q_k^\theta(\mathscr{X},\mathscr{X}^f)$, respectively, by

$$\hat{X}_k^f(\mathscr{X},\mathscr{X}^\theta) = \langle q_k^f(\mathscr{X},\mathscr{X}^\theta),\underline{1}\rangle^{-1} q_k^f(\mathscr{X},\mathscr{X}^\theta), \tag{4.22}$$

$$\hat{X}_k^\theta(\mathscr{X},\mathscr{X}^f) = \langle q_k^\theta(\mathscr{X},\mathscr{X}^f),\underline{1}\rangle^{-1} q_k^\theta(\mathscr{X},\mathscr{X}^f). \tag{4.23}$$

Here $q_k^f(\mathscr{X},\mathscr{X}^\theta)$ is conveniently computed using the following "forward" recursion:

$$q_{k+1}^f(\mathscr{X},\mathscr{X}^\theta) = B^f(y_{k+1},x_{k+1},X_k^\theta)A^f q_k^f(\mathscr{X},\mathscr{X}^\theta), \tag{4.24}$$

where $B^f(y_{k+1},x_{k+1},X_k^\theta) = \operatorname{diag}\left(b_{k+1}^f(1),\ldots,b_{k+1}^f(L_\theta)\right)$, and where $b_{k+1}^f(i) := P\left[y_{k+1} \mid X_{k+1}^f = e_i, x_{k+1}, X_k^\theta\right]$, for $b_{k+1}^f(i)$ given explicitly in (4.26).

Also, $q_k^\theta(\mathscr{X},\mathscr{X}^f)$ is conveniently computed using (4.11) to give the following "forward" recursion:

$$\begin{aligned} q_{k+1}^\theta(\mathscr{X},\mathscr{X}^f) &= B^\theta(y_{k+1},x_{k+1})\left(\sum_{i=1}^{L_f} A^\theta(e_i)X_{k+1}^f(i)\right).q_k^\theta(\mathscr{X},\mathscr{X}^f) \\ &= B^\theta(y_{k+1},x_{k+1})\mathscr{A}^\theta\left[X_{k+1}^f \otimes I_{L_\theta}\right].q_k^\theta(\mathscr{X},\mathscr{X}^f), \end{aligned} \tag{4.25}$$

where $B^\theta(y_{k+1},x_{k+1}) = \operatorname{diag}\left(b_{k+1}^\theta(1),\ldots,b_{k+1}^{f\theta}(L_\theta)\right)$, and where $b_k^\theta(i)$ is given in (4.18). Also, $\mathscr{A}^\theta = \left[A^\theta(e_1),\ldots,A^\theta\left(e_{L_f}\right)\right]$.

We can now write

$$\begin{aligned} b_k^f(i) = &\frac{1}{2\pi\sigma_w^2} \\ &\times \exp\left(-\frac{\left[y_k^R - A_c\left(\cos\left(z_\theta' A^\theta(e_i)X_{k-1}^\theta\right)c_k^R - \sin\left(z_\theta' A^\theta(e_i)X_{k-1}^\theta\right)c_k^I\right)\right]^2}{2\sigma_w^2}\right. \\ &\left. - \frac{\left[y_k^I - A_c\left(\sin\left(z_\theta' A^\theta(e_i)X_{k-1}^\theta\right)c_k^R + \cos\left(z_\theta' A^\theta(e_i)X_{k-1}^\theta\right)c_k^I\right)\right]^2}{2\sigma_w^2}\right). \end{aligned} \tag{4.26}$$

We seek to express the observations, y_k, in terms of the unnormalized conditional estimates, $q_{k-1}^f(\mathscr{X},\mathscr{X}^\theta)$.

Lemma 4.5 *The conditional measurements* $\mathbf{y}_k(\mathscr{X},\mathscr{X}^\theta)$ *are defined by*

$$\begin{aligned} \mathbf{y}_k(\mathscr{X},\mathscr{X}^\theta) = &H_\theta\left[I_{L_\theta} \otimes x_k\right] \\ &\times \mathscr{A}^\theta\left[\left\langle q_{k-1}^f(\mathscr{X},\mathscr{X}^\theta),\underline{1}\right\rangle^{-1} A^f q_{k-1}^f(\mathscr{X},\mathscr{X}^\theta) \otimes I_{L_\theta}\right] X_{k-1}^\theta \\ &+ n_k(\mathscr{X},\mathscr{X}^\theta), \end{aligned} \tag{4.27}$$

where $n_k(\mathscr{X},\mathscr{X}^\theta)$ *is a* $\left(\mathscr{X}_k,\mathscr{X}_{k-1}^\theta,\mathscr{Y}_{k-1}\right)$ *martingale increment.*

Proof Following standard arguments since $q_k^f(\mathscr{X},\mathscr{X}^\theta)$ is measurable with respect to $\{\mathscr{X}_k,\mathscr{X}_{k-1}^\theta,\mathscr{Y}_k\}$, $E\left[w_{k+1}^R \mid \mathscr{Y}_k\right]=0$, $E\left[w_{k+1}^I \mid \mathscr{Y}_k\right]=0$, and $E\left[V_{k+1}^f \mid \mathscr{Y}_k\right]=0$, then

$$\begin{aligned}
&E\left[n_k(\mathscr{X},\mathscr{X}^\theta) \mid \mathscr{X}_k,\mathscr{X}_{k-1}^\theta,\mathscr{Y}_{k-1}\right]\\
&= E\Big[H_\theta\left[I_{L_\theta}\otimes x_k\right]X_k^\theta+\mathbf{w}_k\\
&\qquad -H_\theta[I_{L_\theta}\otimes x_k]\mathscr{A}^\theta\left[\langle q_{k-1}^f(\mathscr{X},\mathscr{X}^\theta),\underline{1}\rangle^{-1}A^f q_{k-1}^f(\mathscr{X},\mathscr{X}^\theta)\otimes I_{L_\theta}\right]X_{k-1}^\theta\\
&\qquad \Big|\ \mathscr{X}_k,\mathscr{X}_{k-1}^\theta,\mathscr{Y}_{k-1}\Big]\\
&= H_\theta\cdot\left[I_{L_\theta}\otimes x_k\right]\\
&\quad\times\Big(\mathscr{A}^\theta\left[\hat{X}_k^f(\mathscr{X},\mathscr{X}^\theta)\otimes I_{L_\theta}\right]X_{k-1}^\theta\\
&\qquad -\mathscr{A}^\theta\left[\langle q_{k-1}^f(\mathscr{X},\mathscr{X}^\theta),\underline{1}\rangle^{-1}A^f q_{k-1}^f(\mathscr{X},\mathscr{X}^\theta)\otimes I_{L_\theta}\right]X_{k-1}^\theta\Big)\\
&= 0.
\end{aligned}$$

□

In practice, however, as noted above, we do not have access to X_{k-1}^θ, but at best its conditional expectation, $q_{k-1}^\theta(\mathscr{X},\mathscr{X}^f)$. Therefore, the conditional measurement for a more useful model generating $\mathbf{y}_k(\mathscr{X})$ (which can be used in practice) does not have a martingale increment noise term, $n_k(\mathscr{X})$. In addition, the covariance matrix, R_n, of $n_k(\mathscr{X})$, is of higher magnitude than that of w_k. The exact form of R_n is, however, too complicated for presentation here, and "application-based" estimates of R_n can be used when implementing these algorithms.

In summary, we have the following lemma:

Lemma 4.6 *The state-space representation* (4.19) *can be reformulated to give the following conditional information-state signal model, with states* $q_k^f(\mathscr{X},\mathscr{X}^\theta)$ *and* $q_k^\theta(\mathscr{X},\mathscr{X}^f)$,

$$\boxed{\begin{aligned}
q_{k+1}^f(\mathscr{X},\mathscr{X}^\theta) &= q_k^f(\mathscr{X},\mathscr{X}^\theta)B^f\left(y_{k+1},x_k,X_k^\theta\right)A^f,\\
q_{k+1}^\theta(\mathscr{X},\mathscr{X}^f) &= q_k^\theta(\mathscr{X},\mathscr{X}^f)B^\theta(y_{k+1},x_{k+1})\\
&\quad\times\mathscr{A}^\theta\left[\langle q_k^f(\mathscr{X},\mathscr{X}^\theta),\underline{1}\rangle^{-1}A^f q_k^f(\mathscr{X},\mathscr{X}^\theta)\otimes I_{L_\theta}\right],\\
x_{k+1} &= Fx_k+v_k,\\
\mathbf{y}_k(\mathscr{X}) &= H_\theta\left[I_{L_\theta}\otimes x_k\right]\mathscr{A}^\theta\left[\frac{A^f q_{k-1}^f(\mathscr{X},\mathscr{X}^\theta)}{\left\langle q_{k-1}^f(\mathscr{X},\mathscr{X}^\theta),\underline{1}\right\rangle}\otimes I_{L_\theta}\right]\\
&\quad\times\frac{q_{k-1}^\theta(\mathscr{X},\mathscr{X}^f)}{\langle q_{k-1}^\theta(\mathscr{X},\mathscr{X}^f),\underline{1}\rangle}+n_k(\mathscr{X}).
\end{aligned}} \tag{4.28}$$

This fading channel FM signal model is now in standard state-space form to allow the application of Kalman filtering.

Higher-Order Message Models

Lemma 4.6 provides insight into methods for coping with higher-order message models, and thus allows us to relax the assumption (4.6). To do this we continue to quantize the range of frequencies, but we allow the model of the frequencies to be in a continuous range and in vector form. Therefore, the state-space representation of the frequency message is no longer a first-order system. Also, the quantization errors are now involved with the phase estimate. This approach allows the message-frequency model to be in a continuous range while still employing the attractive optimal filtering of the HMM filter for the phase. The following state-space model applies,

$$\boxed{\begin{aligned} x^f_{k+1} &= F^f_k x^f_k + v^f_{k+1}, \\ X^\theta_{k+1} &= A^\theta\left(X^f_k(h' x^f_k)\right).X^\theta_k, \\ x_{k+1} &= F x_k + v_{k+1}, \\ \mathbf{y}_k &= H_\theta\left[I_{L_\theta} \otimes x_k\right] X^\theta_k + \mathbf{w}_k, \end{aligned}} \tag{4.29}$$

where F^f_k is the function associated with the dynamics of the frequency message given by the state x^f_k, the scalar message frequency is given by $h' x^f_k$, and X^f_k is the quantized frequency in state-space form.

Following the steps presented above for the signal under the assumption (4.6), an information-state signal model can be generated for this higher-order state space signal model. As with the previous information-state signal model, this higher-order model also results in the EKF/HMM scheme which follows.

6.5 Coupled-Conditional Filters

The algorithm presented here is referred to as the KF/HMM scheme. It consists of a KF for channel estimation, coupled with HMM filters for signal state estimation. These algorithms apply equally to both the QAM and FM signal models presented in the previous two sections.

In the QAM case of (3.26), the HMM estimator for the signal state, $\hat{q}_k$, conditioned on the channel estimate sequence $\{\hat{x}_k\}$, is given by

$$\hat{q}_{k+1}(\hat{x}_k) = B(y_{k+1}, \hat{x}_k) A \hat{q}_k(\hat{x}_{k-1}), \tag{5.1}$$

$$\hat{X}_k(\hat{x}_{k-1}) = \langle q_k(\hat{x}_{k-1}), \underline{1}\rangle^{-1} q_k(\hat{x}_{k-1}). \tag{5.2}$$

In the FM case of (4.28), the conditional HMM estimators for the signal states, $\hat{q}_k^f$, $\hat{q}_k^\theta$, are given by

$$\hat{q}_{k+1}^f(\hat{x}_k, \hat{q}_k^\theta) = B^f(y_{k+1}, \hat{x}_k, \hat{q}_k^\theta) A^f \hat{q}_k^f(\hat{x}_{k-1}, \hat{q}_{k-1}^\theta), \tag{5.3}$$

$$\hat{X}_k^f(\hat{x}_{k-1}, \hat{q}_{k-1}^\theta) = \langle q_k^f(\hat{x}_{k-1}, \hat{q}_{k-1}^\theta), \underline{1}\rangle^{-1} q_k^f(\hat{x}_{k-1}, \hat{q}_{k-1}^\theta), \tag{5.4}$$

$$\hat{q}_{k+1}^\theta(\hat{x}_k, \hat{q}_k^f) = B^\theta(y_{k+1}, \hat{x}_k, \hat{q}_k^f)(A^\theta) \hat{q}_k^\theta(\hat{x}_{k-1}, \hat{q}_{k-1}^f), \tag{5.5}$$

$$\hat{X}_k^\theta(\hat{x}_{k-1}, \hat{q}_{k-1}^f) = \langle q_k^\theta(\hat{x}_{k-1}, \hat{q}_{k-1}^f), \underline{1}\rangle^{-1} q_k^\theta(\hat{x}_{k-1}, \hat{q}_{k-1}^f). \tag{5.6}$$

The Kalman filter equations for the channel parameter, x_k, conditioned on the indicator state estimates, $\hat{X}_{k-1}$ in the QAM case, and $\hat{X}_{k-1}^f$ and $\hat{X}_{k-1}^\theta$ in the FM case, or equivalently on the corresponding information states denoted loosely here as $q_{k-1}, q_{k-1}^f, q_{k-1}^\theta$, are

$$\hat{x}_{k|k} = \hat{x}_{k|k-1} + K_k\left[y_k - H_k' \hat{x}_{k|k-1}\right], \tag{5.7}$$

$$\hat{x}_{k+1|k} = F\hat{x}_{k|k}, \tag{5.8}$$

$$K_k = \Sigma_{k|k-1} H_k \left[H_k' \Sigma_{k|k-1} H_k + R_k\right]^{-1}, \tag{5.9}$$

$$\Sigma_{k|k} = \Sigma_{k|k-1} - \Sigma_{k|k-1}\left[H_k'\Sigma_{k|k-1}H_k + R_k\right]^{-1} H_k'\Sigma_{k|k-1}, \tag{5.10}$$

$$\Sigma_{k+1|k} = F\Sigma_{k|k}F' + Q_k, \tag{5.11}$$

where

$$H_k = \begin{cases} \dfrac{\partial\left(H\,.\,[I_{2^N} \otimes Fx]\,\langle q_{k-1}, \underline{1}\rangle^{-1} A q_{k-1}\right)}{\partial x}\Big|_{x=x_k} \\ \qquad\qquad \text{for QAM model (3.26)} \\ \dfrac{\partial\left(H_\theta\,.\,[I_{L_\theta} \otimes Fx]\,\mathscr{A}^\theta\left[\frac{A^f q_{k-1}^f}{\langle q_{k-1}^f, \underline{1}\rangle} \otimes I_{L_\theta}\right]\frac{q_{k-1}^\theta}{\langle q_{k-1}^\theta, \underline{1}\rangle}\right)}{\partial x}\Big|_{x=x_k} \\ \qquad\qquad \text{for FM model (4.28)} \end{cases} \tag{5.12}$$

and R is the covariance matrix of the noise on the observations $\mathbf{w}$ given in (4.16), Q is the covariance matrix of v, and Σ is the covariance matrix of the channel parameter estimate $\hat{x}$ [x is defined in (3.7)].

Figure 5.1 gives a block diagram for this adaptive HMM scheme, for the QAM model, when Switch 1 is open and Switch 2 is in the top position. This figure is generated from the observation representation (4.15). Further assumptions can be made for simplification if the maximum *a priori* estimate of q_k were used, indicated by having Switch 2 in the lower position. Figure 5.2 gives a block diagram for the adaptive HMM scheme, for the FM model, when Switch 1 and 2 are open.

Figure 5.3 shows the scheme in simplified block form for the case of (3.26). A further suboptimal KF/HMM scheme can be generated by using the state-space signal models (3.26) and (4.28), and estimating the KF conditioned on a maximum *a*

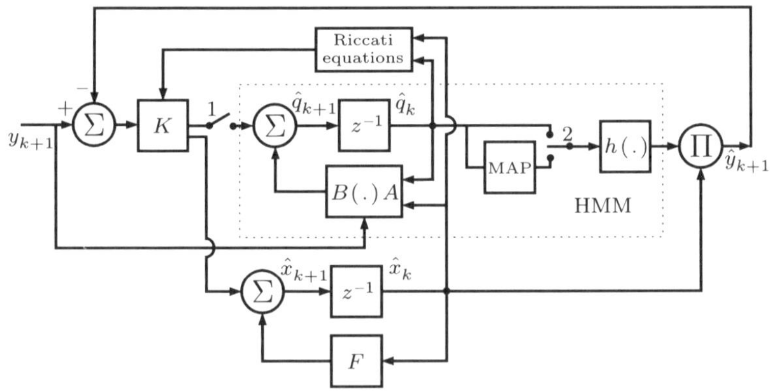

Fig. 5.1 KF/HMM scheme for adaptive HMM filter

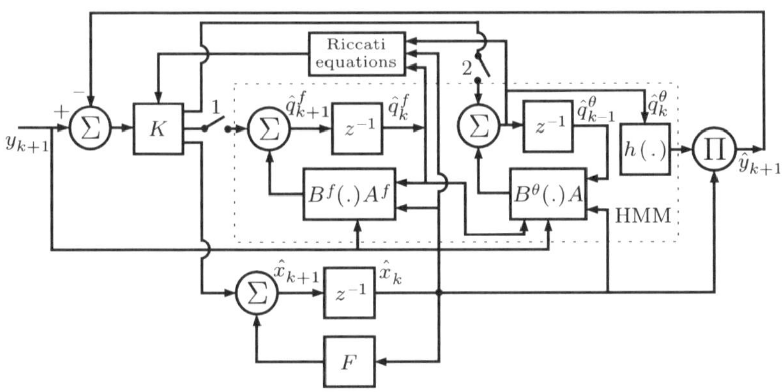

Fig. 5.2 KF/HMM scheme for adaptive HMM filter for second model

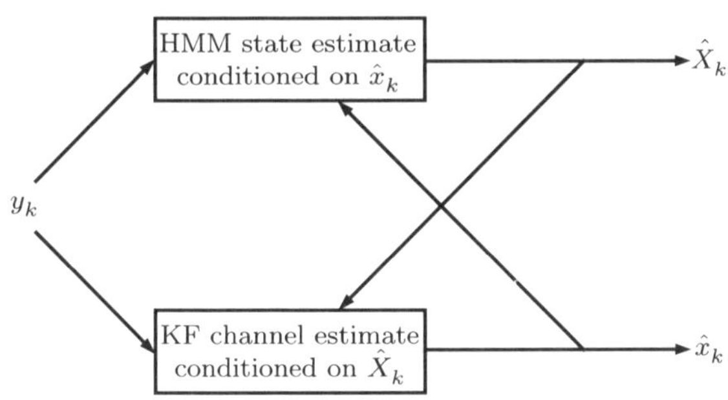

Fig. 5.3 KF/HMM adaptive HMM scheme

priori probability estimates $(\hat{q}^f_k)^{\text{MAP}}$, $(\hat{q}^\theta_k)^{\text{MAP}}$ and $(\hat{q}_k)^{\text{MAP}}$. Figure 5.4 shows this scheme in block form, for the case of (3.26). In fact, hybrid versions can be derived by setting the small-valued elements of $\hat{q}^f_k$, $\hat{q}^\theta_k$, and $\hat{q}_k$ associated with low probabilities, to zero and renormalizing.

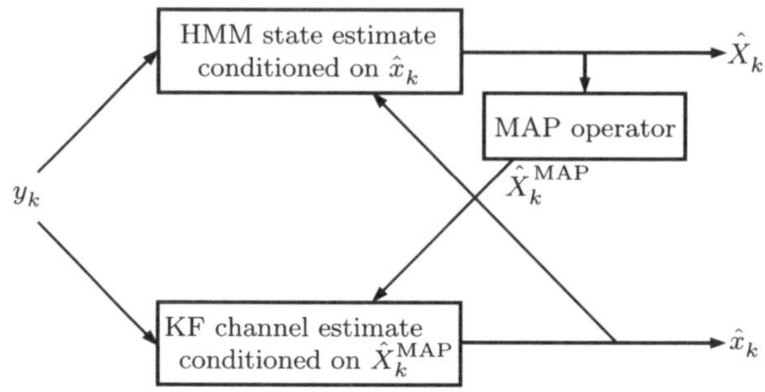

Fig. 5.4 KF/HMM adaptive HMM scheme with MAP approximation

Robustness Issues

Due to the inherently suboptimal nature of these adaptive HMM algorithms, it is necessary to consider robustness issues. The KF/HMM schemes presented above are practical derivatives of EKF/HMM schemes and effectively set the Kalman gain terms associated with the respective q's to zero. This produces decoupled conditional estimates which are then used to condition the other estimates. There is no theory for convergence in these decoupled situations when dealing with time varying parameters.

In an effort to address the inevitable robustness question, we look to the standard procedures from Kalman filtering. A widely practiced method for adding robustness is to model the estimate errors due to incorrect conditioning, as noise in the observations. This procedure can also be used with our adaptive HMM techniques. By adding extra noise to the observation model, the vector of parametrized probability densities (symbol probabilities) will be more uniform. That is, the diagonal "observation" update matrix, $B(.)$, in the "forward" procedure for the discrete-state estimate q_k will place less emphasis on the observations. An additional method for adding robustness to the adaptive HMM scheme is to assume the probability of remaining in the same state is higher than it actually is, that is, by using a more diagonally dominant A. This will also have the effect of placing less importance on the observations through the "forward" procedure for the discrete-state estimate q_k.

These robustness techniques are of course an attempt to counter estimation errors in high noise. They therefore restrict the ability of the estimates to track quickly varying parameters, as the rapid changes will effectively be modeled as noise. There is here, as in all cases, a trade-off to be made between robustness and tracking ability.

Simulation Studies

Example 1 A 16-state QAM signal is generated under assumption (3.11) with parameter values $a_{ii} = 0.95$, $a_{ij} = (1 - a_{ij}) / (N - 1)$ for $i \neq j$, $(z^{(i)})^R = \pm 0.01976 \pm 0.03952$, $(z^{(i)})^I = \pm 0.01976 \pm 0.03952$. The channel characteristics used in this ex-

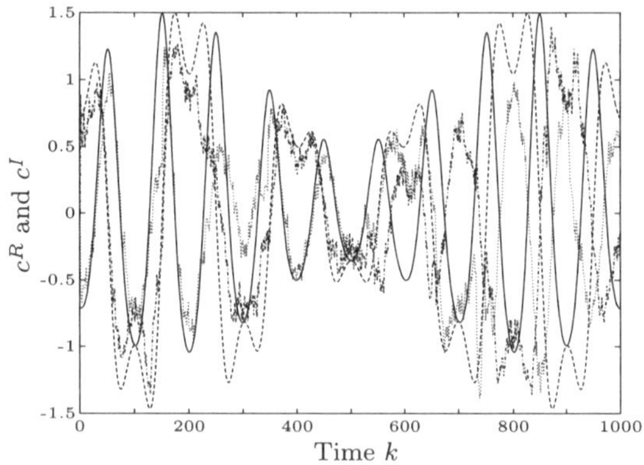

Fig. 5.5 Example 1: QAM $\hat{c}_n^R$ and $\hat{c}_n^I$ for SNR $= 6$ dB

ample is given by

$$\kappa(t) = 1 + 0.5\sin\left(\frac{3\pi t}{1000}\right),$$
$$\phi(t) = 0.75\pi\cos\left(\frac{10\pi t}{1000}\right)$$

and the signal-to-noise ratio associated with the observations in the absence of fading is SNR $= \left(E_b/\sigma_w^2\right) = 6$ dB, where E_b is the energy per bit associated with the transmitted signal. Of course, much lower SNRs can be accommodated in the presence of more slowly varying channels, and it should be noted that the SNR effectively varies as the channel fades. The lowest effective SNR in this example occurs at $k = 500$ where SNR $= 0$ dB.

The case of $a_{ii} = 0.95$ represents a high probability of remaining in the same state. It is known that, in the case of independent data sequences (i.e., $a_{ii} = \frac{1}{N}$ for all i), the matched filter is optimal. In fact, under these conditions, the HMM filter becomes the matched filter. We present the case of a non-i.i.d. data sequence to demonstrate the area where the HMM/KF schemes provide best improvement over conventional approaches.

In this example we demonstrate the KF/HMM scheme of Figure 5.3. The results are presented in Figure 5.5 and Figure 5.6, and show that even though the channel changes quite quickly, good estimates are generated. Figure 5.5 shows the true channel values and the estimated values in real and imaginary format, that is, exactly as estimated from (5.7)–(5.11). Figure 5.6(a) shows the actual channel amplitude κ_k and the estimate of this, generated from the estimates in Figure 5.5. Likewise, Figure 5.6(b) shows the actual channel phase shift ϕ_k and the estimate of this, generated from the estimates in Figure 5.5. Small glitches can be seen in the amplitude and phase estimates at points where tracking is slow and the received channel amplitude is low, but the recovery after this burst of errors seems to be quite good. It is natural that the estimates during these periods be worse, since the noise on the observations is effectively greater when $\kappa_k < 1$, as seen from the signal model (3.20).

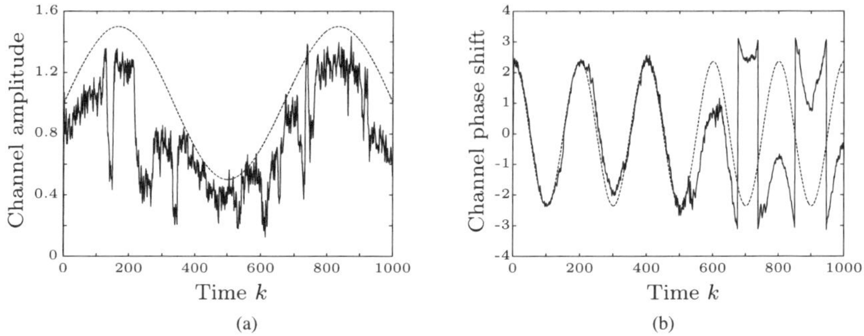

Fig. 5.6 Example 1: QAM $\hat{\kappa}_n$ and $\hat{\phi}_n$ for SNR = 6 dB

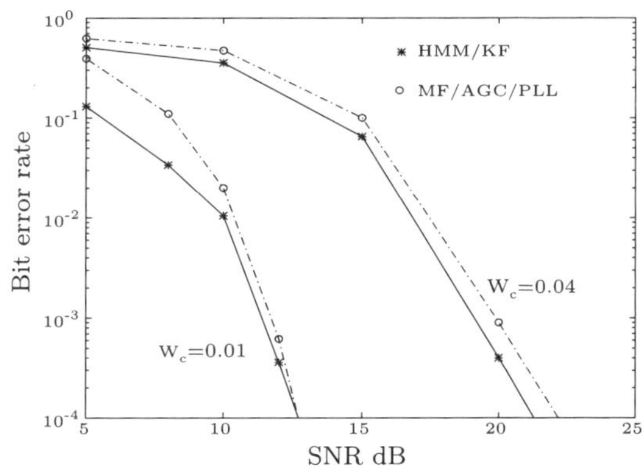

Fig. 5.7 BER vs. SNR for complex-valued channels (16 QAM)

Example 2 In this example, we demonstrate the ability of the HMM/KF adaptive algorithm to demodulate a 16-QAM signal, in the presence of a complex-valued stochastic channel. The channel gain and phase shift variations are given by LPF white Gaussian processes. The variance of the Gaussian process for the channel amplitude is 1, while the variance for the phase shift is 5. In both cases the bandwidth of the LPF is W_c. The results for this example are displayed in Figure 5.7. This example provides a comparison between the HMM/KF scheme and the conventional MF/AGC/PLL scheme.

Example 3 A frequency modulation scheme, under assumption (4.6) with $a^f_{ii} = 0.95, a^f_{ij} = \left(1 - a^f_{ij}\right) / (N-1)$ for $i \neq j$, is generated with $L_f = 16$. (This is equivalent to a 16-state frequency shift-keyed digital scheme.) The value of θ was quantized into $L_\theta = 32$ values, under Lemma 4.2. The signal is of amplitude $A_c = 0.2$. The deterministic channel gives a more rigorous test which is easily repeatable, and allows results to be displayed in a manner that more clearly shows tracking ability of these schemes. The channel characteristics are the same as those used in Example 1. The signal-to-noise ratio associated with the observations in the absence of fading is SNR $= \left(E_b/\sigma^2_w\right) = 2.4$ dB, where E_b is the energy per bit associated with the transmitted signal, if the signal were a 16-FSK digital signal. Of course much lower

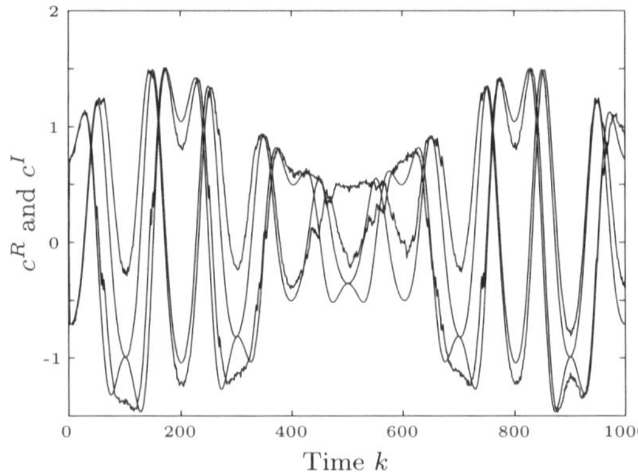

Fig. 5.8 Example 2: FM $\hat{c}_n^R$ and $\hat{c}_n^I$ for SNR = 2.4 dB

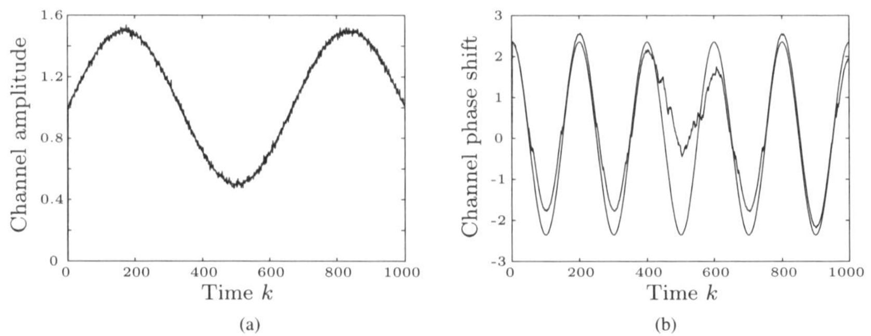

Fig. 5.9 Example 2: FM $\hat{\kappa}_n$ and $\hat{\phi}_n$ for SNR = 2.4 dB

SNRs can be accommodated in the presence of more slowly varying channels, and it should be noted that the SNR effectively varies as the channel fades. The lowest effective SNR in this example occurs at $k = 500$ where SNR = 1.8 dB.

The estimation scheme used here is the decoupled KF/HMM scheme implemented on the FM signal model given in (4.28). The results for the decoupled scheme are presented in Figure 5.8 and Figure 5.9 and show that even though the channel changes quite quickly, good estimates are generated. Figure 5.8 shows the true channel values and the estimated values in real and imaginary format, that is, exactly as estimated from (5.7)–(5.11). Figure 5.9(a) shows the actual channel amplitude κ_k and the estimate of this, generated from the estimates in Figure 5.8. Likewise, Figure 5.9(b) shows the actual channel phase shift ϕ_k and the estimate of this generated from the estimates in Figure 5.8. These results show sudden phase shifts, seen as glitches in the phase estimate in Figure 5.9. These can be any multiple of π/L_f due to the symmetry of the phase quantization. In this case, there is tracking degradation over the period where channel amplitude is less than one. It is natural that the estimates during these periods be worse, since the noise on the observations is effectively greater when $\kappa_k < 1$.

For further simulation studies the reader is directed to Collings and Moore (1994b) where comparisons appear between these KF/HMM schemes and the traditional MF/AGC/PLL approach.

6.6 Notes

The material for this chapter has been drafted in three papers (Collings et al. 1993; Collings and Moore 1994b,a). In turn, these build on earlier work by the authors, on on-line adaptive EM-based schemes in Krishnamurthy and Moore (1993). The EM schemes themselves are studied in Baum et al. (1970) and Dempster et al. (1977). Original work on RPE schemes appears in Moore and Weiss (1979), Ljung (1977), and Ljung and Söderström (1983).

For general background on FM demodulation in extended Kalman filters, see Anderson and Moore (1979). For the case of CPM demodulation in fading channels, see Lodge and Moher (1990).

Part III
Continuous-Time HMM Estimation

Chapter 7
Discrete-Range States and Observations

7.1 Introduction

In this chapter a finite-state space, continuous-time Markov chain hidden in another Markov chain is considered. The state space is taken to be the set of unit vectors $S_X = \{e_i\}$, $e_i = (0,0,\dots,1,\dots,0)'$ of $\mathbb{R}^N$.

Basic martingales associated with Markov chains are identified in Appendix B.

In Section 7.2 we derive finite-dimensional filters and smoothers for various processes related to the Markov chain.

In Section 7.3 optimal estimates of the parameters of the model are obtained via the EM algorithm.

7.2 Dynamics

Consider the Markov process $\{X_t\}$, $t \geq 0$, defined on a probability space $(\Omega, \mathscr{F}, P)$, whose state space is the set, $S_X = \{e_1,\dots,e_N\} \subset \mathbb{R}^N$. Write $p_t^i = P(X_t = e_i)$, $0 \leq i \leq N$. We shall suppose that for some family of matrices A_t, $p_t = \left(p_t^1,\dots,p_t^N\right)'$ satisfies the *forward Kolmogorov equation*

$$\frac{dp_t}{dt} = A_t p_t. \tag{2.1}$$

$A_t = (a_{ij}(t))$, $t \geq 0$, is, therefore, the family of so-called *Q-matrices* of the process. Because A_t is a Q-matrix,

$$a_{ii}(t) = -\sum_{j \neq i} a_{ji}(t). \tag{2.2}$$

The *fundamental transition matrix* associated with A_t will be denoted by $\Phi(t,s)$, so with I the $N \times N$ identity matrix

$$\frac{d\Phi(t,s)}{dt} = A_t\Phi(t,s), \qquad \Phi(s,s) = I, \tag{2.3}$$
$$\frac{d\Phi(t,s)}{ds} = -\Phi(t,s)A_s, \qquad \Phi(t,t) = I.$$

[If A_t is constant $\Phi(t,s) = \exp((t-s)A)$.]

Consider the process in state $x \in S$ at time s and write $X_{s,t}(x)$ for its state at the later time $t \geq s$. Then

$$E_{s,x}[X_t \mid \mathscr{F}_s] = E_{s,x}[X_t \mid X_s] = E_{s,x}[X_{s,t}(x)] = \Phi(t,s)x.$$

Write $\mathscr{F}_t^s$ for the right-continuous, complete filtration generated by $\sigma\{X_r : s \leq r \leq t\}$, and $\mathscr{F}_t = \mathscr{F}_t^0$. We have the following representation result.

Lemma 2.1

$$M_t \stackrel{\triangle}{=} X_t - X_0 - \int_0^t A_r X_r dr$$

is an $\{\mathscr{F}_t\}$ *martingale.*

Proof Suppose $0 \leq s \leq t$. Then

$$\begin{aligned}
E[M_t - M_s \mid \mathscr{F}_s] &= E\left[X_t - X_s - \int_s^t A_r X_r dr \mid \mathscr{F}_s\right] \\
&= E\left[X_t - X_s - \int_s^t A_r X_r dr \mid X_s\right] \\
&= E_{s,X_s}[X_t] - X_s - \int_s^t A_r E_{s,X_s}[X_r] dr \\
&= \Phi(t,s)X_s - X_s - \int_s^t A_r \Phi(r,s) X_s dr = 0
\end{aligned}$$

by (2.3). Therefore, the (special) semimartingale representation of the Markov chain X is:

$$X_t = X_0 + \int_0^t A_r X_r dr + M_t.$$

□

The observed process Y_t has a finite discrete range which is also identified, for convenience, with the set of standard unit vectors of $\mathbb{R}^M$, $S_Y = \{f_1, \ldots, f_M\}$, where $f_i = (0, \ldots, 0, 1, 0, \ldots, 0)'$, $1 \leq i \leq M$. The processes X and Y are not independent; rather, given the evolution of X, then Y is a Markov chain with Q matrix

$$\begin{aligned}
C_t &= C_t(X_t) \qquad\qquad (2.4)\\
&= \sum_{m=1}^N C_m \langle X_t, e_m \rangle, \qquad C_m = \left(c_{ij}^m\right), \ 1 \leq i,j \leq M, \ 1 \leq m \leq N.
\end{aligned}$$

Notation 2.2 *Write* $c_{ij} = c_{ij}(r) = \sum_{m=1}^N \langle X_r, e_m \rangle c_{ij}^m$.

Also write $\mathscr{G}_t^0 = \sigma(X_s, Y_s, 0 \leq s \leq t)$, *and* $\mathscr{G}_t$ *for the* right continuous, *complete filtration generated by* $\mathscr{G}_t^0$.

Lemma 2.3 *In view of the Markov property assumption and* (2.4) $W_t := Y_t - Y_0 - \int_0^t C_r Y_r \, dr$ *is an* $\mathcal{G}_t$*-martingale.*

Proof The proof is left as an exercise. □

Lemma 2.4 *The* predictable quadratic variation *of the process* Y *is given by*

$$\langle Y, Y \rangle_t = \operatorname{diag} \int_0^t C_r Y_r \, dr - \int_0^t (\operatorname{diag} Y_r) C_r' \, dr \\ - \int_0^t C_r (\operatorname{diag} Y_r) \, dr. \tag{2.5}$$

Proof

$$\begin{aligned} Y_t Y_t' &= Y_0 Y_0' + \int_0^t Y_{r-} dY_r' + \int_0^t dY_r Y_{r-}' + [Y, Y]_t \\ &= Y_0 Y_0' + \int_0^t Y_r (C_r Y_r)' \, dr + \int_0^t Y_{r-} dW_r' \\ &\quad + \int_0^t (C_r Y_r) Y_r' \, dr + \int_0^t dW_r Y_{r-}' \\ &\quad + [Y, Y]_t - \langle Y, Y \rangle_t + \langle Y, Y \rangle_t , \end{aligned} \tag{2.6}$$

where $[Y, Y]_t - \langle Y, Y \rangle_t$ is an $\mathcal{G}_t$-martingale. However,

$$Y_t Y_t' = \operatorname{diag} Y_t,$$

$$Y_r (C_r Y_r)' = (\operatorname{diag} Y_r) C_r', \tag{2.7}$$

$$(C_r Y_r) Y_r' = C_r (\operatorname{diag} Y_r). \tag{2.8}$$

We also have

$$Y_t Y_t' = \operatorname{diag} Y_0 + \operatorname{diag} \int_0^t C_r Y_r + \operatorname{diag} W_t. \tag{2.9}$$

$Y_t Y_t'$ is a special semimartingale, so using the uniqueness of its decomposition from (2.6), (2.7), (2.8), and (2.9), we have (2.5). □

The dynamics of our model follow

$$X_t = X_0 + \int_0^t A_r X_r \, dr + V_t, \tag{2.10}$$

$$Y_t = Y_0 + \int_0^t C_r Y_r \, dr + W_t, \tag{2.11}$$

where (2.10) and (2.11) are the semimartingale representations of the processes X_t and Y_t, respectively.

Notation 2.5 *With* $\mathcal{Y}_t^0 = \sigma(Y_s, 0 \le s \le t)$, $\{\mathcal{Y}_t\}$, $t \ge 0$, *is the corresponding right continuous complete filtration. Note* $\mathcal{Y}_t \subset \mathcal{G}_t$, $\forall t \ge 0$. *For* ϕ_t *an integrable and measurable process write* $\hat{\phi}_t$ *for its* $\mathcal{Y}$-optional projection *under* P, *so* $\hat{\phi}_t = E[\phi_t \mid \mathcal{Y}_t]$

a.s. and $\hat{\phi}_t$ is the filtered estimate of ϕ_t. [For a discussion of optional projections see Elliott (1982b).] Optional projections take care of measurability in both t and ω; conditional expectations only concern measurability in ω. Denote by $\mathscr{K}_t^{k\ell}$ the number of jumps of Y from state f_k to state f_ℓ in the interval of time $[0,t]$ with $k \neq \ell$.

Then,

$$\begin{aligned}\langle f_k, Y_{r-}\rangle \langle f_\ell, dY_r\rangle &= \langle f_k, Y_{r-}\rangle \langle f_\ell, Y_r - Y_{r-}\rangle \\ &= \langle f_k, Y_{r-}\rangle \langle f_\ell, Y_r\rangle \\ &= I\left[Y_{r-} = f_k, Y_r = f_\ell\right],\end{aligned}$$

so that for $k \neq \ell$

$$\begin{aligned}\mathscr{K}_t^{k\ell} &= \int_0^t \langle f_k, Y_{r-}\rangle \langle f_\ell, dY_r\rangle \\ &= \int_0^t \langle f_k, Y_{r-}\rangle \langle f_\ell, C_r Y_r\rangle \, dr + \int_0^t \langle f_k, Y_r\rangle \langle f_\ell, dW_r\rangle\end{aligned}$$

using (2.11). This is the semimartingale representation of the process $\mathscr{K}_t^{k\ell}$, $k \neq \ell$. Clearly, $\mathscr{K}_t^{k\ell}$ are $\mathscr{Y}_t$-measurable $\forall t \geq 0$ and have no common jumps for $(k,\ell) \neq (k',\ell')$. Now

$$C_r Y_r = \sum_{i,j=1}^{M} \langle Y_r, f_i\rangle f_j c_{ji},$$

and

$$\langle f_\ell, C_r Y_r\rangle = \sum_{i=1} \langle Y_r, f_i\rangle c_{\ell i},$$

also

$$\langle f_k, Y_r\rangle \sum_{i=1} \langle Y_r, f_i\rangle c_{\ell i} = \langle Y_r, f_k\rangle c_{\ell k}.$$

Since $c_{\ell k} = \sum_{m=1}^{N} \langle X_r, e_m\rangle c_{\ell k}^m$ we have

$$\mathscr{K}_t^{k\ell} = \int_0^t \lambda_r^{k\ell} dr + O_t^{k\ell}, \tag{2.12}$$

where

$$\lambda_r^{k\ell} = \sum_{m=1}^{N} \langle Y_r, f_k\rangle \langle X_r, e_m\rangle c_{\ell k}^m, \tag{2.13}$$

and $O_t^{k\ell}$ are martingales.

Definition 2.6 Suppose K_t is a purely discontinuous, increasing, adapted process on the filtered probability space $(\Omega, \mathscr{F}, P, \mathscr{F}_t)$, $t \geq 0$, all of whose jumps equal $+1$.

Denote the sequence of jump times $S_1, S_2, \ldots$; then K_t is a *standard Poisson process* if the random variables $S_1, S_2 - S_1, \ldots, S_n - S_{n-1}, \ldots$ are exponentially distributed with parameter 1, and are, respectively, independent of $\mathscr{F}_0, \mathscr{F}_{S_1}, \ldots, \mathscr{F}_{S_{n-1}}, \ldots$.

Notation 2.7 *$\underline{1} = (1,1,\ldots,1)' \in \mathbb{R}^M$ will denote the vector all of whose entries are 1. I_M will denote the $M \times M$ identity matrix.*

If $A = (a_{ij})$, $B = (b_{ij})$ are $M \times M$ matrices and $b_{ij} \neq 0$ for all i, j, then $\frac{A}{B}$ will denote the matrix $\left(\frac{a_{ij}}{b_{ij}}\right)$. If $A = (a_{ij})$ and $a = (a_{11}, a_{22}, \ldots, a_{MM})$ then $A_0 = A - \operatorname{diag} a$.

Remarks 2.8 Suppose on $(\Omega, \mathscr{F})$ there is a probability measure $\overline{P}$ and a vector of counting processes

$$N_t = (N_t(1), \ldots, N_t(M))$$

such that under $\overline{P}$ each component $N_t(\ell)$ is a standard Poisson process, and $\{X_t\}$ is an independent Markov chain with Q-matrix A, as in (2.1). Write $\{\mathscr{G}_t\}$ for the filtration generated by N and X. Then $N_t - t\underline{1} = Q_t$ is a $(\mathscr{G}_t, \overline{P})$ martingale.

Define

$$Y_t = Y_0 + \int_0^t \left(I_M - Y_{s-}\underline{1}'\right) dN_s \qquad \text{for } Y_0 \in S_Y. \tag{2.14}$$

Then

$$Y_t = Y_0 + \int_0^t \Pi Y_s \, ds + \int_0^t \left(I_M - Y_{s-}\underline{1}'\right) dQ_s,$$

where Π is the $M \times M$ matrix (π_{ij}) with $\pi_{ij} = 1$ for $i \neq j$ and $\pi_{ii} = -(N-1)$.

Write

$$\overline{N}_t = \int_0^t (I_M - \operatorname{diag} Y_{s-}) \, dY_s.$$

Then $\overline{N}$ is a vector of counting processes. Note that $\overline{N} \neq N$. With

$$D_t := \frac{C_t(X_t)}{\Pi}$$

consider the process

$$\overline{\Lambda}_t = 1 + \int_0^t \overline{\Lambda}_{s-} \left((D_{s-})_0 Y_{s-} - \underline{1}\right)' \left(d\overline{N}_s - \Pi_0 Y_{s-} \, ds\right). \tag{2.15}$$

Then

$$\overline{\Lambda}_t = \exp\left[-\int_0^t (C_s(X_s)_0 - \Pi_0) X_s' \underline{1} \, ds\right] \prod_{0<s\leq t} \left(1 + ((D_{s-})_0 X_{s-} - \underline{1})' \Delta N_s\right). \tag{2.16}$$

The probability measure P can be defined by setting

$$\left.\frac{dP}{d\overline{P}}\right|_{\mathscr{G}_t} = \overline{\Lambda}_t.$$

Then it can be shown that, under P,

$$Y_t - Y_0 - \int_0^t C_r(X_r)Y_r\,dr = W_t \tag{2.17}$$

is a $(\mathscr{G}_t, P)$ martingale. Furthermore, under P, X remains a Markov chain with Q-matrix A.

■

Notation 2.9 *If ϕ_t is an $\mathscr{G}_t$-adapted, integrable process then*

$$\hat{\phi}_t = E\left[\phi_t \mid \mathscr{Y}_t\right] = \frac{\overline{E}\left[\overline{\Lambda}_t \phi_t \mid \mathscr{Y}_t\right]}{\overline{E}\left[\overline{\Lambda}_t \mid \mathscr{Y}_t\right]} := \frac{\sigma_t\left(\phi_t\right)}{\sigma_t\left(1\right)}, \tag{2.18}$$

where $\overline{E}$ denotes expectation w.r.t. $\overline{P}$ and $\sigma_t\left(\phi_t\right)$ is the $\mathscr{Y}$-optional projection of $\Lambda_t \phi_t$ under $\overline{P}$. Consequently, $\sigma_t\left(1\right) = \overline{E}\left[\overline{\Lambda}_t \mid Y_t\right]$ is the $\mathscr{Y}$-optional projection of $\overline{\Lambda}_t$ under $\overline{P}$. Further, if $s \leq t$ we shall write $\sigma_t\left(\phi_s\right)$ for the $\mathscr{Y}$-optional projection of $\overline{\Lambda}_t \phi_s$ under $\overline{P}$, so that

$$\sigma_t\left(\phi_s\right) = \overline{E}\left[\overline{\Lambda}_t \phi_s \mid \mathscr{Y}_t\right] \ a.s.$$

and, loosely we write $\sigma_t\left(\phi_t\right) = \sigma\left(\phi_t\right)$.

To summarize, we have under the probability measure P

$$\boxed{\begin{aligned} X_t &= X_0 + \int_0^t A_r X_r\,dr + V_t, \\ Y_t &= Y_0 + \int_0^t C_r(X_r)Y_r\,dr + W_t. \end{aligned}} \tag{2.19}$$

A_t is a Q-matrix satisfying (2.1) and (2.2). Under the probability measure $\overline{P}$

$$\boxed{\begin{aligned} &X_t = X_0 + \int_0^t A_r X_r\,dr + V_t, \\ &Y_t - Y_0 - \int_0^t \Pi Y_r\,dr \quad \text{is a martingale.} \end{aligned}} \tag{2.20}$$

7.3 A General Finite-Dimensional Filter

Let H_t be a scalar process for simplicity of notation of the form

$$H_t = H_0 + \int_0^t \alpha_r \, dr + \int_0^t \beta_r' dV_r + \int_0^t \delta_r' \, dY_r, \tag{3.1}$$

where α, β, δ are $\mathscr{G}_t$-predictable, square-integrable process of appropriate dimensions. That is, α is scalar, $\beta_r = (\beta_r^1, \ldots, \beta_r^N)' \in \mathbb{R}^N$, and $\delta_r \in \mathbb{R}^M$. Recall from Chapters 2 and 3 that by considering $H_k X_k$ we obtain finite-dimensional estimators for various parameters of our models. The same trick works here with $H_t X_t$.

The signal process X is modeled by the semimartingale

$$X_t = X_0 + \int_0^t A_r X_r \, dr + V_t \tag{3.2}$$

so that

$$\begin{aligned} H_t X_t = {} & H_0 X_0 + \int_0^t \alpha_r X_r \, dr + \int_0^t X_{r-} \beta_r' \, dV_r \\ & + \int_0^t X_{r-} \delta_r' \, dY_r + \int_0^t H_r A_r X_r \, dr \\ & + \int_0^t H_{r-} \, dV_r + \sum_{0<r\leq t} (\beta_r' \Delta X_r) \, \Delta X_r. \end{aligned} \tag{3.3}$$

Note $\Delta X_r \Delta Y_r = 0$ a.s. Now

$$\sum_{0<r\leq t} (\beta_r' \Delta X_r) \, \Delta X_r = \sum_{i,j=1}^N \int_0^t (\beta_r^j - \beta_r^i) \langle X_{r-}, e_i \rangle \langle e_j, dX_r \rangle (e_j - e_i)$$

and using (3.2)

$$\begin{aligned} \sum_{0<r\leq t} (\beta_r' \Delta X_r) \, \Delta X_r = {} & \sum_{i,j=1}^N \int_0^t (\beta_r^j - \beta_r^i) \langle X_{r-}, e_i \rangle \langle e_j, dV_r \rangle (e_j - e_i) \\ & + \sum_{i,j=1}^N \int_0^t \left\langle \beta_t^j X_r - \beta_r^i X_r, e_i \right\rangle a_{ji} \, dr \, (e_j - e_i). \end{aligned}$$

Substituting in (3.3) we have

$$\begin{aligned} H_t X_t = {} & H_0 X_0 + \int_0^t \left[\alpha_r X_r + H_r A_r X_r + \sum_{i,j=1}^N \langle \beta_r^j X_r - \beta_r^i X_r, e_i \rangle (e_j - e_i) a_{ji} \right] dr \\ & + \int_0^t X_{r-} \beta_r' \, dV_r + \int_0^t H_{r-} dV_r + \int_0^t X_{r-} \delta_r' \, dY_r \\ & + \sum_{i,j=1}^N \int_0^t (\beta^j - \beta^i) \langle X_{r-}, e_i \rangle \langle e_j, dV \rangle (e_j - e_i). \end{aligned}$$

We now give a recursive equation for the unnormalized estimate $\sigma(H_tX_t)$.

Theorem 3.1 *Write $c^{\cdot}_{ji} = (c^1_{ji}, c^2_{ji}, \ldots, c^N_{ji})$. A recursive linear stochastic differential equation describing the evolution of the process $\sigma(H_tX_t)$ follows:*

$$\begin{aligned}\sigma(H_tX_t) &= \sigma(H_0X_0) + \int_0^t \sigma(\alpha_r X_r)\,dr + \int_0^t \sigma(H_rA_rX_r)\,dr \\ &\quad + \int_0^t \sum_{i,j=1}^N \left\langle \sigma\left(\beta_r^j X_r - \beta_r^i X_r\right), e_i\right\rangle a_{ji}(r)\,dr\,(e_j - e_i) \qquad (3.4)\\ &\quad + \int_0^t \sigma\left(X_{r-}\delta'_{r-}\left(I - Y_{r-}\underline{1}'\right)\operatorname{diag} dY_r\left((D_{r-})_0 Y_{r-} - \underline{1}\right)\right).\end{aligned}$$

Proof The product $X_tH_t\overline{\Lambda}_t$ is calculated using the product rule for semimartingales

$$\begin{aligned}X_tH_t\overline{\Lambda}_t &= H_0X_0 + \int_0^t H_{r-}X_{r-}\,d\overline{\Lambda}_r + \int_0^t \overline{\Lambda}_{r-}\,dX_rH_r \\ &\quad + \left[XH, \overline{\Lambda}\right]_t. \qquad (3.5)\end{aligned}$$

Here,

$$\left[XH, \overline{\Lambda}\right]_t = \sum_{0<r\le t} X_{r-}\overline{\Lambda}_{r-}\delta'_{r-}\left(I - Y_{r-}\underline{1}'\right)\operatorname{diag}\Delta\overline{N}_r\left((D_{r-})_0 Y_{r-} - \underline{1}\right).$$

Substituting in (3.5) and after some simplification

$$\begin{aligned}X_tH_t\overline{\Lambda}_t &= H_0X_0 + \int_0^t X_{r-}\overline{\Lambda}_{r-}\delta'_{r-}\left(I - Y_{r-}\underline{1}'\right)\operatorname{diag} dY_r\left((D_F)_0 Y_{r-} - \underline{1}\right) \\ &\quad + \int_0^t \overline{\Lambda}_r\left[\alpha_rX_r + A_rH_rX_r + \sum_{i,j=1}^N \left\langle \beta_r^j X_r - \beta_r^i X_r, e_i\right\rangle a_{ji}(r)\,(e_j - e_i)\right] dr \\ &\quad + \int_0^t H_{r-}X_{r-}\overline{\Lambda}_{r-}\left((D_{r-})_0 Y_{r-} - \underline{1}\right)'\left(d\overline{N}_r - \Pi_0 Y_r\,dr\right) \\ &\quad + \text{a dv} - \text{martingale}. \qquad (3.6)\end{aligned}$$

Taking the $\mathscr{Y}$-optional projection under $\overline{P}$ of both sides of (3.6), and using in Lemma 3.2 of Chapter 7 in Wong and Hajek (1985) gives at once (3.4). ☐

The next corollary will be used for the derivation of smoothers.

Corollary 3.2 *The above equation is recursive in t, so for $s \le t$ we have the following form.*

$$\begin{aligned}\sigma(H_tX_t) &= \sigma(X_sH_s) + \int_s^t \sigma(\alpha_r X_r)\,dr + \int_s^t \sigma(A_rH_rX_r)\,dr \\ &\quad + \int_s^t \sum_{i,j=1}^N \left\langle \sigma\left(\beta_r^j X_r - \beta_r^i X_r\right), e_i\right\rangle a_{ji}(r)\,dr\,(e_j - e_i) \qquad (3.7)\end{aligned}$$

$$+\int_s^t \sigma\left(X_{r-}\operatorname{Tr}\left[\delta_{r-}\left(I-Y_{r-}\underline{1}'\right)\left((D_{r-})_0 Y_{r-}-\underline{1})'\right]\right)dY_r$$
$$+\sum_{i=1}^{N}\int_0^t \left\langle \sigma\left(H_{r-}X_{r-}\right),e_i\right\rangle \left(D_{r-}(e_i)_0 Y_{r-}-\underline{1}'\right)\left(d\overline{N}_r-\Pi_0 Y_r dr\right)e_i,$$

where Tr *is the trace. Here, the initial condition is* $\overline{E}\left[\Lambda_s H_s X_s \mid \mathscr{Y}_s\right]$, *which again is a* $\mathscr{Y}_s$*-measurable random variable.*

Specializing Theorem 3.1 and Corollary 3.2, the following finite-dimensional filters and smoothers for processes related to the model are computed.

Notation 3.3 *Write* $\Phi(e_i)=(D_{r-}(e_i)_0 Y_{r-}-\underline{1}'),1\le i\le N.$

Zakai Equation for X

Take $H_t=H_0=1$, $\alpha_r=0$, $\beta_r=0\in\mathbb{R}^N$, $\delta_r=0$. Applying Theorem 3.1 and using Notation 3.3 the unnormalized filter for the conditional distribution of the state process follows:

$$\boxed{\begin{aligned}\sigma\left(X_t\right) &= \sigma\left(X_0\right)+\int_0^t A\sigma\left(X_r\right)dr\\ &\quad +\sum_{i=1}^{N}\int_0^t \langle\sigma(X_{r-}),e_i\rangle\,\Phi(e_i)\left(d\overline{N}_r-\Pi_0 Y_r dr\right)e_i.\end{aligned}} \tag{3.8}$$

This is a single finite-dimensional equation for the unnormalized conditional distribution $\sigma\left(X_t\right)$. Note it is linear in $\sigma\left(X_t\right)$.

The Number of Jumps

For e_i, $e_j\in S$, $i\neq j$, consider the stochastic integral

$$V_t^{ij}=\int_0^t \langle X_{r-},e_i\rangle\left\langle e_j,dV_r\right\rangle.$$

Note the integrand is predictable, so V_t^{ij} is a martingale. Now

$$\begin{aligned}\langle X_{r-},e_i\rangle\left\langle e_j,dX_r\right\rangle &= \langle X_{r-},e_i\rangle\left\langle e_j,X_r-X_{r-}\right\rangle=\langle X_{r-},e_i\rangle\left\langle X_r,e_j\right\rangle\\ &= I\left[X_{r-}=e_i \text{ and } X_r=e_j\right].\end{aligned}$$

Write $\mathscr{J}_t^{ij}$ for the number of jumps from e_i to e_j in the time interval $[0,t]$. Then using (2.10) we obtain

$$\begin{aligned}\mathscr{J}_t^{ij} &= \int_0^t \langle X_{r-}, e_i\rangle \langle e_j, dX_r\rangle = \int_0^t \langle X_{r-}, e_i\rangle \langle e_j, A_r X_r\rangle dr + V_t^{ij} \\ &= \int_0^t \langle X_{r-}, e_i\rangle a_{ji}\, dr + V_t^{ij}. \end{aligned} \tag{3.9}$$

This is the semimartingale decomposition of $\mathscr{J}_t^{ij}$. To obtain the Zakai filter for $\mathscr{J}_t^{ij}$, take $H_t = \mathscr{J}_t^{ij}$, $H_0 = 0$, $\alpha_r = \langle X_r, e_i\rangle a_{ji}$, $\beta_r = \langle X_r, e_i\rangle e_j$ $\delta_r = 0$. Then it is seen that:

$$\boxed{\begin{aligned}\sigma\left(\mathscr{J}_t^{ij} X_t\right) &= \int_0^t \left(A_r \sigma\left(\mathscr{J}_r^{ij} X_r\right) + \langle \sigma(X_r), e_i\rangle a_{ji}(r) e_j\right) dr \\ &+ \int_0^t \sum_{i=1}^N \left\langle \sigma\left(Y_{r-}^{ij} X_{r-}\right), e_i\right\rangle \Phi(e_i) \left(d\overline{N}_r - \Pi_0 Y_r dr\right) e_i. \end{aligned}} \tag{3.10}$$

Taking the inner product with $\underline{1} = (1, \ldots, 1)$ gives the finite-dimensional filter $\sigma(\mathscr{J}_t^{ij})$ for the number of transitions in the interval of time 0 to t. This quantity will be used later for the estimation of the probability transitions a_{ji}.

Occupation Time

The time spent by the process X in state e_i is given by

$$\mathscr{O}_t^i = \int_0^t \langle X_r, e_i\rangle dr, 1 \le i \le N.$$

A recursive finite-dimensional filter for this process is needed with (3.10) in order to estimate a_{ji}. Take

$$H_t = \mathscr{O}_t^i, H_0 = 0, \alpha_r = \langle X_r, e_i\rangle, \beta_r = 0 \in \mathbb{R}^N, \delta_r = 0.$$

Substituting in Theorem 3.1, and using Notation 3.3 we have

$$\boxed{\begin{aligned}\sigma\left(\mathscr{O}_t^i X_t\right) &= \int_0^t \left(\langle \sigma(X_r), e_i\rangle e_i + A\sigma\left(\mathscr{O}_r^i X_r\right)\right) dr \\ &+ \int_0^t \sum_{i=i}^N \left\langle \sigma\left(\mathscr{O}_{r-}^i X_{r-}\right), e_i\right\rangle \Phi(e_i) \left(d\overline{N}_r - \Pi_0 Y_r dr\right) e_i. \end{aligned}} \tag{3.11}$$

Together with the filter for $\sigma(X_t)$ we have a finite-dimensional unnormalized filter for $\sigma\left(\mathscr{O}_t^i X_t\right)$, $1 \le i \le N$. Taking the inner product with $\underline{1}$ gives $\sigma\left(\mathscr{O}_t^i\right)$.

Drift Coefficients

In the next section we will see that the estimation of the $c_{\ell k}^m$'s, in the entries of the Q-matrix C_r of the observation process, involves the filtered estimates of the processes

$$\mathscr{A}^m_{k\ell,t} = \int_0^t \langle X_{r-}, e_m\rangle \, d\mathscr{K}^{k\ell}_r \quad \text{and} \quad \mathscr{T}^{km}_t = \int_0^t \langle Y_r, f_k\rangle \langle X_r, e_m\rangle \, dr.$$

The process $\mathscr{A}^m_{k\ell,t}$ increases only when the Y process jumps from f_k to f_ℓ and the X_{r-} process is in state e_m. The $\mathscr{T}^{km}_t$ process measures the total time up to time t for which X is in state e_m and simultaneously Y is in state f_k. Apply Theorem 3.1 to

$$\mathscr{A}^m_{k\ell,t} = \int_0^t \langle X_{r-}, e_m\rangle \left(d\mathscr{K}^{k\ell}_r - \lambda^{k\ell}_r \, dr\right) + \int_0^t \langle X_r, e_m\rangle \lambda^{k\ell}_r \, dr$$

taking $H_t = \mathscr{A}^m_{k\ell,t}$, $H_0 = 0$, $\alpha_r = \langle X_r, e_m\rangle \lambda^{k\ell}_r$, $\delta^{k\ell}_r = \langle X_r, e_m\rangle$, and $\beta_r = 0$, $\langle X_r, e_m\rangle \lambda^{k\ell}_r = \langle X_r, e_m\rangle \langle Y_r, f_k\rangle \sum_{\alpha=1}^N \langle X_r, e_\alpha\rangle c^\alpha_{\ell k} = \langle Y_r, f_k\rangle \langle X_r, e_m\rangle c^m_{\ell k}$, and $\delta^{k\ell}_r X_r = \langle X_r, e_m\rangle X_r = \langle X_r, e_m\rangle e_m$. Write $\Psi(e_i) = (I - Y_r \underline{1}')\Phi(e_i)$. Then the Zakai equation for $\mathscr{A}^m_{k\ell,t} X_t$ is

$$\begin{aligned}
\sigma\left(\mathscr{A}^m_{k\ell,t} X_t\right) &= \int_0^t \left(\langle \sigma(X_r), e_m\rangle \langle Y_r, f_k\rangle c^m_{\ell k} e_m + A_r \sigma\left(\mathscr{A}^m_{k\ell,r} X_r\right)\right) dr \\
&\quad + \int_0^t \sum_{i=1}^N \langle \sigma(X_{r-}), e_m\rangle \Psi(e_i) dY_r e_i \\
&\quad + \int_0^t \left[\sum_{i=1}^N \left\langle \sigma\left(\mathscr{A}^m_{k\ell,r-} X_{r-}\right), e_i\right\rangle \Phi(e_i)\right] \left(d\overline{N}_r - \Pi_0 Y_r dr\right) e_i.
\end{aligned} \tag{3.12}$$

Similarly, with $H_t = \mathscr{T}^{km}_t$, $H_0 = 0$, $\alpha_r = \langle Y_r, f_k\rangle \langle X_r, e_m\rangle$, $\beta_r = 0 \in \mathbb{R}^N$, $\delta^{ij}_r = 0$ for $i, j = 1, \ldots, M$, we have

$$\begin{aligned}
\sigma\left(\mathscr{T}^{km}_t X_t\right) &= \int_0^t \left(\langle \sigma(X_r), e_m\rangle \langle Y_r, f_k\rangle e_m + A_r \sigma\left(\mathscr{T}^{km}_r X_r\right)\right) dr \\
&\quad + \int_0^t \sum_{i=1}^N \left\langle \sigma\left(\mathscr{T}^{km}_{r-} X_{r-}\right), e_i \right\rangle \Phi(e_i) \left(d\overline{N}_r - \Pi_0 Y_r dr\right) e_i.
\end{aligned} \tag{3.13}$$

Smoother for the State

For the smoothed estimates of X_s given $\mathscr{Y}_t$, $s \le t$, take $H_t = H_s = \langle X_s, e_i\rangle$, $s \le t$, $\alpha_r = 0$, $\beta_r = 0 \in \mathbb{R}^N$, $\delta_r = \{\delta^{ij}_r\} = 0$ and apply Corollary 3.2.

$$\begin{aligned}
&\sigma_t\left(\langle X_s, e_i\rangle X_t\right) \\
&= \sigma_s\left(\langle X_s, e_i\rangle X_s\right) + \int_s^t A_r \sigma\left(\langle X_s, e_i\rangle X_r\right) dr \\
&\quad + \int_s^t \sum_{j=1}^N \langle \sigma\left(\langle X_s, e_i\rangle X_{r-}\right), e_j\rangle \Phi(e_j) \left(d\overline{N}_r - \Pi_0 Y_r\right) dr\, e_j.
\end{aligned} \tag{3.14}$$

This is a single equation, finite-dimensional filter, for $\sigma_t(\langle X_s, e_i\rangle X_t) = \overline{E}[\Lambda_t \langle X_s, e_i\rangle X_t \mid \mathscr{Y}_t]$, driven by the $\mathscr{K}^{ij}$'s. Taking the inner product with $\underline{1}$ gives $\sigma_t(\langle X_s, e_i\rangle)$.

Smoother for the Number of Jumps

Take $H_t = H_s = \mathscr{J}_s^{ij}$, $s \le t$, $\alpha_r = 0$, $\beta_r = 0$, and $\delta_r = 0$ in Corollary 3.2:

$$\begin{aligned}\sigma\left(\mathscr{J}_s^{ij} X_t\right) &= \sigma\left(\mathscr{J}_s^{ij} X_s\right) + \int_s^t A_r \sigma\left(\mathscr{J}_s^{ij} X_r\right) dr \\ &\quad + \int_s^t \sum_{k=1}^N \left\langle \sigma\left(\mathscr{J}_{r-}^{ij} X_{r-}\right), e_k \right\rangle \Phi(e_k) \left(d\overline{N}_r - \Pi_0 Y_r dr\right) e_k.\end{aligned} \tag{3.15}$$

Smoother for the Occupation Time

Finite-dimensional smoothers are obtained for $\mathscr{O}_s^i$ by taking $H_t = H_s = \mathscr{O}_s^i$ for $s \le t$ and $\alpha_r = 0$, $\beta_r = 0$, $\delta_r = 0$. Applying Corollary 3.2 gives:

$$\begin{aligned}\sigma\left(\mathscr{O}_s^i X_t\right) &= \sigma\left(\mathscr{O}_s^i X_s\right) + \int_s^t \left(A_r \sigma\left(\mathscr{O}_s^i X_r\right) + \langle \sigma(X_r), e_i\rangle e_i\right) dr \\ &\quad + \int_s^t \sum_{k=1}^N \left\langle \sigma\left(\mathscr{O}_{r-}^i X_{r-}\right), e_k \right\rangle \Phi(e_k) \left(d\overline{N}_r - \Pi_0 Y_r dr\right) e_k.\end{aligned} \tag{3.16}$$

Smoother for $\mathscr{A}_{k\ell,s}^m$ and $\mathscr{T}_s^{km}$

It is left as an exercise to show that the processes $\mathscr{A}_{k\ell,s}^m$ and $\mathscr{T}_s^{km}$ have the following finite-dimensional unnormalized smoothers:

$$\begin{aligned}&\sigma\left(\mathscr{A}_{k\ell,s}^m X_t\right) \\ &\quad = \sigma\left(\mathscr{A}_{k\ell,s}^m X_s\right) + \int_s^t A_r \sigma\left(\mathscr{A}_{k\ell,s}^m X_r\right) dr \\ &\qquad + \int_s^t \langle \sigma(X_r), e_m\rangle \langle Y_r, f_k\rangle c_{\ell k}^m e_m\, dr \\ &\qquad + \int_s^t \sum_{i=1}^N \langle \sigma(X_{r-}), e_m\rangle \Psi(E_i) dY_r e_i \\ &\qquad + \int_s^t \left[\sum_{i=1}^N \left\langle \sigma\left(\mathscr{A}_{k\ell,r-}^m X_{r-}\right), e_i\right\rangle \Phi(e_i)\right] \left(d\overline{N}_r - \Pi_0 Y_r dr\right) e_i,\end{aligned} \tag{3.17}$$

$$\begin{aligned}
&\sigma\left(\mathcal{T}_s^{km} X_t\right) \\
&\quad = \sigma\left(\mathcal{T}_s^{km} X_s\right) + \int_s^t \sum_{i=1}^N \left\langle \sigma\left(\mathcal{T}_{r-}^{km} X_{r-}\right), e_i \right\rangle \Phi(e_i) \left(d\overline{N}_r - \Pi_0 Y_r dr\right) e_i \\
&\qquad + \int_s^t \left(\langle \sigma(X_r), e_m \rangle \langle Y_r, f_k \rangle e_m + A_r \sigma\left(\mathcal{T}_s^{km} X_r\right)\right) dr.
\end{aligned} \tag{3.18}$$

7.4 Parameter Estimation

Suppose, as above, that X_t, $t \geq 0$, is a Markov chain with state space $S_X = \{e_1, \ldots, e_N\}$ and *Q-matrix generator* $A = \{a_{ij}\}$. Then

$$X_t = X_0 + \int_0^t AX_r \, dr + V_t. \tag{4.1}$$

Again, suppose X_t is observed through the process Y with representation

$$Y_t = Y_0 + \int_0^t C_r Y_r \, dr + W_t, \tag{4.2}$$

where C_r is as given in Equation (2.4). The above model, therefore, is determined by the set of parameters

$$\theta := \left\{a_{ij}, c_{\ell k}^m, \ 1 \leq i, j, \ell, m \leq N, \ 1 \leq k \leq M\right\}.$$

Suppose the model is first determined by a set of parameters

$$\theta := \left\{a_{ij}, \ c_{\ell k}^m, \ 1 \leq i, j, \ell, m \leq N, \ 1 \leq k \leq M\right\}$$

and we wish to determine a new set $\hat{\theta} = \left(\hat{a}_{ij}, \ \hat{c}_{\ell k}^m, \ 1 \leq i, j, \ell, m \leq N, \ 1 \leq k \leq M\right)$ which maximizes the log-likelihood defined below. Write $P_{\hat{\theta}}$ and P_θ for their respective probability measures. From (3.9) and (2.12) we have, under P_θ, that

$$\begin{aligned}
\mathcal{J}_t^{ij} &= \int_0^t \langle X_r, e_i \rangle \, a_{ji} \, dr + V_t^{ij}, \\
\mathcal{K}_t^{k\ell} &= \int_0^t \langle Y_r, f_k \rangle \sum_{m=1}^N \langle X_r, e_m \rangle \, c_{\ell k}^m \, dr + O_t^{k\ell}.
\end{aligned}$$

To change, or modify, the intensities of the counting processes $\mathcal{J}_t^{ij}$ and $\mathcal{K}_t^{k\ell}$, that is, to change a_{ji} to $\hat{a}_{ji}$ and $c_{k\ell}^m$ to $\hat{c}_{k\ell}^m$, $m = 1, \ldots, N$ respectively, we must introduce the Radon-Nikodym derivatives $L_t^{ij,k\ell}$ (Brémaud 1981), given by

$$\begin{aligned}
L_t^{ij,k\ell} &= \exp\Bigg[\int_0^t \log\left(\frac{\lambda_r^{k\ell}}{\hat{\lambda}_r^{k\ell}}\right) d\mathscr{K}_r^{k\ell} - \int_0^t (\lambda_r^{k\ell} - \hat{\lambda}_r^{k\ell})dr \\
&\qquad + \int_0^t \log\left(\frac{a_{ji}}{\hat{a}_{ji}}\right) d\mathscr{J}_r^{ij} - \int_0^t (a_{ji} - \hat{a}_{ji}) \langle X_r, e_i\rangle\, dr\Bigg] \\
&= \left(\frac{a_{ji}}{\hat{a}_{ji}}\right)^{\mathscr{J}_t^{ij}} \exp\Bigg[-\int_0^t (a_{ji} - \hat{a}_{ji}) \langle X_r, e_i\rangle\, dr \\
&\qquad + \int_0^t \log\left(\frac{\lambda_r^{k\ell}}{\hat{\lambda}_r^{k\ell}}\right) d\mathscr{K}_r^{k\ell} - \int_0^t (\lambda_r^{k\ell} - \hat{\lambda}_r^{k\ell})dr\Bigg],
\end{aligned}$$

where

$$\hat{\lambda}_r^{k\ell} = \langle Y_r, f_k\rangle \sum_{m=1}^{N} \langle X_r, e_m\rangle\, \hat{c}_{\ell k}^m.$$

Clearly the martingales V^{ij}, $V^{i'j'}$, $O^{k\ell}$, and $O^{k'\ell'}$ are *orthogonal* for $(i,j) \neq (i',j')$ and $(k,\ell) \neq (k',\ell')$. Consequently, to change all the a_{ji} to $\hat{a}_{ji}$ and to change all $c_{\ell k}^m$ to $\hat{c}_{\ell k}^m$, $m = 1,\dots,N$, we should define for $i \neq j$ and $k \neq \ell$

$$\left.\frac{dP_{\hat{\theta}}}{dP_\theta}\right|_{\mathscr{G}_t} = L_t := \prod_{i,j=1}^{N} \prod_{k,\ell=1}^{M} L_t^{ij,k\ell}.$$

The log likelihood is, therefore

$$\begin{aligned}
\log \left.\frac{dP_{\hat{\theta}}}{dP_\theta}\right|_{\mathscr{G}_t} &= \log L_t \\
&= \sum_{i,j=1}^{N} \left\{ \mathscr{J}_t^{ij} \log \frac{\hat{a}_{ji}}{a_{ji}} + \int_0^t (a_{ji} - \hat{a}_{ji}) \langle X_r, e_i\rangle\, dr \right\} \\
&\quad + \sum_{k,\ell=1}^{M} \Bigg\{ \int_0^t \log\left(\frac{\sum_{m=1}^N \langle X_r, e_m\rangle\, \hat{c}_{\ell k}^m}{\sum_{m=1}^N \langle X_r, e_m\rangle\, c_{\ell k}^m}\right) d\mathscr{K}_r^{k\ell} \\
&\quad + \int_0^t \langle Y_r, f_k\rangle \sum_{m=1}^{N} \langle X_r, e_m\rangle (c_{\ell k}^m - \hat{c}_{\ell k}^m)\, dr \Bigg\}.
\end{aligned} \tag{4.3}$$

Now, note that

$$\log\left(\frac{\sum_{m=1}^N \langle X_r, e_m\rangle\, \hat{c}_{\ell k}^m}{\sum_{m=1}^N \langle X_r, e_m\rangle\, c_{\ell k}^m}\right) = \sum_{m=1}^{N} \langle X_r, e_m\rangle \log \frac{\hat{c}_{\ell k}^m}{c_{\ell k}^m}$$

so that, taking the conditional expectation of (4.3), we obtain

$$
\begin{aligned}
E\left[\log \frac{dP_{\hat{\theta}}}{dP_{\theta}} \,\middle|\, \mathscr{Y}_t\right] = & \sum_{i,j=1}^{N} \left(\hat{\mathscr{J}}_t^{ij} \log \hat{a}_{ji} - \hat{a}_{ji} \hat{\mathscr{O}}_t^i\right) \\
& + \sum_{k,\ell=1}^{M} \left\{ \sum_{m=1}^{N} (\log \hat{c}_{\ell k}^m) \widehat{\int_0^t \langle X_r, e_m\rangle \, d\mathscr{K}_r^{k\ell}} \right. \\
& \left. - \sum_{m=1}^{N} \hat{c}_{\ell k}^m \widehat{\int_0^t \langle Y_r, f_k\rangle \langle X, e_m\rangle \, dr} \right\} + \widehat{R}(\theta),
\end{aligned}
\tag{4.4}
$$

where $\hat{R}(\theta) = E[R(\theta) \mid \mathscr{Y}_t]$ does not involve any of the parameters of $\hat{\theta}$. Therefore, the unique maximum of (4.4) over $\hat{\theta}$ occurs at the value of $\hat{\theta}$ obtained by equating to zero the partial derivatives of (4.4) in $\hat{a}_{ji}$ and $\hat{c}_{\ell k}^m$, yielding

$$
\boxed{\hat{a}_{ji} = \frac{\sigma\left(\mathscr{J}_t^{ij}\right)}{\sigma\left(\mathscr{O}_t^i\right)}}
\tag{4.5}
$$

and

$$
\boxed{\hat{c}_{\ell k}^m = \frac{\sigma\left(\int_0^t \langle X_r, e_m\rangle \, d\mathscr{K}_r^{k\ell}\right)}{\sigma\left(\int_0^t \langle Y_r, f_k\rangle \langle X_r, e_m\rangle \, dr\right)} = \frac{\sigma\left(\mathscr{A}_{k\ell,t}^m\right)}{\sigma\left(\mathscr{T}_t^{km}\right)}}
\tag{4.6}
$$

[from Bayes' formula (2.18)]. The family of log-likelihoods is improving and so converges (Dembo and Zeitouni 1986).

7.5 Problems and Notes

Problems

1. Show that the Markov chain X_t has the martingale representation

$$
X_t = \Phi(t,0)\left(X_0 + \int_0^t \Phi(r,0)^{-1} \, dV_r\right),
$$

   where $\Phi(.,.)$ is the fundamental transition matrix (see Appendix B).
2. Prove Lemma 2.3.
3. Show that finite-dimensional smoothers for the processes $\mathscr{A}_{k\ell,s}^m$ and $\mathscr{T}_s^{km}$ described in Section 7.3 are given by (3.17) and (3.18), respectively.

Notes

In previous chapters we obtained finite-dimensional filters and smoothers for a discrete-time Markov chain observed in Gaussian noise. In addition to the filters for the state, finite-dimensional filters and smoothers were obtained for the number of jumps from one state to another of the occupation time in any state, and aslo of a process related to the observations.

In this chapter, the situation considered is that of an unobserved, continuous-time Markov chain which influences the behavior of a second process; in fact, the terms in the intensity, or Q-matrix of the second observed process, are functions of the first, Again, new finite-dimensional filters and smoothers are obtained for quantities analogous to those mentioned above. Particularly interesting are estimates for the joint occupation times of the X and Y processes. Using the expectation maximization (EM) algorithm these estimates are then used to update and improve the parameters of the model. Using the smoothers, filters and possibly new data, the model can be repeatedly revised towards optimality. We, therefore, have an adaptive, self-tuning model.

The results are similar in spirit to work in Davis, Kailath and Segall (1975), van Schuppen (1977), and Boel, Varaiya and Wong (1975). However, these works do not exploit the change of measure. Other novel features of the present chapter are the use again of the idempotent property of the signal process to find a closed-form filter for HX, and the finite-dimensional filters for the number of jumps and occupation times, and the use of the EM algorithm to reestimate the parameters of our model.

Chapter 8
Markov Chains in Brownian Motion

8.1 Introduction

In this chapter, a continuous-time, finite-state Markov chain is observed through a *Brownian motion* with drift. The filtered estimate of the state is the *Wonham filter* (1965). The smoothed estimate of the state is given in Clements and Anderson (1975). A finite-dimensional filter for the number of jumps $\mathcal{J}_t^{ij}$ was obtained by Dembo and Zeitouni (1986) and Zeitouni and Dembo (1988), and used to estimate the parameters of the Markov chain and the observation process. However, this estimation also involves $\mathcal{O}_t^i$ and $\mathcal{T}_t^i$ for which finite-dimensional filters are not given in Zeitouni and Dembo (1988). Our filters allow, therefore, the application of the EM algorithm, an extension of the discrete-time Baum-Welch algorithm (Dembo and Zeitouni 1986; Zeitouni and Dembo 1988). Unlike the Baum-Welch method our equations are recursive and can be implemented by the usual methods of discretization; no backward estimates are required.

Section 8.2 introduces the model. Sections 8.3–8.4 cover the filtering and smoothing of the various processes related to the Markov chain. In Section 8.6 finite-dimensional *predictors* for the various processes are derived. Finally, in Section 8.7 we obtain a finite-dimensional filter for a *non-Markov* multivariate jump process with, *almost surely* (i.e., for "almost" all its *sample paths*), finitely many jumps in any finite-time interval. Some elementary introduction to the concept of *random measures* would be helpful to the understanding of this section, which can be omitted on a first reading.

8.2 The Model

Suppose, that X_t, $t \geq 0$, is a Markov chain defined on a probability space $(\Omega, \mathcal{F}, P)$ with state space $S = \{e_1, e_2, \ldots, e_N\}$. As in Chapter 7, X_t has a semimartingale representation

$$X_t = X_0 + \int_0^t A_r X_r \, dr + V_t. \tag{2.1}$$

The process X_t is not observed directly; rather we suppose there is a (scalar) observation process given by

$$y_t = \int_0^t c(X_r)\,dr + w_t. \tag{2.2}$$

(The extension to vector processes y is straightforward.) Here, w_t is a *standard Brownian motion* on $(\Omega, \mathscr{F}, P)$ which is independent of X_t. Because X takes values in S the function c is given by a vector $c = (c_1, c_2, \ldots, c_N)'$, so that $c(X) = \langle X, c\rangle$ where $\langle \cdot, \cdot \rangle$ denotes the scalar product in $\mathbb{R}^N$. Write

$$\mathscr{G}_t^0 = \sigma\{X_s, y_s : s \le t\}, \qquad \mathscr{Y}_t^0 = \sigma\{y_s : s \le t\},$$

and $\{\mathscr{G}_t\}$, $\{\mathscr{Y}_t\}$, $t \ge 0$, for the corresponding right-continuous, complete filtrations. Note $\mathscr{Y}_t \subset \mathscr{G}_t$ for all t.

We are going to derive finite-dimensional filters and smoothers similar to the ones in Chapter 7. For this we introduce the probability measure $\overline{P}$ by putting

$$\left.\frac{d\overline{P}}{dP}\right|_{\mathscr{G}_t} = \Lambda_t = \exp\left(-\int_0^t \langle X_r, c\rangle\,dw_r - \frac{1}{2}\int_0^t \langle X_r, c\rangle^2\,dr\right).$$

Now Λ_t is a martingale under P and

$$\Lambda_t = 1 - \int_0^t \Lambda_r \langle X_r, c\rangle\,dw_r.$$

By Girsanov's theorem (Elliott 1982b), y is a standard Brownian motion under $\overline{P}$. Define the process $\overline{\Lambda}_t$ by

$$\overline{\Lambda}_t = 1 + \int_0^t \overline{\Lambda}_r \langle X_r, c\rangle\,dy_r \tag{2.3}$$

so that $\overline{\Lambda}_t = \exp\left(\int_0^t \langle X_r, c\rangle\,dy_r - \frac{1}{2}\int_0^t \langle X_r, c\rangle^2\,dr\right)$ and $\overline{\Lambda}_t\Lambda_t = 1$. Note $\overline{\Lambda}_t$ is an $\mathscr{G}$-martingale under $\overline{P}$. However, it is under P that (2.2) holds and so has the form of the observation process influenced by the Markov chain.

In the sequel, we work with a Markov chain X_t and a standard Brownian motion y_t defined on $\{\Omega, \mathscr{F}, \overline{P}\}$. The measure P is defined by putting $\left(dP/d\overline{P}\right)|_{\mathscr{F}_t} = \overline{\Lambda}_t$.

8.3 A General Finite-Dimensional Filter

Consider again a scalar process H_t of the form

$$H_t = H_0 + \int_0^t \alpha_r\,dr + \int_0^t \beta_r'\,dV_r + \int_0^t \delta_r\,dw_r, \tag{3.1}$$

where α, β, δ are $\mathscr{F}$-predictable, square-integrable processes of appropriate dimensions. That is, α_r and δ_r are real and β_r is an N-dimensional vector.

Write $C = \operatorname{diag} c$ for the matrix with diagonal entries $c_1, c_2, \dots, c_N$. Using the product rule for semimartingales

$$\begin{aligned} H_t X_t &= H_0 X_0 + \int_0^t \alpha_r X_r \, dr + \int_0^t \beta_r' X_{r-} \, dV_r \\ &+ \int_0^t \delta_r X_{r-} \, dw_r + \int_0^t H_r A X_r \, dr + \int_0^t H_{r-} \, dV_r \\ &+ \sum_{0<r\le t} (\beta_r' \Delta X_r) \, \Delta X_r. \end{aligned} \tag{3.2}$$

Here

$$\begin{aligned} \sum_{0<r\le t} (\beta_r' \Delta X_r) \, \Delta X_r &= \sum_{i,j=1}^N \int_0^t (\beta_r^j - \beta_r^i) \langle X_{r-}, e_i \rangle \langle e_j, dV_r \rangle (e_j - e_i) \\ &+ \sum_{i,j=1}^N \int_0^t \langle \beta_r^j X_r - \beta_r^i X_r, e_i \rangle a_{ji} \, dr \, (e_j - e_i). \end{aligned}$$

In the last integral we have replaced X_{r-} by X_r. Substituting in (3.2) we have

$$\begin{aligned} H_t X_t &= H_0 X_0 + \int_0^t \alpha_r X_r \, dr + \int_0^t \beta_r X_{r-} \, dV_r \\ &+ \int_0^t \delta_r X_{r-} \, dw_r + \int_0^t H_r A X_r \, dr + \int_0^t H_{r-} \, dV_r \\ &+ \sum_{i,j=1}^N \int_0^t (\beta_r^j - \beta_r^i) \langle X_{r-}, e_i \rangle \langle e_j, dV_r \rangle (e_j - e_i) \\ &+ \sum_{i,j=1}^N \int_0^t \langle \beta_r^j X_r - \beta_r^i X_r, e_i \rangle a_{ji} \, dr \, (e_j - e_i). \end{aligned}$$

Remark 3.1 As in previous chapters, $\sigma(H_t X_t) = \overline{E}\left[\overline{\Lambda}_t H_t X_t \mid \mathscr{Y}_t\right]$.

■

Theorem 3.2 *The recursive equation for the unnormalized estimate $\sigma(H_t X_t)$ is given by the following linear equation:*

$$\begin{aligned} \sigma(H_t X_t) &= \sigma(H_0 X_0) + \int_0^t \sigma(\alpha_r X_r) \, dr + \int_0^t A \sigma(H_r X_r) \, dr \\ &+ \int_0^t \sum_{i,j=1}^N \langle \sigma(\beta_r^j X_r - \beta_r^i X_r), e_i \rangle a_{ji} \, dr \, (e_j - e_i) \\ &+ \int_0^t (\sigma(\delta_r X_r) + C \sigma(H_r X_r)) \, dy_r. \end{aligned} \tag{3.3}$$

Proof

$$\begin{aligned}\overline{\Lambda}_t H_t X_t &= H_0 X_0 + \int_0^t \alpha_r \overline{\Lambda}_r X_r\, dr + \int_0^t \beta_r \overline{\Lambda}_{r-} X_{r-}\, dV_r \\ &+ \int_0^t \delta_r \overline{\Lambda}_{r-} X_{r-}\, dw_r + \int_0^t \overline{\Lambda}_r H_r A X_r\, dr + \int_0^t \overline{\Lambda}_{r-} H_{r-}\, dV_r \\ &+ \sum_{i,j=1}^N \int_0^t \left(\beta_r^j - \beta_r^i\right) \left\langle \overline{\Lambda}_{r-} X_{r-}, e_i \right\rangle \left\langle e_j, dV_r \right\rangle (e_j - e_i) \\ &+ \sum_{i,j=1}^N \int_0^t \left\langle \beta_r^j \overline{\Lambda}_r X_r - \beta_r^i \overline{\Lambda}_r X_r, e_i \right\rangle a_{ij}\, dr\, (e_j - e_i) \\ &+ \sum_{i=1}^N \int_0^t \left\langle \overline{\Lambda}_r X_r, e_i \right\rangle c_i H_r\, dy_r\, e_i + \sum_{i=1}^N \int_0^t \left\langle \overline{\Lambda}_r X_r, e_i \right\rangle \delta_r\, dr\, c_i e_i.\end{aligned}$$

Under $\overline{P}$ y is a standard Brownian motion. Conditioning each side on $\mathcal{Y}_t$ under $\overline{P}$, and using the Fubini Theorem of Wong and Hajek (1985), the result follows. □

The Zakai equation is recursive, so for $s \leq t$ we have the following form:

Corollary 3.3

$$\begin{aligned}\sigma(H_t X_t) &= \sigma(H_s X_s) + \int_s^t \sigma(\alpha_r X_r)\, dr + \int_s^t A\sigma(H_r X_r)\, dr \\ &+ \sum_{i,j=1}^N \int_s^t \left\langle \sigma\left(\left(\beta_r^j - \beta_r^i\right) X_r\right), e_i \right\rangle a_{ji}\, dr\, (e_j - e_i) \\ &+ \int_s^t \left[\sigma(\delta_r X_r) + C\sigma(H_r X_r)\right] dy_r. \qquad (3.4)\end{aligned}$$

Here, the initial condition is $\overline{E}\left[\overline{\Lambda}_s H_s X_s \mid \mathcal{Y}_s\right]$, again a $\mathcal{Y}_s$-measurable random variable.

8.4 States, Transitions, and Occupation Times

We now obtain particular finite-dimensional filters and smoothers, in their unnormalized (Zakai) form by specializing the result of Section 8.3.

The State

Take $H_t = H_0 = 1$, $\alpha_r = 0$, $\beta_r = 0 \in \mathbb{R}^N$, $\delta_r = 0$. Applying Theorem 3.2 we obtain a single, finite-dimensional equation for the unnormalized conditional distribution $\sigma(X_t)$:

$$\sigma(X_t) = \overline{E}\left[\overline{\Lambda}_t X_t \mid \mathcal{Y}_t\right]$$

which is

$$\boxed{\sigma(X_t) = \hat{X}_0 + \int_0^t A\sigma(X_r)\,dr + \int_0^t C\sigma(X_r)\,dy_r.} \tag{4.1}$$

For the smoothed estimates of X_s given $\mathscr{Y}_t$, $s \le t$, take

$$H_t = H_s = \langle X_s, e_i \rangle, \qquad s \le t, \qquad \alpha_r = 0, \qquad \beta_r = 0, \qquad \delta_r = 0$$

and apply Corollaries 3.3. [Rather than taking $H_t = \langle X_s, e_i \rangle$ and estimating $P(X_s = e_i \mid \mathscr{Y}_t) = \pi_t(\langle X_s, e_i \rangle)$ we could consider all states of X_s simultaneously by taking $H_t = X_s$, $s \le t$; however, the product $H_t X_t$ would then have to be interpreted as a tensor, or Kronecker, product $H_t X_t'$.] Substituting $H_t = \langle X_s, e_i \rangle$, $s \le t$, in (3.4) gives

$$\boxed{\begin{aligned}\sigma_t(\langle X_s, e_i \rangle X_t) = \sigma_s(\langle X_s, e_i \rangle X_s) &+ \int_s^t A\sigma_r(\langle X_s, e_i \rangle X_r)\,dr \\ &+ \int_s^t C\sigma_r(\langle X_s, e_i \rangle X_r)\,dy_r.\end{aligned}} \tag{4.2}$$

This is a single-equation finite-dimensional filter for $\sigma_t(\langle X_s, e_i \rangle X_t) = \overline{E}\left[\overline{\Lambda}_t \langle X_s, e_i \rangle X_t \mid \mathscr{Y}_t\right]$ driven by y. Taking the inner product with $\underline{1}$ gives $\sigma_t(\langle X_s, e_i \rangle)$.

The Number of Jumps

In Appendix B it is shown that the number of jumps from e_i to e_j in the time interval $[0,t]$ is given by:

$$\mathscr{J}_t^{ij} = \int_0^t \langle X_{r-}, e_i \rangle a_{ji}\,dr + V_t^{ij}. \tag{4.3}$$

To obtain the unnormalized filter equation, take $H_t = \mathscr{J}_t^{ij}$, $H_0 = 0$, $\alpha_r = \langle X_r, e_i \rangle a_{ji}$, $\beta_r = \langle X_r, e_i \rangle e_j$, $\delta_r = 0$.

The Zakai equation for $\sigma(\mathscr{J}_t^{ij} X_t)$ is obtained by substituting in Theorem 3.2:

$$\boxed{\begin{aligned}\sigma(\mathscr{J}_t^{ij} X_t) = &\int_0^t \langle \sigma(X_r), e_i \rangle a_{ji} e_j\,dr \\ &+ \int_0^t A\sigma(\mathscr{J}_r^{ij} X_r)\,dr + \int_0^t C\sigma(\mathscr{J}_r^{ij} X_r)\,dy_r.\end{aligned}} \tag{4.4}$$

The smoothed estimate of $\mathscr{J}_s^{ij}$ given $\mathscr{Y}_t$, $s \le t$, is obtained from Corollary 3.3 by taking $H_t = H_s = \mathscr{J}_s^{ij}$, $s \le t$, $\alpha_r = 0$, $\beta_r = 0$ and $\delta_r = 0$. Then, from (3.4) we have the finite-dimensional Zakai form of the smoother

$$\boxed{\sigma(\mathscr{J}_s^{ij} X_t) = \sigma(\mathscr{J}_s^{ij} X_s) + \int_s^t A\sigma(\mathscr{J}_s^{ij} X_r)\,dr + \int_s^t C\sigma(\mathscr{J}_s^{ij} X_r)\,dy_r.} \tag{4.5}$$

The Occupation Time

The time spent by the process in state e_i is given by

$$\mathcal{O}_t^i = \int_0^t \langle X_r, e_i \rangle \, dr, \qquad 1 \le i \le N.$$

Take

$$H_t = \mathcal{O}_t^i, \qquad H_0 = 0, \qquad \alpha_r = \langle X_r, e_i \rangle,$$
$$\beta_r = 0 \in \mathbb{R}^N, \qquad \delta_r = 0.$$

Substituting again in Theorem 3.2 gives the Zakai equation

$$\boxed{\sigma\left(\mathcal{O}_t^i X_t\right) = \int_0^t \langle \sigma(X_r), e_i \rangle e_i \, dr + \int_0^t A\sigma\left(\mathcal{O}_r^i X_r\right) dr + \int_0^t \left(C\sigma\left(\mathcal{O}_r^i X_r\right)\right) dy_r.} \tag{4.6}$$

The finite-dimensional unnormalized smoother is obtained for $\mathcal{O}_s^i$ by taking $H_t = H_s = \mathcal{O}_s^i$ for $s \le t$ and $\alpha_r = 0$, $\beta_r = 0$, $\delta_r = 0$. Applying Corollary 3.3 gives

$$\boxed{\sigma\left(\mathcal{O}_s^i X_t\right) = \sigma\left(\mathcal{O}_s^i X_s\right) + \int_s^t A\sigma\left(\mathcal{O}_s^i X_r\right) dr + \int_s^t \left(C\sigma\left(\mathcal{O}_s^i X_r\right)\right) dy_r.} \tag{4.7}$$

The Drift Coefficient

In the next section we shall see that the estimation of the drift coefficient $c = (c_1, c_2, \ldots, c_N)'$ of the observation process involves the filtered estimate of the processes

$$\mathcal{T}_t^i = \int_0^t \langle X_r, e_i \rangle \, dy_r = \int_0^t c_i \langle X_r, e_i \rangle \, dr + \int_0^t \langle X_r, e_i \rangle \, dw_r.$$

Taking $H_t = \mathcal{T}_t^i$, $H_0 = 0$, $\alpha_r = c_i \langle X_r, e_i \rangle$, $\beta_r = 0$ and $\delta_r = \langle X_r, e_i \rangle$ we shall apply Theorem 3.2, noting again that $X_r \alpha_r = c_i \langle X_r, e_i \rangle X_r = c_i \langle X_r, e_i \rangle e_i$ and $X_r \delta_r = X_r \langle X_r, e_i \rangle = \langle X_r, e_i \rangle e_i$. The Zakai equation here is

$$\boxed{\begin{aligned}\sigma\left(\mathcal{T}_t^i X_t\right) &= c_i \int_0^t \langle \sigma(X_r), e_i \rangle e_i \, dr + \int_0^t A\sigma\left(\mathcal{T}_r^i X_r\right) dr \\ &\quad + \int_0^t \left(C\sigma\left(\mathcal{T}_r^i X_r\right) + \langle \sigma(X_r), e_i \rangle e_i\right) dy_r.\end{aligned}}$$

Taking $H_t = \mathcal{T}_s^i$ for $s \le t$, $\alpha_r = 0$, $\beta_r = 0$ and $\delta_r = 0$, we obtain from Corollary 3.3 the following finite-dimensional unnormalized smoother:

$$\boxed{\sigma\left(\mathcal{T}_s^i X_t\right) = \sigma\left(\mathcal{T}_s^i X_s\right) + \int_s^t A\sigma\left(\mathcal{T}_s^i X_r\right) dr + \int_s^t C\sigma\left(\mathcal{T}_s^i X_r\right) dy_r.} \tag{4.8}$$

Remark 4.1 In all the above smoothing equations when we take an inner product with $\underline{1}$ the integral involving A will vanish because $\sum_{j=1}^{N} a_{ji} = 0$. ∎

8.5 Parameter Estimation

This problem is nicely discussed in Dembo and Zeitouni (1986) and Zeitouni and Dembo (1988). We first review their formulation in our setting.

Suppose, as above, that X_t, $t \geq 0$, is a Markov chain with representation (2.1). Again, suppose X_t is observed through the process (2.2).

The above model, therefore, is determined by the set of parameters

$$\theta := (a_{ij},\ 1 \leq i, j \leq N,\ c_i,\ 1 \leq i \leq N).$$

Suppose the model is first determined by a set of parameters

$$\theta = (a_{ij},\ c_i,\ 1 \leq i, j \leq N)$$

and we wish to determine a new set

$$\hat{\theta} = (\hat{a}_{ij},\ \hat{c}_i,\ 1 \leq i, j \leq N)$$

which maximizes the log-likelihood defined below. Write P_θ and $P_{\hat{\theta}}$ for their respective probability measures.

To change all the a_{ji} to $\hat{a}_{ji}$ (see Chapter 7) and to change the c_i to $\hat{c}_i$, we should define

$$\left.\frac{dP_{\hat{\theta}}}{dP_\theta}\right|_{\mathscr{F}_t} := \prod_{i \neq j} L_t^{ij} \exp\left\{ \int_0^t \langle X_r, \hat{c} - c\rangle\, dy_r - \frac{1}{2}\int_0^t \left(\langle X_r, \hat{c}\rangle^2 - \langle X_r, c\rangle^2\right) dr \right\}. \tag{5.1}$$

Now recall that $\langle X_r, c\rangle = \sum_{i=1}^{N} \langle X_r, e_i\rangle c_i$, $\langle X_r, \hat{c}\rangle^2 = \sum_{i=1}^{N} \langle X_r, e_i\rangle c_i^2$, etc. Therefore, taking the log and the conditional expectation on $\mathscr{Y}_t$ of both sides of (5.1), we obtain using the notation of Section 8.4:

$$E\left[\log \frac{dP_{\hat{\theta}}}{dP_\theta} \,\Big|\, \mathscr{Y}_t\right] = \sum_{\substack{i,j=1 \\ \neq j}}^{N} \left(\hat{\mathscr{J}}_t^{ij} \log \hat{a}_{ji} - a_{ji} \hat{\mathscr{O}}_t^i\right) + \sum_{i=1}^{N} \left(c_i \hat{\mathscr{T}}_t^i - \tfrac{1}{2} g_i^2 \hat{\mathscr{O}}_t^i\right) + \hat{R}(\theta). \tag{5.2}$$

Here again $\hat{R}(\theta)$ is independent of $\hat{\theta}$.

The unique maximum of (5.2) over $\hat{\theta}$, obtained by equating to zero the partial derivatives of (5.2) in $\hat{a}_{ji}$ and $\hat{c}_i$, is therefore, given by

$$\boxed{\hat{a}_{ji} = \frac{\sigma\left(\mathscr{J}_t^{ij}\right)}{\sigma\left(\mathscr{O}_t^i\right)},} \tag{5.3}$$

$$\boxed{\hat{c}_i = \frac{\sigma\left(\mathscr{T}_t^i\right)}{\sigma\left(\mathscr{O}_t^i\right)}.} \tag{5.4}$$

This parameter set gives $P_{\hat{\theta}}$, the next probability measure in the sequence of steps in the EM procedure.

The sequence of log-likelihoods constructed this way is increasing and so converges. The convergence of the sequence of θ is discussed in Dembo and Zeitouni (1986) and Zeitouni and Dembo (1988).

8.6 Finite-Dimensional Predictors

The State

The prediction problem discusses the derivation of an equation for $\pi_s(X_t) := E[X_t \mid \mathscr{Y}_s]$, $0 \le s \le t$. For fixed t, $\pi_s(X_t)$ is a $\mathscr{Y}_s$-martingale. Consider for fixed t, the $\mathscr{G}_s$-martingale

$$\eta_s := E[X_t \mid \mathscr{G}_s] = \Phi(t,s)X_s. \tag{6.1}$$

Bayes' theorem implies

$$\pi_s(X_t) = E[X_t \mid \mathscr{Y}_s] = \frac{\overline{E}\left[X_t\overline{\Lambda}_t \mid \mathscr{Y}_s\right]}{\overline{E}\left[\overline{\Lambda}_s \mid \mathscr{Y}_s\right]} = \frac{\sigma_s(X_t)}{\sigma_s(1)}, \qquad \text{say,} \tag{6.2}$$

where $\overline{\Lambda}_t$ is given by (2.3). We are interested in deriving a recursive equation for $\overline{E}\left[\overline{\Lambda}_t X_t \mid \mathscr{Y}_s\right] := \sigma_s(X_t)$. Using the product rule, with $C = \operatorname{diag}(c)$ as before,

$$X_t\overline{\Lambda}_t = X_0 + \int_0^t \overline{\Lambda}_r C X_r \, dy_r + \int_0^t \overline{\Lambda}_r A_r X_r \, dr + \int_0^t \overline{\Lambda}_r \, dV_r. \tag{6.3}$$

Lemma 6.1 *The solution of* (6.3) *is given by*

$$X_t\overline{\Lambda}_t = \Phi(t,0)\left\{X_0 + \int_0^t \Phi(r,0)^{-1}\overline{\Lambda}_r C X_r \, dy_r + \int_0^t \Phi(r,0)^{-1}\overline{\Lambda}_r \, dV_r\right\}. \tag{6.4}$$

Proof Here, (6.4) is obtained by variation of constants, or alternatively differentiating (6.4) yields (6.3). □

Now, using a result of Wong and Hajek (1985) to exchange conditioning and integration in $\overline{E}\left[\overline{\Lambda}_s X_s \mid \mathscr{Y}_s\right]$, the desired Zakai recursion follows:

$$\boxed{\sigma_s(X_t) = \Phi(t,0)\,\sigma(X_0) + \int_0^s \Phi(t,r)\,C\sigma(X_r)\,dy_r.} \tag{6.5}$$

Together with (4.1), Equation (6.5) which is driven by y, provides a finite-dimensional predictor for the unnormalized estimate of X.

Number of Jumps and Occupation Time

The number of jumps from e_i to e_j in the time interval $[0,t]$ is (see Appendix B)

$$\mathscr{J}_t^{ij} = \int_0^t \langle X_{r-}, e_i\rangle\, a_{ji}(r)\,dr + \int_0^t \langle X_{r-}, e_i\rangle \left\langle e_j, dV_r\right\rangle. \tag{6.6}$$

Denote $\eta_s^{ij} = E\left[\mathscr{J}_t^{ij} \mid \mathscr{G}_s\right]$; then we have the following:

Theorem 6.2

$$\begin{aligned}\eta_s^{ij} &= \left\langle \int_0^t \Phi(r,0)\,X_0 a_{ji}(r)\,dr, e_i\right\rangle + \left\langle \int_0^s \langle X_{r-}, e_i\rangle\, dV_r, e_j\right\rangle \\ &\quad + \left\langle \int_0^s \int_u^t \Phi(r,u)\,a_{ji}(r)\,dr\,dV_u, e_i\right\rangle.\end{aligned} \tag{6.7}$$

Proof From (6.6) we write

$$\begin{aligned}\eta_s^{ij} &= E\left[\mathscr{J}_t^{ij} \mid \mathscr{G}_s\right] \\ &= E\left[\int_0^t \langle X_{r-}, e_i\rangle\, a_{ji}(r)\,dr \,\middle|\, \mathscr{G}_s\right] + \int_0^s \langle X_{r-}, e_i\rangle \left\langle e_j, dV_r\right\rangle.\end{aligned} \tag{6.8}$$

Now, using the martingale form, $X_t = \Phi(t,0)\left(X_0 + \int_0^t \Phi(r,0)^{-1}\,dV_r\right)$

$$\begin{aligned}&E\left[\int_0^t \langle X_{r-}, e_i\rangle\, a_{ji}(r)\,dr \,\middle|\, \mathscr{G}_s\right] \\ &\quad = \left\langle \int_0^t \Phi(r,0)\,X_0, e_i\right\rangle a_{ji}(r)\,dr \\ &\qquad + E\left[\left\langle \int_0^t \Phi(r,0) \int_0^r \Phi^{-1}(u,0)\,dV_u, e_i\right\rangle a_{ji}(r)\,dr \,\middle|\, \mathscr{G}_s\right] \\ &\quad = \left\langle \int_0^t \Phi(r,0)\,X_0, e_i\right\rangle a_{ji}(r)\,dr \qquad (6.9) \\ &\qquad + \left\langle \int_0^s \int_u^t \Phi(r,u)\,a_{ji}(r)\,dr\,dV_u, e_i\right\rangle. \qquad (6.10)\end{aligned}$$

Substituting (6.9) into (6.8) gives (6.7). □

We shall first obtain the following finite-dimensional, recursive filter for $\widehat{\eta_s^{ij}X_s} = E\left[\mathscr{J}_t^{ij}X_s \mid \mathscr{G}_s\right]$:

Theorem 6.3

$$\begin{aligned}\widehat{\eta_s^{ij}X_s} = \widehat{\eta_0^{ij}X_0} &+ \int_0^s \langle \hat{X}_r, e_i\rangle a_{ji}(r)\,dr\,(e_j - e_i) + \int_0^s A_r\widehat{\eta_r^{ij}X_r}\,dr \\ &+ \Big\langle \int_0^s \int_r^t \Phi(u,r)\,a_{ji}(u)\,du \\ &\qquad \times \sum_{\substack{k,\ell \\ k\neq\ell}} \langle \hat{X}_r, e_k\rangle a_{\ell k}(r)\,dr\,(e_\ell - e_k), e_i \Big\rangle (e_\ell - e_k) \\ &+ \int_0^s \left(\widehat{CX_r\eta_r^{ij}} - \langle \hat{X}_r, c\rangle \widehat{X_r\eta_r^{ij}}\right)\left(dy_r - \langle \hat{X}_r, c\rangle dr\right),\end{aligned} \tag{6.11}$$

with a Zakai form

$$\boxed{\begin{aligned}\sigma\left(\eta_s^{ij}X_s\right) = \sigma\left(\eta_0^{ij}X_0\right) &+ \int_0^s \langle \sigma(X_r), e_i\rangle a_{ji}(r)\,dr\,(e_j - e_i) \\ &+ \int_0^s A_r\sigma\left(\eta_r^{ij}X_r\right)dr \\ &+ \Big\langle \sum_{k,\ell} \Gamma_{\ell k}^{ji}(s,t), e_i \Big\rangle (e_\ell - e_k) \\ &+ \int_0^s C\sigma\left(\eta_r^{ij}X_r\right)dy_r.\end{aligned}} \tag{6.12}$$

Here $\Gamma_{\ell k}^{ji}(s,t) = \int_0^s \int_r^t \Phi(u,r)\,a_{ji}(u)\,du\,\langle \sigma(X_r), e_k\rangle a_{\ell k}(r)\,dr\,(e_\ell - e_k)$.

Proof Using the Itô product rule for (4.1) and (6.7) we have

$$\eta_s^{ij}X_s = \eta_0^{ij}X_0 + \int_0^s \eta_r^{ij}A_rX_r\,dr + \left[\eta^{ij}, X\right]_s + \mathscr{G}_s\text{-martingale}.$$

Here,

$$\left[\eta^{ij}, X\right]_s = \sum_{0<r\leq s} \Delta\eta_r^{ij}\Delta X_r$$

and

$$\begin{aligned}\Delta\eta_r^{ij}\Delta X_r &= \langle X_{r-}, e_i\rangle \langle e_j, \Delta X_r\rangle \Delta X_r \\ &\quad + \Big\langle \int_r^t \Phi(u,r)\,a_{ji}(u)\,du\Delta X_r, e_i \Big\rangle \Delta X_r \\ &= \langle X_{r-}, e_i\rangle \langle X_r, e_j\rangle (e_j - e_i) \\ &\quad + \Big\langle \int_r^t \Phi(u,r)\,a_{ji}(u)\,du\end{aligned}$$

$$
\begin{aligned}
&\times \sum_{\substack{k,\ell \\ k\neq\ell}} \langle X_{r-}, e_k\rangle \langle X_r, e_\ell\rangle (e_\ell - e_k), e_i \Big\rangle (e_\ell - e_k) \\
&= \langle X_{r-}, e_i\rangle \langle e_j, dX_r\rangle (e_j - e_i) \\
&\quad + \Big\langle \int_r^t \Phi(u,r)\, a_{ji}(u)\, du \\
&\qquad \times \sum_{\substack{k,\ell \\ k\neq\ell}} \langle X_{r-}, e_k\rangle \langle e_\ell, dX_r\rangle (e_\ell - e_k), e_i \Big\rangle (e_\ell - e_k).
\end{aligned}
$$

Hence

$$
\begin{aligned}
\eta_s^{ij} X_s &= \eta_0^{ij} X_0 + \int_0^s \eta_r^{ij} A_r X_r\, dr + \int_0^s \langle X_r, e_i\rangle a_{ji}(r)\, dr\, (e_j - e_i) \\
&\quad + \Big\langle \int_0^s \Big[\int_r^t \Phi(u,r)\, a_{ji}(u)\, du \\
&\qquad \times \sum_{k,\ell} \langle X_r, e_k\rangle a_{\ell k}(r) \Big] dr\, (e_i - e_k), e_i \Big\rangle (e_\ell - e_k) \\
&\quad + \mathscr{G}_s\text{-martingale.}
\end{aligned} \tag{6.13}
$$

Taking the $\mathscr{Y}_s$-optional projection on both sides of (6.13) gives

$$
\begin{aligned}
\widehat{\eta_s^{ij} X_s} &= \widehat{\eta_0^{ij} X_0} + \int_0^s \langle \hat{X}_r, e_i\rangle a_{ji}(r)\, dr\, (e_j - e_i) + \int_0^s A_r \widehat{\eta_r^{ij} X_r}\, dr \\
&\quad + \Big\langle \int_0^s \int_r^t \Phi(u,r)\, a_{ji}(u)\, du \\
&\qquad \times \sum_{\substack{k,\ell \\ k\neq\ell}} \langle \hat{X}_r, e_k\rangle a_{\ell k}(r)\, dr\, (e_\ell - e_k), e_i \Big\rangle (e_\ell - e_k) \\
&\quad + \int_0^s \gamma_r\, d\nu_r,
\end{aligned} \tag{6.14}
$$

where $\nu_r := y_r - \int_0^r \langle \hat{X}_r, c\rangle dr$ is the *innovation* process and γ_r is a square-integrable $\mathscr{Y}_s$-predictable process which we will identify using special semimartingale representations. Multiplying together (2.2) and (6.14) and taking the $\mathscr{Y}_s$-optional projection gives us:

$$
\begin{aligned}
\widehat{\eta_s^{ij} X_s y_s} &= \int_0^s \eta_r^{ij} X_r \widehat{\langle X_r, c\rangle}\, dr + \int_0^s y_r A_r \widehat{\eta_r^{ij} X_r}\, dr \\
&\quad + \int_0^s y_r \langle \hat{X}_r, e_i\rangle a_{ji}(r)\, dr\, (e_j - e_i) \\
&\quad + \int_0^s y_r \Big\langle \int_r^t \Phi(u,r)\, a_{ji}(u)\, du
\end{aligned}
$$

$$\times \sum_{\substack{k,\ell \\ k\neq\ell}} \langle \hat{X}_r, e_k \rangle a_{\ell k}(r)\, dr\, (e_\ell - e_k), e_i \Big\rangle (e_\ell - e_k)$$

$$+\, \mathscr{Y}_s\text{-martingale.} \tag{6.15}$$

However, multiplying together the *innovation* form $y_t = \int_0^t \langle \hat{X}_r, c \rangle dr + v_t$ of (2.2) and (6.14) we also have

$$\begin{aligned}
\widehat{\eta_s^{ij} X_s y_s} &= \int_0^s \langle \hat{X}_r, c \rangle \widehat{\eta_r^{ij} X_r}\, dr \\
&+ \int_0^s y_r \langle \hat{X}_r, e_i \rangle a_{ji}(r)\, dr\, (e_j - e_i) + \int_0^s y_r A_r \eta_r^{ij} X_r\, dr \\
&+ \int_0^s y_r \Big\langle \int_r^t \Phi(u,r)\, a_{ji}(u)\, du \\
&\qquad \times \sum_{\substack{k,\ell \\ k\neq\ell}} \langle \hat{X}_r, e_k \rangle a_{\ell k}(r)\, dr\, (e_\ell - e_k), e_i \Big\rangle (e_\ell - e_k) \\
&+ \int_0^s \gamma_r\, dr.
\end{aligned} \tag{6.16}$$

Equating the bounded variation terms in (6.15) and (6.16) gives

$$\gamma_r = \widehat{C\eta_r^{ij} X_r} - \langle \hat{X}_r, c \rangle \widehat{\eta_r^{ij} X_r}. \tag{6.17}$$

This together with (6.14) gives (6.11). Equation (6.12) is easily obtained using the change of measure and Bayes' rule. □

Taking the inner product of (6.11) and (6.12) with the vector $\underline{1} = (1, 1, \dots, 1)$ we obtain the normalized and unnormalized predictors for the number of jumps, that is,

$$\Big\langle \widehat{\eta_s^{ij} X_s}, \underline{1} \Big\rangle = \widehat{\eta_s^{ij}} \qquad \text{and} \qquad \langle \sigma(\eta_s^{ij} X_s), \underline{1} \rangle = \sigma(\eta_s^{ij}).$$

With similar arguments we have predictors for the occupation time $\mathscr{O}_t^i := \int_0^t \langle X_r, e_i \rangle dr$. If we denote $\Gamma_s^i := E\left[\mathscr{O}_t^i \mid \mathscr{G}_s\right]$, then similarly to Theorem 6.2,

$$\Gamma_s^i = \Big\langle \int_0^s \Phi(r,0) X_0\, dr, e_i \Big\rangle + \Big\langle \int_0^s \int_r^t \Phi(r,u)\, du M_u, e_i \Big\rangle.$$

In this case

$$\begin{aligned}
\widehat{\Gamma_s^i X_s} &= \widehat{\Gamma_0^i X_0} + \int_0^s A_r \widehat{\Gamma_r^i X_r}\, dr \\
&+ \Big\langle \int_0^s \int_r^t \Phi(u,r)\, a_{ji}(u)\, du
\end{aligned}$$

$$\times \sum \langle \hat{X}_r, e_k \rangle a_{\ell k}(r)\, dr\, (e_\ell - e_k), e_i \Big\rangle (e_\ell - e_k)$$

$$+ \int_0^s \left(C\widehat{\Gamma^i_r X_r} - \langle \hat{X}_r, c\rangle \widehat{\Gamma^i_r X_r} \right) \left(dy_r - \langle \hat{X}_r, c\rangle dr \right), \tag{6.18}$$

with a Zakai form

$$\boxed{\begin{aligned} \sigma\left(\Gamma^i_s X_s\right) &= \sigma\left(\Gamma^i_0 X_0\right) + \int_0^s A_r \sigma\left(\Gamma^i_r X_r\right) dr \\ &\quad + \left\langle \sum_{k,\ell} \Gamma^{ji}_{\ell k}(s,t), e_i \right\rangle (e_\ell - e_k) \\ &\quad + \int_0^s C\sigma\left(X_r \Gamma^i_r\right) dy_r. \end{aligned}} \tag{6.19}$$

Here $\Gamma^{ji}_{\ell k}(s,t) = \int_0^s \int_r^t \Phi(u,r)\, a_{ji}(u)\, du \,\langle \sigma(X_r), e_k \rangle a_{\ell k}(r)\, dr\, (e_\ell - e_k)$. Taking the inner product of (6.18) and (6.19) with the vector $\underline{1} = (1, \ldots, 1)$ gives the desired predictors for the occupation time at any state e_i up to time t.

8.7 A Non-Markov Finite-Dimensional Filter

Consider a process S_t, $t \geq 0$, with right-constant sample paths defined on a probability space $(\Omega, \mathscr{F}, P)$. Its state space is an arbitrary finite set $\hat{S} = \{s_1, \ldots, s_N\}$. By considering the functions $\phi_k(s_i)$, defined so that $\phi_k(s_i) = 0$ if $i \neq k$ and $\phi_k(s_k) = 1$, and writing $X_t := (\phi_1(S_t), \ldots, \phi_N(S_t))$ we see that equivalently we can consider a process X_t, $t \geq 0$, whose state space is the set $S = \{e_1, \ldots, e_N\}$ of unit (column) vectors $e_i = (0, \ldots, 1, 0, \ldots, 0)'$ of $\mathbb{R}^N$. $\{\mathscr{F}_t\}$ will be the right-continuous complete filtration generated by X.

Suppose that in any finite-time interval X_t has, almost surely, only finitely many jumps (this is implied by the boundedness of a_{ii} defined below), and write $T_k(\omega)$, $k \in N$, for the kth jump time. $\delta_{T_k(\omega)}(dr)$ will denote the unit mass at time $T_k(\omega)$ and, if $X_{T_k(\omega)} = e_{i_k(\omega)}$, write $\delta_{i_k(\omega)}(i)$ for the function which is 1 if $i = i_k(\omega)$ and 0 otherwise. X_t is a *multivariate jump process* and we can write

$$\begin{aligned} X_t &= X_0 + \sum_{0<r\leq t} \Delta X_r, \qquad \text{where } \Delta X_r = X_r - X_{r-} \\ &= X_0 + \int_0^t \sum_i (e_i - X_{r-}) \left(\sum_k \delta_{T_k(\omega)}(dr)\, \delta_{i_k(\omega)}(i) \right). \end{aligned} \tag{7.1}$$

The *random measure* $\sum_k \delta_{T_k(\omega)}(dr)\, \delta_{i_k(\omega)}(i)$ picks out the jump size and the jump times and, following Jacod (1979), has a *predictable compensator* $\nu^p(dr, e_i)$. In turn, ν^p factors into its *Lévy system* :

$$\nu^p(dr, e_i) = \lambda(e_i, X_{r-}, r, \omega)\, d\Lambda(r, X_{r-}, \omega).$$

Roughly speaking, $d\Lambda(r, X_{r-}, \omega)$ is the conditional probability that the next jump occurs at time r given the previous history of the process. (In the Markov case the conditioning is only on the event that the process is in state X_{r-} just before r). $\lambda(e_i, X_{r-}, r, \omega)$, defined for $e_i \neq X_{r-}$, is the conditional probability that the process jumps from X_{r-} to $X_r = e_i$, given that there is a jump at time r and given the previous history of the process.

The measure $d\Lambda$ is a nonnegative random measure on $\mathbb{R}^+$. For simplicity we shall suppose it is *absolutely continuous* with respect to *Lebesgue measure*, so that there is a predictable Radon-Nikodym derivative $a(r, X_{r-}, \omega)$ such that

$$d\Lambda(r, X_{r-}, \omega) = a(r, X_{r-}, \omega)\, dr.$$

Writing $a(e_i, X_{r-}, r, \omega) = \lambda(e_i, X_{r-}, r, \omega)\, a(r, X_{r-}, \omega)$ for $e_i \neq X_{r-}$ we see

$$\sum_{e_i \neq X_{r-}} a(e_i, X_{r-}, r, \omega) = a(r, X_{r-}, \omega).$$

If we define a matrix $A(r, \omega) = (a_{ij}(r, \omega))$ by putting

$$\begin{aligned} a_{ij}(r, \omega) &= a(e_j, e_i, r, \omega), \qquad && i \neq j, \\ a_{ii}(r, \omega) &= -a(r, e_i, \omega), && 1 \leq i \leq N, \end{aligned}$$

then (with obvious notation),

$$\sum_i (e_i - X_{r-})\, a_{iX_{r-}} = A(r, \omega) X_{r-}$$

and the representation (7.1) can be written

$$\begin{aligned} X_t = X_0 &+ \int_0^t \sum_i (e_i - X_{r-}) \left(\sum_k \delta_{T_k(\omega)}(dr)\, \delta_{i_k(\omega)}(i) - a_{iX_{r-}} \right) dr \\ &+ \int_0^t A(r, \omega) X_{r-}\, dr. \end{aligned} \tag{7.2}$$

Note X is a Markov process if and only if the elements of A are deterministic.

The decomposition (7.2) expresses X as the sum of a martingale V and a predictable process of *integrable variation*. The martingale V is

$$\begin{aligned} V_t &= \int_0^t \sum_i (e_i - X_{r-}) \left(\sum_k \delta_{T_k(\omega)}(dr)\, \delta_{i_k(\omega)}(i) - a_{iX_{r-}}\, dr \right) \\ &= X_t - X_0 - \int_0^t A(r, \omega) X_{r-}\, dr && (7.3) \\ &= X_t - X_0 - \int_0^t A(r, \omega) X_r\, dr, && (7.4) \end{aligned}$$

because for almost all ω $X_r(\omega) = X_{r-}(\omega)$ except for countably many r. (This observation will be used to equate similar integrals below.)

For unit vectors $e_i, e_j \in S$, $i \neq j$, consider the stochastic integral

$$V_t^{ij} = \int_0^t \langle e_i, X_{r-} \rangle \langle e_j, dV_r \rangle . \tag{7.5}$$

Note the integrand is predictable, so V^{ij} is a martingale. Now

$$\begin{aligned} \langle e_i, X_{r-} \rangle \langle e_j, dX_r \rangle &= \langle e_i, X_{r-} \rangle \langle e_j, \Delta X_r \rangle \\ &= I(X_{r-} = e_i,\, X_r = e_j) . \end{aligned} \tag{7.6}$$

Substituting (7.3) in (7.5)

$$V_t^{ij} = \int_0^t \langle e_i, X_{r-} \rangle \langle e_j, dX_r \rangle - \int_0^t \langle e_i, X_{r-} \rangle \langle e_j, A(r,\omega) X_{r-} \rangle dr.$$

Write $\mathscr{J}_t^{ij}$ for the number of jumps of process X from e_i to e_j up to time t. Then using (7.6)

$$\begin{aligned} V_t^{ij} &= \mathscr{J}_t^{ij} - \int_0^t \langle X_{r-}, e_i \rangle a_{ji}(r,\omega)\, dr \\ &= \mathscr{J}_t^{ij} - \int_0^t \langle X_r, e_i \rangle a_{ji}(r,\omega)\, dr \end{aligned}$$

(again because $X_r = X_{r-}$ a.s. except for countably many r). Therefore, we can write

$$\mathscr{J}_t^{ij} = \int_0^t \langle X_r, e_i \rangle a_{ji}(r,\omega)\, dr + V_t^{ij} . \tag{7.7}$$

Observation Process

Suppose $\{1,2,\dots,N\} = A(1) \cup A(2) \cup \cdots \cup A(d)$ where $A(k) \cap A(\ell) = \emptyset$ if $k \neq \ell$. For a set $A(k)$ write $I(A(k))$ for the vector $\sum_{i \in A(k)} e_i$. For $k, \ell \in \{1,\dots,d\}$, $k \neq \ell$, define

$$Y_t^{k\ell} = \sum_{(i,j) \in A(k) \times A(\ell)} \mathscr{J}_t^{ij}$$

so from (7.6)

$$Y_t^{k\ell} = \int_0^t \langle I(A(k)), X_{r-} \rangle \langle I(A(\ell)), dX_r \rangle .$$

Substituting from (7.3) we obtain, similarly to (7.7),

$$Y_t^{k\ell} = \int_0^t h^{k\ell}(r)\, dr + Q_t^{k\ell},$$

where

$$h^{k\ell}(r) = \sum_{(i,j)\in A(k)\times A(\ell)} \langle X_r, e_i\rangle a_{ji}(r,\omega)$$
$$\text{and } Q_t^{k\ell} = \sum_{(i,j)\in A(k)\times A(\ell)} V_t^{ij}.$$

The observation process will be a set of processes of the form $Y_t^{k\ell}$. That is, we suppose there is a collection of pairs (k_1,ℓ_1), (k_2,ℓ_2), ..., (k_p,ℓ_p) with k_i, $\ell_i \in \{1,\dots,d\}$, $k_i \neq \ell_i$, and $(k_i,\ell_i) \neq (k_j,\ell_j)$ for $i \neq j$. For simplicity we shall write Y^j for $Y^{k_j\ell_j}$, etc., so the observation process is

$$Y_t := \left(Y_t^1, Y_t^2, \dots, Y_t^p\right)',$$
$$\text{where } Y_t^j = \int_0^t h^j(r)\,dr + Q_t^j. \tag{7.8}$$

Note that if $i \neq j$ then Y^i jumps at different times to Y^j, so the martingales Q^i and Q^j are orthogonal.

Write $\{\mathscr{Y}_t\}$ for the right-continuous, complete filtration generated by Y. Our objective is to obtain a recursive equation for $\hat{X}_t = E[X_t \mid \mathscr{Y}_t]$.

Definition 7.1 The innovation process is

$$\tilde{Q}_t := (\tilde{Q}_t^1, \dots, \tilde{Q}_t^p)'$$

where

$$\tilde{Q}_t^j = Y_t^j - \int_0^t \hat{h}^j(r)\,dr.$$

Simple arguments, using again Fubini's theorem, show that $\tilde{Q}^j$ is a $\mathscr{Y}$-martingale. We can, therefore, write

$$Y_t^j = \int_0^t \hat{h}^j(r)\,dr + \tilde{Q}_t^j. \tag{7.9}$$

Similar calculations, again using Fubini's theorem and (7.4), show that the process

$$\tilde{V}_t := \hat{X}_t - \hat{X}_0 - \int_0^t \widehat{A(r,\omega)X_r}\,dr$$

is an $\mathbb{R}^N$-valued martingale with respect to the $\mathscr{Y}$-filtration. Because $\mathscr{Y}_0$ is the trivial σ-field, $\{\Omega,\phi\}$, $\hat{X}_0$ is a constant vector, the initial distribution of X.

From the representation result for martingales with respect to a multivariate jump process $\tilde{V}$ (Brémaud 1981) can be represented as

$$\tilde{V}_t = \int_0^t \gamma_r\,d\tilde{Q}_r$$
$$= \sum_{j=1}^p \int_0^t \gamma_r^j\,d\tilde{Q}_r^j.$$

Here γ is a $\mathscr{Y}$-predictable $N \times p$ matrix valued process with columns

$$\gamma_r^j = \left(\gamma_r^{1j}, \dots, \gamma_r^{Nj}\right)', \qquad j = 1, \dots, p.$$

We, therefore, have

$$\hat{X}_t = \hat{X}_0 + \int_0^t A\widehat{(r,\omega)X_r}\, dr + \int_0^t \gamma_r\, d\tilde{Q}_r. \tag{7.10}$$

Theorem 7.2

$$\gamma_r^j = \hat{h}^j(r)^{-1} I_{\hat{h}^j(r)\neq 0} \left\{ \sum_{(\sigma,\rho)\in A(k_j)\times A(\ell_j)} \widehat{\langle e_\sigma, X_r\rangle a_{\rho\sigma}}(r,\omega) e_\rho - \hat{h}^j(r)\hat{X}_r \right\}.$$

Proof The filtering problem will be solved if we determine the elements of γ. This we do by calculating $\hat{X}_t Y_t'$ two ways; in fact we calculate the jth column $\langle \hat{X}_t, Y_t^j \rangle$ in two ways. Now

$$X_t Y_t^j = \int_0^t X_{r-}\, dY_r^j + \int_0^t dX_r Y_{r-}^j + \sum_{0<r\le t} \Delta X_r \Delta Y_r^j$$

and from (7.4) this is

$$\begin{aligned} &= \int_0^t X_{r-}\, dY_r^j + \int_0^t A(r,\omega) X_r Y_{r-}\, dr \\ &\quad + \int_0^t dV_r Y_{r-} + \sum_{0<r\le t} (X_r - X_{r-}) \Delta Y_r^j. \end{aligned} \tag{7.11}$$

Note

$$\sum_{0<r\le t} X_{r-} \Delta Y_r^j = \int_0^t X_{r-}\, dY_r^j.$$

Also

$$\Delta Y_r^j = \langle I(A(k_j)), X_{r-}\rangle \langle I(A(\ell_j)), X_r\rangle$$

and

$$X_r \langle e_\rho, X_r\rangle = e_\rho \langle e_\rho, X_r\rangle, 1 \le \rho \le N,$$

so

$$X_r \Delta Y_r^j = \sum_{\rho\in A(\ell_j)} \langle I(A(k_j)), X_{r-}\rangle e_\rho \langle e_\rho, X_r\rangle.$$

Therefore,

$$\begin{aligned} \sum_{0<r\le t} \Delta X_r \Delta Y_r^j &= \sum_{0<r\le t} (X_r - X_{r-}) \Delta Y_r^j \\ &= \sum_{\rho\in A(\ell_j)} \langle I(A(k_j)), X_{r-}\rangle e_\rho \langle e_\rho, X_r\rangle - \int_0^t X_{r-}\, dY_r^j \end{aligned}$$

$$= \sum_{(\sigma,\rho)\in A(k_j)\times A(\ell_j)} \int_0^t \langle e_\sigma, X_r\rangle a_{\rho\sigma}(r,\omega)\, dr\, e_\rho$$
$$+ \sum_{(\sigma,\rho)\in A(k_j)\times A(\ell_j)} Q_t^{\sigma\rho} e_\rho - \int_0^t X_{r-}\, dY_r^j.$$

Substituting in (7.11) we see

$$\begin{aligned} X_t Y_t^j &= \int_0^t A(r,\omega) X_r Y_{r-}^j\, dr + \int_0^t dV_r\, Y_{r-} \\ &+ \sum_{(\sigma,\rho)\in A(k_j)\times A(e_j)} \int_0^t \langle e_\sigma, X_r\rangle a_{\rho\sigma}(r,\omega)\, dr\, e_\rho \\ &+ \sum_{(\sigma,\rho)\in A(k_j)\times A(\ell_j)} Q_t^{\sigma\rho} e_\rho. \end{aligned} \tag{7.12}$$

Taking the $\mathscr{Y}$-optional projection of each side of (7.12) and using the fact that

$$\widehat{A(r,\omega) X_r Y_{r-}^j} = A(r,\omega)\widehat{X_r} Y_{r-}^j$$

we have

$$\begin{aligned} \hat{X}_t Y_t^j &= \int_0^t A(r,\omega) X_r Y_{r-}^j\, dr \\ &+ \sum_{(\sigma,\rho)\in A(k_j)\times A(\ell_j)} \int_0^t \widehat{\langle e_\sigma, X_r\rangle a_{\rho\sigma}}(r,\omega) dr\, e_\rho + \overline{H}_t, \end{aligned} \tag{7.13}$$

where $\overline{H}$ is a square-integrable $\mathscr{Y}$-martingale. However, from (7.9) and (7.10) we also have

$$\begin{aligned} \hat{X}_t Y_t^j &= \int_0^t \hat{X}_{r-} dY_r^j + \int_0^t d\hat{X}_r Y_{r-}^j + \sum_{0<r\le t} \Delta\hat{X}_r \Delta Y_r^j \\ &= \int_0^t \hat{X}_{r-} \hat{h}^j(r)\, dr + \int_0^t \hat{X}_{r-}\, d\tilde{Q}_r^j \\ &+ \int_0^t A(r,\omega) X_r Y_{r-}^j\, dr + \int_0^t \gamma_r Y_{r-}^j\, d\tilde{Q}_r + \sum_{0<r\le t} \gamma_r^j \Delta Y_r^j \end{aligned}$$

because the martingales $\tilde{Q}^i$ and $\tilde{Q}^j$ jump at different times if $i \neq j$. That is,

$$\hat{X}_t Y_t^j = \int_0^t \hat{X}_r \hat{h}^j(r)\, dr + \int_0^t A(r,\omega) X_r Y_{r-}^j\, dr + \int_0^t \gamma_r^j \hat{h}^j(r)\, dr + H_t^2, \tag{7.14}$$

where H^2 is a square-integrable $\mathscr{Y}$-martingale. Now $\hat{X}_t Y_t^j$ is a special semimartingale so the decompositions (7.13) and (7.14), into the sum of a martingale and a predictable bounded variation process, must be the same. Therefore, equating the bounded variation processes we see

$$\gamma_r^j = \hat{h}^j(r)^{-1} I_{\hat{h}^j(r)\neq 0}\left\{ \sum_{(\sigma,\ell)\in A(k_j)\times A(\ell_j)} \widehat{\langle e_\sigma, X_r\rangle a_{\rho\sigma}}(r,\omega) e_\rho - \hat{h}^j(r)\hat{X}_r \right\}.$$

□

Corollary 7.3 *If the entries $a_{ij}(t,\omega)$ of the matrix $A(t,\omega)$ are adapted to the $\mathscr{Y}$-filtration, then*

$$\widehat{\langle e_\sigma, X_r\rangle a_{\rho\sigma}}(r,\omega) = \langle e_\sigma, \hat{X}_r\rangle a_{\rho\sigma}(r,\omega)$$

so

$$\hat{h}^j(r) = \sum_{(\sigma,\rho)\in A(k_j)\times A(\ell_j)} \langle e_\sigma, \hat{X}_r\rangle a_{\rho\sigma}(r,\omega)$$

and $\hat{X}_r$ is given by a finite-dimensional filter. That is,

$$\begin{aligned} \hat{X}_t = \hat{X}_0 &+ \int_0^t A(r,\omega)\hat{X}_r\,dr \\ &+ \int_0^t \left(\sum_{j=1}^p \hat{h}^j(r)^{-1} I_{\hat{h}^j(r)\neq 0} \right. \\ &\qquad \left. \times \left\{ \sum_{(\sigma,\rho)\in A(k_j)\times A(\ell_j)} \langle e_\sigma, \hat{X}_r\rangle a_{\rho\sigma}(r,\omega)\, e_\rho - \hat{h}^j(r)\hat{X}_r \right\} \right) d\tilde{Q}_r^j. \end{aligned}$$

A Zakai Equation

To derive the Zakai form of this filter suppose the $a_{\sigma\rho}(t,\omega)$ are adapted to the $\mathscr{Y}$-filtration and suppose there is a $k > 0$ such that $a_{\sigma\rho}(t,\omega) \geq k$ for $(\sigma,\rho) \in A(k_j)\times A(\ell_j)$ and all j.

Define a new measure $\overline{P}$, on $(\Omega, \mathscr{F}, P)$ by putting $E\left[d\overline{P}/dP \mid \mathscr{G}_t\right] = \Lambda_t$ where Λ is the martingale

$$\Lambda_t = 1 + \sum_{j=1}^p \int_0^t \Lambda_{r-}\left(h^j(r-)^{-1} - 1\right) dQ_r^j.$$

Then under $\overline{P}$ the components Y^j of Y are independent Poisson processes. Consequently define $\overline{Q}_t^j = Y_t^j - t$ and write $\overline{Q}_t = \left(\overline{Q}^1, \ldots, \overline{Q}^p\right)'$. Consider the $\left(\overline{P}, \mathscr{F}\right)$ martingale

$$\overline{\Lambda}_t := 1 + \sum_{j=1}^p \int_0^t \overline{\Lambda}_{r-}\left(h^j(r-) - 1\right) d\overline{Q}_r^j.$$

Then it is easily checked that

$$\Lambda_t \overline{\Lambda}_t = 1 \qquad \text{a.s.}$$

We take $\overline{P}$ as the reference probability and compute expectations under $\overline{P}$. However, it is under P that

$$Y_t^j = \int_0^t h^j(r-)\,dr + Q_t^j, \qquad j = 1, \dots, p.$$

Write $\Pi\left(\overline{\Lambda}_t\right)$ for the $\mathscr{Y}$-optional projection of $\overline{\Lambda}$ under $\overline{P}$. Then for each $t \geq 0$

$$\Pi\left(\overline{\Lambda}_t\right) = \overline{E}\left[\overline{\Lambda}_t \mid \mathscr{Y}_t\right] \qquad \text{a.s.}$$

Furthermore, it can be shown that

$$\Pi\left(\overline{\Lambda}_t\right) = 1 + \sum_{j=1}^{p} \int_0^t \lambda_r^j \, d\overline{Q}_r^j,$$

where $\lambda_r^j = \Pi\left(\overline{\Lambda}_{r-}\right)\left(\hat{h}^j(r-) - 1\right)$.

Write

$$\begin{aligned}\sigma(X_t) &= \overline{E}\left[\overline{\Lambda}_t X_t \mid \mathscr{Y}_t\right] \qquad \text{a.s.}\\ &= q_t, \qquad \text{say.}\end{aligned}$$

Also, $\sigma(1) = \Pi\left(\overline{\Lambda}_t\right)$. q_t is an unnormalized conditional expectation of X_t given $\mathscr{Y}_t$ because

$$\hat{X}_t = E\left[X_t \mid \mathscr{Y}_t\right] = \frac{\sigma(X_t)}{\sigma(1)} = \frac{q_t}{\Pi\left(\overline{\Lambda}_t\right)}.$$

Therefore,

$$\boxed{\begin{aligned} q_t &= \hat{X}_0 + \int_0^t A_r q_{r-}\,dr \\ &\quad + \sum_{j=1}^{p} \sum_{(\sigma,\rho)\in A(k_j)\times A(\ell_j)} \int_0^t \left[\langle e_\sigma, q_{r-}\rangle a_{\rho\sigma}(r,\omega) e_\rho - q_{r-}\right] d\overline{Q}_r^j. \end{aligned}} \tag{7.15}$$

Note again this equation is linear in q.

8.8 Problems and Notes

Problems

1. Fill in the proof of Theorem 3.2
2. Derive Equation (6.12).
3. Fill in the details in the derivation of Zakai Equation (7.15).

4. Prove that the solution of (6.3) is given by

$$X_t\overline{\Lambda}_t = \Phi(t,0)\left\{X_0 + \int_0^t \Phi(r,0)^{-1}\overline{\Lambda}_r C X_r\, dy_r + \int_0^t \Phi(r,0)^{-1}\overline{\Lambda}_r\, dV_r\right\}.$$

Notes

The literature on semimartingales and stochastic integrals is found in many textbooks and monographs (Elliott 1982b; Wong and Hajek 1985; Chung and Williams 1990).

In Dembo and Zeitouni (1986) $\hat{\mathscr{T}}_t^i$ is written as $\int_0^t E\left[\langle X_r, g\rangle \mid \mathscr{Y}_t\right] dy_r$, a non-adapted stochastic integral which is not defined, at least in Dembo and Zeitouni (1986). Also, it is not clear the reference to Yao (1985) in Zeitouni and Dembo (1988) provides a finite-dimensional filter for $\hat{\mathscr{O}}_t^i$ in the general case.

However, the results of Section 8.4 above give explicit finite-dimensional filters (and smoothers) for $\hat{\mathscr{J}}_t^{ij}$, $\hat{\mathscr{O}}_t^i$, and $\hat{\mathscr{T}}_t^i$, $1 \le i, j \le N$.

Part IV
Two-Dimensional HMM Estimation

Chapter 9
Hidden Markov Random Fields

9.1 Discrete Signal and Observations

In the previous chapters we used ideas and techniques to solve filtering and estimation problems for dynamical systems evolving in one-dimensional (discrete or continuous) time. Here, working again with the reference probability technique, we discuss similar problems for "discrete-time" *random fields*, that is, sets of random variables taking their indices from unordered countable sets, such as, for example, *lattices* in Euclidean spaces. Our goal is to derive algorithms that could be useful to restore, or filter, noisy or [as expressed by Besag (1986)] "dirty images."

We shall be working under the assumption that the random fields are *Markov random fields*. For that, consider a lattice L of points and consider a system of neighborhoods $\mathscr{N} = \{N_\ell, \ell \in L\}$ over L, such that each N_ℓ consists of a certain number $|N_\ell|$ of points of L, not including ℓ. Denote $X(N_\ell) = \{X_k, k \in N_\ell\}$. Then $\{X_\ell, \ell \in L\}$ is a Markov random field if $P[X_\ell = x \mid X_k, k \neq \ell, k \in L] = P[X_\ell = x \mid X(N_\ell)]$; that is, the dependence between the random variable is, for each X_ℓ, determined only by random variables in its neighborhood N_ℓ.

A random field X on a lattice L is considered. At each point ℓ of the lattice X_ℓ takes some value. The random field X is not directly observed; rather there is a noisy observation process y which is a function of X and, in the "blurred" case, of some of the neighbors of X. We, therefore, have a hidden Markov random field, HMRF.

Conditions are given which ensure X is a Markov random field. The problems discussed here are the following: given a set of observations $\{y_\ell, \ell \in L\}$, determine the most likely signal $\{X_\ell, \ell \in L\}$ and, also, determine the parameters of the model, that is, the transition probabilities of the Markov random field X and the observations y. The technique used is that of a measure transformation which changes all the signal, X_ℓ, and observation, y_ℓ, random variables into independent, identically, uniformly distributed random variables. The use of this measure change is equivalent to employing a form of *Bayes' Theorem*; however, to exhibit analogies with the rest of the book we choose to introduce a new probability measure. The lattice L could be the set of pixel locations in some image. We first discuss the case where the observation variables y take values in a discrete set. In Section 9.2 we considered the situation where y is scalar and the signal X is observed in Gaussian noise. In Section 9.3, both the signal X and observations y take continuous values.

The Markov Random Field

Consider a finite lattice L. (In particular, L could consist of a grid or array of points in $\mathbb{R}^d$.) Associated with L we suppose there is a discrete Markov field X_ℓ, $\ell \in L$, with a finite-state space S defined on a probability space $(\Omega, \mathscr{F}, P)$. Without loss of generality we can suppose that S consists of the set $S_X = \{e_1, \dots, e_M\}$ of standard unit vectors in $\mathbb{R}^M$ for some positive integer M. Then $X_\ell \in S$ for each $\ell \in L$. We shall suppose that each point $\ell \in L$ has a neighborhood N_ℓ consisting of points of L different from ℓ. The number of points in N_ℓ may vary for ℓ on or near the boundary of L. Write $\overline{N}_\ell = N_\ell \cup \ell$ and $|N_\ell|$ for the number of points in N_ℓ.

Given the state $X_\ell = x_\ell$ and given $X(N_\ell)$ we suppose the site $\ell \in L$ has an *energy* proportional to

$$b(x_\ell, X(N_\ell)).$$

Here $b(\cdot,\cdot)$ is a positive function defined on $S \times S^{|N_\ell|}$. We suppose a probability measure P is defined on the finite-state space $\Omega = S^L$ of this discrete random field by setting

$$P(x) = \frac{\prod_{\ell \in L} b(x_\ell, X(N_\ell))}{Z} \tag{1.1}$$

for $x = \{x_\ell, \ell \in L\} \in S^L$. Here $Z = \sum_{x^* \in S^L} \prod_{\ell \in L} b\left(x^*_\ell, x^*(N_\ell)\right)$ is a normalizing constant.

We shall assume in the sequel that, for each $\ell \in L$,

$$b(x_\ell, X(N_\ell)) = \prod_{n \in N_\ell} a^{x_\ell}(x_n)\, a(x_\ell). \tag{1.2}$$

Here each $a^{\cdot}(\cdot)$ is a positive function on S^2 and $a(\cdot)$ is a positive function on S.

Remark 1.1 Note our model generalizes the *Ising* (Kinderman and Snell 1980) situation where, for each $\ell \in L$ and $x \in S^L$, a function U_ℓ is considered where, for constants J, m, H,

$$U_\ell(x) = -\frac{J}{2} \sum_{n \in N_\ell} \sigma_\ell(x)\, \sigma_n(x) - mH\sigma_\ell(x).$$

In this Ising case $S = \{e_1, e_2\}$ and $\sigma_\ell(x) = +1$ if $x_\ell = e_1$ and $\sigma_\ell(x) = -1$ if $x_\ell = e_2$. Then, for $x = \{x_\ell, \ell \in L\}$

$$e^{-U_\ell(x)} = \prod_{n \in N_\ell} a^{x_\ell}(x_n)\, a(x_\ell),$$

where $a^{x_\ell}(x_n) = \exp\left(-\frac{J}{2}\sigma_\ell(x)\, \sigma_n(x)\right)$ and $a(x_\ell) = \exp(-mH\sigma_\ell(x_\ell))$.

∎

Lemma 1.2 *With P defined on $\Omega = S^L$ by (1.1), if assumption (1.2) is satisfied, the random field X satisfies the Markov property*

$$P[X_\ell = e_m \mid X_k, k \neq \ell, k \in L] = P[X_\ell = e_m \mid X(N_\ell)] = \frac{\prod_{n \in N_\ell} a^{x_n}(e_m) a^m(x_n) a(e_m)}{\sum_{p=1}^M \prod_{n \in N_\ell} a^{x_n}(e_p) a^p(x_n) a(e_p)}.$$

Proof

$$\begin{aligned} &P[X_\ell = e_m \mid X_k, k \in L, k \neq \ell] \\ &= \frac{P[X_\ell = e_m, X_k = x_k, k \neq \ell]}{\sum_{p=1}^M P[X_k = x_k, X_\ell = e_p, k \neq \ell]} \\ &= \frac{\begin{array}{r}\prod_{\substack{k \in L \\ k \notin \bar{N}_\ell}} \prod_{\alpha \in N_k} a^{x_k}(x_\alpha) a(x_k) \prod_{n \in N_\ell} a^m(x_n) a(e_m) \\ \times \prod_{\substack{\beta \in N_n \\ \beta \neq \ell}} a^{x_n}(x_\beta) a^{x_n}(e_m) a(x_n)\end{array}}{\begin{array}{r}\sum_{p=1}^M \prod_{\substack{k \in L \\ k \notin \bar{N}_\ell}} \prod_{\alpha \in N_k} a^{x_k}(x_\alpha) a(x_k) \\ \times \prod_{n \in N_\ell} a^p(x_n) a(e_p) \prod_{\substack{\beta \in N_n \\ \beta \neq \ell}} a^{x_n}(x_\beta) a^{x_n}(e_p) a(x_n)\end{array}} \\ &= \frac{\prod_{n \in N_\ell} a^m(x_n) a^{x_n}(e_m) a(e_m)}{\sum_{p=1}^M \prod_{n \in N_\ell} a^p(x_n) a^{x_n}(e_p) a(e_p)}. \end{aligned}$$

□

Remark 1.3 We shall assume in the sequel that assumption (1.2) is satisfied and that P is defined by (1.1). Note this implies the Markov field is *homogeneous in space*; that is, the transition probabilities depend only on the neighbors and not on the location (though for sites ℓ on or near a boundary, $|N_\ell|$ may vary). Nonhomogeneous random fields can be discussed using the measure change method; however, parameter estimation is more difficult. Write $\mathscr{F}_L$ for the complete σ-field on $\Omega = S^L$ generated by X. Write A_ℓ for the $M \times M^{|N_\ell|}$ matrix of probability transitions

$$\begin{aligned} &a^m(x_1, \ldots, x_{|N_\ell|}) \\ &= P[X_\ell = e_m \mid X(N_\ell)] \qquad (1.3) \\ &= \frac{\prod_{n \in N_\ell} a^{x_n}(e_m) a^m(x_n) a(e_m)}{\sum_{p=1}^M \prod_{n \in N_\ell} a^{x_n}(e_p) a^p(x_n) a(e_p)}, \qquad x_i \in S = \{e_1, \ldots, e_M\}. \end{aligned}$$

■

Now $\mathscr{F}_{L-\{\ell\}}$ is the σ-field generated by all the X_k, with the exception of X_ℓ. With $\otimes$ denoting the tensor product we have:

Lemma 1.4 *The signal process X has the representation*

$$X_\ell = A_\ell \left(\otimes_{n \in N_\ell} X_n \right) + V_\ell, \qquad (1.4)$$

where V_ℓ satisfies

$$E\left[V_\ell \mid \mathscr{F}_{L-\{\ell\}}\right] = 0.$$

Proof

$$\begin{aligned} E\left[V_\ell \mid \mathscr{F}_{L-\{\ell\}}\right] &= E\left[X_\ell - A_\ell\left(\otimes_{n\in N_\ell} X_n\right) \mid \mathscr{F}_{L-\{\ell\}}\right] \\ &= E\left[X_\ell \mid \mathscr{F}_{L-\{\ell\}}\right] - A_\ell\left(\otimes_{n\in N_\ell} X_n\right) \\ &= A_\ell\left(\otimes_{n\in N_\ell} X_n\right) - A_\ell\left(\otimes_{n\in N_\ell} X_n\right) \\ &= 0. \end{aligned}$$

□

The Observation Process

The process X is not observed directly. Rather, we observe the values Y_ℓ, $\ell \in L$, of another process Y which, without loss of generality, we identify with the standard unit vectors $f_1, \ldots, f_K$ of $\mathbb{R}^K$ for some suitable positive integer K. Write

$$c_{km} = P\left[Y_\ell = f_k \mid X_\ell = e_m\right] \tag{1.5}$$

and C for the $K \times M$ matrix $\{c_{km}\}$, $k = 1, \ldots, K$, $m = 1, \ldots, M$, so that $E\left[Y_\ell \mid X_\ell\right] = CX_\ell$. Write $\mathscr{Y}_L = \sigma\{Y_\ell, \ell \in L\}$ and assume that

$$E\left[Y_\ell \mid \mathscr{F}_L, \mathscr{Y}_{L-\{\ell\}}\right] = E\left[Y_\ell \mid X_\ell\right]. \tag{1.6}$$

Lemma 1.5 *For $\ell \in L$*

$$Y_\ell = CX_\ell + W_\ell$$

where W_ℓ satisfies $E\left[W_\ell \mid \mathscr{Y}_{L-\{\ell\}}\right] = 0$.

Proof

$$\begin{aligned} E\left[W_\ell \mid \mathscr{Y}_{L-\{\ell\}}\right] &= E\left[Y_\ell - CX_\ell \mid \mathscr{Y}_{L-\{\ell\}}\right] \\ &= E\left[E\left[Y_\ell - CX_\ell \mid \mathscr{F}_L, \mathscr{Y}_{L-\{\ell\}}\right] \mid \mathscr{Y}_{L-\{\ell\}}\right]. \end{aligned}$$

Using (1.6) this is

$$\begin{aligned} &= E\left[E\left[Y_\ell \mid X_\ell\right] - CX_\ell \mid \mathscr{Y}_{L-\{\ell\}}\right] \\ &= E\left[CX_\ell - CX_\ell \mid \mathscr{Y}_{L-\{\ell\}}\right] \\ &= 0. \end{aligned}$$

□

In summary, we have the discrete hidden Markov random field (HMRF) model

$$\boxed{\begin{aligned} X_\ell &= A_\ell\left(\otimes_{n\in N_\ell} X_n\right) + V_\ell, \\ Y_\ell &= CX_\ell + W_\ell, \end{aligned}} \tag{1.7}$$

where A_ℓ is the $M \times M^{|N_\ell|}$ matrix of probability transitions given by (1.3) and C is the $K \times M$ transition matrix with entries given by (1.5). The values V_ℓ and W_ℓ are such that $E\left[V_\ell \mid \mathscr{F}_{L-\{\ell\}}\right] = 0$ and $E\left[W_\ell \mid \mathscr{Y}_{L-\{\ell\}}\right] = 0$.

Change of Measure for the Y Process

Assume that $c^k(X_\ell) = E\left[\langle Y_\ell, f_k\rangle \mid X_\ell\right] > 0$ for $1 \le k \le K$ and all the values $e_1, \dots, e_M$ taken on by X_ℓ for $\ell \in L$. Write

$$\lambda_\ell = \frac{1}{K} \prod_{k=1}^{K} \left(\frac{1}{c^k(X_\ell)}\right)^{\langle Y_\ell, f_k\rangle}$$

and

$$\Lambda_L = \prod_{\ell \in L} \lambda_\ell.$$

Lemma 1.6 $E\left[\lambda_\ell \mid \mathscr{F}_L, \mathscr{Y}_{L-\{\ell\}}\right] = 1.$

Proof

$$\begin{aligned} E\left[\lambda_\ell \mid \mathscr{F}_L, \mathscr{Y}_{L-\{\ell\}}\right] &= \frac{1}{K} E\left[\prod_{k=1}^{K}\left(\frac{1}{c^k(x_\ell)}\right)^{\langle Y_\ell, f_k\rangle} \mid \mathscr{F}_L, \mathscr{Y}_{L-\{\ell\}}\right] \\ &= \frac{1}{K}\sum_{k=1}^{K} \frac{1}{c^k(x_\ell)} E\left[\langle Y_\ell, f_k\rangle \mid \mathscr{F}_L, \mathscr{Y}_{L-\{\ell\}}\right]. \end{aligned}$$

Using assumption (1.6) this is

$$= \frac{1}{K} E\left[\sum_{k=1}^{K} \frac{c^k(x_\ell)}{c^k(x_\ell)}\right] = 1.$$

□

Using Lemma 1.6 and repeated conditioning we see that $E[\Lambda_L] = 1$.

Now construct a new probability measure $\overline{P}$ on $\left(S^L, \mathscr{F}_L \vee \mathscr{Y}_L\right)$ by setting $d\overline{P}/dP = \left(d\overline{P}/dP\right)|_{\mathscr{F}_L \vee \mathscr{Y}_L} = \Lambda_L$.

Theorem 1.7 *Under $\overline{P}$ the random variables Y_ℓ, $\ell \in L$, are i.i.d. and uniformly distributed over $\{f_1, \dots, f_K\}$.*

Proof Denote by $\overline{E}$ the expectation under $\overline{P}$. Using a version of Bayes' Theorem we see

$$\begin{aligned} &\overline{E}\left[\langle Y_\ell, f_k\rangle \mid \mathscr{F}_L, \mathscr{Y}_{L-\{\ell\}}\right] \\ &\quad = \frac{E\left[\langle Y_\ell, f_k\rangle \Lambda_L \mid \mathscr{F}_L, \mathscr{Y}_{L-\{\ell\}}\right]}{E\left[\Lambda_L \mid \mathscr{F}_L, \mathscr{Y}_{L-\{\ell\}}\right]} \end{aligned}$$

$$= \frac{\Lambda_{L-\{\ell\}} E\left[\langle Y_\ell, f_k\rangle K^{-1} c^k(X_\ell)^{-1} \mid \mathscr{F}_L, \mathscr{Y}_{L-\{\ell\}}\right]}{\Lambda_{L-\{\ell\}} E\left[\lambda_\ell \mid \mathscr{F}_L, \mathscr{Y}_{L-\{\ell\}}\right]}.$$

From Lemma 1.6 this is

$$\begin{aligned}
&= E\left[\langle Y_\ell, f_k\rangle \frac{1}{K}\frac{1}{c^k(X_\ell)} \mid \mathscr{F}_L, \mathscr{Y}_{L-\{\ell\}}\right] \\
&= \frac{1}{K}\frac{1}{c^k(X_\ell)} E\left[\langle Y_\ell, f_k\rangle \mid X_\ell\right] \\
&= \frac{1}{K}.
\end{aligned}$$

□

Remark 1.8 Under $\overline{P}$, $X_\ell = A\left(\otimes_{n\in N_\ell} X_n\right) + V_\ell$. Note that under $\overline{P}$ the Y_ℓ are, in particular, independent of the X_ℓ. Write $\overline{\lambda}_\ell = \lambda_\ell^{-1}$ and $\overline{\Lambda}_L = \prod_{\ell\in L}\overline{\lambda}_\ell$. It can be shown, as in Lemma 1.6, that $\overline{E}\left[\overline{\lambda}_\ell \mid \mathscr{F}_L, \mathscr{Y}_{L-\{\ell\}}\right] = 1$ and $\overline{E}\left[\overline{\Lambda}_L\right] = 1$. Set $dP/d\overline{P} = \overline{\Lambda}_L$. Then under P, $P\left[Y_k = f_k \mid \mathscr{F}_L, \mathscr{Y}_{L-\{\ell\}}\right] = c^k(X_\ell)$.

■

Change of Measure for the Signal Process

In this section we start with the measure $\overline{P}$ on $\left(S^L, \mathscr{F}_L \vee \mathscr{Y}_L\right)$, so the random variables Y_ℓ are i.i.d. and uniformly distributed, and the distribution of $X \in \Omega = S^L$ is defined by (1.1). Assume that $a^m(X_\ell) > 0$, $a(X_\ell) > 0$ for all $\ell \in L$ and $m = 1, \dots, M$.

Suppose that $\hat{\phi}(\cdot)$ is the uniform probability distribution on S, so that $\hat{\phi}(e_i) = \frac{1}{M}$, $1 \le i \le M$. Write

$$\hat{\Phi}(x_\ell, \ell \in L) = \prod_{\ell\in L}\hat{\phi}(x_\ell) = M^{-L}.$$

Define $\Gamma_L = \hat{\Phi}(x)/P(x)$ where $P(x)$ is given by (1.1). Note $E[\Gamma_L] = 1$, and a new measure $\hat{P}$ can be defined on $\left(S^L, \mathscr{F}_L \vee \mathscr{Y}_L\right)$ by putting

$$\frac{d\hat{P}}{d\overline{P}} = \Gamma_L.$$

Lemma 1.9 *Under $\hat{P}$ the random variables X_ℓ, $\ell \in L$, are i.i.d., with uniform distribution over S.*

Proof Denote by $\hat{E}$ the expectation under $\hat{P}$. Using again a version of Bayes' Theorem we have

$$\hat{E}\left[\langle X_\ell, e_m\rangle \mid \mathscr{F}_{L-\ell}\right] = \frac{\overline{E}\left[\langle X_\ell, e_m\rangle \Gamma_\ell \mid \mathscr{F}_{L-\ell}\right]}{\overline{E}\left[\Gamma_L \mid \mathscr{F}_{L-\ell}\right]}.$$

Cancelling, and leaving in the expectation only terms involving X_ℓ this is

$$= \frac{\overline{E}\left[\frac{\langle X_\ell, e_m\rangle \hat{\phi}(X_\ell)}{\prod_{n\in N_\ell} a^{X_n}(e_m) a^m(X_n) a(e_m)} \mid \mathscr{F}_{L-\ell}\right]}{\overline{E}\left[\frac{\hat{\phi}(X_\ell)}{\prod_{n\in N_\ell} a^{X_n}(X_\ell) a^{X_\ell}(X_n) a(X_\ell)} \mid \mathscr{F}_{L-\ell}\right]}$$

$$= \frac{\frac{1}{\prod_{n\in N_\ell} a^{x_n}(e_m) a^m(x_n) a(e_m)} \overline{E}\left[\langle X_\ell, e_m\rangle \mid \mathscr{F}_{L-\ell}\right]}{\sum_{p=1}^{M} \frac{1}{\prod_{n\in N_\ell} a^{x_n}(e_p) a^p(x_n) a(e_p)} \overline{E}\left[\langle X_\ell, e_p\rangle \mid \mathscr{F}_{L-\ell}\right]}.$$

Writing $\pi(p) = \prod_{n\in N_\ell} a^{x_n}(e_p) a^p(x_n) a(e_p)$ this is

$$= \frac{\frac{1}{\pi(m)} \cdot \frac{\pi(m)}{\sum_{p=1}^{M} \pi(p)}}{\sum_{p=1}^{M} \frac{1}{\pi(p)} \cdot \frac{\pi(p)}{\sum_{k=1}^{M} \pi(k)}} = \frac{1}{M}.$$

□

Now it is possible to start with a probability measure $\hat{P}$ on $\left(S^L, \mathscr{F}_L \vee \mathscr{Y}_L\right)$ such that the X_ℓ, are also i.i.d. random variables uniformly distributed over $S = \{e_1, \dots, e_M\}$. Then, given functions $\{a^m(x_n), a(e_m)\}$ for $m = 1, \dots, M$ and $n \in N_\ell$, $\ell \in L$, we can construct a probability measure $\overline{P}$ such that

$$\overline{P}[X_\ell = e_m \mid X_k, k \neq \ell] = \frac{\pi(m)}{\sum_{k=1}^{M} \pi(k)}. \tag{1.8}$$

We shall adopt here the convention $0^0 = 1$. In fact, the probability measure $\overline{P}$ is defined by setting $d\overline{P}/d\hat{P} = \overline{\Gamma}_L$ where

$$\overline{\Gamma}_L = \Gamma_L^{-1} = \frac{P(x_\ell)}{\hat{\Phi}(x_\ell)}.$$

Lemma 1.10 *Under the probability measure $\overline{P}$, Equation* (1.8) *holds.*

Proof The proof is left as an exercise. □

We see, therefore, that we can start with a probability measure $\hat{P}$ on $\left(S^L, \mathscr{F}_L \vee \mathscr{Y}_L\right)$ such that both processes X_ℓ, $\ell \in L$ and Y_ℓ, $\ell \in L$ are i.i.d. and uniformly distributed over $\{e_1, \dots, e_M\}$ and $\{f_1, \dots, f_K\}$, respectively. To retrieve the situation of Lemmas 1.2 and 1.5 we define a probability measure P by setting

$$\frac{dP}{d\hat{P}} = \frac{dP}{d\overline{P}} \frac{d\overline{P}}{d\hat{P}} := \overline{\Lambda}_L \overline{\Gamma}_L.$$

Recursion

We shall work under the probability measure $\hat{P}$, so that X_ℓ, $\ell \in L$, and Y_ℓ, $\ell \in L$, are are i.i.d. and uniformly distributed over $\{e_1, \dots, e_M\}$ and $\{f_1, \dots, f_K\}$, respectively. Using a version of Bayes' Theorem we write

$$E\left[\langle X_\ell, e_m\rangle \mid \mathcal{Y}_L\right] = \frac{\hat{E}\left[\langle X_\ell, e_m\rangle \overline{\Lambda}_L \overline{\Gamma}_L \mid \mathcal{Y}_L\right]}{\hat{E}\left[\overline{\Lambda}_L \overline{\Gamma}_L \mid \mathcal{Y}_L\right]}.$$

Notation 1.11 *Write $q_\ell(e_m)$, $m = 1, \dots, M$ for the unnormalized conditional distribution $\hat{E}\left[\langle X_\ell, e_m\rangle \overline{\Lambda}_L \overline{\Gamma}_L \mid \mathcal{Y}_L\right]$.*

Theorem 1.12 *For $\ell \in L$ and $m = 1, \dots, M$ a recursivelike equation for the unnormalized conditional distribution of one single random variable X_ℓ given all the observation $\mathcal{Y}_L$ is as follows:*

$$\boxed{q_\ell(e_m) = \frac{K}{M} c^{Y_\ell}(e_m)\, \beta_{L-\ell,L}(e_m),} \tag{1.9}$$

where

$$\boxed{\beta_{L-\ell,L}(e_m) = \hat{E}\left[\overline{\Lambda}_{L-\ell}\overline{\Gamma}_L (X_\ell = e_m) \mid X_\ell = e_m, \mathcal{Y}_L\right].} \tag{1.10}$$

Proof

$$\begin{aligned}
&\hat{E}\left[\langle X_\ell, e_m\rangle \overline{\Lambda}_L \overline{\Gamma}_L \mid \mathcal{Y}_L\right] \\
&\quad = \hat{E}\left[\langle X_\ell, e_m\rangle K c^{Y_\ell}(e_m) \overline{\Lambda}_{L-\ell} \overline{\Gamma}_L \mid \mathcal{Y}_L\right] \\
&\quad = K c^{Y_\ell}(e_m) \hat{E}\left[\langle X_\ell, e_m\rangle \hat{E}\left[\overline{\Lambda}_{L-\ell} \overline{\Gamma}_L \mid X_\ell = e_m, \mathcal{Y}_L\right] \mid \mathcal{Y}_L\right] \\
&\quad = \frac{K}{M} c^{Y_\ell}(e_m) \hat{E}\left[\overline{\Lambda}_{L-\ell} \overline{\Gamma}_L \mid X_\ell = e_m, \mathcal{Y}_L\right] \\
&\quad = \frac{K}{M} c^{Y_\ell}(e_m)\, \beta_{L-\ell,L}(e_m).
\end{aligned}$$

□

Theorem 1.13 *Write*

$$\begin{aligned}
&\beta_{L-\{\ell_1,\ell_2,\dots,\ell_p\}}(e_{m_1}, e_{m_2}, \dots, e_{m_p}) \\
&\quad = \hat{E}\left[\overline{\Lambda}_{L-\{\ell_1,\ell_2,\dots,\ell_p\}} \overline{\Gamma}_L \mid X_{\ell_1} = e_{m_1}, \dots, X_{\ell_p} = e_{m_p}, \mathcal{Y}_L\right].
\end{aligned}$$

Then $\beta_{L-\ell,L}(e_m)$, $m = 1, \dots, M$, satisfies the following "backward recursive" equation for any $\ell^* \neq \ell$; $\ell, \ell^* \in L$

$$\beta_{L-\ell,L}(e_m) = \frac{K}{M}\sum_{k=1}^{M} c^{Y_{\ell^*}}(e_k)\,\beta_{L-\{\ell,\ell^*\},L}(e_m,e_k)$$

and

$$\beta_{L-L,L}\left(e_{k_1},\ldots,e_{k_L}\right) = \overline{\Gamma}_L\left(e_{k_1},\ldots,e_{k_L}\right).$$

Proof

$$\begin{aligned}
&\beta_{L-\{\ell\},L}(e_m)\\
&\quad = \hat{E}\left[\overline{\Lambda}_{L-\ell}\overline{\Gamma}_L \mid X_\ell = e_m, \mathscr{Y}_L\right]\\
&\quad = \hat{E}\left[Kc^{Y_\ell^*}(X_{\ell^*})\,\hat{E}\left[\overline{\Lambda}_{L-\{\ell,\ell^*\}}\overline{\Gamma}_L \mid X_\ell = e_m, X_{\ell^*}, \mathscr{Y}_L\right] \mid X_\ell = e_m, \mathscr{Y}_L\right]\\
&\quad = K\sum_{k=1}^{M} c^{Y_{\ell^*}}(e_k)\,\beta_{L-\{\ell,\ell^*\},L}(e_m,e_k)\,\hat{E}\left[\langle X_{\ell^*}, e_k\rangle\right]\\
&\quad = \frac{K}{M}\sum_{k=1}^{M} c^{Y_{\ell^*}}(e_k)\,\beta_{L-\{\ell,\ell^*\},L}(e_m,e_k),
\end{aligned}$$

where

$$\beta_{L-\{\ell,\ell^*\},L}(e_m,e_k) = \hat{E}\left[\overline{\Lambda}_{L-\{\ell,\ell^*\}}\overline{\Gamma}_L \mid X_\ell = e_m, X_{\ell^*} = e_k, \mathscr{Y}_L\right].$$

Finally in the recursion we would have

$$\begin{aligned}
&\beta_{\{r\},L}\left(e_{k_i}, i \neq r, i \in L\right)\\
&\quad = \hat{E}\left[Kc^{Y_r}(X_r)\,\overline{\Gamma}\left(e_{k_i}, i \neq r, i \in L\right) \mid X_i = e_{k_i}, i \neq r, \mathscr{Y}_L\right]\\
&\quad = K\sum_{p=1}^{M} c^{Y_r}(e_p)\,\overline{\Gamma}\left(e_{k_i}, i \neq r, e_p\right)\hat{E}\left[\langle X_r, e_p\rangle\right]\\
&\quad = \frac{K}{M}\sum_{p=1}^{M} c^{Y_r}(e_p)\,\overline{\Gamma}\left(e_{k_i}, i \neq r, e_p\right)\\
&\quad = \frac{K}{M}\sum_{p=1}^{M} c^{Y_r}(e_p)\,\beta_{L-L,L}\left(e_{k_1},\ldots,e_{k_L}\right),
\end{aligned}$$

and

$$\beta_{L-L,L}\left(e_{k_1},\ldots,e_{k_L}\right) = \overline{\Gamma}_L\left(e_{k_1},\ldots,e_{k_L}\right).$$

□

The next result gives the normalized conditional distribution of the whole signal given the observation.

Theorem 1.14 *Let $X = (X_\ell, \ell \in L)$ and $x \in S^L$, then the conditional distribution of the hidden signal given the observations is given by the following expression*:

$$\boxed{P\left[X = x \mid \mathscr{Y}_L\right] = \frac{\prod_{\ell\in L}\prod_{n\in N_\ell} a^{x_\ell}(x_n)\,a(x_\ell)\,c^{Y_\ell}(x_\ell)}{\sum_{x^*\in S^L}\prod_{\ell\in L}\prod_{n\in N_\ell} a^{x_\ell^*}(x_n^*)\,a\left(x_\ell^*\right)c^{Y_\ell}\left(x_\ell^*\right)}.} \tag{1.11}$$

Proof $P[X=x \mid \mathscr{Y}_L] = E\left[\prod_{\ell\in L}\langle X_\ell, x_\ell\rangle \mid \mathscr{Y}_L\right]$. Using Bayes' Theorem and the independence assumption under $\hat{P}$ this is

$$
\begin{aligned}
&= \frac{\hat{E}\left[\prod_{\ell\in L}\langle X_\ell, x_\ell\rangle \overline{\Lambda}_L \overline{\Gamma}_L \mid \mathscr{Y}_L\right]}{\hat{E}\left[\overline{\Lambda}_L \overline{\Gamma}_L \mid \mathscr{Y}_L\right]} \\
&= \frac{\hat{E}\left[\prod_{\ell\in L}\langle X_\ell, x_\ell\rangle M^L K^L \prod_{\ell\in L}\prod_{n\in N_\ell} a^{x_\ell}(x_n)\, a(x_\ell)\, c^{Y_\ell}(x_\ell)\right]}{\hat{E}\left[M^L K^L \prod_{\ell\in L}\prod_{k=1}^{M}\left(\prod_{n\in N_\ell} a^k(X_n)\, a(e_k)\right)^{\langle X_\ell, e_k\rangle} c^{Y_\ell}(X_\ell)\right]} \\
&= \frac{\prod_{\ell\in L}\prod_{n\in N_\ell} a^{x_\ell}(x_n)\, a(x_\ell)\, c^{Y_\ell}(x_\ell)\, \hat{E}\left[\langle X_\ell, x_\ell\rangle\right]}{\sum_{x'\in S^L}\prod_{\ell\in L}\prod_{n\in N_\ell} a^{x'_\ell}(x'_n)\, a(x'_\ell)\, c^{Y_\ell}(x'_\ell)\, \hat{E}\left[\langle X_\ell, x'_\ell\rangle\right]}
\end{aligned}
$$

and the result follows because $\hat{E}\left[\langle X_\ell, x_\ell\rangle\right] = \frac{1}{M}$. □

Remark 1.15 Quantity (1.11) is a function of the hidden signal x. For any possible signal $x = (x_\ell, \ell\in L)$ write

$$
\begin{aligned}
\mathscr{L}(x) &= \log \prod_{\ell\in L}\prod_{n\in N_\ell} a^{x_\ell}(x_n)\, a(x_\ell)\, c^{Y_\ell}(x_\ell) \qquad (1.12)\\
&= \sum_{\ell\in L}\sum_{n\in N_\ell} \log a^{x_\ell}(x_n) + \sum_{\ell\in L}\log c^{Y_\ell}(x_\ell) + \sum_{\ell\in L}\log a(x_\ell).
\end{aligned}
$$

For any component x_{ℓ_1} of x there is at least one possible value $e_{\ell_1}\in S$ of x_{ℓ_1} for which

$$\mathscr{L}\left(x \mid e_{\ell_1}\right) = \max_{x_{\ell_1}\in S}\mathscr{L}(x).$$

Here the maximization takes place with all components of x fixed, with the exception of x_{ℓ_1}, and $\left(x \mid e_{\ell_1}\right)$ denotes the signal x modified so that e_{ℓ_1} occurs in the ℓ_1 location. Suppose x_{ℓ_2} is a component of $\left(x \mid e_{\ell_1}\right)$. Then there is at least one possible value $e_{\ell_2}\in S$ of x_{ℓ_2} for which

$$\mathscr{L}\left(x \mid e_{\ell_1}, e_{\ell_2}\right) = \max_{x_{\ell_2}\in S}\mathscr{L}\left(x \mid e_{\ell_1}\right).$$

Again, $\left(x \mid e_{\ell_1}, e_{\ell_2}\right)$ denotes the signal $\left(x \mid e_{\ell_1}\right)$ modified so that e_{ℓ_2} occurs in the ℓ_2 location. This procedure can be repeated. Note that we obtain a monotonic increasing sequence of log-likelihoods

$$\mathscr{L}(x) \le \mathscr{L}\left(x \mid e_{\ell_1}\right) \le \mathscr{L}\left(x \mid e_{\ell_1}, e_{\ell_2}\right) \le \cdots \le \mathscr{L}\left(x \mid e_{\ell_1}, e_{\ell_2}, \ldots, e_{\ell_p}\right)$$

and so the sequence $\left(x \mid e_{\ell_1}\right)$, $\left(x \mid e_{\ell_1}, e_{\ell_2}\right)$, ..., $\left(x \mid e_{\ell_1}, \ldots, e_{\ell_p}\right)$ provides better and better estimates of the signal, given the observations.

■

In the next paragraph an alternative method of maximizing $\mathscr{L}(x)$ is proposed.

Maximum A Posterior (MAP) Distribution of the Image

Remark 1.16 Again consider quantity (1.12).

$$\mathscr{L}(x) = \sum_{\ell \in L} \sum_{n \in N_\ell} \log a^{x_\ell}(x_n) + \sum_{\ell \in L} \log c^{Y_\ell}(x_\ell) + \sum_{\ell \in L} \log a(x_\ell). \tag{1.13}$$

Given the observations Y, we suppose that each pixel $\ell \in L$ has an independent probability distribution $p(\ell) = (p_1(\ell), p_2(\ell), \dots, p_N(\ell))$ over S. Write $p = (p(\ell), \ell \in L)$ and E_p for the expectation under p. We now wish to determine the distribution p which maximizes $E_p \mathscr{L}(x \mid y)$. That is, we wish to maximize $E_p \mathscr{L}(x)$ subject to the constraints

$$p_j(\ell) \geq 0 \qquad \forall \ell \in L \text{ and } j = 1, \dots, M$$

and

$$\sum_{j=1}^{M} p_j(\ell) = 1 \qquad \forall \ell \in L.$$

To effect this, consider real variables $\rho_j(\ell)$ such that $\rho_j^2(\ell) = p_j(\ell)$. Then we require $\Sigma \rho_j(\ell)^2 = 1, \forall \ell \in L$.

■

Write

$$L(\rho, \lambda) = E_p \mathscr{L}(x) + \sum_{\ell \in L} \lambda_\ell \left(\sum_{j=1}^{M} \rho_j(\ell)^2 - 1 \right).$$

Write $k(\ell)$ for the set of k such that $\ell \in N_k$. Differentiating $L(\rho, \lambda)$ w.r.t. $\rho_j(\ell)$ and λ_ℓ gives a sparse system of $(M+1)L$ equations with $(M+1)L$ unknowns

$$\begin{aligned}
\frac{\partial L(p, \lambda)}{\partial \rho_1(1)} = 2\rho_1(1) \Bigg(& \sum_{n \in N_1} \sum_{j=1}^{M} \log a^{e_1}(e_j) \rho_j^2(n) \\
& + \sum_{k \in k(1)} \sum_{j=1}^{M} \log a^{e_1}(e_j) \rho_j^2(k) \\
& + \log c^{Y_1}(e_1) + \log a(e_1) + \lambda_1 \Bigg) = 0 \\
\frac{\partial L(p, \lambda)}{\partial \rho_2(1)} = 2\rho_2(1) \Bigg(& \sum_{n \in N_1} \sum_{j=1}^{M} \log a^{e_2}(e_j) \rho_j^2(n) \\
& + \sum_{k \in k(1)} \sum_{j=1}^{M} \log a^{e_1}(e_j) \rho_j^2(k) \\
& + \log c^{Y_1}(e_2) + \log a(e_2) + \lambda_1 \Bigg) = 0 \\
& \vdots
\end{aligned}$$

$$\frac{\partial L}{\partial \rho_N(1)} = 2\rho_N(1)\left(\sum_{n\in N_1}\sum_{j=1}^{M} \log a^{e_M}(e_j)\,\rho_j^2(n) + \sum_{k\in k(1)}\sum_{j=1}^{M} \log a^{e_1}(e_j)\,\rho_j^2(k) + \log c^{Y_1}(e_M) + \log a(e_M) + \lambda_1\right) = 0$$

$$\frac{\partial L}{\partial \rho_1(2)} = 2\rho_1(2)\left(\sum_{n\in N_2}\sum_{j=1}^{M} \log a^{e_1}(e_j)\,\rho_j^2(n) + \sum_{k\in k(2)}\sum_{j=1}^{M} \log a^{e_2}(e_j)\,\rho_j^2(k) + \log c^{Y_2}(e_1) + \log a(e_1) + \lambda_2\right) = 0$$

$$\vdots$$

$$\frac{\partial L}{\partial \rho_N(2)} = 2\rho_N(2)\left(\sum_{n\in N_2}\sum_{j=1}^{M} \log a^{e_M}(e_j)\,\rho_j^2(n) + \sum_{k\in k(2)}\sum_{j=1}^{M} \log a^{e_2}(e_j)\,\rho_j^2(k) + \log c^{Y_2}(e_M) + \log a(e_M) + \lambda_2\right) = 0$$

$$\vdots$$

$$\frac{\partial L}{\partial \rho_1(L)} = 2\rho_1(L)\left(\sum_{n\in N_L}\sum_{j=1}^{M} \log a^{e_1}(e_j)\,\rho_j^2(n) + \sum_{k\in k(2)}\sum_{j=1}^{M} \log a^{e_2}(e_j)\,\rho_j^2(k) + \log c^{Y_L}(e_1) + \log a(e_1) + \lambda_L\right) = 0$$

$$\vdots$$

$$\frac{\partial L}{\partial \rho_N(L)} = 2\rho_N(L)\left(\sum_{n\in N_L}\sum_{j=1}^{M} \log a^{e_M}(e_j)\,\rho_j^2(n) + \sum_{k\in k(L)}\sum_{j=1}^{M} \log a^{e_L}(e_j)\,\rho_j^2(k) + \log c^{Y_L}(e_M) + \log a(e_M) + \lambda_L\right) = 0$$

$$\frac{\partial L}{\partial \lambda_1} = \sum_{j=1}^{M} \rho_j^2(1) - 1 = 0$$

$$\vdots$$

$$\frac{\partial L}{\partial \lambda_L} = \sum_{j=1}^{M} \rho_j^2(L) - 1 = 0.$$

Once a candidate for the critical value $\hat{p} = (\hat{p}_j(\ell),\ j = 1,\ldots,M,\ \ell \in L)$ has been found, an estimate for $\hat{x} = (\hat{x}_\ell, \ell \in L)$ is obtained by choosing, for each $\ell \in L$, the state e_i in S corresponding to the maximum value of $\hat{p}(\ell) = (\hat{p}_1(\ell),\ldots,\hat{p}_M(\ell))$.

An advantage of this procedure is that it simultaneously estimates maximal values of x_ℓ for all pixels $\ell \in L$, thus avoiding the iterative procedures of the *ICM* or *simulated annealing* (Ripley 1988).

Estimation

We now estimate the parameters in our model, namely, the entries of the matrices A and C in the hidden Markov field X and the observation process Y, respectively. To simplify discussion we shall consider only the transition matrix A corresponding to nonboundary points ℓ of L, so that $|N_\ell|$ is constant. The transition matrix for points on or near the boundary can be estimated in a similar way. Write

$$\begin{aligned} P[X_\ell = e_m \mid X_n = e_{k_n}, n \in N_\ell] &= a^m\left(e_{k_1},\ldots,e_{k_N}\right) \\ &= \frac{\pi(m)}{\sum_{p=1}^{M} \pi(p)}, \qquad m = 1,\ldots,M \end{aligned}$$

and $N = |N_\ell|$ for the fixed number of neighbors of points in N_ℓ. Also write $P[Y_\ell = f_k \mid X_\ell = e_m] = c_{km}$, $k = 1,\ldots,K$; $m = 1,\ldots,M$ for the entries of matrix C. These parameters are subject to the constraints:

$$\sum_{m=1}^{M} a^m\left(e_{k_1},\ldots,e_{k_N}\right) = 1, \tag{1.14}$$

$$\sum_{k=1}^{K} c_{km} = 1. \tag{1.15}$$

Assume a prior set of parameters

$$\left\{a^m\left(e_{k_1},\ldots,e_{k_N}\right); c_{km}; 1 \le k_i, m \le M, 1 \le k \le K\right\}.$$

We shall determine a new set of parameters

$$\left\{\hat{a}^m\left(e_{k_1},\ldots,e_{k_N}\right); \hat{c}_{km}, 1 \le k_i, m \le M, 1 \le k \le K\right\}$$

which maximizes the *pseudo-log-likelihood* defined below. Consider first the parameters c_{km}. Define

$$\lambda_\ell = \prod_{k=1}^{K}\prod_{m=1}^{M}\left(\frac{\hat{c}_{km}}{c_{km}}\right)^{\langle X_\ell, e_m\rangle\langle Y_\ell, f_k\rangle}$$

and

$$\Lambda_L = \prod_{\ell\in L}\lambda_\ell.$$

It can be shown that $E[\Lambda_L] = 1$. Define a new probability measure $\hat{P}$ on $(\Omega, \mathscr{F}_L \vee \mathscr{Y}_L)$ by putting $d\hat{P}/dP = \Lambda_L$.

Lemma 1.17 *Under the probability measure* $\hat{P}$

$$\hat{P}\left[Y_\ell = f_k \mid \mathscr{F}_L, \mathscr{Y}_{L-\{\ell\}}\right] = \hat{c}_{km} \text{ on the set } [X_\ell = e_m].$$

Proof Assume $X_\ell = e_m$; then by a version of Bayes' Theorem we can write

$$\begin{aligned}\hat{P}\left[Y_\ell = f_k \mid \mathscr{F}_L, \mathscr{Y}_{L-\{\ell\}}\right] &= \hat{E}\left[\langle Y_\ell, f_k\rangle \mid \mathscr{F}_L, \mathscr{Y}_{L-\{\ell\}}\right]\\ &= \frac{E\left[\langle Y_\ell, f_k\rangle \Lambda_L \mid \mathscr{F}_L, \mathscr{Y}_{L-\{\ell\}}\right]}{E\left[\Lambda_L \mid \mathscr{F}_L, \mathscr{Y}_{L-\{\ell\}}\right]}\\ &= \frac{E\left[\langle Y_\ell, f_k\rangle \lambda_\ell \mid \mathscr{F}_L, \mathscr{Y}_{L-\{\ell\}}\right]}{E\left[\lambda_\ell \mid \mathscr{F}_L, \mathscr{Y}_{L-\{\ell\}}\right]}.\end{aligned}$$

Now

$$\begin{aligned}E\left[\lambda_\ell \mid \mathscr{F}_L, \mathscr{Y}_{L-\{\ell\}}\right] &= E\left[\prod_{k=1}^{K}\left(\frac{\hat{c}_{km}}{c_{km}}\right)^{\langle Y_\ell, f_k\rangle} \mid \mathscr{F}_L, \mathscr{Y}_{L-\{\ell\}}\right]\\ &= \sum_{k=1}^{K}\frac{\hat{c}_{km}}{c_{km}} E\left[\langle Y_\ell, f_k\rangle \mid \mathscr{F}_L, \mathscr{Y}_{L-\{\ell\}}\right]\end{aligned}$$

in view of assumption (1.6) this is

$$\begin{aligned}&= \sum_{k=1}^{K}\frac{\hat{c}_{km}}{c_{km}} E\left[\langle Y_\ell, f_k\rangle \mid X_\ell = e_m\right]\\ &= \sum \frac{\hat{c}_{km}}{c_{km}} c_{km} = \sum_{k=1}^{K}\hat{c}_{km} = 1.\end{aligned}$$

Therefore

$$\begin{aligned}\hat{P}\left[Y_\ell = f_k \mid \mathscr{F}_L, \mathscr{Y}_{L-\{\ell\}}\right] &= E\left[\langle Y_\ell, f_k\rangle \lambda_\ell \mid \mathscr{F}_L, \mathscr{Y}_{L-\{\ell\}}\right]\\ &= E\left[\langle Y_\ell, f_k\rangle \frac{\hat{c}_{km}}{c_{km}} \mid \mathscr{F}_L, \mathscr{Y}_{L-\{\ell\}}\right]\\ &= \frac{\hat{c}_{km}}{c_{km}} E\left[\langle Y_\ell, f_k\rangle \mid X_\ell = e_m\right] = \hat{c}_{km}.\end{aligned}$$

□

Theorem 1.18 *The maximum likelihood estimate of c_{km} given the observations Y_ℓ, $\ell \in L$, is given by*

$$\boxed{\hat{c}_{km} = \frac{\sum_{\ell\in L} \langle Y_\ell, f_k\rangle q_\ell(e_m)}{\sum_{\ell\in L} q_\ell(e_m)}.} \tag{1.16}$$

Where $q_\ell(e_m)$ is the unnormalized conditional distribution given by Theorem 1.12.

Proof

$$\begin{aligned} &E[\log \Lambda_L \mid \mathscr{Y}_L] \\ &= E\left[\sum_{k=1}^{K}\sum_{m=1}^{M}\sum_{\ell\in L} \langle X_\ell, e_m\rangle \langle Y_\ell, f_k\rangle (\log \hat{c}_{km} - \log c_{km}) \mid \mathscr{Y}_L\right] \\ &= \sum_k\sum_m\sum_\ell \langle Y_\ell, f_k\rangle E[\langle X_\ell, e_m\rangle \mid \mathscr{Y}_L] \log \hat{c}_{km} + R, \end{aligned} \tag{1.17}$$

where R is independent of $\hat{c}_{km}$. Now the $\hat{c}_{km}$ satisfy constraint (1.15)

$$\sum_{k=1}^{K} \hat{c}_{km} = 1$$

with dynamic form $\sum_{\ell\in L}\sum_{m=1}^{M}\sum_{k=1}^{K} \langle X_\ell, e_m\rangle \hat{c}_{km} = L$ or conditional form

$$\sum_{m=1}^{M}\sum_{k=1}^{K}\sum_{\ell\in L} E[\langle X_\ell, e_m\rangle \mid \mathscr{Y}_L] \hat{c}_{km} = L. \tag{1.18}$$

We wish, therefore, to choose $\hat{c}_{km}$ to maximize the quantity (1.17) subject to the constraint (1.18). Using the Lagrange multiplier technique this is

$$\hat{c}_{km} = \frac{\sum_{\ell\in L} \langle Y_\ell, f_k\rangle E[\langle X_\ell, e_m\rangle \mid \mathscr{Y}_L]}{\sum_{\ell\in L} E[\langle X_\ell, e_m\rangle \mid \mathscr{Y}_L]}$$

or

$$\boxed{\hat{c}_{km} = \frac{\sum_{\ell\in L} \langle Y_\ell, f_k\rangle q_\ell(e_m)}{\sum_{\ell\in L} q_\ell(e_m)},} \tag{1.19}$$

where $q_\ell(e_m)$ is the unnormalized conditional distribution given by Theorem 1.12. □

Consider now the parameters $a^m(e_{k_1}, \ldots, e_{k_N})$ in the matrix A. To replace these parameters by $\hat{a}^m(e_{k_1}, \ldots, e_{k_N})$ we consider, the pseudo-likelihood

$$\Gamma_L = \prod_{\ell\in L}\prod_{m=1}^{M}\prod_{n\in N_\ell}\prod_{k_n=1}^{M} \left(\frac{\hat{a}^m(e_{k_1}, \ldots, e_{k_N})}{a^m(e_{k_1}, \ldots, e_{k_N})}\right)^{\langle X_\ell, e_m\rangle \prod_{n\in N_\ell}\langle X_n, e_{k_n}\rangle}.$$

Then

$$
\begin{aligned}
&E\left[\log \Gamma_L \mid \mathscr{Y}_L\right] \\
&= \sum_{\ell \in L} \sum_{m=1}^{M} \sum_{n \in N_\ell} \sum_{k_n=1}^{M} E\left[\langle X_\ell, e_m\rangle \prod_{n \in N_\ell} \langle X_n, e_{k_n}\rangle \mid \mathscr{Y}_L\right] \\
&\quad \times \log \hat{a}^m\left(e_{k_1}, \ldots, e_{k_N}\right) + \overline{R},
\end{aligned} \tag{1.20}
$$

where $\overline{R}$ is independent of $\hat{a}^m\left(e_{k_1}, \ldots, e_{k_N}\right)$. We have the constraint

$$
\sum_{m=1}^{M} \hat{a}^m\left(e_{k_1}, \ldots, e_{k_N}\right) = 1
$$

with dynamic form

$$
\sum_{\ell \in L} \sum_{m=1}^{M} \sum_{n \in N_\ell} \sum_{k_n=1}^{M} \hat{a}^m\left(e_{k_1}, \ldots, e_{k_N}\right) \prod_{n \in N_\ell} \langle X_n, e_{k_n}\rangle = L
$$

and conditional form

$$
\sum_{\ell \in L} \sum_{m=1}^{M} \sum_{n \in N_\ell} \sum_{k_n=1}^{M} \hat{a}^m\left(e_{k_1}, \ldots, e_{k_N}\right) E\left[\prod_{n \in N_\ell} \langle X_n, e_{k_n}\rangle \mid \mathscr{Y}_L\right] = L. \tag{1.21}
$$

We wish therefore to choose $\hat{a}^m\left(e_{k_1}, \ldots, e_{k_N}\right)$ to maximize the conditional pseudo-log-likelihood (1.20) subject to the constraint (1.21). Using again the Lagrange multiplier technique we obtain

$$
\hat{a}^m\left(e_{k_1}, \ldots, e_{k_N}\right) = \frac{\sum_{\ell \in L} E\left[\langle X_\ell, e_m\rangle \prod_{n \in N_\ell} \langle X_n, e_{k_n}\rangle \mid \mathscr{Y}_L\right]}{\sum_{\ell \in L} E\left[\prod_{n \in N_\ell} \langle X_n, e_{k_m}\rangle \mid \mathscr{Y}_L\right]}.
$$

From Bayes' Theorem this is

$$
= \frac{\sum_{\ell \in L} \hat{E}\left[\langle X_\ell, e_m\rangle \prod_{n \in N_\ell} \langle X_n, e_{k_n}\rangle \overline{\Lambda}_L \overline{\Gamma}_L \mid \mathscr{Y}_L\right]}{\sum_{\ell \in L} \hat{E}\left[\prod_{n \in N_\ell} \langle X_n, e_{k_n}\rangle \overline{\Lambda}_L \overline{\Gamma}_L \mid \mathscr{Y}_L\right]}.
$$

Using Notation 1.11 we can write

$$
\boxed{\hat{a}^m\left(e_{k_1}, \ldots, e_{k_N}\right) = \frac{\sum_{\ell \in L} q_{\overline{N}_\ell}\left(e_m, e_{k_n}, n \in N_\ell\right)}{\sum_{\ell \in L} q_{N_\ell}\left(e_{k_n}, n \in N_\ell\right)},} \tag{1.22}
$$

where $\overline{N}_\ell = N_\ell \cup \ell$ and as in Theorems 1.12 and 1.13

$$
\begin{aligned}
&q_{\overline{N}_\ell}\left(e_m, e_{k_n}, n \in N_\ell\right) \\
&= \left(\frac{K}{M}\right)^{\overline{N}_\ell} \prod_{n \in N_\ell} c^{Y_n}\left(e_{k_n}\right) c^{Y_\ell}\left(e_m\right) \beta_{L-\overline{N}_\ell, L}\left(e_m, e_{k_n}, n \in N_\ell\right),
\end{aligned}
$$

$$q_{N_\ell}(e_{k_n}, n \in N_\ell) = \left(\frac{K}{M}\right)^{N_\ell} \prod_{n \in N_\ell} c^{Y_n}(e_{k_n}) \beta_{L-N_\ell,L}(e_{k_n}, n \in N_\ell)$$

are unnormalized conditional distributions.

A Blurred Observation Process

In this section the Markov random field X is still given by

$$X_\ell = A_\ell\left(\otimes_{n \in N_\ell} X_n\right) + V_\ell \tag{1.23}$$

but the observation process Y will express some "blurring" of the signal X:

$$Y_\ell = C_\ell\left(\otimes_{n \in \overline{N}_\ell} X_n\right) + W_\ell, \tag{1.24}$$

where C_ℓ is the $K \times M^{|\overline{N}_\ell|}$ matrix of probability transitions

$$c^k\left(x_1, \ldots, x_{|\overline{N}_\ell|}\right) = P\left[Y_\ell = f_k \mid X_n = x_n, n \in \overline{N}_\ell\right], \qquad x_i \in S,$$

and $Y_\ell \in \{f_1, \ldots, f_K\}$,

$$E\left[Y_\ell \mid X\left(\overline{N}_\ell\right)\right] = C_\ell\left(\otimes_{n \in \overline{N}_\ell} X_n\right).$$

Remark 1.19 All the above discussions go through with minor changes in the proofs. We shall mention only the most relevant results. Note that an assumption like (1.2) is not needed for the $c^m\left(X\left(\overline{N}_\ell\right)\right)$. Furthermore, the neighborhood system N_ℓ used in (1.24) can differ from the one used in the dynamics of (1.23). Theorem 1.14 reads as:

$$\boxed{P[X = x \mid \mathscr{Y}_L] = \frac{\prod_{\ell \in L} \prod_{n \in N_\ell} a^{x_\ell}(x_n)\, a(x_\ell)\, c^{Y_\ell}\left(x_m, m \in \overline{N}_\ell\right)}{\sum_{x' \in S^L} \prod_{\ell \in L} \prod_{n \in N_\ell} a^{x'_\ell}(x'_n)\, a\left(x'_\ell\right) c^{Y_\ell}\left(x'_m, m \in \overline{N}_\ell\right)}.} \tag{1.25}$$

■

The MAP method of the previous section and the discussion in Remark 1.15 apply here. The maximum (pseudo) log-likelihood estimates of $c^k\left(e_{j_1}, \ldots, e_{j_{\overline{N}}}\right)$ and $a^m\left(e_{k_1}, \ldots, e_{k_N}\right)$ are given by

$$\boxed{c'^k\left(e_{j_1}, \ldots, e_{j_{\overline{N}}}\right) = \frac{\sum_{\ell \in L} \langle Y_\ell, f_k \rangle\, q_{\overline{N}_\ell}\left(e_{j_n}, n \in \overline{N}_\ell\right)}{\sum_{\ell \in L} q_{\overline{N}_\ell}\left(e_{j_n}, n \in \overline{N}_\ell\right)}} \tag{1.26}$$

and

$$\boxed{a^{m\prime}(e_{k_1},\dots,e_{k_N}) = \frac{\sum_{\ell\in L} q_{\overline{N}_\ell}(e_m, e_{k_n}, n\in N_\ell)}{\sum_{\ell\in L}\sum_{i=1}^{M} q_{\overline{N}_\ell}(e_i, e_{k_n}, n\in N_\ell)}.} \tag{1.27}$$

Here

$$\begin{aligned} q_{\overline{N}_\ell}(x_n, n\in \overline{N}_\ell) &= \frac{K}{M^{\overline{N}_\ell}} c^{Y_\ell}(x_n, n\in \overline{N}_\ell)\,\beta_{L-\{\ell\}}(x_n, n\in \overline{N}_\ell), \\ \beta_{L-\{\ell\}}(x_n, n\in \overline{N}_\ell) &= \hat{E}\left[\overline{\Lambda}_{L-\{\ell\}}\overline{\Gamma}_L \mid X_n = x_n, n\in \overline{N}_\ell, \mathcal{Y}_L\right], \\ \beta_{L-L}(x_n, n\in L) &= \overline{\Gamma}_L(x_n, n\in L) \end{aligned}$$

and the Radon-Nikodym derivative $\overline{\Lambda}_L$ is defined as

$$\overline{\Lambda}_L = \prod_{\ell\in L} K\left(\prod_{k=1}^{K} c^k\left(X\left(\overline{N}_\ell\right)\right)\right)^{\langle Y_\ell, f_k\rangle}.$$

9.2 HMRF Observed in Gaussian Noise

The process X is again described by

$$X_\ell = A_\ell\left(\otimes_{n\in N_\ell} X_n\right) + V_\ell. \tag{2.1}$$

We shall suppose the y process is scalar. (The case of a vector observation process can be treated similarly.) Further, we suppose the real-valued y process has the form

$$y_\ell = g\left(\otimes_{n\in\overline{N}_\ell} X_n\right) + \sigma\left(\otimes_{n\in\overline{N}_\ell} X_n\right) w_\ell. \tag{2.2}$$

Here the w_ℓ, $\ell\in L$, are i.i.d. $N(0,1)$ random variables. Because $X_\ell \in S$ for $\ell\in L$ the functions g and σ are determined by vectors $\left(g_1,\dots,g_{M^{|\overline{N}_\ell|}}\right)$ and $\left(\sigma_1,\dots,\sigma_{M^{|\overline{N}_\ell|}}\right)$, with $\sigma_i > 0$, for $i = 1,\dots,M^{|\overline{N}_\ell|}$, respectively. Then $g\left(\otimes_{n\in\overline{N}_\ell} X_n\right) = \left\langle g, \otimes_{n\in\overline{N}_\ell} X_n\right\rangle$ and $\sigma\left(\otimes_{n\in\overline{N}_\ell} X_n\right) = \left\langle \sigma, \otimes_{n\in\overline{N}_\ell} X_n\right\rangle$ where $\langle\ ,\ \rangle$ denotes the usual inner product in $\mathbb{R}^{M^{|\overline{N}_\ell|}}$.

Note that a different neighborhood system $\widehat{N}_\ell$ could be introduced in the observations. In image processing observations of the form (2.2) are said to be *blurred*.

Changes of Measure

Starting with the y process, define

$$\begin{aligned} \lambda_\ell &= \lambda_\ell\left(\otimes_{n\in\overline{N}_\ell} X_n, w_\ell\right) \\ &= \left\langle \sigma, \otimes_{n\in\overline{N}_\ell} X_n\right\rangle \\ &\quad \times \exp\left\{-\tfrac{1}{2}\left(\langle g, \otimes_{n\in\overline{N}_\ell} X_n\rangle + \langle \sigma, \otimes_{n\in\overline{N}_\ell} X_n\rangle w_\ell\right)^2 + \tfrac{1}{2} w_\ell^2\right\} \end{aligned}$$

and

$$\Lambda_L = \prod_{\ell \in L} \lambda_\ell.$$

Using repeated conditioning we see that $E[\Lambda_L] = 1$. A new probability measure $\overline{P}$ on $(S^L, \mathscr{F}_L \vee \mathscr{Y}_L)$ is obtained by setting $d\overline{P}/dP = \Lambda_L$.

Theorem 2.1 *Under the probability measure $\overline{P}$ the random variables y_ℓ, $\ell \in L$, are i.i.d. $N(0,1)$.*

Proof With $\overline{E}$ for the expectation under $\overline{P}$. Then for any integrable $f : \mathbb{R} \to \mathbb{R}$ and using a version of Bayes' Theorem

$$\begin{aligned}\overline{E}\left[f(y_\ell) \mid \mathscr{F}_L, \mathscr{Y}_{L-\{\ell\}}\right] &= \frac{E\left[f(y_\ell)\Lambda_L \mid \mathscr{F}_L, \mathscr{Y}_{L-\{\ell\}}\right]}{E\left[\Lambda_L \mid \mathscr{F}_L, \mathscr{Y}_{L-\{\ell\}}\right]} \\ &= \frac{E\left[f(y_\ell)\lambda_\ell \mid \mathscr{F}_L, \mathscr{Y}_{L-\{\ell\}}\right]}{E\left[\lambda_\ell \mid \mathscr{F}_L, \mathscr{Y}_{L-\{\ell\}}\right]}.\end{aligned}$$

Now

$$\begin{aligned}&E\left[\lambda_\ell \mid \mathscr{F}_L, \mathscr{Y}_{L-\{\ell\}}\right] \\ &\quad= \int_{-\infty}^{+\infty} \frac{1}{\sqrt{2\pi}} \exp\left\{-\tfrac{1}{2}\left(\langle g, \otimes x_n\rangle + \langle \sigma, \otimes x_n\rangle w\right)^2\right\} \langle \sigma, \otimes_{n \in \overline{N}_\ell} x_n\rangle dw.\end{aligned}$$

The values of g, σ, and the x_n are known, so after a change of variable this integral equals 1. Hence,

$$\begin{aligned}&\overline{E}\left[f(y_\ell) \mid \mathscr{F}_L, \mathscr{Y}_{L-\{\ell\}}\right] \\ &\quad= \frac{1}{\sqrt{2\pi}} \int_{-\infty}^{+\infty} \Big\{ f(y_\ell) \exp\left[-\tfrac{1}{2}\left(\langle g, \otimes x_n\rangle + \langle \sigma, \otimes x_n\rangle w_\ell\right)^2\right] \\ &\qquad\qquad \times \langle \sigma, \otimes_{n \in \overline{N}_\ell} x_n\rangle \Big\} dw_\ell \\ &\quad= \frac{1}{\sqrt{2\pi}} \int_{-\infty}^{+\infty} f(y_\ell) \exp\left(-\tfrac{1}{2} y_\ell^2\right) dy_\ell\end{aligned}$$

and the result follows. □

Now, starting from the probability measure $\overline{P}$, we define another probability measure $\hat{P}$ such that under $\hat{P}$ the X_ℓ, $\ell \in L$, are i.i.d. random variables with uniform distribution over $\{e_1, \ldots, e_M\}$ and the y_ℓ are i.i.d. with normal density $(1/\sqrt{2\pi}) \exp\{-\frac{1}{2} y^2\}$. We start with $\hat{P}$ on $(\Omega, \mathscr{F}_L \vee \mathscr{Y}_L)$. To return to the real-world situation, set

$$\frac{dP}{d\hat{P}} = \frac{dP}{d\overline{P}} \frac{d\overline{P}}{d\hat{P}} = \overline{\Lambda}_L \overline{\Gamma}_L. \tag{2.3}$$

Here $\overline{\Lambda}_L$ is the inverse of Λ_L so that

$$\begin{aligned}\overline{\Lambda}_L &= \prod_{\ell\in L} \lambda_\ell^{-1} \\ &= \prod_{\ell\in L} \left\langle \sigma, \otimes_{n\in\overline{N}_\ell} X_n \right\rangle^{-1} \exp\left[-\tfrac{1}{2}w_\ell^2 + \tfrac{1}{2}\left(\langle g, \otimes X_n\rangle + \langle \sigma, \otimes X_n\rangle w_\ell\right)^2\right].\end{aligned}$$

Lemma 2.2 *Under the probability measure P defined by* (2.3) *the random variables*

$$w_\ell := \frac{y_\ell - \left\langle g, \otimes_{n\in\overline{N}_\ell} X_n\right\rangle}{\left\langle \sigma, \otimes_{n\in\overline{N}_\ell} X_n\right\rangle}, \qquad \ell \in L,$$

are i.i.d. and normally distributed with density $\left(1/\sqrt{2\pi}\right)\exp\left(-\frac{1}{2}w^2\right)$.

Proof Let $f:\mathbb{R}\to\mathbb{R}$ be any integrable function. Then

$$\begin{aligned}E\left[f(w_\ell) \mid \mathscr{F}_L, \mathscr{Y}_{L-\{\ell\}}\right] &= \frac{\hat{E}\left[f(w_\ell)\overline{\Lambda}_L\overline{\Gamma}_L \mid \mathscr{F}_L, \mathscr{Y}_{L-\{\ell\}}\right]}{\hat{E}\left[\overline{\Lambda}_L\overline{\Gamma}_L \mid \mathscr{F}_L, \mathscr{Y}_{L-\{\ell\}}\right]} \\ &= \frac{\overline{\Gamma}_L\overline{\Lambda}_{L-\{\ell\}}\hat{E}\left[f(w_\ell)\overline{\lambda}_\ell \mid \mathscr{F}_L, \mathscr{Y}_{L-\{\ell\}}\right]}{\overline{\Gamma}_L\overline{\Lambda}_{L-\{\ell\}}\hat{E}\left[\overline{\lambda}_\ell \mid \mathscr{F}_L, \mathscr{Y}_{L-\{\ell\}}\right]}.\end{aligned}$$

Now as in the proof of Theorem 2.1 $\hat{E}\left[\overline{\lambda}_\ell \mid \mathscr{F}_L, \mathscr{Y}_{L-\{\ell\}}\right] = 1$ so that

$$\begin{aligned}&E\left[f(w_\ell) \mid \mathscr{F}_L, \mathscr{Y}_{L-\{\ell\}}\right] \\ &\quad = \hat{E}\left[f(w_\ell)\overline{\lambda}_\ell \mid \mathscr{F}_L, \mathscr{Y}_{L-\{\ell\}}\right] \\ &\quad = \int_{-\infty}^{+\infty} \frac{f(w_\ell)\exp\left\{-\frac{1}{2}w_\ell^2 + \frac{1}{2}\left(g(\otimes x_n) + \sigma(\otimes x_n)w_\ell\right)^2\right\}\exp\left(-\frac{1}{2}y_\ell^2\right)}{\sigma(\otimes x_n)\sqrt{2\pi}}\,dy_\ell.\end{aligned}$$

Since $y_\ell = g(\otimes X_n) + \sigma(\otimes X_n)w_\ell$ and w_ℓ is the only unknown random variable, this is

$$= \int_{-\infty}^{+\infty} f(w_\ell)\exp\left(-\frac{1}{2}w_\ell^2\right)\frac{dw_\ell}{\sqrt{2\pi}}.$$

The result follows. □

Signal Estimation

Theorem 2.3 *Let* $X = (X_\ell, \ell\in L)$ *and* $x \in S^L$ *be any signal. Then the conditional probability distribution of the signal is*:

$$\boxed{P\left[X = x \mid \mathscr{Y}_L\right] = \frac{\Psi(x,y)}{\sum_{x^*\in S^L}\Psi(x^*,y)},} \tag{2.4}$$

where

$$\Psi(x,y) = \prod_{\ell \in L} \langle \sigma, \otimes_{n \in \overline{N}_\ell} x_n \rangle^{-1} \exp\left\{ -\frac{1}{2} \left(\frac{y_\ell - \langle g, \otimes_{n \in \overline{N}_\ell} x_n \rangle}{\langle \sigma, \otimes_{n \in \overline{N}_\ell} x_n \rangle} \right)^2 \right\}$$
$$\times \prod_{n \in N_\ell} a^{x_\ell}(x_n)\, a(x_\ell).$$

Proof Here, $P[X = x \mid \mathscr{Y}_L] = E[\prod_{\ell \in L} \langle X_\ell, x_\ell \rangle \mid \mathscr{Y}_L]$. Using a version of Bayes' Theorem and the independence assumption under $\hat{P}$ this is then

$$= \frac{\hat{E}\left[\prod_{\ell \in L} \langle X_\ell, X_\ell \rangle \overline{\Lambda}_\ell \overline{\Gamma}_L \mid \mathscr{Y}_L\right]}{\hat{E}\left[\overline{\Lambda}_L \overline{\Gamma}_L \mid \mathscr{Y}_L\right]}$$
$$= \frac{\Psi(x,y) M^L \hat{E}\left[\langle X_\ell, x_\ell \rangle\right]}{\sum_{x' \in S^L} \Psi(x',y) M^L \hat{E}\left[\langle X_\ell, x'_\ell \rangle\right]}.$$

The result follows because $\hat{E}\left[\langle X_\ell, x_\ell \rangle\right] = \frac{1}{M}$. □

Remark 2.4 Quantity (2.4) is a function of the hidden signal x. For any possible signal $x = (x_\ell, \ell \in L)$, given $y = (y_\ell, \ell \in L)$, write

$$\mathscr{L}(x) = \log \Psi(x,y)$$
$$= \log \prod_{\ell \in L} \langle \sigma, \otimes_{n \in \overline{N}_\ell} x_n \rangle^{-1} \exp\left\{ -\frac{1}{2} \left(\frac{y_\ell - \langle g, \otimes_{n \in \overline{N}_\ell} x_n \rangle}{\langle \sigma, \otimes_{n \in \overline{N}_\ell} x_n \rangle} \right)^2 \right\}$$
$$\times \prod_{n \in N_\ell} a^{x_\ell}(x_n)\, a(x_\ell)$$
$$= \sum_{\ell \in L} \log \langle \sigma, \otimes_{n \in \overline{N}_\ell} x_n \rangle^{-1} - \frac{1}{2} \sum_{\ell \in L} \left(\frac{y_\ell - \langle g, \otimes x_n \rangle}{\langle \sigma, \otimes_{n \in \overline{N}_\ell} x_n \rangle} \right)^2$$
$$+ \sum_{\ell \in L} \sum_{n \in N_\ell} \log a^{x_\ell}(x_n) + \sum_{\ell \in L} \log a(x_\ell).$$

The maximization techniques used to estimate the signal x which were discussed for $\mathscr{L}(x)$ in Section 9.1 can be applied to the present model. ■

Notation 2.5 *Write* $q_{\overline{N}_\ell}(x_n, n \in \overline{N}_\ell)$ *for the unnormalized conditional distribution* $\hat{E}\left[\prod_{n \in \overline{N}_\ell} \langle X_n, x_n \rangle \overline{\Lambda}_L \overline{\Gamma}_L \mid \mathscr{Y}_L\right]$.

Lemma 2.6 *We have the following relation:*

$$q_{\overline{N}_\ell}(x_n, n \in \overline{N}_\ell) = \frac{\langle \sigma, \otimes_{n \in \overline{N}_\ell} x_n \rangle^{-1}}{M^{|\overline{N}_\ell|}} \exp\left\{ -\frac{1}{2} \left(\frac{y_\ell - \langle g, \otimes_{n \in \overline{N}_\ell} x_n \rangle}{\langle \sigma, \otimes_{n \in \overline{N}_\ell} x_n \rangle} \right)^2 + \frac{1}{2} y_\ell^2 \right\}$$
$$\times \beta_{L - \{\ell\}}(x_n, n \in \overline{N}_\ell),$$

where

$$\beta_{L-\{\ell\}}\left(x_n, n \in \overline{N}_\ell\right) = \hat{E}\left[\overline{\Lambda}_{L-\{\ell\}}\overline{\Gamma}_L \mid X_n = x_n, n \in \overline{N}_\ell, \mathscr{Y}_L\right].$$

Proof

$$\begin{aligned}
&\hat{E}\left[\prod_{n\in\overline{N}_\ell}\langle X_n, x_n\rangle\,\overline{\Lambda}_L\overline{\Gamma}_L \,\middle|\, \mathscr{Y}_L\right] \\
&\quad = \hat{E}\left[\prod_{n\in\overline{N}_\ell}\langle X_n, x_n\rangle\,\overline{\lambda}_\ell\hat{E}\left[\overline{\Lambda}_{L-\{\ell\}}\overline{\Gamma}_L \mid X_n = x_n, n\in\overline{N}_\ell, \mathscr{Y}_L\right] \,\middle|\, \mathscr{Y}_L\right].
\end{aligned}$$

Writing

$$\hat{E}\left[\Lambda_{L-\{\ell\}}\overline{\Gamma}_L \mid X_n = x_n, n \in \overline{N}_\ell, \mathscr{Y}_L\right] := \beta_{L-\{\ell\}}\left(x_n, n \in \overline{N}_\ell\right)$$

and substituting for

$$\begin{aligned}
\overline{\lambda}_\ell &= \overline{\lambda}_\ell\left(X_n, n \in \overline{N}_\ell, y_\ell\right) \\
&= \langle \sigma, \otimes X_n\rangle^{-1}\exp\left\{-\frac{1}{2}\left(\frac{y_\ell - \langle g, \otimes X_n\rangle}{\langle \sigma, \otimes X_n\rangle}\right)^2 + \frac{1}{2}y_\ell^2\right\}
\end{aligned}$$

the result follows using the independence assumption under the probability measure $\hat{P}$. □

Write

$$\begin{aligned}
&\beta_{L-\{\ell_1,\ell_2,\dots,\ell_p\}}\left(x_{n_1}, n_1 \in \overline{N}_{\ell_1}; x_{n_2}, n_2 \in \overline{N}_{\ell_2}; \dots; x_{n_p}, n_p \in \overline{N}_{\ell_p}\right) \\
&\quad = \hat{E}\Big[\overline{\Lambda}_{L-\{\ell_1,\dots,\ell_p\}}\overline{\Gamma}_L \\
&\qquad\qquad \Big| X_{n_1} = x_{n_1}, n_1 \in \overline{N}_{\ell_1}; \dots; X_{n_p} = x_{n_p}, n_p \in \overline{N}_{\ell_p}, \mathscr{Y}_L\Big].
\end{aligned}$$

Then using double conditioning as in the proof of Lemma 2.6 we see that $\beta_{L-\{\ell\}}(x_n, n \in \overline{N}_\ell)$ satisfies the following "backward recursive" equation for any $\ell^* \neq \ell$; $\ell, \ell^* \in L$:

$$\begin{aligned}
\beta_{L-\{\ell\}}\left(x_n, n \in \overline{N}_\ell\right) = &\sum_{S^{|\overline{N}_{\ell^*}|}} \left\langle \sigma, \otimes_{n^*\in\overline{N}_{\ell^*}} x_{n^*}\right\rangle^{-1} \\
&\times \exp\left\{-\frac{1}{2}\left(\frac{y_{\ell^*} - \left\langle g, \otimes_{n^*\in\overline{N}_{\ell^*}} x_n\right\rangle}{\left\langle \sigma, \otimes_{n^*\in\overline{N}_{\ell^*}} x_{n^*}\right\rangle}\right)^2 + \frac{1}{2}y_{\ell^*}^2\right\} \\
&\times \beta_{L-\{\ell,\ell^*\}}\left(x_{n^*}, n^* \in \overline{N}_\ell \cup \overline{N}_{\ell^*}\right)
\end{aligned}$$

and

$$\beta_{L-L}\left(x_n, n \in L\right) = \overline{\Gamma}_L\left(x_n, n \in L\right).$$

Parameter Estimation

We only need to estimate the parameters in the observation process described by the equation

$$y_\ell = \langle g, \otimes_{n\in\overline{N}_\ell} X_n \rangle + \langle \sigma, \otimes_{n\in\overline{N}_\ell} X_n \rangle w_\ell,$$

where $g = (g_1, \ldots, g_{M^{|\overline{N}_\ell|}}) \in \mathbb{R}^{M^{|\overline{N}_\ell|}}$ and $\sigma = (\sigma_1, \ldots, \sigma_{M^{|\overline{N}_\ell|}}) \in \mathbb{R}_+^{M^{|\overline{N}_\ell|}}$. We remark again, the neighborhoods N_ℓ used in the definition of y_ℓ could differ from those used in X_ℓ. To simplify the discussion we shall consider only nonboundary points of L, so that $|N_\ell|$ is constant.

To replace the parameters by $\hat{g}_1, \ldots, \hat{g}_{M^{\overline{N}}}$ and $\hat{\sigma}_1, \ldots, \hat{\sigma}_{M^{\overline{N}}}$ we must consider the Radon-Nikodym derivative

$$\Lambda_L = \prod_{\ell\in L} \frac{\langle \sigma, \otimes_{n\in\overline{N}_\ell} X_n \rangle \exp\left\{ -\frac{1}{2} \left(\frac{y_\ell - \langle \hat{g}, \otimes_{n\in\overline{N}_\ell} X_n \rangle}{\langle \hat{\sigma}, \otimes_{n\in\overline{N}_\ell} X_n \rangle} \right)^2 \right\}}{\langle \hat{\sigma}, \otimes_{n\in\overline{N}_\ell} X_n \rangle \exp\left\{ -\frac{1}{2} \left(\frac{y_\ell - \langle g, \otimes_{n\in\overline{N}_\ell} X_n \rangle}{\langle \sigma, \otimes_{n\in\overline{N}_\ell} X_n \rangle} \right)^2 \right\}}.$$

Now

$$\begin{aligned}
\log \Lambda_L &= \sum_{\ell\in L} \left(-\log\langle \hat{\sigma}, \otimes_{n\in\overline{N}_\ell} X_n \rangle - \frac{1}{2} \left(\frac{y_\ell - \langle \hat{g}, \otimes_{n\in\overline{N}_\ell} X_n \rangle}{\langle \sigma, \otimes_{n\in\overline{N}_\ell} X_n \rangle} \right)^2 \right) + R(\sigma, g) \\
&= -\sum_{\ell\in L} \sum_{k_1,\ldots,k_{\overline{N}}=1}^{M} \prod_{n\in\overline{N}_\ell} \langle X_n, e_{k_n} \rangle \left[\log \hat{\sigma}_{k_1,\ldots,k_{\overline{N}}} + \frac{\left(y_\ell - \hat{g}_{k_1,\ldots,k_{\overline{N}}} \right)^2}{2\hat{\sigma}^2_{k_1,\ldots,k_{\overline{N}}}} \right] \\
&\quad + R(\sigma, g),
\end{aligned}$$

where $\hat{\sigma}_{k_1,\ldots,k_{\overline{N}}} = \langle \hat{\sigma}, e_{k_1} \otimes \cdots \otimes e_{k_{\overline{N}}} \rangle$, $\hat{g}_{k_1,\ldots,k_{\overline{N}}} = \langle \hat{g}, e_{k_1} \otimes \cdots \otimes e_{k_{\overline{N}}} \rangle$ and $R(\sigma, g)$ is independent of $\hat{\sigma}$ and $\hat{g}$. Then

$$\begin{aligned}
E[\log \Lambda_L \mid \mathscr{Y}_L] &= -\sum_{\ell\in L} \sum_{k_1,\ldots,k_{\overline{N}}=1}^{M} E\left[\prod_{n\in\overline{N}_\ell} \langle X_n, e_{k_n} \rangle \mid \mathscr{Y}_L \right] \\
&\qquad \times \left[\log \hat{\sigma}_{k_1,\ldots,k_{\overline{N}}} + \frac{\left(y_\ell - \hat{g}_{k_1,\ldots,k_{\overline{N}}} \right)^2}{2\hat{\sigma}^2_{k_1,\ldots,k_{\overline{N}}}} \right],
\end{aligned}$$

$$\begin{aligned}
\frac{\partial}{\partial \hat{\sigma}_{k_1,\ldots,k_{\overline{N}}}} E[\log \Lambda_L \mid \mathscr{Y}_L] &= -\sum_{\ell\in L} E\left[\prod_{n\in\overline{N}_\ell} \langle X_n, e_{k_n} \rangle \mid \mathscr{Y}_L \right] \\
&\qquad \times \left[\frac{1}{\hat{\sigma}_{k_1,\ldots,k_{\overline{N}}}} - \frac{\left(y_\ell - \hat{g}_{k_1,\ldots,k_{\overline{N}}} \right)^2}{\hat{\sigma}^3_{k_1,\ldots,k_{\overline{N}}}} \right] = 0
\end{aligned}$$

and the optimal estimate of $\hat{\sigma}$ is

$$\hat{\sigma}_{k_1,\dots,k_{\overline{N}}} = \frac{\sum_{\ell\in L} E\left[\prod_{n\in\overline{N}_\ell}\langle X_n, e_{k_n}\rangle \mid \mathscr{Y}_L\right]\left(y_\ell - \hat{g}_{k_1,\dots,k_{\overline{N}}}\right)^2}{\sum_{\ell\in L} E\left[\prod_{n\in\overline{N}_\ell}\langle X_n, e_{k_n}\rangle \mid \mathscr{Y}_L\right]}.$$

From Bayes' Theorem this

$$\hat{\sigma}_{k_1,\dots,k_{\overline{N}}} = \frac{\sum_{\ell\in L} \hat{E}\left[\prod_{n\in\overline{N}_\ell}\langle X_n, e_{k_n}\rangle \overline{\Lambda}_L\overline{\Gamma}_L \mid \mathscr{Y}_L\right]\left(y_\ell - \hat{g}_{k_1,\dots,k_{\overline{N}}}\right)^2}{\sum_{\ell\in L} \hat{E}\left[\prod_{n\in\overline{N}_\ell}\langle X_n, e_{k_n}\rangle \overline{\Lambda}_L\overline{\Gamma}_L \mid \mathscr{Y}_L\right]}.$$

Using Notation 2.5 we can write

$$\boxed{\hat{\sigma}_{k_1,\dots,k_{\overline{N}}} = \frac{\sum_{\ell\in L} q_{\overline{N}_\ell}\left(e_{k_n}, n\in\overline{N}_\ell\right)\left(y_\ell - \hat{g}_{k_1,\dots,k_{\overline{N}}}\right)^2}{\sum_{\ell\in L} q_{\overline{N}_\ell}\left(e_{k_n}, n\in\overline{N}_\ell\right)}.} \tag{2.5}$$

The optimal estimate of $\hat{g}_{k_1,\dots,k_{\overline{N}}}$ is a solution of $\frac{\partial E[\log\Lambda_L \mid \mathscr{Y}_L]}{\partial \hat{g}_{k_1,\dots,k_{\overline{N}}}} = 0$ and using the same argument we obtain

$$\boxed{\hat{g}_{k_1,\dots,k_{\overline{N}}} = \frac{\sum_{\ell\in L} q_{\overline{N}_\ell}\left(e_{k_n}, n\in\overline{N}_\ell\right) y_\ell}{\sum_{\ell\in L} q_{\overline{N}_\ell}\left(e_{k_n}, n\in\overline{N}_\ell\right)},} \tag{2.6}$$

where $q_{\overline{N}_\ell}$ is the unnormalized conditional distribution given in Lemma 2.6.

9.3 Continuous-State HMRF

Signal and Observations

A random field $X_\cdot$ on a lattice L is considered which at each point ℓ of the lattice can take any real value. The actual value X_ℓ is related to the values of X_n for n in some neighborhood N_ℓ of ℓ, plus some (additive) noise. Here the set of observations y_ℓ, $\ell\in L$, are also real-valued and involve additive noise. Our development focuses on the situation where the transitions in the Markov random field $X_\cdot$ are related to Gaussian densities, and the noise in the observations y_ℓ is also Gaussian. In this case the MAP estimate of the signal, given the observations, is the solution of a *sparse* set of linear equations. The observed process y is assumed to satisfy the dynamics

$$y_\ell = \sum_{n\in\overline{N}_\ell} c_n x_n + w_\ell, \tag{3.1}$$

where $\overline{N}_\ell = N_\ell \cup \ell$, the coefficients c_ℓ, $\ell\in L$, are real numbers and the w_ℓ, $\ell\in L$, are independent random variables with positive densities ψ_ℓ. The hidden signal is described by a set of real random variables X_ℓ, $\ell\in L$, with joint probability density

$$\Phi\left(x_{\ell}, \ell \in L\right)=\frac{\prod_{\ell \in L} \prod_{\ell_n \in N_{\ell}} \exp \left\{\frac{-\left|x_{\ell}-a_{\ell_n} x_{\ell_n}\right|^2}{2 \sigma^2}\right\}}{Z}, \tag{3.2}$$

where Z is a normalizing constant.

Lemma 3.1 *The conditional density of X_ℓ given all the other X_ℓ's is given by*

$$\Phi_{\ell}\left(x_{\ell} \mid x_k, k \neq \ell\right)=\frac{\prod_{n \in N_{\ell}} \exp \left\{\frac{-\left|x_{\ell}-a_n x_n\right|^2-\left|x_n-a_{\ell} x_{\ell}\right|^2}{2 \sigma^2}\right\}}{\int_{\mathbb{R}} \prod_{n \in N_{\ell}} \exp \left\{\frac{-\left|x_{\ell}-a_n x_n\right|^2-\left|x_n-a_{\ell} x_{\ell}\right|^2}{2 \sigma^2}\right\} d x_{\ell}}.$$

Proof

$$\begin{aligned}
\Phi_{\ell}\left(x_{\ell} \mid x_k, k \neq \ell\right) & =\frac{\Phi\left(x_k, k \in L\right)}{\int_{\mathbb{R}} \Phi\left(x_{\ell}, x_k, k \neq \ell\right) d x_{\ell}} \\
& =\frac{\prod_{k \in L} \prod_{n \in N_k} \exp \left(\frac{1}{2 \sigma^2}\right)\left\{-\left|x_k-a_n x_n\right|^2\right\}}{\int_{\mathbb{R}} \prod_{\ell \in L} \prod_{n \in N_k} \exp \left(\frac{1}{2 \sigma^2}\right)\left\{-\left|x_k-a_n x_n\right|^2\right\} d x_{\ell}}.
\end{aligned}$$

After cancellation we have only terms involving x_ℓ; this, therefore, is

$$=\frac{\prod_{n \in N_{\ell}} \exp \left(\frac{1}{2 \sigma^2}\right)\left\{-\left|x_{\ell}-a_n x_n\right|^2-\left|x_n-a_{\ell} x_{\ell}\right|^2\right\}}{\int_{\mathbb{R}} \prod_{n \in N_{\ell}} \exp \left(\frac{1}{2 \sigma^2}\right)\left\{-\left|x_{\ell}-a_n x_n\right|^2-\left|x_n-a_{\ell} x_{\ell}\right|^2\right\} d x_{\ell}}.$$

□

Changes of Measure

Consider first the y process. Define

$$\Lambda_L=\prod_{\ell \in L} \frac{\psi_{\ell}\left(y_{\ell}\right)}{\psi_{\ell}\left(w_{\ell}\right)}.$$

A new probability measure $\overline{P}$ on $\mathscr{F}_L \vee \mathscr{Y}_L$ is obtained if we set $d\overline{P}/dP=\Lambda_L$. Under $\overline{P}$ the random variables y_ℓ, $\ell \in L$ are independent with densities ψ_ℓ. Now consider the signal X. Let $\hat{\phi}(x)$ be any positive probability density defined over $\mathbb{R}$, and write $\hat{\Phi}(x_\ell, \ell \in L)=\prod_{\ell \in L} \hat{\phi}(x_\ell)$. Define

$$\Gamma_L=\frac{\hat{\Phi}\left(x_{\ell}, \ell \in L\right)}{\Phi\left(x_{\ell}, \ell \in L\right)},$$

where $\Phi(x_\ell, \ell \in L)$ is given by (3.2). Define a measure $\hat{P}$ on $\mathscr{F}_L \vee \mathscr{Y}_L$ by putting

$$\frac{d\hat{P}}{d\overline{P}}=\Gamma_L.$$

Lemma 3.2 *Under $\hat{P}$, the random variables X_ℓ, $\ell \in L$, are i.i.d. with density $\hat{\phi}$.*

Proof Suppose $f : \mathbb{R} \to \mathbb{R}$ is any integrable function. Then

$$\hat{E}\left[f(X_\ell) \mid \mathscr{F}_{L-\{\ell\}}\right] = \frac{\overline{E}\left[\Gamma_L f(X_\ell) \mid \mathscr{F}_{L-\{\ell\}}\right]}{\overline{E}\left[\Gamma_L \mid \mathscr{F}_{L-\{\ell\}}\right]}.$$

Leaving under the expectation only terms involving X_ℓ this is

$$\frac{\overline{E}\left[\frac{\hat{\phi}(X_\ell) f(X_\ell)}{\prod_{n \in N_\ell} \exp \frac{1}{2\sigma^2}\left\{-|X_\ell - a_n x_n|^2 - |x_n - a_\ell X_\ell|^2\right\}} \mid \mathscr{F}_{L-\{\ell\}}\right]}{\overline{E}\left[\frac{\hat{\phi}(X_\ell)}{\prod_{n \in N_\ell} \exp \frac{1}{2\sigma^2}\left\{-|X_\ell - a_n x_n|^2 - |x_n - a_\ell X_\ell|^2\right\}} \mid \mathscr{F}_{L-\{\ell\}}\right]}.$$

Using Lemma 3.1 and writing

$$\pi(x_\ell) = \prod_{n \in N_\ell} \exp\left\{\frac{-|x_\ell - a_n x_n|^2 - |x_n - a_\ell x_\ell|^2}{2\sigma^2}\right\}$$

this is

$$= \frac{\int_{\mathbb{R}} \frac{\hat{\phi}(x_\ell) f(x_\ell)}{\pi(x_\ell)} dx_\ell \frac{\pi(x_\ell)}{\int_{\mathbb{R}} \pi(x_\ell) dx_\ell}}{\int_{\mathbb{R}} \frac{\hat{\phi}(x_\ell)\pi(x_\ell)}{\pi(x_\ell) \int_{\mathbb{R}} \pi(x_\ell)} dx_\ell}$$

$$= \int_{\mathbb{R}} \hat{\phi}(x_\ell) f(x_\ell) dx_\ell = \hat{E}\left[f(X_\ell)\right],$$

and the result follows. □

Estimation

We shall work under $\hat{P}$, so that X and y are two sets of independent random variables with respective probability density functions $\hat{\phi}$ and ψ_ℓ. In this section we choose ψ_ℓ to be the normal density $\left(2\pi\rho^2\right)^{-1/2} \exp\left(-x^2/2\rho^2\right)$.

Notation 3.3 *Let $x = (x_\ell, \ell \in L)$ be any value of the hidden signal X and write*

$$E\left[\prod_{\ell \in L} I(X_\ell \in dx_\ell) \mid \mathscr{Y}_L\right] := q_L(x)$$

for the conditional probability density function of X given the observation y.

Theorem 3.4 *If the probability density function of X is as given by* (3.2) *then*

$$\boxed{q_L(x) = \frac{\Gamma(x,y)}{\int_{\mathbb{R}^L} \Gamma(x,y)\, dx}.} \tag{3.3}$$

Here $\Gamma(x,y) = \prod_{\ell\in L}\prod_{n\in N_\ell}\exp\left\{\frac{-|x_\ell - a_n x_n|^2}{2\sigma^2} - \frac{\left(y_\ell - \sum_{m\in\overline{N}_\ell} c_m x_m\right)^2}{2\rho^2}\right\}$.

Proof For any arbitrary integrable function $f : \mathbb{R}^L \to \mathbb{R}$

$$E\left[f(X) \mid \mathscr{Y}_L\right] = \frac{\hat{E}\left[f(X)\overline{\Lambda}_L\overline{\Gamma}_L \mid \mathscr{Y}_L\right]}{\hat{E}\left[\overline{\Lambda}_L\overline{\Gamma}_L \mid \mathscr{Y}_L\right]}$$

$$= \frac{\hat{E}\left[\frac{f(X)\prod_{\ell\in L}\prod_{n\in N_\ell}\exp\left\{\frac{-|X_\ell - a_n X_n|^2}{2\sigma^2} - \frac{\left(y_\ell - \sum_{m\in\overline{N}_\ell} c_m X_m\right)^2}{2\rho^2}\right\}}{\hat{\Phi}(X)} \,\middle|\, \mathscr{Y}_L\right]}{\hat{E}\left[\frac{\prod_{\ell\in L}\prod_{n\in N_\ell}\exp\left\{\frac{-|X_\ell - a_n X_n|^2}{2\sigma^2} - \frac{\left(y_\ell - \sum_{m\in\overline{N}_\ell} c_m X_m\right)^2}{2\rho^2}\right\}}{\hat{\Phi}(X)} \,\middle|\, \mathscr{Y}_L\right]}.$$

Under $\hat{P}$, X has the probability density function $\hat{\Phi}$. Therefore, this is

$$\frac{\int_{\mathbb{R}^L} f(x)\prod_{\ell\in L}\prod_{n\in N_\ell}\exp\left\{\frac{-|x_\ell - a_n x_n|^2}{2\sigma^2} - \frac{\left(y_\ell - \sum_{m\in\overline{N}_\ell} c_m x_m\right)^2}{2\rho^2}\right\}dx}{\int_{\mathbb{R}^L}\prod_{\ell\in L}\prod_{n\in N_\ell}\exp\left\{\frac{-|x_\ell - a_n x_n|^2}{2\sigma^2} - \frac{\left(y_\ell - \sum_{m\in\overline{N}_\ell} c_m x_m\right)^2}{2\rho^2}\right\}dx}. \tag{3.4}$$

On the other hand we have

$$E\left[f(X) \mid \mathscr{Y}_L\right] = \int_{\mathbb{R}^L} f(x)\,q_L(x)\,dx. \tag{3.5}$$

Equations (3.4) and (3.5) yield at once (3.3). □

Theorem 3.5 *Suppose A is the $L\times L$ matrix such that the ℓth row has the only nonzero entries*

$$A_{\ell,\ell} = \frac{2c_\ell^2}{\rho^2} + \frac{v_\ell + a_\ell^2}{\sigma^2}$$

$$A_{\ell,n} = \frac{2c_n c_\ell}{\rho^2} - \frac{2a_n}{\sigma^2} \qquad \text{if } n \in N_\ell,$$

and

$$A_{\ell,m} = \frac{c_m c_\ell}{\rho^2} \qquad \text{if } m \in N_n,\ n \in N_\ell,\ l \neq m.$$

Here v_ℓ is the cardinality of N_ℓ. Also B is a matrix with nonzero entries c_n/ρ^2 for $n \in \overline{N}_\ell$. If A is nonsingular the maximum conditional likelihood of the signal $x = (x_\ell, \ell \in L)$, given the observation $y = (y_\ell, \ell \in L)$, is given by $x_M = A^{-1}By$.

Proof For any $x \in \mathbb{R}^L$ write

$$\begin{aligned}
\mathscr{L}(x) &= \log \prod_{\ell \in L} \prod_{n \in N_\ell} \exp\left\{ \frac{-|x_\ell - a_n x_n|^2}{2\sigma^2} \right\} \\
&\quad \times \prod_{\ell \in L} \exp\left\{ \frac{\left(y_\ell - \sum_{m \in \overline{N}_\ell} c_m x_m\right)^2}{2\rho^2} \right\} \\
&= -\left(\sum_{\ell \in L} \sum_{n \in N_\ell} \frac{1}{2\sigma^2} (x_\ell - a_n x_n)^2 + \sum_{\ell \in L} \frac{1}{2\rho^2} \left(y_\ell - \sum_{m \in \overline{N}_\ell} c_m x_m \right)^2 \right).
\end{aligned}$$

Now

$$\begin{aligned}
-\frac{\partial}{\partial x_\ell} \mathscr{L}(x) &= \sum_{n \in N_\ell} \frac{1}{\sigma^2} (x_\ell - a_n x_n) + \frac{c_\ell}{\rho^2} \sum_{m \in \overline{N}_\ell} c_m x_m - \frac{c_\ell}{\rho^2} y_\ell \\
&\quad - \frac{a_\ell}{\sigma^2} \sum_{n \in N_\ell} (x_n - a_\ell x_\ell) - \sum_{n \in N_\ell} \frac{c_\ell}{\rho^2} \left(y_n - \sum_{m \in \overline{N}_n} c_m x_m \right) \\
&= \left(\frac{v_\ell}{\sigma^2} + \frac{2c_\ell^2}{\rho^2} + \frac{a_\ell^2}{\sigma^2} \right) x_\ell + \sum_{n \in N_\ell} \left(\frac{2c_n c_\ell}{\rho^2} - \frac{2a_n}{\sigma^2} \right) x_n \\
&\quad + \sum_{n \in N_\ell} \sum_{m \in N_n, m \neq \ell} \frac{c_\ell c_m}{\rho^2} x_m - \frac{c_\ell}{\rho^2} y_\ell - \sum_{n \in N_\ell} \frac{c_\ell y_n}{\rho^2}.
\end{aligned}$$

Setting $(\partial/\partial x_\ell)\mathscr{L}(x) = 0$ for $\ell = 1, \ldots, L$ yields a system of L linear equations in matrix notation

$$Ax = By.$$

Hence

$$x_M = A^{-1} By,$$

where $B = \operatorname{diag}\left(\frac{c_1}{\rho^2}, \frac{c_2}{\rho^2}, \ldots, \frac{c_L}{\rho^2}\right)$. □

Remark 3.6 The form of the matrix A depends on how the lattice and neighborhood system are defined. Simple cases indicate A is in general nonsingular. The measure change method will work with other forms of density for the signal X and observations y, though perhaps with not quite such explicit results. It is of interest that Theorem 3.5 describes the MAP estimate for the signal without techniques such as *simulated annealing*.

■

9.4 Example: A Mixed HMRF

The signal we consider here could be from a region of distant space. The atmosphere causes distortion and blurring of the picture. Another possible source of noise is inherent to the recording device itself. The scene is partitioned into a lattice L of pixels or *frames* and all the quantities of interest will be indexed by L. We shall assume that at each frame or pixel $\ell \in L$ the signal consists of essentially two components: the luminescence, which is represented by a continuous real random variable X_ℓ; and the number of stars, Z_ℓ, which is represented by a discrete random variable. The observed information at each pixel ℓ is given by a continuous real random variable $y_{1,\ell}$ and a Poisson random variable $y_{2,\ell}$. All random variables are defined initially on a probability space $(\Omega, \mathscr{F}, P)$. Write $\mathscr{F}_L = \sigma(Z_\ell, \ell \in L)$, $\mathscr{G}_L = \sigma(X_\ell, \ell \in L)$ and $\mathscr{Y}_L = \sigma\left(y_{1,\ell}, y_{2,\ell}, \ell \in L\right)$ for the complete σ-fields generated by the two components of the signal process and the two components of the observation process respectively.

Write $\overline{N}_\ell = N_\ell \cup \{\ell\}$ and $X\left(\overline{N}_\ell\right) = \left\{X_k, k \in \overline{N}_\ell\right\}$ and $Z\left(\overline{N}_\ell\right) = \left\{Z_k, k \in \overline{N}_\ell\right\}$. Given $Z\left(\overline{N}_\ell\right)$ and $X\left(\overline{N}_\ell\right)$ we suppose there is an "energy" at $\ell \in L$ proportional to

$$\Phi_\ell\left(Z\left(\overline{N}_\ell\right), X\left(\overline{N}_\ell\right)\right) = \prod_{n \in N_\ell} \exp\{-\mu(Z_n, X_n, X_\ell)\} \left(\frac{\mu(Z_n, X_n, X_\ell)}{Z_\ell!}\right)^{Z_\ell} \times \exp\left\{\frac{-|X_\ell - a_n(Z_n, Z_\ell) X_n|}{2\sigma^2}\right\}^2. \tag{4.1}$$

We suppose a probability measure P is defined on the state space $\Omega = \mathbb{N}^L \times \mathbb{R}^L$ by setting

$$P[Z = z, X = x] = C^{-1} \prod_{\ell \in L} \Phi_\ell\left(z\left(\overline{N}_\ell\right), x\left(\overline{N}_\ell\right)\right) := \Phi_L(z, x) \tag{4.2}$$

for $(z, x) \in \mathbb{N}^L \times \mathbb{R}^L$. Here $C = \sum_{z \in \mathbb{N}^L} \int_{\mathbb{R}^L} \Phi(z, x)\, dx$ is a normalizing constant. In view of (4.1) and (4.2) we have

$$\begin{aligned} P&\left[Z_\ell = z_\ell, X_\ell = x_\ell \mid \mathscr{F}_{L-\{\ell\}}, \mathscr{G}_{L-\{\ell\}}\right] \\ &:= \Psi\left(z_\ell, x_\ell \mid \mathscr{F}_{L-\{\ell\}}, \mathscr{G}_{L-\{\ell\}}\right) \\ &= P[Z_\ell = z_\ell, X_\ell = x_\ell \mid Z(N_\ell), X(N_\ell)] \\ &= \frac{\Pi(z_\ell, x_\ell)}{\sum_{k=0}^{\infty} \int_{\mathbb{R}} \Pi(k, x)\, dx}. \end{aligned} \tag{4.3}$$

Here

$$\Pi(z_\ell, x_\ell) = \prod_{n \in N_\ell} \exp\left\{-\mu(z_n, x_n, x_\ell) - \mu(z_\ell, x_\ell, x_n) - |x_\ell - a_n(z_n, z_\ell) x_n|^2 - |x_n - a_\ell(z_\ell, z_n) x_\ell|^2 \left(\frac{\mu(z_n, x_n, x_\ell)}{Z_\ell!}\right)^{Z_\ell} (\mu(z_\ell, x_\ell, x_n))^{Z_n}\right\}. \tag{4.4}$$

That is, (X_ℓ, Z_ℓ) is a Markov random field.

We suppose that our *mixed HMRF* is described by

$$\boxed{\begin{array}{l} X_\ell,\ \ell \in L,\ Z_\ell,\ \ell \in L, \\ y_{1,\ell} = h(Z_\ell, X_\ell) + w_\ell, \\ y_{2,\ell} = \nu(Z_\ell, X_\ell) + m_\ell. \end{array}} \tag{4.5}$$

Here $\nu(\cdot)$ is the intensity of the Poisson process $y_{2,\ell}$; w_ℓ are independent real-valued random variables with density functions ψ_ℓ and;

$$E\left[m_\ell \mid \mathscr{F}_L, \mathscr{G}_L, \mathscr{Y}_L - \{y_{2,\ell}\}\right] = 0$$

and X and Z have the joint probability distribution given by (4.2).

We shall define a new probability measure $\hat{P}$ on $(\Omega, \mathscr{F}_L \vee \mathscr{G}_L \vee \mathscr{Y}_L)$ such that the four processes X_ℓ, Z_ℓ, $y_{1,\ell}$ and $y_{2,\ell}$ become four sets of independent random variables which are, in particular, independent of each other. Let $\hat{\phi}(x)$ be any positive probability density defined over $\mathbb{R}$ (e.g., a standard normal), and write

$$\hat{\Phi}_L(x_\ell, \ell \in L) = \prod_{\ell \in L} \hat{\phi}(x_\ell).$$

Assuming that all quantities of interest are positive, define

$$\Delta_L = \frac{1}{e^{2L}} \frac{\hat{\Phi}_L(X)}{\Phi_L(X,Z)} \prod_{\ell \in L} \frac{\psi_\ell(y_{1,\ell})}{\psi_\ell(w_\ell)} \frac{\exp(\nu(Z_\ell, X_\ell))}{(\nu(Z_\ell, X_\ell))^{y_{2,\ell}} Z_\ell!}$$

for $X = (X_\ell, \ell \in L)$, $Z = (Z_\ell, \ell \in L)$.

Lemma 4.1 $E[\Delta_L] = 1$.

Proof

$$\begin{aligned} E[\Delta_L] &= E\left[E\left[\Delta_L \mid \mathscr{Y}_L - \{y_{1,\ell}\}, \mathscr{F}_L, \mathscr{G}_L\right]\right] \\ &= E\left[\frac{\hat{\Phi}_L(X)}{e^{2L}\Phi_L(X,Z)} \prod_{\ell \in L} \frac{\exp(\nu(Z_\ell, X_\ell))}{(\nu(Z_\ell, X_\ell))^{y_{2,L}} Z_\ell!} \int_{\mathbb{R}} \psi_\ell(y_{1,\ell})\, dy_{1,\ell}\right]. \end{aligned}$$

The integral is equal to 1 and proceeding in the same manner with $y_{2,\ell}$ this becomes

$$\begin{aligned} &= E\left[\frac{\hat{\Phi}_L(X)}{\Phi_L(X,Z)\, e^L} \frac{1}{\prod_{\ell \in L} Z_\ell!}\right] \\ &= \frac{1}{e^L} \sum_{(k_\ell, \ell \in L) \in \mathbb{N}^L} \int_{\mathbb{R}^L} \frac{\hat{\Phi}_L(x)}{\Phi_L(x,k)} \frac{\Phi_L(x,k)}{\prod k_\ell!} dx \\ &= \frac{e^L}{e^L} = 1. \end{aligned}$$

□

A new probability measure $\hat{P}$ on $(\Omega, \mathscr{F}_L \vee \mathscr{G}_L \vee \mathscr{Y}_L)$ is obtained if we set $d\hat{P}/dP = \Delta_L$.

Theorem 4.2 *Under the probability measure $\hat{P}$ the process $\{Z_\ell, \ell \in L\}$, $\{X_\ell, \ell \in L\}$, $\{y_{1,\ell}, \ell \in L\}$, and $\{y_{2,\ell}\}$ are four sets of independent random variables which are, in particular, independent of each other. Moreover, Z_ℓ and $y_{2,\ell}$ are Poisson random variables with intensities equal to 1, X_ℓ has probability density function $\hat{\phi}_\ell$, and $y_{1,\ell}$ has probability density function ψ_ℓ.*

Proof Let f, g from $\mathbb{R} \to \mathbb{R}$ be any integrable functions and ξ, η any summable functions from $\mathbb{N} \to \mathbb{R}$. Then using a version of Bayes' Theorem we write

$$\begin{aligned}
&\hat{E}\left[f(X_\ell)\, g(y_{1,\ell})\, \xi(Z_\ell)\, \eta(y_{2,\ell}) \mid \mathscr{F}_{L-\{\ell\}}, \mathscr{G}_{L-\{\ell\}}, \mathscr{Y}_{L-\{\ell\}}\right] \\
&\quad = \frac{E\left[f(X_\ell)\, g(y_{1,\ell})\, \xi(Z_\ell)\, \eta(y_{2,\ell})\, \Delta_L \mid \mathscr{F}_{L-\{\ell\}}, \mathscr{G}_{L-\{\ell\}}, \mathscr{Y}_{L-\{\ell\}}\right]}{E\left[\Delta_L \mid \mathscr{F}_{L-\{\ell\}}, \mathscr{G}_{L-\{\ell\}}, \mathscr{Y}_{L-\{\ell\}}\right]} \\
&\quad := \frac{\langle f(X_\ell)\, g(y_{1,\ell})\, \xi(Z_\ell)\, \eta(y_{2,\ell}), \Delta_L \rangle}{\langle 1, \Delta_L \rangle} \\
&\quad = \frac{\langle f(X_\ell)\, g(y_{1,\ell})\, \xi(Z_\ell)\, \eta(y_{2,\ell}), \delta_\ell \rangle}{\langle 1, \delta_\ell \rangle}.
\end{aligned}$$

Here

$$\delta_\ell = \frac{\hat{\phi}(X_\ell)}{Z_\ell!} \frac{\psi_\ell(y_{1,\ell})}{\psi_\ell(w_\ell)} \frac{\exp(\nu(Z_\ell, X_\ell))}{(\nu(Z_\ell, X_\ell))^{y_{2,\ell}}\, \Pi(Z_\ell, X_\ell)}$$

and $\Pi(Z_\ell, X_\ell)$ is given by (4.3). Using repeated conditioning as in the proof of Lemma 4.1, we see that

$$\begin{aligned}
&\langle 1, \delta_\ell \rangle = \left[\sum_{k=0}^{\infty} \int_{\mathbb{R}} \Pi(x, k)\, dx\right]^{-1}, \\
&\langle f(X_\ell)\, g(y_{1,\ell})\, \xi(Z_\ell)\, \eta(y_{2,\ell}), \delta_\ell \rangle \\
&\quad = \hat{E}[f(X_\ell)]\, \hat{E}[g(y_{1,\ell})]\, \hat{E}[\xi(Z_\ell)]\, \hat{E}[\eta(y_{2,\ell})]\, \langle 1, \delta_\ell \rangle,
\end{aligned}$$

and X_ℓ, Z_ℓ, $y_{1,\ell}$, and $y_{2,\ell}$ have the stated probability distributions. □

Conditional Distribution of the Scene

We shall work under the probability measure $\hat{P}$.

Notation 4.3 *Let $k = (k_\ell, \ell \in L) \in \mathbb{N}^L$ and $x = (x_\ell, \ell \in L) \in \mathbb{R}^L$ be any value of the signal components and write*

$$E\left[\prod_{\ell \in L} I(X_\ell \in dx_\ell, Z_\ell = k_\ell) \mid \mathscr{Y}_L\right] := q_L(k, x)$$

for the conditional probability density distribution function of the signal given the observation y.

Theorem 4.4

$$\boxed{q_L(k,x) = \frac{\Upsilon(k,x,y)}{\sum_{k'\in\mathbb{N}^L}\int_{\mathbb{R}^L}\Upsilon(k',x',y)\,dx'}.} \tag{4.6}$$

Here

$$\Upsilon(k,x,y) = \Phi_L(k,x)\prod_{\ell\in L}\psi_\ell\left(y_{1,\ell}-h(k_\ell,x_\ell)\right)\left(\nu(k_\ell,x_\ell)\right)^{y_{2,\ell}}\exp\{-\nu(k_\ell,x_\ell)\}.$$

Proof For any integrable function $f:\mathbb{N}^L\times\mathbb{R}^L\to\mathbb{R}$ and by a version of Bayes' Theorem we write:

$$E\left[f(Z,X)\mid\mathscr{Y}_L\right] = \frac{\hat{E}\left[f(Z,X)\overline{\Delta}_L\mid\mathscr{Y}_L\right]}{\hat{E}\left[\overline{\Delta}_L\mid\mathscr{Y}_L\right]}$$

using the independence and distribution assumptions under $\hat{P}$; after simplification this is equal to

$$\frac{\begin{aligned}\sum_{k\in\mathbb{N}^L}\int_{\mathbb{R}^L}\Big[f(k,x)\Phi_L(k,x)\prod_{\ell}\in L\psi_\ell\left(y_{1,\ell}-h(k_\ell,x_\ell)\right)\left(\nu(k_\ell,x_\ell)\right)^{y_{2,\ell}}\\ \times\exp\left(\nu(k_\ell,x_\ell)\right)\Big]dx\end{aligned}}{\begin{aligned}\sum_{k^*\in\mathbb{N}^L}\int_{\mathbb{R}^L}\Big[\Phi_L(k^*,x^*)\prod_{\ell}\in L\psi_\ell\left(y_{1,\ell}-h\left(k_\ell^*,x_\ell^*\right)\right)\left(\nu\left(k_\ell^*,x_\ell^*\right)\right)^{y_{2,\ell}}\\ \times\exp\left(\nu\left(k_\ell^*,x_\ell^*\right)\right)\Big]dx^*\end{aligned}}. \tag{4.7}$$

On the other hand we have

$$E\left[f(Z,X)\mid\mathscr{Y}_L\right] = \sum_{k\in\mathbb{N}^L}\int_{\mathbb{R}^L}f(x,k)\,q_L(x,k)\,dx. \tag{4.8}$$

Comparing (4.7)–(4.8) yields at once (4.6). □

Maximum A Posteriori Distribution of the Scene

To obtain the maximum posterior estimate of the scene, given the observations y, the values of $(k,x)=(k_\ell,x_\ell,\ell\in L)$ which maximize $q_L(k,x)$ given by (4.6) could be obtained. However, the k_ℓ take integer values, and it is difficult to find the maximizing values of (k,x). Procedures discussed in the literature (Ripley 1988) include the ICM (iterated conditional modes), or *simulated annealing*. As in Section 9.1 we propose an alternative method which leads to a sparse system of equations.

Write

$$\begin{aligned}\mathscr{L}(k,x\mid y) = \log\Phi_L(k,x) &+ \sum_{\ell\in L}\log\psi_\ell\left(y_{1,\ell}-h(k_\ell,x_\ell)\right)\\ &+\sum_{\ell\in L}\left(y_{2,\ell}-1\right)\nu(k_\ell,x_\ell).\end{aligned} \tag{4.9}$$

It is sufficient to seek values of $\left(\hat{k},\hat{x}\right)$ which maximize $\mathscr{L}(k,x \mid y)$. For any positive integer N and any pixel $\ell \in L$ suppose there is a probability distribution $p(\ell) = (p_1(\ell), p_2(\ell), \ldots, p_N(\ell), p_{N+1}(\ell))$ which assigns a probability $p_i(\ell)$ to the integer i, $1 \leq i \leq N$, and $p_{N+1}(\ell)$ to the integers greater than N. Write $p = (p(\ell), \ell \in L)$ for the corresponding distribution and E_p for the expectation. Consider

$$\mathscr{L}(p,x \mid y) = E_p\left[\mathscr{L}(k,x \mid y)\right]. \tag{4.10}$$

Both variables p and x are now continuous and we propose to investigate the critical points $(\hat{p},\hat{x})$ of $\mathscr{L}(p,x \mid y)$ subject to the constraints

$$p_j(\ell) \geq 0, \qquad \forall \ell \in L,\ 1 \leq j \leq N+1, \tag{4.11}$$

and

$$\sum_{j=1}^{N+1} p_j(\ell) = 1, \qquad \forall \ell \in L.$$

Once a candidate for $\left(\hat{p},\hat{x}\right)$ has been found an estimate for $\left(\hat{k},\hat{x}\right)$ is obtained by choosing, for each $\ell \in L$, the integer i corresponding to the maximum value of $\hat{p}(\ell)$. In case this procedure gives $i = N+1$ for a large number of pixels ℓ perhaps initially a larger value of N should be chosen. To effect this, consider real variables $\rho_j(\ell)$ such that $\rho_j^2(\ell) = p_j(\ell)$. Then we require

$$\sum_{j=1}^{N+1} \rho_j^2(\ell) = 1, \qquad \forall \ell \in L. \tag{4.12}$$

Write

$$L(\rho,\lambda) = E_p \mathscr{L}(k,x \mid y) + \sum_{\ell \in L} \lambda_\ell \left(\sum_{j=1}^{N+1} \rho_j^2(\ell) - 1 \right).$$

Differentiating $L(\rho,\lambda)$ w.r.t. to $\rho_j(\ell)$, x_ℓ and λ_ℓ gives a sparse system of $(N+3)L$ equations for the critical values $(\hat{\rho},\hat{x})$.

Again an advantage of this procedure is that it attempts to find simultaneously maximal values of $k(\ell)$ and $x(\ell)$ for all pixels $\ell \in L$, thus avoiding the iterative procedures of the ICM or *simulated annealing*.

9.5 Problems and Notes

Problems

1. Show that under the probability measure $\overline{P}$ defined in Theorem 1.7 the random variables X_ℓ, $\ell \in L$ satisfy the Markov random field property given in Lemma 1.2.
2. Establish the result of Lemma 1.10

3. Explain whether the algorithm described in Remark 1.15 converges necessarily to a global maximum or not.
4. In the blurred model described in Section 9.1 derive (1.25), (1.26), and (1.27).
5. Discuss the MAP estimation of the signal X for the HMRF observed in Gaussian noise in Section 9.2.

Notes

The literature on image processing is extensive. Important contributions include Besag (1986), Geman and Geman (1984), Qian and Titterington (1990), and Ripley (1988).

Part V
HMM Optimal Control

Chapter 10
Discrete-Time HMM Control

10.1 Control of Finite-State Processes

Discrete-time control problems are treated, for example, in Kumar and Varaiya (1986a) and Caines (1988). Here we discuss the discrete-time, *partially observed control* problem using the reference probability. The reference probability is constructed explicitly, and the role of the dynamics in the separated problem clarified. The unnormalized conditional probabilities, which describe the state of the process given the observations, play the role of *information states*, and the control problem can be recast as a *fully observed optimal control* problem. *Dynamic programming* results and *minimum principles* are obtained, in terms of *separated controls*, and *adjoint processes* are described for each model.

Dynamics

We use the notation of Chapter 2.

The state and observation processes are as described in Chapter 2 by the equations

$$\begin{aligned} X_{k+1} &= AX_k + V_{k+1}, \\ Y_{k+1} &= CX_k + W_{k+1}. \end{aligned}$$

Now, we suppose that the transition matrix $A(.)$ in the chain X depends on a control parameter u taking values in some measurable space U. At time k the control u_k is to be $\mathscr{Y}_k$-measurable. Write $\underline{U}(k)$ for the set of such controls and $\underline{U}(h,h+\ell) = \underline{U}(h) \cup \underline{U}(h+1) \cup \cdots \cup \underline{U}(h+\ell)$. For $u = (u_0, \dots, u_{K-1}) \in \underline{U}(0, K-1)$, with $u_i \in \underline{U}(i)$, where K is the finite horizon, X will denote the corresponding process. We suppose there is a probability $\overline{P}$ on $(\Omega, \mathscr{G}_k)$ such that under $\overline{P}$ the Y_ℓ are i.i.d. random variables uniformly distributed over the set of unit standard vectors $\{f_1, \dots, f_M\}$ of

$\mathbb{R}^M$. A new probability measure P^u is defined by putting

$$\left.\frac{dP^u}{d\overline{P}}\right|_{\mathscr{G}_k} = \overline{\Lambda}_k^u := \prod_{\ell=1}^{k}\left(\sum_{i=1}^{M}\left(Mc_\ell^i\right)\langle Y_\ell, f_i\rangle\right).$$

Recall from the notation of Chapter 2 that

$$c_{k+1} = E\left[Y_{k+1} \mid \mathscr{G}_k\right] = CX_k$$

and $c_{k+1}^i = \langle c_{k+1}, f_i\rangle = E\left[Y_{k+1}^i \mid \mathscr{G}_k\right] = \langle CX_k, f_i\rangle$. Then under P^u, let us write the model as

$$\begin{aligned} X_{k+1} &= A(u_k)X_k + V_{k+1}, \\ Y_{k+1} &= CX_k + W_{k+1}. \end{aligned}$$

From Equation (2.6.1) the recursion equation for the unnormalized distribution again has the form

$$q_{k+1}^u = \overline{E}\left[\overline{\Lambda}_{k+1}^u X_{k+1} \mid \mathscr{Y}_{k+1}\right] = M\sum_{j=1}^{N}\left\langle q_k^u, e_j\right\rangle a_j^u \prod_{i=1}^{M} c_{ij}^{Y_{k+1}^i}. \tag{1.1}$$

The initial value q_0 is the distribution of X_0.

Equation (1.1) describes the observable dynamics of a separated problem. q_{k+1}^u is an *information state*. That is, if we know q_k^u, $Y_0, \dots, Y_{k+1}$ and u_k, Equation (1.1) enables us to determine q_{k+1}^u. The information state q_k^u is a positive (not necessarily normalized) measure on $S = \{e_1, \dots, e_N\}$. As in Chapter 2 we work under $\overline{P}$, so the Y's remain i.i.d. and uniformly distributed. A more general model would allow the entries of C to be u-dependent.

Cost

Suppose there is a cost associated with the process of the form

$$J(X_0, u) = \sum_{k=0}^{K}\langle X_k, \ell_k(u)\rangle, \qquad \text{for } u = (u_0, \dots, u_K) \in \underline{U}(0, K). \tag{1.2}$$

Note the cost is, without loss of generality, linear in X.

Here, for each $k \in \{0, 1, \dots, K\}$ and $u \in U$, $\ell_k(u) \in \mathbb{R}^N$. Then the expected cost if control u is used is

$$\begin{aligned} V_0(u) &= E\left[J(X_0, u)\right] \\ &= \overline{E}\left[\overline{\Lambda}_K^u\left(\sum_{k=0}^{K}\langle X_k, \ell_k(u_k)\rangle\right)\right] \end{aligned}$$

$$= \overline{E}\left[\sum_{k=0}^{K}\left\langle \overline{E}\left[\overline{\Lambda}_k^u X_k \mid \mathscr{Y}_k\right], \ell_k(u_k)\right\rangle\right]$$

$$= \overline{E}\left[\sum_{k=0}^{K}\langle q_k^u, \ell_k(u_k)\rangle\right]. \tag{1.3}$$

We see that the cost is expressed in terms of the information states given recursively by (1.1).

If q is such an information state at time k then the expected remaining cost corresponding to the control $u \in \underline{U}(k,K)$ is given by

$$V_k(q,u) = \overline{E}\left[\sum_{j=k}^{K}\langle q_j^u, \ell_j(u_j)\rangle \mid q_k = q\right]. \tag{1.4}$$

For $0 \le k \le K$ the cost process is defined as the essential infimum under $\overline{P}$

$$V(k,q) = \bigwedge_{u\in\underline{U}(k,K-1)} V_k(q,u).$$

For $k = K$ set $V(K,q) = \overline{E}\left[\langle q, \ell_K(u)\rangle\right]$. The following dynamic programming result is then established:

Lemma 1.1

$$\boxed{V(k,q) = \bigwedge_{u\in\underline{U}(k)} \overline{E}\left[\langle q, \ell_k(u)\rangle + V\left(k+1, q_{k+1}^u\right) \mid q_k = q\right].} \tag{1.5}$$

Proof

$$V(k,q) = \bigwedge_{u\in\underline{U}(k,K-1)} V_k(q,u) = \bigwedge_{u\in\underline{U}(k)}\ \bigwedge_{v\in\underline{U}(k+1,K-1)} V_k(q,u).$$

Using (1.4) and double conditioning this is:

$$= \bigwedge_{u\in\underline{U}(k)}\ \bigwedge_{v\in\underline{U}(k+1,K-1)} \overline{E}\left[\overline{E}\left[\langle q, \ell_k(u)\rangle + \sum_{j=k+1}^{K}\left\langle q_j^u, \ell_j(u)\right\rangle \,\middle|\, \mathscr{Y}_{k+1}\right] \,\middle|\, q_k = q\right]$$

$$= \bigwedge_{u\in\underline{U}(k)}\left\{\overline{E}\left[\langle q, \ell_k(u)\rangle \mid q_k = q\right] + \bigwedge_{v\in\underline{U}(k+1,K-1)} \overline{E}\left[\overline{E}\left[\sum_{j=k+1}^{K}\left\langle q_j^u, \ell_j(u)\right\rangle \,\middle|\, \mathscr{Y}_{k+1}\right] \,\middle|\, q_k = q\right]\right\}.$$

Using the lattice property for the controls [see Lemma 16.14 of Elliott (1982b)], the inner minimization and first expectation can be interchanged, so this is

$$= \bigwedge_{u \in \underline{U}(k)} \Bigg\{ \overline{E}\left[\langle q, \ell_k(u)\rangle \mid q_k = q\right] + \overline{E}\left[\bigwedge_{v \in \underline{U}(k+1,K-1)} \overline{E}\left[\sum_{j=1}^{K} \langle q_j^u, \ell_j(u)\rangle \,\Big|\, \mathcal{Y}_{k+1}\right] \,\Big|\, q_k = q\right]\Bigg\}$$

$$= \bigwedge_{u \in \underline{U}(k)} \overline{E}\left[\langle q, \ell_k(u)\rangle + V\left(k+1, q_{k+1}^u\right) \mid q_k = q\right].$$

□

Definition 1.2 A control $u \in \underline{U}(0, K-1)$ is said to be *separated* if u_k depends on $(y_0, \ldots, y_k)$ only through the information states q_k^u. Write $\underline{U}_s(0, K-1)$ for the set of separated controls.

Lemma 1.3

$$V(k,q) = \bigwedge_{u \in \underline{U}_s(k,K-1)} V_k(q,u). \tag{1.6}$$

Proof The proof will use backward induction in k. Clearly

$$\begin{aligned} V(K,q) &= \bigwedge_{u \in \underline{U}(K)} V_K(q,u) = \bigwedge_{u \in \underline{U}(K)} \overline{E}\left[\langle q, \ell_K(u)\rangle\right] \\ &= \bigwedge_{u \in \underline{U}_s(K)} \overline{E}\left[\langle q, \ell_K(u)\rangle\right] \end{aligned}$$

and the result holds for $k = K$. Then from Lemma 1.1

$$V(k,q) = \bigwedge_{u \in \underline{U}(k)} \overline{E}\left[\langle q, \ell_k(u)\rangle + V\left(k+1, q_{k+1}^u\right) \mid q_k = q\right].$$

It is clear that a minimizing u_k (or a sequence of minimizing u_k's) depends only on the information state $q_k = q$. Therefore,

$$\begin{aligned} V(k,q) &= \bigwedge_{u \in \underline{U}_s(k)} \overline{E}\left[\langle q, \ell_k(u)\rangle + \bigwedge_{v \in \underline{U}_s(k+1,K-1)} V_{k+1}\left(q_{k+1}^u, v\right) \,\Big|\, q_k = q\right] \\ &= \bigwedge_{u \in \underline{U}_s(k,K-1)} V_k(q,u). \end{aligned}$$

□

A minimum principle has the following form:

Theorem 1.4 *Suppose u^* is a separated control such that, for each positive measure q on $\{e_1,\dots,e_N\}$, $u_k^*(q)$ achieves the minimum in (1.5). Then $V_k(q,u^*) = V(k,q)$ and u^* is an optimal control.*

Proof We again use backward induction in k. Clearly

$$V_K(q,u^*) = \overline{E}\left[\langle q, \ell_K(u^*)\rangle\right] = V(K,q).$$

Suppose the result holds for $k+1, k+2, \dots, K$. Then

$$\begin{aligned} V_k(q,u_k^*) &= \overline{E}\left[\langle q, \ell_k(u^*)\rangle + V_{k+1}\left(q_{k+1}^{u^*}, u^*\right) \mid q_h = q\right] \\ &= \overline{E}\left[\langle q, \ell_k(u^*)\rangle + V\left(k+1, q_{k+1}^{u^*}\right) \mid q_k = q\right] \\ &= V(k,q). \end{aligned}$$

Now for any other $u \in \underline{U}(0,K)$,

$$V_k(q,u^*) = V(k,q) \le V_k(q,u),$$

and, in particular, $V_0(q,u^*) \le V_0(q,u)$, so u^* is optimal. □

Adjoint Process

For simplicity suppose the cost is purely terminal at the final time K, so

$$J(X_0,u) = \langle X_K, \ell_K(u_K)\rangle,$$

where $u_K \in \underline{U}(K)$. Consequently, $\ell_K(u_K)$ is $\mathscr{Y}_K$ measurable and so a function of $Y_1,\dots,Y_K$. Then, as in (1.3),

$$\begin{aligned} V_0(u) &= E\left[\langle X_K, \ell_K(u_K)\rangle\right] \\ &= \overline{E}\left[\langle q_K^u, \ell_K(u_K)\rangle\right]. \end{aligned}$$

Theorem 1.5 *Define $\eta_K^u = \ell_K(u_K)$ and, if $\eta_{k+1}^u = \eta_{k+1}^u(Y_1,\dots,Y_{k+1})$ has been defined, set*

$$\boxed{\eta_k^u = \eta_k^u(Y_1,\dots,Y_k) = \sum_{j=1}^{N}\sum_{i=1}^{M} \left\langle \eta_{k+1}^u(f_i), a_j^u\right\rangle c_{ij} e_j,} \tag{1.7}$$

where $\eta_{k+1}^u(f_i) = \eta_{k+1}^u(Y_1,\dots,Y_k,f_i)$. Then η_k^u is the adjoint process such that

$$\overline{E}\left[\langle \ell_K(u), q_K^u\rangle \mid \mathscr{Y}_k\right] = \langle \eta_k^u, q_k^u\rangle.$$

Proof Again we use backward induction

$$\overline{E}\left[\langle \ell_K(u), q_K^u\rangle \mid \mathscr{Y}_K\right] = \langle \ell_K(u), q_K^u\rangle = \langle \eta_K^u, q_K^u\rangle.$$

So the result holds for $k = K$. Suppose $\eta^u_{k+1} = \eta^u_{k+1}(Y_1, \dots, Y_k, Y_{k+1})$ has been defined. Then

$$
\begin{aligned}
&\overline{E}\left[\left\langle \eta^u_{k+1}, q^u_{k+1}\right\rangle \mid \mathscr{Y}_k\right] \\
&= \overline{E}\left[\left\langle \eta^u_{k+1}, M\sum_{j=1}^{N}\left\langle q^u_k, e_j\right\rangle a^u_j \prod_{i=1}^{M} c_{ij}^{Y^i_{k+1}}\right\rangle \mid \mathscr{Y}_k\right] \\
&= M\sum_{j=1}^{N}\left\langle q^u_k, e_j\right\rangle \overline{E}\left[\left\langle \eta^u_{k+1}, a^u_j\right\rangle \left\langle Y_{k+1}, c_j\right\rangle \mid \mathscr{Y}_k\right] \\
&= \left\langle q^u_k, \sum_{j=1}^{N}\sum_{i=1}^{M}\left\langle \eta^u_{k+1}(f_i), a^u_j\right\rangle c_{ij} e_j \right\rangle = \left\langle q^u_k, \eta^u_k\right\rangle
\end{aligned}
$$

and the result follows. □

Remark 1.6 Notice that the adjoint process is given by finite-dimensional linear equations.

■

The Dependent Case

When the noises in the signal and observations processes are not independent as in Section 2.10, the observable dynamics of the separated control problem are given in Lemma 2.10.4 as

$$
\tilde{q}^u_{k+1} = \sum_{r=1}^{M}\sum_{j=1}^{N}\left\langle \tilde{q}^u_k, e_j\right\rangle s^u_{r\cdot j}\left\langle y_{k+1}, f_r\right\rangle.
$$

Here $s_{rij} = P(Y_{k+1} = f_r, X_{k+1} = e_i \mid X_k = e_j)$ and $s_{r\cdot j}$ is the vector $\left(s_{r1j}, \dots, s_{rNj}\right)$. The initial value $\tilde{q}_0$ is the (normalized) distribution of X_0. The above analysis goes through and the dynamic programming results are exactly as before. The adjoint process $\tilde{\eta}^u_k$ is given by

$$
\tilde{\eta}^u_k = \sum_{r=1}^{k}\sum_{j=1}^{N}\left\langle \tilde{\eta}^u_{k+1}(f_r), s^u_{r\cdot j}\right\rangle e_j. \tag{1.8}
$$

Markov Chain in Gaussian Noise

The model

$$
\begin{aligned}
X_{k+1} &= AX_k + V_{k+1}, \\
y_{k+1} &= c(X_k) + \sigma(X_k) w_{k+1},
\end{aligned}
$$

where y is real-valued and the w_k are i.i.d. $N(0,1)$ random variables, is discussed in Chapter 3 where we derive various estimators of quantities related to the state and observations processes. The recursive equation giving updates of the unnormalized conditional distribution of the state is reported here for convenience:

$$\gamma_{k+1}(X_{k+1}) = \sum_{i=1}^{N} \left\langle \gamma_k(X_k), \Gamma^i(y_{k+1}) \right\rangle a_i,$$

where

$$\Gamma^i(y_k) = \left[\frac{\phi\left(\frac{y_k - c_i}{\sigma_i}\right)}{\sigma_i \phi(y_k)} \right] e_i.$$

The initial value q_0 is the distribution of X_0. The above discussion goes through. The adjoint process is given by

$$\boxed{\eta_k^u = \sum_{i=1}^{N} \frac{e_i}{\sigma_i} \int_{-\infty}^{+\infty} \left\langle \eta_{k+1}^u, A(u) e_i \right\rangle \phi_{k+1}\left(\frac{y - c_i}{\sigma_i}\right) dy.} \tag{1.9}$$

We recall that ϕ_k is the positive densities of w_k, and the w_k form a sequence of independent random variables. The c_i and σ_i are the components of the functions c and σ, respectively.

10.2 More General Processes

Dynamics

Consider a finite-time horizon control problem and, for simplicity, suppose the noise is additive in the state and observation processes. All processes are defined initially on a probability space $(\Omega, \mathcal{F}, P)$.

The state process $\{x_k\}$, $k = 0, 1, \dots, K$, takes values in $\mathbb{R}^d$ and has dynamics

$$x_{k+1} = A_k(x_k, u_k) + v_{k+1}. \tag{2.1}$$

We suppose the initial density $\pi_0(z)$ of x_0 is known.

The observation process $\{y_k\}$, $k = 0, 1, \dots, K$, takes values in $\mathbb{R}^m$ and has dynamics

$$y_{k+1} = C_k(x_k) + w_{k+1}. \tag{2.2}$$

We suppose $y_0 = 0 \in \mathbb{R}^m$. For $0 \le k \le K$ write $y^k = \{y_0, y_1, \dots, y_k\}$. $\{\mathcal{G}_k\}$ is the complete filtration generated by x and y. $\{\mathcal{Y}_k\}$ is the complete filtration generated by y.

The noise in the state process is a sequence $\{v_k\}$, $1 \le k \le K$, of independent $\mathbb{R}^d$-valued random variables having densities ψ_k. The noise in the observation process is a sequence $\{w_k\}$, $1 \le k \le K$, of independent $\mathbb{R}^m$-valued random variables having positive densities ϕ_k, $\phi_k(b) > 0$, for all $b \in \mathbb{R}^m$. The parameter u_k in (2.1) represents the control variable, and takes values in a set $U \subset \mathbb{R}^p$. At time k, u_k is $\mathscr{Y}_k$ measurable, that is, u_k is a function of y^k. For $0 \leq k < K$ write $\underline{U}(k)$ for the set of such control functions and

$$\underline{U}(k, k+\ell) = \underline{U}(k) \cup \underline{U}(k+1) \cup \cdots \cup \underline{U}(k+\ell).$$

For $u \in \underline{U}(0, K-1)$, X will denote the trajectory $(x_0, X_1, X_2, \dots, X_K)$ determined by (2.1), and a sequence $v_1, \dots, v_K$ of noise terms.

Unnormalized Densities

Suppose we have an equivalent probability measure $\overline{P}$ on $(\Omega, \mathscr{G}_K)$ such that under $\overline{P}$ $\{y_k\}$ is a sequence of independent random variables having positive densities ϕ_k and for any $u \in \underline{U}(0, K-1)$, X_k satisfies the dynamics in (2.1).

Suppose $u \in \underline{U}(0, K-1)$. Define

$$\overline{\Lambda}_k^u = \prod_{\ell=1}^{k} \frac{\phi_\ell\left(y_\ell - C_{\ell-1}(X_{\ell-1})\right)}{\phi_\ell(y_\ell)}.$$

Then a probability P^u can be defined by setting the restriction of $dP^u/d\overline{P}$ to $\mathscr{G}_K$ equal to $\overline{\Lambda}_K^u$. It is under P^u that the state and observation processes have the form (2.1) and (2.2).

Write $q_k^u(z)$ for the unnormalized conditional density such that

$$\overline{E}\left[\overline{\Lambda}_k^u I(X_k \in dz) \mid \mathscr{Y}_k\right] = q_k^u(z)\, dz.$$

The normalized conditional density $p_k^u(z)$ is then given by:

$$p_k^u(z) = \frac{q_k^u(z)}{\int_{\mathbb{R}^d} q_k^u(\xi)\, d\xi},$$

and for any Borel test function f

$$E^u\left[f(X_{k+1}) \mid \mathscr{Y}_{k+1}\right] = \int_{\mathbb{R}^d} f(z)\, p_{k+1}^u(z)\, dz.$$

Similarly to Theorem 4.4.8 we have the following recurrence relation for q_k^u:

Theorem 2.1

$$\boxed{\begin{aligned} q_{k+1}^u(z) &= \phi_{k+1}(y_{k+1})^{-1} \\ &\quad \times \int_{\mathbb{R}^d} \psi_{k+1}(z - A_k(\xi, u_k))\, \phi_{k+1}(y_{k+1} - C_k(\xi))\, q_k^u(\xi)\, d\xi. \end{aligned}} \tag{2.3}$$

Remark 2.2 This equation describes the observable dynamics of a separated problem. $q^u_{k+1}(\cdot)$ is an "information state" in the sense of Kumar and Varaiya (1986a). That is, if we know $q^u_k(\cdot)$, y^{k+1} and u_k, Equation (2.3) enables us to determine $q^u_{k+1}(\cdot)$.

■

The initial information state q_0 is just π_0, the (normalized) density of x_0. Note that, even if π_0 is a unit mass at a particular x_0, $q^u_1(z) = \phi_1(y_1)^{-1}\psi_1(z - A_0(x_0,u_0)) \times \phi_1(y_1 - C_0(x_0))$, and the consequent terms $q^u_2, q^u_3, \dots$ follow from Equation (2.3).

Cost

Suppose, given x_0 and $u \in \underline{U}(0, K-1)$, the cost function associated with the problem is of the form

$$J(x_0,u) = \sum_{k=0}^{K-1} \ell_k(X_k,u_k) + \ell_K(X_K). \tag{2.4}$$

Then the expected cost, if control u is used and the density of x_0 is $\pi_0(\cdot)$, is

$$V_0(\pi_0,u) = E\left[J(x_0,u)\right]. \tag{2.5}$$

This can be expressed

$$\begin{aligned}
V_0(\pi_0,u) &= \overline{E}\left[\overline{\Lambda}^u_K\left(\sum_{k=0}^{K-1} \ell_k(X_k,u_k) + \ell_K(X_K)\right)\right] \\
&= \sum_{k=0}^{K-1} \overline{E}\left[\overline{\Lambda}^u_k \ell_k(X_k,u_k)\right] + \overline{E}\left[\overline{\Lambda}^u_K \ell_K(X_K)\right] \\
&= \overline{E}\left[\sum_{k=0}^{K-1} \langle \ell_k(z,u_k), q^u_k(z)\rangle + \langle \ell_K(z), q^u_K(z)\rangle\right] \\
&= \overline{E}\left[\overline{E}\left[\sum_{k=0}^{K-1} \langle \ell_k(z,u_k), q^u_k(z)\rangle + \langle \ell_K(z), q^u_K(z)\rangle \mid \mathscr{Y}_K\right]\right],
\end{aligned}$$

where, for example, we write

$$\begin{aligned}
\langle \ell_k(z,u_k), q^u_k(z)\rangle &= \int_{\mathbb{R}^d} \ell_k(z,u_k)\, q^u_k(z)\, dz \\
&= \overline{E}\left[\overline{\Lambda}^u_k \ell_k(X_k,u_k) \mid \mathscr{Y}_k\right].
\end{aligned}$$

Remark 2.3 We have seen the information state at time k belongs to the set S of positive measures $q(\cdot)$ on $\mathbb{R}^d$. Note the probability measures are a subset of S.

■

Here S is an infinite-dimensional space. A metric can be defined on S using the L^1 norm, so that for $q^1(\cdot), q^2(\cdot) \in S$

$$d\left(q^1,q^2\right) = \left\|q^1 - q^2\right\| = \int_{\mathbb{R}^d} \left|q^1(z) - q^2(z)\right| dz.$$

Any $q \in S$ can be normalized to give a probability measure $\pi(q) = q(\cdot)/\|q\|$.

Consider the process starting from some intermediate time k, $0 \le k \le K$, from some state $q(\cdot) \in S$. Then, for $u \in \underline{U}(k, K-1)$

$$q_{k+1}^{u}(z) = \phi_{k+1}(y_{k+1})^{-1} \times \int_{\mathbb{R}^d} \psi_{k+1}(z - A_k(\xi, u_k))\, \phi_{k+1}(y_{k+1} - C_k(\xi))\, q(\xi)\, d\xi. \tag{2.6}$$

The remaining information states $q_n^u(\cdot)$, $k+1 < n \le K$, are obtained from (2.3).

The expected cost accumulated, starting from state $q(\cdot) \in S$ and using control $u \in \underline{U}(k, K-1)$ is, therefore,

$$V_k(q, u) = \overline{E}\left[\sum_{j=k}^{K-1} \langle \ell_j(z, u_j), q_j^u(z) \rangle + \langle \ell_K(z), q_K^u(z) \rangle \mid q_k = q \right]. \tag{2.7}$$

The problem is now in a separated form. The filtering recursively determines the unnormalized, conditional probabilities which are the information states, $q_k^u(\cdot)$. These evolve according to the linear dynamics (2.3), and the cost is expressed linearly in terms of these information states.

For $0 \le k \le K$ the cost process is again the essential infimum

$$V(k, q) = \bigwedge_{u \in \underline{U}(k, K-1)} V_k(q, u).$$

The dynamic programming identity and the minimum principle have the same forms as in Lemma 1.1 and Theorem 3.3 and their proofs are left as an exercise.

The Adjoint Process

Consider any control $u \in \underline{U}(0, K-1)$. We shall suppose for simplicity of notation that the cost is purely terminal at the final time K, so

$$J(x_0, u) = \ell_K(X_K).$$

Then

$$\begin{aligned} V(\pi_0, u) &= E[\ell_K(X_K)] \\ &= \overline{E}[\langle \ell_K(z), q_K^u(z) \rangle]. \end{aligned}$$

Theorem 2.4 *There is a process $\eta_k^u(z, y^k)$, adapted to $\mathscr{Y}_k$, such that for $0 \le k \le K$*

$$\overline{E}[\langle \ell_K(z), q_K^u(z) \rangle \mid \mathscr{Y}_k] = \langle \eta_k^u(z, y^k), q_k^u(z) \rangle.$$

Further, η_k^u *evolves in reverse time so that*

$$\boxed{\begin{aligned}\eta_k^u(\xi, y^k) = \int_{\mathbb{R}^d}\int_{\mathbb{R}^m}\Big[\eta_{k+1}^u(z, y^k, y)\,\phi_{k+1}(y - C_k(\xi))\\ \times \psi_{k+1}(z - A_k(\xi, u_k))\Big]dz\,dy.\end{aligned}} \tag{2.8}$$

Proof Again we use backward induction. Define $\eta_K^u(z, y^K) = \ell_K(z)$ so

$$\begin{aligned}\overline{E}[\langle \ell_K(z), q_K^u(z)\rangle \mid \mathscr{Y}_K] &= \langle \ell_K(z), q_K^u(z)\rangle\\ &= \langle \eta_K^u(z, y^K), q_K^u(z)\rangle.\end{aligned}$$

Suppose $\eta_{k+1}^u(z, y^{k+1})$ has been defined. Then

$$\langle \eta_{k+1}^u(z, y^{k+1}), q_{k+1}^u(z)\rangle = \int_{\mathbb{R}^d}\eta_{k+1}^u(z, y^{k+1})q_{k+1}^u(z)\,dz$$

and

$$\begin{aligned}&\overline{E}[\langle \eta_{k+1}^u(z, y^{k+1}), q_{k+1}^u(z)\rangle \mid \mathscr{Y}_k]\\ &= \int_{\mathbb{R}^d}\int_{\mathbb{R}^d}\int_{\mathbb{R}^m}\Big[\eta_{k+1}^u(z, y^k, y_{k+1})\,\phi_{k+1}(y_{k+1})^{-1}\,\phi_{k+1}(y_{k+1} - C_k(\xi))\\ &\qquad \times \psi_{k+1}(z - A_k(\xi, u_k))\,q_k^u(\xi)\,\phi_{k+1}(y_{k+1})\Big]dz\,d\xi\,dy_{k+1}\\ &= \langle \eta_k^u(\xi, y^k), q_k^u(\xi)\rangle,\end{aligned}$$

where

$$\begin{aligned}\eta_k^u(\xi, y^k) = \int_{\mathbb{R}^d}\int_{\mathbb{R}^m}\Big[\eta_{k+1}^u(z, y^k, y)\,\phi_{k+1}(y - C_k(\xi))\\ \times \psi_{k+1}(z - A_k(\xi, u_k))\Big]dz\,dy.\end{aligned}$$

□

Remark 2.5 Note in particular

$$\begin{aligned}V(\pi_0, u) &= \overline{E}[\langle \ell_K(z), q_K^u(z)\rangle]\\ &= \overline{E}[\langle \eta_0^u(\xi, y_0), \pi_0(\xi)\rangle]\\ &= \overline{E}[\langle \eta_k^u(\xi, y^k), q_k^u(\xi)\rangle].\end{aligned}$$

■

Remark 2.6 Note again that the adjoint process derives from a linear equation. This is now infinite-dimensional for other than the discrete-state case.

■

Parameter Estimation and Dual Control

Suppose we have a situation where the model contains an unknown parameter θ, which we also wish to estimate. That is, suppose the state dynamics and observation processes are of the form:

$$x_{k+1} = a_k(x_k, u_k, \theta) + v_{k+1}, \tag{2.9}$$
$$y_{k+1} = c_k(x_k, \theta) + w_{k+1}, \qquad 0 \le k \le K. \tag{2.10}$$

Here θ takes values in some measure space (Θ, β, λ), with λ a probability measure. Θ could be a (subset of a) Euclidean space.

For example, a simple case would be (one-dimensional) linear dynamics and observations of the form

$$\begin{aligned} x_{k+1} &= \theta^1 x_k + \theta^2 u_k + v_{k+1}, \\ y_{k+1} &= \theta^3 x_k + w_{k+1}. \end{aligned}$$

The analysis of the previous sections goes through, taking the θ^i to be additional state variables. The unnormalized conditional density $q_k^u(z, \theta_1, \theta_2, \theta_3)$ is defined by

$$\begin{aligned} &\overline{E}\left[\overline{\Lambda}_k^u I(X_k \in dz) I\left(\theta^1 \in d\theta_1\right) I\left(\theta^2 \in d\theta_2\right) I\left(\theta^3 \in d\theta_3\right) \mid \mathscr{Y}_k\right] \\ &\quad = q_k^u(z, \theta_1, \theta_2, \theta_3)\, dz\, d\theta_1\, d\theta_2\, d\theta_3, \end{aligned}$$

and the recursive Equations (2.3) and dynamic programming results are exactly as before.

10.3 A Dependent Case

Controlled Dynamics

All processes are defined initially on a probability space $(\Omega, \mathscr{F}, P)$. Suppose $\{x_\ell\}$, $\ell \in \mathbb{N}$, is a discrete-time stochastic process taking values in some Euclidean space $\mathbb{R}^d$. We suppose that x_0 has a known distribution $\pi_0(x)$. Here $\{v_\ell\}$, $\ell \in \mathbb{N}$, will be a sequence of independent, $\mathbb{R}^k$-valued, random variables with probability measures $d\psi_\ell$ and $\{w_\ell\}$ a sequence of $\mathbb{R}^m$-valued, random variables with positive probability density functions ϕ_ℓ. The parameter u_k represents the control variable, and takes values in a set $U \subset \mathbb{R}^r$. For $k \in \mathbb{N}$, $a_k : \mathbb{R}^d \times \mathbb{R}^k \times \mathbb{R}^m \times U \to \mathbb{R}^d$ are measurable functions, and we suppose for $k \ge 0$ that

$$x_{k+1} = a_{k+1}(x_k, v_{k+1}, w_{k+1}, u_k). \tag{3.1}$$

The signal process X is not observed directly; rather we suppose there is an observation process $\{y_\ell\}$, $\ell \in \mathbb{N}$, related to the signal and taking values in some Euclidean space $\mathbb{R}^p$ and for $k \in \mathbb{N}$, $c_k : \mathbb{R}^d \times \mathbb{R}^m \to \mathbb{R}^p$ are measurable functions such that

$$y_{k+1} = c_{k+1}(x_k, w_{k+1}). \tag{3.2}$$

At time k, u_k is $\mathscr{Y}_k$ measurable, that is, u_k is a function of $y^k = \{y_0, y_1, \dots, y_k\}$. We shall consider a finite-time horizon control problem so that $0 \le k \le K-1$ from now on.

We assume here that for each $0 \le k \le K-1$, there is an inverse map $d_k : \mathbb{R}^d \times \mathbb{R}^d \times \mathbb{R}^m \times U \to \mathbb{R}^k$ such that if (3.1) holds

$$v_{k+1} = d_{k+1}(x_{k+1}, x_k, w_{k+1}, u_k). \tag{3.3}$$

Note this is the case if $x_{k+1} = \tilde{a}_{k+1}(x_k, w_{k+1}, u_k) + v_{k+1}$.

We require d_k to be differentiable in the first-variable for $0 \le k \le K-1$. We also assume that for each $0 \le k \le K-1$, there is an inverse map $g_k : \mathbb{R}^p \times \mathbb{R}^d \to \mathbb{R}^m$ such that if (3.2) holds

$$w_{k+1} = g_{k+1}(y_{k+1}, x_k).$$

Again, this is the case if $y_{k+1} = \tilde{c}_{k+1}(x_k) + w_{k+1}$. Finally, we require

$$C_{k+1}(x_k, w_{k+1}) = \left.\frac{\partial c_{k+1}(x_k, w)}{\partial w}\right|_{w=w_{k+1}},$$

$$G_{k+1}(y_{k+1}, x_k) = \left.\frac{\partial g_{k+1}(y, x_k)}{\partial y}\right|_{y=h_{k+1}}$$

to be nonsingular for $0 \le k \le K-1$.

Separation of the Problem

Suppose $u \in \underline{U}(0, K-1)$. Define

$$\lambda_\ell^u = \frac{\phi_\ell(y_\ell)}{\phi_\ell(g_\ell(y_\ell, X_{\ell-1}))} \left(\frac{G\ell(y_\ell, x_{\ell-1})^{-1}}{\partial y} \right)^{-1}$$

and

$$\Lambda_k^u = \prod_{\ell=1}^{k} \lambda_\ell^u.$$

Then a probability measure $\overline{P}^u$ can be defined by setting the restriction of $d\overline{P}^u / dP^u$ to $\mathscr{F}_K$ equal to Λ_K^u. It can be shown that under $\overline{P}^u$:

1. $\{y_k\}$ is a sequence of independent random variables having positive densities ϕ_k,
2. X_k satisfies the dynamics

$$X_{k+1} = a_{k+1}(X_k, v_{k+1}, g_k(y_{k+1}, X_k), u_k), \qquad 0 \le k \le K-1.$$

We now suppose that our starting, or reference, measure is $\overline{P}^u$ on $(\Omega, \mathscr{F}_K)$ and, proceeding in an inverse manner, we define

$$\overline{\lambda}_\ell^u = \frac{\phi_\ell(g_\ell(y_\ell, X_{\ell-1}))}{\phi_\ell(y_\ell)} \frac{C_\ell(X_{\ell-1}, w_\ell)^{-1}}{w_\ell}$$

and $\overline{\Lambda}_k^u = \prod_{\ell=1}^k \overline{\lambda}_\ell^u$. Then the probability measure P^u, under which (3.1) and (3.2) hold, is obtained by setting $\left(dP^u / d\overline{P}^u\right)|_{\mathscr{G}_K} = \overline{\Lambda}_K^u$.

Write $d\alpha_k^u(x)$ for the unnormalized conditional probability measure such that

$$\overline{E}^u\left[\overline{\Lambda}_k^u I(x_k \in dx) \mid \mathscr{Y}_k\right] := d\alpha_k^u(x).$$

Theorem 3.1 *For $1 \leq k \leq K$, a recursion for $d\alpha_k^u(x)$ is given by*

$$d\alpha_k^u(x) = \int_{\mathbb{R}^d} \Phi_k(x, y_k, z)\, d\psi_k(d_k(x, z, g_k(y_k, z), u_{k-1}))\, d\alpha_{k-1}^u(z). \tag{3.4}$$

Here

$$\begin{aligned}\Phi_k(x, y_k, z) &= \frac{1}{\phi_k(y_k)} \phi_k(g_k(y_k, z)) C_k(z, g_k(y_k, z))^{-1} \\ &\quad \times \left| \frac{\partial d_k(x, z, g_k(y_k, z), u_{k-1})}{\partial x} \right|.\end{aligned}$$

Proof The proof is similar to that of Theorem 4.4.2. □

Cost

Suppose, given x_0 and $u \in \underline{U}(0, K-1)$, the cost function associated with the problem is of the form

$$J(x_0, u) = \sum_{k=0}^{K-1} \ell_k(X_k, u_k) + \ell_K(X_K)$$

with expected cost

$$\begin{aligned}V_0(\pi_0, u) &= E^u[J(x_0, u)] \\ &= \overline{E}^u\left[\overline{\Lambda}_K^u \sum_{k=0}^{K-1} \ell_k(X_k, u_k) + \overline{\Lambda}_K^u \ell_K(X_K)\right] \\ &= \overline{E}^u\left[\sum_{k=0}^{K-1} \int_{\mathbb{R}^d} \ell_k(z, u_k)\, d\alpha_k^u(z) + \int_{\mathbb{R}^d} \ell_K(z)\, d\alpha_K^u(z)\right].\end{aligned}$$

If we write

$$\langle \ell_k(z,u_k), d\alpha_k^u(z)\rangle = \int_{\mathbb{R}^d} \ell_k(z,u_k)\, d\alpha_k^u(z),$$

this is

$$= \overline{E}^u\left[\sum_{k=0}^{K-1} \langle \ell_k(z,u_k), d\alpha_k^u(z)\rangle + \langle \ell_K(z), d\alpha_K^u(z)\rangle\right].$$

The information states $d\alpha_k^u$ are positive measures on $\mathbb{R}^d$. Consider the process starting from some intermediate time k, $0 \le k \le K$, from some state $d\alpha$. Then the expected remaining cost, corresponding to $u \in \underline{U}(k, K-1)$ is given by

$$V_k(d\alpha, u) = \overline{E}^u\Bigg[\sum_{j=k}^{K-1} \langle \ell_j(z,u_j), d\alpha_j^u(z)\rangle$$
$$+ \langle \ell_K(z), d\alpha_K^u(z)\rangle \;\Big|\; d\alpha_k^u = d\alpha\Bigg].$$

For $0 \le k \le K$ the cost process is defined as the essential minimum

$$V(k, d\alpha) = \bigwedge_{u \in \underline{U}(k,K-1)} V_k(d\alpha, u). \tag{3.5}$$

For $k = K$ set $V(K, d\alpha) = \overline{E}^u[\langle \ell_K(z), d\alpha(z)\rangle]$. The following dynamic programming result is then established.

Lemma 3.2 *For $0 \le k \le K-1$ and $d\alpha$ a positive measure on $\mathbb{R}^d$*

$$V(k, d\alpha) = \bigwedge_{u \in \underline{U}(k)} \overline{E}^u\left[\langle \ell_k(z,u_k), d\alpha(z)\rangle + V\left(k+1, d\alpha_{k+1}^u\right) \mid d\alpha_k = d\alpha\right]. \tag{3.6}$$

A dynamic programming result can be obtained as in earlier sections, and a minimum principle has the following form.

Theorem 3.3 *Suppose u^* is a separated control such that, for each positive measure $d\alpha$ on $\mathbb{R}^d$, $u_k^*(d\alpha)$ achieves the minimum in* (1.5). *Then $V_k(d\alpha, u^*) = V(k, d\alpha)$, and u^* is an optimal control.*

Proof The proof will use backward induction on k. Clearly

$$V_K(d\alpha, u^*) = \overline{E}^u[\langle \ell_K(z), d\alpha(z)\rangle] = V(K, d\alpha).$$

Suppose the result holds for $k+1, k+2, \ldots, K$. Then

$$\begin{aligned} V_k(d\alpha, u_k^*) &= \overline{E}^{u^*}\left[\langle \ell_k(z,u^*), d\alpha(z)\rangle + V_{k+1}(d\alpha_{k+1}^{u^*}, u^*) \;\big|\; d\alpha_k = d\alpha\right] \\ &= \overline{E}^{u^*}\left[\langle \ell_k(z,u^*), d\alpha(z)\rangle + V(k+1, d\alpha_{k+1}^{u^*}) \;\big|\; d\alpha_k = d\alpha\right] \\ &= V(k, d\alpha). \end{aligned}$$

Now for any other $u \in \underline{U}(0, K-1)$,

$$V_k(d\alpha, u^*) = V(k, d\alpha) \leq V_k(d\alpha, u),$$

and, in particular, $V_0(d\alpha, u^*) \leq V_0(d\alpha, u)$, so u^* is optimal. □

The Adjoint Process

Consider any control $u \in \underline{U}(0, K-1)$. We shall suppose for simplicity of notation that the cost is purely terminal at the final time K, so

$$J(x_0, u) = \ell_K(X_K).$$

Then

$$\begin{aligned} V(\pi_0, u) &= E^u[\ell_K(X_K)] \\ &= \overline{E}^u[\langle \ell_K(x), d\alpha_K^u(x)\rangle]. \end{aligned}$$

Theorem 3.4 *There is a process $\beta_k^u(x, y^k)$, adapted to $\mathscr{Y}_k$, such that for $0 \leq k \leq K$*

$$\overline{E}^u[\langle \ell_K(x), d\alpha_K^u(x)\rangle \mid \mathscr{Y}_k] = \langle \beta_k^u(x, y^k), d\alpha_k^u(x)\rangle.$$

Further, β_k^u evolves in reverse time so that

$$\begin{aligned} \beta_k^u(z, y^k) = \int_{\mathbb{R}^p} \Bigg[\int_{\mathbb{R}^d} & \beta_{k+1}^u(x, y^k, y_{k+1}) \phi_{k+1}(g_{k+1}(y_{k+1}, z))\, d\psi_{k+1} \\ & \times (d_{k+1}(x, z, g_{k+1}(y_{k+1}, z), u_k)) C_{k+1}(z, g_{k+1}(y_{k+1}, z))^{-1} \\ & \times \left| \frac{\partial d_{k+1}(x, z, g_{k+1}(y_{k+1}, z), u_k)}{\partial x} \right| \Bigg] dy_{k+1}. \end{aligned}$$

Proof The proof is left as an exercise. □

10.4 Problems and Notes

Problems

1. Derive the dynamic programming result (1.8).
2. Derive the adjoint process given by (1.9) for the control of a Markov chain in Gaussian noise.
3. Obtain the unnormalized recursive density and the dynamic programming results for the model given by (2.9) and (2.10) in the dual control section.
4. Derive the adjoint process given in Theorem 3.4.

Notes

In this chapter discrete-time, partially observed control problems are discussed by explicitly constructing a reference probability under which the observations are independent. Using the unnormalized conditional probabilities as information states, the problems are treated in separated form. Dynamic programming and minimum principle results are obtained. The idea of measure change from earlier chapters has again been exploited.

Chapter 11
Risk-Sensitive Control of HMM

11.1 Introduction

We saw in Chapter 10 that a stochastic control problem was solved using an information state, that is, an unnormalized conditional density. This problem was of the *risk-neutral* type. In this chapter we will solve a *risk-sensitive* stochastic control problem. It turns out, surprisingly, that the appropriate information state is *not* an unnormalized conditional density; rather, the information state also depends on the cost function.

Risk-sensitive problems involve an exponential cost function, which reflects the controllers' aversion to risk. A risk-sensitive controller is more conservative than the risk-neutral controller discussed in Chapter 10, since the cost function penalizes large values—a manifestation of the exponential function. The risk-sensitive stochastic control problem is formulated and solved in Section 11.2.

Interestingly, the risk-sensitive problem is closely related to H^∞, or robust, control. Actually, it incorporates features of both H^∞ and H_2 (i.e., risk-neutral) control. The relationship is due to the fact that H^∞ problems can be formulated in terms of dynamic games. In the case of linear systems with quadratic cost functions, the solution of the risk-sensitive problem coincides with that of a dynamic game; in the general nonlinear case, one must employ an asymptotic (small noise) limit to make the connection.

Accordingly, to point out this connection (and to emphasize the role of risk-sensitivity), we include a small noise parameter ε and a risk-sensitive parameter $\mu > 0$ in our formulation. Note, however, that the solution to the risk-sensitive problem in no way depends on asymptotic limits.

In Section 11.3, we briefly explain the connection with H^∞ dynamic games, and in Section 11.4 the relationship between risk-sensitive and risk-neutral (or H_2) problems is explained. Finally, we give an example in Section 11.5 for which the information state is finite-dimensional.

11.2 The Risk-Sensitive Control Problem

Dynamics

On a probability space $(\Omega, \mathscr{F}, P^u)$ we consider a risk-sensitive stochastic control problem for the discrete-time system

$$x_{k+1}^{\varepsilon} = a(x_k^{\varepsilon}, u_k) + v_{k+1}^{\varepsilon} \in \mathbb{R}^n, \tag{2.1}$$
$$y_{k+1}^{\varepsilon} = c(x_k^{\varepsilon}) + w_{k+1}^{\varepsilon} \in \mathbb{R}, \tag{2.2}$$

on the finite-time interval $k = 0, 1, 2, \ldots, K$. The process $x^{\varepsilon} \in \mathbb{R}$ represents the state of the system, and is not directly measured. The process $y^{\varepsilon} \in \mathbb{R}$ is measured and is called the observation process. This observation process can be used to select the control actions u_k. The values $\mathscr{G}_k$ and $\mathscr{Y}_k$ denote the complete filtrations generated by $\left(x_\ell^{\varepsilon}, y_\ell^{\varepsilon}, 0 \le \ell \le k\right)$ and $\left\{y_\ell^{\varepsilon}, 0 \le \ell \le k\right\}$, respectively.

We assume:

1. x_0^{ε} has density $\rho(x) = (2\pi)^{-n/2} \exp\left(-\frac{1}{2}|x|^2\right)$.
2. $\left\{v_k^{\varepsilon}\right\}$ is an $\mathbb{R}^n$-valued i.i.d. noise sequence with density
$$\psi^{\varepsilon}(v) = (2\pi\varepsilon)^{-n/2} \exp\left(-\tfrac{1}{2\varepsilon}|v|^2\right).$$
3. $y_0^{\varepsilon} = 0$.
4. The set of random variables $\left\{w_k^{\varepsilon}\right\}$ is a real-valued i.i.d. noise sequence with density
$$\phi^{\varepsilon}(w) = (2\pi\varepsilon)^{-1/2} \exp\left(-\tfrac{1}{2\varepsilon}|w|^2\right),$$
independent of x_0^{ε} and $\left\{v_k^{\varepsilon}\right\}$.
5. The function $a(.) \in C^1(\mathbb{R}^n \times \mathbb{R}^m, \mathbb{R}^n)$ is bounded and Lipschitz continuous, uniformly in $(x, u) \in \mathbb{R}^n \times U$.
6. The controls u_k take values in $U \subset \mathbb{R}^m$, assumed compact, and are $\mathscr{Y}_k$-measurable. We write $\underline{U}(k, \ell)$ for the set of such control processes defined on the interval $k, \ldots, \ell$.
7. The function $c(.) \in C(\mathbb{R}^n)$ is bounded and uniformly continuous.

The probability measure P^u can be defined in terms of an equivalent reference measure $\overline{P}$. Under $\overline{P}$, $\left\{y_k^{\varepsilon}\right\}$ is i.i.d. with density ϕ^{ε}, independent of $\left\{x_k^{\varepsilon}\right\}$, and x^{ε} satisfies (2.1). For $u \in \underline{U}(0, K-1)$,

$$\left.\frac{dP^u}{d\overline{P}}\right|_{\mathscr{G}_k} = \overline{\Lambda}_k^{\varepsilon} = \prod_{\ell=1}^{k} \overline{\lambda}^{\varepsilon}\left(x_{\ell-1}^{\varepsilon}, y_\ell^{\varepsilon}\right),$$

where

$$\overline{\lambda}^{\varepsilon}(x, y) := \frac{\phi(y - c(x))}{\phi(y)} = \exp\left(-\frac{1}{\varepsilon}\left[\frac{1}{2}|c(x)|^2 - c(x)y\right]\right).$$

Cost Function

The cost function is defined for admissible $u \in \underline{U}(0, K-1)$ and $\mu > 0$ by

$$J^{\mu,\varepsilon}(x_0, u) = \left[\exp\left(\frac{\mu}{\varepsilon}\left\{\sum_{k=0}^{K-1} L(x_k^\varepsilon, u_k) + \Phi(x_K^\varepsilon)\right\}\right)\right],$$

$$V^{\mu,\varepsilon}(\pi_0, u) = E\left[J^{\mu,\varepsilon}(x_0, u)\right], \tag{2.3}$$

where the density on x_0 is π_0.

The *partially observed risk-sensitive stochastic control problem* is to find $u^* \in \underline{U}(0, K-1)$ such that

$$V^{\mu,\varepsilon}(\pi_0, u^*) = \bigwedge_{u \in \underline{U}(0,K-1)} V^{\mu,\varepsilon}(\pi_0, u).$$

Here, we assume:

1. $L \in C(\mathbb{R}^n \times \mathbb{R}^m)$ is nonnegative, bounded and uniformly continuous. uniformly in $(x, u) \in \mathbb{R}^n \times U$.
2. $\Phi \in C(\mathbb{R}^n)$ is nonnegative, bounded, and uniformly continuous.

Remark 2.1 The assumptions 1–9 are stronger than necessary. For example, the boundedness assumption for a can be replaced by a linear growth condition. In addition, a "diffusion" coefficient can be inserted into the system. Other choices for the initial density ρ are possible.

■

The parameters $\mu > 0$ and $\varepsilon > 0$ are measures of risk sensitivity and noise variance. In view of our assumptions, the cost function is finite for all $\mu > 0$, $\varepsilon > 0$. For risk-sensitive problems the cost is of an exponential form.

In terms of the reference measure, the cost can be expressed as

$$V^{\mu,\varepsilon}(\pi_0, u) = \overline{E}\left[\overline{\Lambda}_K^\varepsilon \exp\left(\frac{\mu}{\varepsilon}\left\{\sum_{k=0}^{K-1} L(x_k^\varepsilon, u_k) + \Phi(x_K^\varepsilon)\right\}\right)\right]. \tag{2.4}$$

Information State

We consider the space $L^\infty(\mathbb{R}^n)$ and its dual $L^{\infty *}(\mathbb{R}^n)$, which includes $L^1(\mathbb{R}^n)$. We will denote the natural bilinear pairing between $L^\infty(\mathbb{R}^n)$ and $L^{\infty *}(\mathbb{R}^n)$ by $\langle \tau, \beta \rangle$ for $\tau \in L^{\infty *}(\mathbb{R}^n)$, $\beta \in L^\infty(\mathbb{R}^n)$. In particular, for $\alpha \in L^1(\mathbb{R}^n)$ and $\beta \in L^\infty(\mathbb{R}^n)$ we have

$$\langle \alpha, \beta \rangle = \int_{\mathbb{R}^n} \alpha(x) \beta(x) \, dx.$$

We now define an information state process $q_k^{\mu,\varepsilon} \in L^{\infty *}(\mathbb{R}^n)$ by

$$\int_{\mathbb{R}^n} \eta(x)\, q_k^{\mu,\varepsilon}(x)\, dx = \langle q_k^{\mu,\varepsilon}, \eta \rangle \tag{2.5}$$

$$= \overline{E}\left[\eta(x_k^\varepsilon) \exp\left(\frac{\mu}{\varepsilon} \sum_{\ell=0}^{k-1} L(x_\ell^\varepsilon, u_\ell) \right) \overline{\Lambda}_k^\varepsilon \,\middle|\, \mathscr{Y}_k \right]$$

for all test functions η in $L^\infty(\mathbb{R}^n)$, for $k = 1, \ldots, K$ and $q_0^{\mu,\varepsilon} = \rho \in L^1(\mathbb{R}^n)$. We note that this information state is similar to that of Chapter 10, but it now includes part of the exponential cost function. We introduce the bounded linear operator $\Sigma^{\mu,\varepsilon}$: $L^\infty(\mathbb{R}^n) \to L^\infty(\mathbb{R}^n)$ defined by

$$\Sigma^{\mu,\varepsilon}(u,y)\,\beta(\xi) := \exp\left(\frac{\mu}{\varepsilon} L(\xi,u)\right) \lambda^\varepsilon(\xi,y) \int_{\mathbb{R}^n} \psi^\varepsilon(z - a(\xi,u))\, \beta(z)\, dz. \tag{2.6}$$

The bounded linear operator $\Sigma^{\mu,\varepsilon *} : L^{\infty *}(\mathbb{R}^n) \to L^{\infty *}(\mathbb{R}^n)$ adjoint to $\Sigma^{\mu,\varepsilon}$ is defined by

$$\langle \Sigma^{\mu,\varepsilon *} \tau, \eta \rangle = \langle \tau, \Sigma^{\mu,\varepsilon} \eta \rangle$$

for all $\tau \in L^{\infty *}(\mathbb{R}^n)$, $\eta \in L^\infty(\mathbb{R}^n)$.

The following theorem establishes that $q_k^{\mu,\varepsilon}$ is in $L^1(\mathbb{R}^n)$ and its evolution is governed by the operator $\Sigma^{\mu,\varepsilon *}$, and for $\alpha \in L^1(\mathbb{R}^n)$, $\eta \in L^\infty(\mathbb{R}^n)$, we have

$$\Sigma^{\mu,\varepsilon *}(u,y)\,\alpha(z) = \int_{\mathbb{R}^n} \psi^\varepsilon(z - a(\xi,u)) \exp\left(\frac{\mu}{\varepsilon} L(\xi,u)\right) \lambda^\varepsilon(\xi,y)\, \alpha(\xi)\, d\xi. \tag{2.7}$$

Note that $q_k^{\mu,\varepsilon}$ is an unnormalized "density" for the state which also depends on the cost.

Similarly to Theorem 10.2.1, we have the following result.

Theorem 2.2 *The information state $q_k^{\mu,\varepsilon}$ satisfies the recursion*

$$\boxed{\begin{aligned} q_k^{\mu,\varepsilon} &= \Sigma^{\mu,\varepsilon *}(u_{k-1}, y_k^\varepsilon)\, q_{k-1}^{\mu,\varepsilon}, \\ q_0^{\mu,\varepsilon} &= \rho. \end{aligned}} \tag{2.8}$$

Further, $q_k^{\mu,\varepsilon} \in L^1(\mathbb{R}^n)$ since $\rho \in L^1(\mathbb{R}^n)$ and $\Sigma^{\mu,\varepsilon *}$ maps $L^1(\mathbb{R}^n)$ into $L^1(\mathbb{R}^n)$.

Proof The proof is left as an exercise. □

Remark 2.3 When $L \equiv 0$, the recursion (2.8) reduces to the Duncan-Mortensen-Zakai equation for the unnormalized conditional density given in Theorems 10.2.1 and 10.3.1. See also Kumar and Varaiya (1986b). ■

Next, we define an adjoint operator $\Sigma^{\mu,\varepsilon *} : C_b(\mathbb{R}^n) \to C_b(\mathbb{R}^n)$. [Actually it maps $L^\infty(\mathbb{R}^n)$ to $L^\infty(\mathbb{R}^n)$, but we need only $C_b(\mathbb{R}^n)$ for our purpose here where $C_b(\mathbb{R}^n) := \{q \in C(\mathbb{R}^n) : |q(x)| \le C,$ for some $C \ge 0\}$.] This is the analog of the recursion of

Theorem 10.3.4. The operator $\Sigma^{\mu,\varepsilon}$ actually maps $C_b(\mathbb{R}^n)$ into $C_b(\mathbb{R}^n)$. Then we can define a process $\beta_k^{\mu,\varepsilon} \in C_b(\mathbb{R}^n)$ by

$$\boxed{\begin{aligned} \beta_{k-1}^{\mu,\varepsilon} &= \Sigma^{\mu,\varepsilon}(u_{k-1}, y_k^{\varepsilon})\,\beta_k^{\mu,\varepsilon}, \\ \beta_K^{\mu,\varepsilon} &= \exp\left(\tfrac{\mu}{\varepsilon}\Phi\right). \end{aligned}} \tag{2.9}$$

It is straightforward to establish the adjoint relationships

$$\begin{aligned} \langle \Sigma^{\mu,\varepsilon *}\alpha, \beta \rangle &= \langle \alpha, \Sigma^{\mu,\varepsilon}\beta \rangle, \\ \langle q_k^{\mu,\varepsilon}, \beta_k^{\mu,\varepsilon} \rangle &= \langle q_{k-1}^{\mu,\varepsilon}, \beta_{k-1}^{\mu,\varepsilon} \rangle, \end{aligned} \tag{2.10}$$

for all $\alpha \in L^1(\mathbb{R}^n)$, $\beta \in C_b(\mathbb{R}^n)$, and all k.

Alternate Representation of the Cost

Following Bensoussan and van Schuppen (1985), we define for $u \in \underline{U}(0, K-1)$

$$K^{\mu,\varepsilon}(\pi_0, u) = \overline{E}\left[\left\langle q_K^{\mu,\varepsilon}, \exp\left(\frac{\mu}{\varepsilon}\Phi\right)\right\rangle\right], \tag{2.11}$$

a cost function associated with the new "state" process $q_k^{\mu,\varepsilon}$.

Theorem 2.4 *We have for all* $u \in \underline{U}(0, K-1)$

$$J^{\mu,\varepsilon}(u) = K^{\mu,\varepsilon}(u). \tag{2.12}$$

Proof By (2.5),

$$\begin{aligned} K^{\mu,\varepsilon}(\pi_0, u) &= \overline{E}\left[\overline{E}\left[\exp\left(\frac{\mu}{\varepsilon}\Phi(x_K^{\varepsilon})\right)\exp\left(\frac{\mu}{\varepsilon}\sum_{k=0}^{K-1} L(x_k^{\varepsilon}, u_k)\right)\overline{\Lambda}_K^{\varepsilon}\,\middle|\, \mathscr{Y}_K\right]\right] \\ &= \overline{E}\left[\exp\left(\frac{\mu}{\varepsilon}\left\{\sum_{k=0}^{K-1} L(x_k^{\varepsilon}, u_k) + \Phi(x_K^{\varepsilon})\right\}\right)\overline{\Lambda}_K^{\varepsilon}\right] \\ &= V^{\mu,\varepsilon}(\pi_0, u) \end{aligned}$$

using (2.4). □

We now define an alternate but equivalent stochastic control problem with complete state information. Under the measure P^u, consider the state process $q_k^{\mu,\varepsilon}$ governed by (2.8) and the cost $K^{\mu,\varepsilon}(\pi_0, u)$ given by (2.11). The new problem is to find $u^* \in \underline{U}(0, K-1)$ minimizing $K^{\mu,\varepsilon}$.

Let $\underline{U}_s(k, \ell)$ denote the set of control processes defined on the interval $k, \ldots, \ell$ which are adapted to $\sigma(q_j^{\mu,\varepsilon}, k \le j \le \ell)$. Such policies are called separated (Kumar and Varaiya, 1986b).

Dynamic Programming

The alternate stochastic control problem can be solved using dynamic programming. Consider now the state $q^{\mu,\varepsilon}$ on the interval $k,\ldots,K$ with initial condition $q_k^{\mu,\varepsilon} = q \in L^1(\mathbb{R}^n)$:

$$\boxed{\begin{aligned} &q_\ell^{\mu,\varepsilon} = \Sigma^{\mu,\varepsilon *}(u_{\ell-1}, y_\ell^\varepsilon)\, q_{\ell-1}^{\mu,\varepsilon}, \qquad k+1 \le \ell \le K, \\ &q_k^{\mu,\varepsilon} = q. \end{aligned}} \tag{2.13}$$

The corresponding value function for this control problem is defined for $q \in L^1(\mathbb{R}^n)$ by

$$\begin{aligned} V^{\mu,\varepsilon}(k,q) &= \bigwedge_{u \in \underline{U}(k,K-1)} \overline{E}\left[\langle q_k^{\mu,\varepsilon}, \beta_k^{\mu,\varepsilon}\rangle \mid q_k^{\mu,\varepsilon} = q\right] \\ &= \bigwedge_{u \in \underline{U}(k,K-1)} \overline{E}\left[\left\langle q_k^{\mu,\varepsilon}, \exp\left(\frac{\mu}{\varepsilon}\Phi\right)\right\rangle\right]. \end{aligned} \tag{2.14}$$

Note that this function is expressed in terms of the adjoint process $\beta_k^{\mu,\varepsilon}$, given by (2.9). See Theorem 10.2.4.

The following result is the analog of Lemma 10.1.1.

Theorem 2.5 (Dynamic programming equation) *The value function $V^{\mu,\varepsilon}$ satisfies the recursion*

$$\boxed{\begin{aligned} &V^{\mu,\varepsilon}(k,q) = \bigwedge_{u_k \in \underline{U}(k,k)} \overline{E}\left[V^{\mu,\varepsilon}\left(k+1, \Sigma^{\mu,\varepsilon *}\left(u_k, y_{k+1}^\varepsilon\right) q\right)\right] \\ &V^{\mu,\varepsilon}(q,K) = \left\langle q, \exp\left(\frac{\mu}{\varepsilon}\Phi\right)\right\rangle. \end{aligned}} \tag{2.15}$$

Proof

$$\begin{aligned} &V^{\mu,\varepsilon}(k,q) \\ &= \bigwedge_{u \in \underline{U}(k,k)} \bigwedge_{v \in \underline{U}(k+1,K-1)} \overline{E}\left[\langle q_k^{\mu,\varepsilon}, \Sigma^{\mu,\varepsilon}\left(u_k, y_{k+1}^\varepsilon\right) \beta_{k+1}^{\mu,\varepsilon}\rangle \mid q_k^{\mu,\varepsilon} = q\right] \\ &= \bigwedge_{u \in \underline{U}(k,k)} \bigwedge_{v \in \underline{U}(k+1,K-1)} \overline{E}\Big[\overline{E}\left[\langle \Sigma^{\mu,\varepsilon *}\left(u_k, y_{k+1}^\varepsilon\right) q_k^{\mu,\varepsilon},\ \beta_{k+1}^{\mu,\varepsilon}\rangle \mid \mathcal{Y}_{k+1}\right] \mid q_k^{\mu,\varepsilon} = q\Big] \\ &= \bigwedge_{u \in \underline{U}(k,k)} \overline{E}\Big[\bigwedge_{v \in \underline{U}(k+1,K-1)} \overline{E}\left[\langle \Sigma^{\mu,\varepsilon *}\left(u_k, y_{k+1}^\varepsilon\right) q_k^{\mu,\varepsilon},\ \beta_{k+1}^{\mu,\varepsilon}\rangle \mid \mathcal{Y}_{k+1}\right] \Big| q_k^{\mu,\varepsilon} = q\Big] \\ &= \bigwedge_{u \in \underline{U}(k,k)} \overline{E}\Big[\bigwedge_{v \in \underline{U}(k+1,K-1)} \overline{E}\left[\langle q_{k+1}^{\mu,\varepsilon}, \beta_{k+1}^{\mu,\varepsilon}\rangle \mid q_{k+1}^{\mu,\varepsilon} = \Sigma^{\mu,\varepsilon *}\left(u_k, y_{k+1}^\varepsilon\right) q\right]\Big] \\ &= \bigwedge_{u \in \underline{U}(k,k)} \overline{E}\left[V^{\mu,\varepsilon}\left(k+1, \Sigma^{\mu,\varepsilon *}\left(u_k, y_{k+1}^\varepsilon\right) q\right)\right]. \end{aligned}$$

The interchange of minimization and conditional expectation is justified because of the lattice property of the set of controls (Elliott 1982b, Chapter 16). □

Theorem 2.6 (Verification) *Suppose that $u^* \in \underline{U}_s(0,K-1)$ is a policy such that, for each $k=0,\dots,K-1$, $u_k^* = \overline{u}_k^*\left(q_k^{\mu,\varepsilon}\right)$, where $\overline{u}_k^*(q)$ achieves the minimum in* (2.15). *Then $u^* \in \underline{U}(0,K-1)$ and is an optimal policy for the partially observed risk-sensitive stochastic control problem (Section* 11.2*).*

Proof The proof is similar to Theorem 10.1.4. □

Remark 2.7 As in Chapter 10, the significance of Theorem 2.6 is that it establishes the optimal policy of the risk-sensitive stochastic control problem as a *separated* policy through the process $q_k^{\mu,\varepsilon}$ which serves as an "information state" (Kumar and Varaiya 1986b).

■

11.3 Connection with $\mathbf{H}^\infty$ Control

In this section we explain the connection of the risk-sensitive problem with H^∞ control through dynamic games. Complete details can be found in James (1992).

Information State

For $\gamma \in G := \left\{\gamma \in \mathbb{R}^2 : \gamma_1 > 0, \gamma_2 \geq 0\right\}$ define

$$\mathscr{D}^\gamma := \left\{p \in C(\mathbb{R}^n) : p(x) \leq -\gamma_1 |x|^2 + \gamma_2\right\},$$
$$\mathscr{D} := \left\{p \in C(\mathbb{R}^n) : p(x) \leq -\gamma_1 |x|^2 + \gamma_2, \text{ for some } \gamma \in G\right\}.$$

We equip these spaces with the topology of uniform convergence on compact subsets. In the sequel, $B(x,\alpha) \subset \mathbb{R}^p$ denotes the open ball centered at $x \in \mathbb{R}^p$ of radius $\alpha > 0$.

The inner product $\langle .,. \rangle$ is replaced by the "sup pairing"

$$(p,q) := \sup_{x \in \mathbb{R}^n} \{p(x) + q(x)\}. \tag{3.1}$$

This is defined for $p \in \mathscr{D}$, $q \in C_b(\mathbb{R}^n)$, and arises naturally in view of the Varadhan-Laplace lemma (James 1992). In fact,

$$\lim_{\varepsilon \to 0} \frac{\varepsilon}{\mu} \log \left\langle \exp\left(\frac{\mu}{\varepsilon} p\right), \exp\left(\frac{\mu}{\varepsilon} q\right)\right\rangle = (p,q) \tag{3.2}$$

[uniformly on compact subsets of $\mathscr{D}^\gamma \times C_b(\mathbb{R}^n)$, for each $\gamma \in G$].

Define operators $\Gamma^{\mu *}: \mathscr{D} \to \mathscr{D}$, and $\Gamma^{\mu}: C_b(\mathbb{R}^n) \to C_b(\mathbb{R}^n)$ by

$$\begin{aligned}\Gamma^{\mu *}(u,y)\,p(z) &:= \sup_{\xi\in\mathbb{R}^n}\left\{L(\xi,u) - \frac{1}{2\mu}|z-a(\xi,u)|^2\right.\\ &\qquad\left. -\frac{1}{\mu}\left[\frac{1}{2}|c(\xi)|^2 - c(\xi)y\right] + p(\xi)\right\},\\ \Gamma^{\mu}(u,y)\,q(\xi) &:= \sup_{z\in\mathbb{R}^n}\left\{-\frac{1}{2\mu}|z-a(\xi,u)|^2 + q(z)\right\}\\ &\qquad + L(\xi,u) - \frac{1}{\mu}\left[\frac{1}{2}|c(\xi)|^2 - c(\xi)y\right].\end{aligned} \tag{3.3}$$

With respect to the "sup pairing" $(\cdot,\cdot)$, these operators satisfy:

$$(\Gamma^{\mu *}p, q) = (p, \Gamma^{\mu} q). \tag{3.4}$$

Also, $\Gamma^{\mu *}(u,y): \mathscr{D}^{\gamma} \to \mathscr{D}$ is continuous for each $\gamma \in G$; in fact, the map $(u,y,p) \mapsto \Gamma^{\mu *}(u,y)\,p$, $U \times \mathbb{R} \times \mathscr{D}^{\gamma} \to \mathscr{D}$ is continuous.

The next theorem is a logarithmic limit result for the information state and its dual, stated in terms of operators (i.e., semigroups).

Theorem 3.1 *We have*

$$\begin{aligned}\lim_{\varepsilon\to 0}\frac{\varepsilon}{\mu}\log\Sigma^{\mu,\varepsilon *}(u,y)\exp\left(\frac{\mu}{\varepsilon}p\right) &= \Gamma^{\mu *}(u,y)\,p,\\ \lim_{\varepsilon\to 0}\frac{\varepsilon}{\mu}\log\Sigma^{\mu,\varepsilon}(u,y)\exp\left(\frac{\mu}{\varepsilon}q\right) &= \Gamma^{\mu}(u,y)\,q\end{aligned} \tag{3.5}$$

in $\mathscr{D}$ uniformly on compact subsets of $U \times \mathbb{R} \times \mathscr{D}^{\gamma}$ for each $\gamma \in G$, and respectively, in $C_b(\mathbb{R}^n)$ uniformly on compact subsets of $U \times \mathbb{R} \times C_b(\mathbb{R}^n)$.

Proof From (2.7), we have

$$\begin{aligned}&\frac{\varepsilon}{\mu}\log\Sigma^{\mu,\varepsilon *}(u,y)\exp\left(\frac{\mu}{\varepsilon}p\right)z\\ &\quad= \frac{\varepsilon}{\mu}\log\int_{\mathbb{R}^n}\exp\frac{\mu}{\varepsilon}\left(-\frac{1}{2\mu}|z-a(\xi,u)|^2 - \frac{n\varepsilon}{2\mu}\log(2\pi\varepsilon)\right.\\ &\qquad\qquad\left. -\frac{1}{\mu}\left[\frac{1}{2}|c(\xi)|^2 - c(\xi)y\right] + L(\xi,u) + p(\xi)\right)d\xi.\end{aligned}$$

Therefore,

$$\begin{aligned}&\lim_{\varepsilon\to 0}\frac{\varepsilon}{\mu}\log\Sigma^{\mu,\varepsilon *}(u,y)\exp\left(\frac{\mu}{\varepsilon}p\right)z\\ &\quad= \sup_{\xi\in\mathbb{R}^n}\left\{L(\xi,u) - \frac{1}{2\mu}|z-a(\xi,u)|^2 - \frac{1}{\mu}\left[\frac{1}{2}|c(\xi)|^2 - c(\xi)y\right] + p(\xi)\right\}\\ &\quad= \Gamma^{\mu *}(u,y)\,p(z)\end{aligned}$$

uniformly in $a = (z,u,y,p) \in A$, where $A = B(0,R) \times U \times B(0,R) \times K$, and $K \subset \mathscr{D}^\gamma$ is compact. This proves the first part of (3.5). The second part is proven similarly. □

Risk-Sensitive Value Function

We now give a result which provides a dynamic programming result for the small noise limit of the risk-sensitive value function $V^{\mu,\varepsilon}$. The proof can be found in James, Baras and Elliott (1993).

Theorem 3.2 *The function $W^\mu(p,k)$ defined for $p \in \mathscr{D}$ by*

$$W^\mu(k,p) := \lim_{\varepsilon \to 0} \frac{\varepsilon}{\mu} \log V^{\mu,\varepsilon}\left(\exp\left(\frac{\mu}{\varepsilon}p\right),k\right) \tag{3.6}$$

exists [i.e., the sequence converges uniformly on compact subsets of $\mathscr{D}^\gamma$ $(\gamma \in G)$], is continuous on $\mathscr{D}^\gamma$ $(\gamma \in G)$, and satisfies the recursion

$$\boxed{\begin{aligned} W^\mu(k,p) &= \inf_{u\in U}\sup_{y\in\mathbb{R}}\left\{W^\mu\left(k+1,\Gamma^{\mu*}(u,y)p,\right) - \frac{1}{2\mu}|y|^2\right\}, \\ W^\mu(K,p) &= (p,\Phi). \end{aligned}} \tag{3.7}$$

Remark 3.3 In James et al. (1993) $W^\mu(p,k)$ is interpreted as the optimal cost function (upper value) for a deterministic *dynamic game* problem. Also, the limit $\lim_{\varepsilon\to 0}\frac{\varepsilon}{\mu}\log q_k^{\mu,\varepsilon} = p_k^\mu$ is the information state for the game. This is the most important contribution of James et al. (1993).

■

11.4 Connection with $\mathbf{H_2}$ or Risk-Neutral Control

In this section we explain how a risk-neutral stochastic control problem is obtained if in the risk-sensitive stochastic control problem the risk-sensitivity parameter μ tends to zero. In this limiting case, the index expressed as a power series expansion is seen to approximate its first two terms and thus be effectively the exponent itself, so the limiting connection to risk-neutral, i.e., H_2, control is not surprising.

Information State

Similarly to Chapter 10 we now define the bounded linear operator $\Sigma^\varepsilon : L^1(\mathbb{R}^n) \to L^1(\mathbb{R}^n)$ and $\Sigma^\varepsilon : L^\infty(\mathbb{R}^n) \to L^\infty(\mathbb{R}^n)$ by:

$$\Sigma^\varepsilon(u,y)q(z) := \int_{\mathbb{R}^n} \psi^\varepsilon(z - a(\xi,u))\lambda^\varepsilon(\xi,y)q(\xi)\,d\xi,$$

$$\Sigma^{\varepsilon}(u,y)\beta(\xi) := \int_{\mathbb{R}^n} \psi^{\varepsilon}(z - a(\xi,u))\beta(\xi)\, dz\lambda^{\varepsilon}(\xi,y). \tag{4.1}$$

These operators are adjoint with respect to the inner product $\langle \cdot, \cdot \rangle$.

Theorem 4.1 *We have*

$$\begin{aligned} \lim_{\mu \to 0} \Sigma^{\mu,\varepsilon}(u,y)\, q &= \Sigma^{\varepsilon}(u,y)\, q, \\ \lim_{\mu \to 0} \Sigma^{\mu,\varepsilon}(u,y)\, \beta &= \Sigma^{\varepsilon}(u,y)\, \beta \end{aligned} \tag{4.2}$$

uniformly on bounded subsets of $U \times \mathbb{R} \times L^1(\mathbb{R}^n)$ *[respectively,* $U \times \mathbb{R} \times L^\infty(\mathbb{R}^n)$*].*

Proof This result follows simply from the definitions in (2.7) and (4.1). □

Next, we define a process $q_k^{\varepsilon} \in L^1(\mathbb{R}^n)$ and its dual (w.r.t. $\langle \cdot, \cdot \rangle$) β_k^{ε} by the recursion

$$\begin{cases} q_k^{\varepsilon} = \Sigma^{\varepsilon}\left(u_{k-1}, y_k^{\varepsilon}\right) q_{k-1}^{\varepsilon}, \\ q_0^{\varepsilon} = \rho; \end{cases} \tag{4.3}$$

$$\begin{cases} \beta_{k-1}^{\varepsilon} = \Sigma^{\varepsilon}\left(u_{k-1}, y_k^{\varepsilon}\right) \beta_k^{\varepsilon}, \\ \beta_K^{\varepsilon} = 1. \end{cases} \tag{4.4}$$

A Risk-Neutral Control Problem

We are, therefore, in the situation of Chapter 10. We again consider the discrete-time stochastic system in (2.1) and (2.2), and formulate a *partially observed risk-neutral stochastic control problem* with cost

$$V^{\varepsilon}(\pi_0, u) = E^u \left[\sum_{k=0}^{K-1} L(x_k^{\varepsilon}, u_k) + \Phi(x_K^{\varepsilon}) \right] \tag{4.5}$$

defined for $u \in \underline{U}(0, K-1)$, where $\underline{U}(0, K-1)$, etc., are as defined above. This cost function is finite for all $\varepsilon > 0$.

We quote the following result from Lemma 10.1.3, which establishes that the optimal policy is separated through the information state q_k^{ε} satisfying (4.4).

Theorem 4.2 *The unnormalized conditional density* q_k^{ε} *is an information state for the risk-neutral problem, and the value function defined for* $q \in L^1(\mathbb{R}^n)$ *by*

$$W^{\varepsilon}(k,q) = \bigwedge_{u \in \underline{U}_s(k,K-1)} \overline{E} \left[\sum_{\ell=k}^{K-1} \langle q_\ell^{\varepsilon}, L(\cdot, u) \rangle + \langle q_K^{\varepsilon}, \Phi \rangle \mid q_k^{\varepsilon} = q \right] \tag{4.6}$$

satisfies the dynamic programming equation

$$\boxed{\begin{aligned} W^{\varepsilon}(k,q) &= \inf_{u\in U} \overline{E}\left[\langle q, L(\cdot,u)\rangle + W^{\varepsilon}\left(k+1, \Sigma^{\varepsilon}\left(u, y^{\varepsilon}_{k+1}\right) q\right)\right], \\ W^{\varepsilon}(K,q) &= \langle q, \Phi\rangle . \end{aligned}} \tag{4.7}$$

If $u^* \in \underline{U}_s(0, K-1)$ *is a policy such that, for each* $k = 0, \ldots, K-1$, $u^*_k = \overline{u}^*_k\left(q^{\varepsilon}_k\right)$, *where* $\overline{u}^*_k(q)$ *achieves the minimum in (4.7), then* $u^* \in \underline{U}(0, K-1)$ *and is an optimal policy for the partially observed risk-neutral problem.*

Remark 4.3 The function $W^{\varepsilon}(q,k)$ depends continuously on $q \in L^1(\mathbb{R}^n)$. ■

Risk-Sensitive Value Function

The next theorem evaluates the small risk limit of the risk-sensitive stochastic control problem. Note that normalization of the information state is required. The limit operation picks out the supremum on the right side.

Theorem 4.4 *We have*

$$\lim_{\mu\to 0} \frac{\varepsilon}{\mu} \log \frac{V^{\mu,\varepsilon}(k,q)}{\langle q,1\rangle} = \frac{W^{\varepsilon}(k,q)}{\langle q,1\rangle} \tag{4.8}$$

uniformly on bounded subsets of $L^1(\mathbb{R}^n)$.

Proof 1. We claim that

$$V^{\mu,\varepsilon}(k,q) = \langle q,1\rangle + \frac{\mu}{\varepsilon} W^{\varepsilon}(k,q) + o(\mu) \tag{4.9}$$

as $\mu \to 0$ uniformly on bounded subsets of $L^1(\mathbb{R}^n)$.

For $k = K$,

$$\begin{aligned} V^{\mu,\varepsilon}(K,q) &= \left\langle q, \exp\left(\frac{\mu}{\varepsilon}\Phi\right)\right\rangle \\ &= \langle q,1\rangle + \frac{\mu}{\varepsilon}\langle q,\Phi\rangle + o(\mu) \\ &= \langle q,1\rangle + \frac{\mu}{\varepsilon} W^{\varepsilon}(K,q) + o(\mu) \end{aligned}$$

as $\mu \to 0$, uniformly on bounded subsets of $L^1(\mathbb{R}^n)$.

Assume now that (4.9) is true for $k+1, \ldots, K$. Then

$$\begin{aligned} &V^{\mu,\varepsilon}(k,q;u) \\ &\quad := \overline{E}\left[V^{\mu,\varepsilon}\left(k+1, \Sigma^{\mu,\varepsilon}\left(u, y^{\varepsilon}_{k+1}\right) q\right)\right] \end{aligned}$$

$$
\begin{aligned}
&= \int_{\mathbb{R}} \phi^{\varepsilon}(y) V^{\mu,\varepsilon}(k+1, \Sigma^{\mu,\varepsilon}(u,y) q)\, dy \\
&= \int_{\mathbb{R}^n} \int_{\mathbb{R}} \left[\phi^{\varepsilon}(y - c(\xi)) \exp\left(\frac{\mu}{\varepsilon} L(\xi,u)\right) \right. \\
&\qquad\qquad \left. \times V^{\mu,\varepsilon}\left(k+1, \frac{\Sigma^{\mu,\varepsilon}(u,y) q}{\langle \Sigma^{\mu,\varepsilon}(u,y) q, 1\rangle}\right) q(\xi) \right] d\xi\, dy \\
&= \int_{\mathbb{R}^n} \int_{\mathbb{R}} \left\{ \phi^{\varepsilon}(y - c(\xi)) \left[1 + \frac{\mu}{\varepsilon} L(\xi,u) + o(\mu) \right] \right. \\
&\qquad\qquad \left. \times \left[1 + \frac{\mu}{\varepsilon} W^{\varepsilon}\left(k+1, \frac{\Sigma^{\mu,\varepsilon}(u,y) q}{\langle \Sigma^{\mu,\varepsilon}(u,y) q, 1\rangle}\right) + o(\mu) \right] q(\xi) \right\} d\xi\, dy \\
&= \langle q, 1\rangle + \frac{\mu}{\varepsilon} \left\{ \langle q, L(\cdot, u)\rangle \right. \\
&\qquad\qquad + \int_{\mathbb{R}^n} \int_{\mathbb{R}} \left[\phi^{\varepsilon}(y - c(\xi)) W^{\varepsilon}\left(k+1, \frac{\Sigma^{\varepsilon}(u,y) q}{\langle \Sigma^{\varepsilon}(u,y) q, 1\rangle}\right) \right. \\
&\qquad\qquad\qquad \left. \left. \times q(\xi) \right] d\xi\, dy \right\} + o(\mu) \\
&= \langle q, 1\rangle + \frac{\mu}{\varepsilon} \left\{ \langle q, L(\cdot, u)\rangle + \int_{\mathbb{R}} \phi^{\varepsilon}(y) W^{\varepsilon}(k+1, \Sigma^{\varepsilon}(u,y) q)\, dy \right\} + o(\mu)
\end{aligned}
$$

as $\mu \to 0$ uniformly on bounded subsets of $U \times L^1(\mathbb{R}^n)$. Thus, using the continuity of $(q,u) \mapsto V^{\mu,\varepsilon}(k,q;u)$,

$$
\begin{aligned}
V^{\mu,\varepsilon}(k,q) &= \bigwedge_{u \in U} V^{\mu,\varepsilon}(q,k;u) \\
&= \langle q, 1\rangle + \frac{\mu}{\varepsilon} \bigwedge_{u \in U} \left\{ \langle q, L(\cdot, u)\rangle \right. \\
&\qquad\qquad \left. + \int_{\mathbb{R}} \phi^{\varepsilon}(y) W^{\varepsilon}(k+1, \Sigma^{\varepsilon}(u,y) q)\, dy \right\} + o(\mu) \\
&= \langle q, 1\rangle + \frac{\mu}{\varepsilon} W^{\varepsilon}(k,q) + o(\mu),
\end{aligned}
$$

uniformly on bounded subsets of $L^1(\mathbb{R}^n)$, proving (4.9).

2. To complete the proof, note that (4.9) implies

$$
\frac{V^{\mu,\varepsilon}(k,q)}{\langle q, 1\rangle} = 1 + \frac{\mu}{\varepsilon} \frac{W^{\varepsilon}(k,q)}{\langle q, 1\rangle} + o(\mu)
$$

and hence

$$
\frac{\varepsilon}{\mu} \log \frac{V^{\mu,\varepsilon}(k,q)}{\langle q, 1\rangle} = \frac{W^{\varepsilon}(k,q)}{\langle q, 1\rangle} + o(1)
$$

as $\mu \to 0$, uniformly on bounded subsets of $L^1(\mathbb{R}^n)$. □

Remark 4.5 We conclude from Theorems 4.2 and 4.4 that the small risk limit of the partially observed stochastic risk-sensitive problem is a partially observed stochastic risk-neutral problem.

■

11.5 A Finite-Dimensional Example

In this section we present a discrete-time analogue of Bensoussan and Elliott (1995). Suppose $\{\Omega, \mathscr{F}, P\}$ is a probability space with a complete filtration $\{\mathscr{G}_k\}$, $k \in \mathbb{N}$, on which are given two sequences of independent random variables x_k and y_k, having normal densities $\psi_k = N(0, Q_k)$ and $\phi_k = N(0, R_k)$ where Q_k and R_k are $n \times n$ and $m \times m$ positive definite matrices for all $k \in \mathbb{N}$.

Note that, unlike earlier sections we *start* with a measure P under which the x_k and y_k processes are independent random variables, that is, P is already playing the role of $\overline{P}$. The Λ we define below, therefore, plays the role of $\overline{\Lambda}$. A measure P^u is then defined, so that under P^u the x_k and y_k processes satisfy the dynamics in (5.1) and (5.2).

Let U be a nonempty subset of $\mathbb{R}^p$. $\mathscr{Y}_0$ is the trivial σ-field and for $k \geq 1$ $\mathscr{Y}_k = \sigma\{y_\ell, \ell \leq k\}$ is the complete filtration generated by y. The admissible controls u are the set of U-valued $\{\mathscr{Y}_k\}$-adapted processes, that is an admissible control u is a sequence $(u_0, \ldots, u_k, \ldots)$ where u_k is $\mathscr{Y}_k$ measurable.

Write $U(k, \ell)$ for the set of such control processes defined on the interval $k, \ldots, \ell$.

Consider the following functions

$$\begin{aligned} A_k(u) :&\quad U \times \mathbb{N} \to L(\mathbb{R}^n, \mathbb{R}^n) \qquad \text{(the space of } n \times n \text{ matrices)} \\ B_k(u) :&\quad U \times \mathbb{N} \to \mathbb{R}^n \\ C_k(u) :&\quad U \times \mathbb{N} \to L(\mathbb{R}^m, \mathbb{R}^m). \end{aligned}$$

In the sequel, for any admissible control u, we shall write $A_k(u)$ for $A_k(u_k)$, and so on.

Define

$$\begin{aligned} \Lambda^u_{k+1} &= \Lambda_{k+1} \\ &= \prod_{\ell=0}^{k} \frac{\psi_{\ell+1}(x_{\ell+1} - A_\ell(u)x_\ell - B_\ell(u))\,\phi_{\ell+1}(y_{\ell+1} - C_\ell(u)x_{\ell+1}(u))}{\psi_{\ell+1}(x_{\ell+1})\,\phi_{\ell+1}(y_{\ell+1})}. \end{aligned}$$

Then Λ_k is an $\mathscr{G}_k$ martingale and $E[\Lambda_k] = 1$. A new probability measure can be defined by putting $\left(dP^u/dP\right)|_{\mathscr{G}_k} = \Lambda_k$.

Define the processes $v_k = v^u_k$ and $w_k = w^u_k$ by

$$\begin{aligned} v_{k+1} &= x_{k+1} - A_k(u_k)x_k - B_k(u_k), \\ w_k &= y_k - C_k(u_k)x_k. \end{aligned}$$

Then under P^u, v_k and w_k are two sequences of independent, normally distributed random variables with densities ϕ_k and ψ_k, respectively. Therefore, under P^u

$$x_{k+1} = A_k(u_k)x_k + B_k(u_k) + v_{k+1} \tag{5.1}$$
$$y_k = C_k(u_k)x_k + w_k. \tag{5.2}$$

Cost

Consider the following mappings

$$\begin{aligned} &M(.) : U \to L(\mathbb{R}^n, \mathbb{R}^n) \qquad \text{(the space of } n \times n \text{ matrices)} \\ &m(.) : U \to \mathbb{R}^n \\ &N(.) : U \to \mathbb{R} \\ &\Phi(.) : \mathbb{R}^n \to \mathbb{R}. \end{aligned}$$

As in Section 11.2, for any admissible control u and real number θ we consider the expected exponential risk sensitive cost

$$\begin{aligned} V_1(u) &= \theta E^u D_r^u \exp(\Phi(x_T)) \\ &= \theta E\left[\Lambda_T D_r^u \exp(\Phi(x_T))\right], \end{aligned} \tag{5.3}$$

where

$$D_{0,k}^u = D_k^u = \exp\left(\theta \sum_{\ell=1}^{k} \left[\langle M(u_{\ell-1})x_\ell, x_\ell\rangle + \langle m(u_{\ell-1}), x_\ell\rangle + N(u_{\ell-1})\right]\right). \tag{5.4}$$

To simplify notation, in the sequel we suppress the time index on u.

Note here that θ is playing the role of the $\frac{\mu}{\varepsilon}$ of Section 11.4.

Finite-Dimensional Information States

Notation 5.1 *For any admissible control u consider the measure*

$$q_k^u(x)\,dx := E\left[\Lambda_k^u D_k^u I(x_k \in dx) \mid \mathscr{Y}_k\right]. \tag{5.5}$$

Then q_k^u satisfies the following forward recursion

$$\begin{aligned} q_{k+1}^u(x) = &\frac{\phi_{k+1}(y_{k+1} - C_{k+1}(u)x)}{\phi_{k+1}(y_{k+1})} \\ &\times \exp\{\theta(\langle M(u)x, x\rangle + \langle m(u), x\rangle + N(u))\} \\ &\times \int_{\mathbb{R}^n} \psi_{k+1}(x - A_{k+1}(u)z - B_{k+1}(u))\, q_k(z)\,dz. \end{aligned} \tag{5.6}$$

The linearity of the dynamics and the fact that v_k and w_k are independent and normally distributed implies that $q_k^u(x)$ is an unnormalized normal density which we write as

$$q_k^u(x) = Z_k(u)\exp\left\{-\tfrac{1}{2}(x-\hat{x}_k(u))'\Sigma_k^{-1}(u)(x-\hat{x}_k(u))\right\} \tag{5.7}$$

$\Sigma_k^{-1}(u)$, $\hat{x}_k(u)$ and $Z_k(u)$ are given by the following algebraic recursions:

$$\boxed{\begin{aligned}
&\left(\Sigma_{k+1}^{-1}(u)\hat{x}_{k+1}(u)\right)\\
&\qquad = C'_{k+1}(u)R_{k+1}^{-1}y_{k+1} + Q_{k+1}^{-1}A'_k(u)\overline{\Sigma}_k(u)\left(\Sigma_k^{-1}(u)\hat{x}_k(u)\right)\\
&\qquad\quad + \left(I - Q_{k+1}^{-1}A_k(u)\overline{\Sigma}_k(u)A'_k(u)\right)Q_{k+1}^{-1}B_k(u) + \theta m(u),\\
&\Sigma_{k+1}^{-1}(u) = -2\theta M(u) + C'_{k+1}(u)R_{k+1}^{-1}C_{k+1}(u)\\
&\qquad\quad + Q_{k+1}^{-1}\left\{I - A_k(u)\overline{\Sigma}_k(u)A'_k(u)Q_{k+1}^{-1}\right\},\\
&Z_{k+1}(u) = Z_k(u)\left|Q_{k+1}\right|^{-\frac{1}{2}}\left|\overline{\Sigma}_k(u)\right|^{1/2}\\
&\qquad\quad \times\exp\left(-\frac{1}{2}\left\{b_k(u) - \hat{x}'_{k+1}(u)\Sigma_{k+1}^{-1}(u)\hat{x}_{k+1}(u)\right\}\right).
\end{aligned}} \tag{5.8}$$

Here

$$\begin{aligned}
\overline{\Sigma}_k(u) &= \left[A'_k(u)Q_{k+1}^{-1}A_k(u) + \Sigma_k^{-1}(u)\right]^{-1}, \text{ is a symmetric matrix}\\
b_k(u) &= -2\theta N(u) + B'_k(u)Q_{k+1}^{-1}\left\{I - A_k(u)\overline{\Sigma}_k A'_k(u)Q_{k+1}^{-1}\right\}B_k(u)\\
&\quad + \hat{x}'_k(u)\Sigma_k^{-1}(u)\left\{I - \overline{\Sigma}_k(u)\right\}\Sigma_k^{-1}(u)\hat{x}_k\\
&\quad - 2B'_k(u)Q_{k+1}^{-1}A_k(u)\overline{\Sigma}_k(u)\Sigma_k^{-1}(u)\hat{x}_k(u), \text{ is a scalar}
\end{aligned}$$

so exhibiting the recursive nature of (5.8). The symbol $|.|$ denotes the determinant of a matrix.

The recursions for $\hat{x}$, Σ and Z are algebraic and involve no integration.

For this partially observed stochastic control problem the information state $q_k^u(x)$ (which is, in general, a measure-valued process) is determined by the three finite-dimensional parameters $\hat{x}_k(u)$, $\Sigma_k(u)$, and $Z_k(u)$. These parameters can be considered as the state ξ of the process: $\xi_k^u = (\hat{x}_k(u), \Sigma_k(u), Z_k(u))$, and we can write

$$q_k^u(x) = q_k(\xi_k^u, x) = Z_k(u)\exp\left\{-\tfrac{1}{2}(x-\hat{x}_k(u))'\Sigma_k^{-1}(u)(x-\hat{x}_k(u))\right\}.$$

For integrable $f(x)$ write

$$\langle q_k(\xi_k^u), f\rangle = \int_{\mathbb{R}^n} q_k(\xi_k^u, x)f(x)\,dx,$$

which in fact equals $E\left[\Lambda_k^u D_{0,k}^u f(x_k) \mid \mathscr{Y}_k\right]$.

A Separation Principle

For any admissible control u, we saw that the expected total cost is

$$\begin{aligned}
V_1(u) &= E\left[\Lambda_T^u \exp\left(\theta\left\{\sum_{k=1}^{T}\Big[\langle M(u)x_k, x_k\rangle + \langle m(u), x_k\rangle + N(u)\Big] + \Phi(x_T)\right\}\right)\right] \\
&= E\left[\Lambda_T^u D_{0,T}^u \exp(\theta\Phi(x_T))\right] \\
&= E\left[E\left[\Lambda_T^u D_{0,T}^u \exp(\theta\Phi(x_T)) \mid \mathscr{Y}_T\right]\right] \\
&= E\left[\int_{\mathbb{R}^n} \exp(\theta\Phi(x))\, q_T(x)\, dx\right] \\
&= E\left[\langle q_T(\xi_T^u, x), \exp(\theta\Phi(x))\rangle\right].
\end{aligned}$$

Adjoint Process

Now for any k, $0 < k < T$ write $\Lambda_{k,T} = \prod_{k=k}^{T} \lambda_k$,

$$\begin{aligned}
V_1(u) &= E\left[\Lambda_k^u \Lambda_{k+1,T}^u D_{0,k}^u D_{k+1,T}^u \exp(\theta\Phi(x_T))\right] \\
&= E\left[\Lambda_k^u D_{0,k}^u E\left[\Lambda_{k+1,T}^u D_{k+1,T}^u \exp(\theta\Phi(x_T)) \mid x_0, \dots, x_k, \mathscr{Y}_T\right] \mid \mathscr{Y}_T\right].
\end{aligned}$$

Write $\beta_k^u(x_k) = E\left[\Lambda_{k+1,T}^u D_{k+1,T}^u \exp(\theta\Phi(x_T)) \mid x_k, \mathscr{Y}_T\right]$ where, using the Markov property of x, the conditioning involves only x_k. Note that $\beta_T(x_T) = \exp\Phi(x_T)$. Therefore

$$\begin{aligned}
V_1(u) &= E\left[\Lambda_k^u D_{0,k}^u \beta_k^u(x_k)\right] \\
&= E\left[E\left[\Lambda_k^u D_{0,k}^u \beta_k^u(x_k) \mid \mathscr{Y}_k\right]\right] \\
&= E\left[\langle q_k(\xi_k^u), \beta_k^u\rangle\right].
\end{aligned}$$

Note this decomposition is independent of k, so

$$\begin{aligned}
V_1(u) &= E\left[\langle \pi_0, \beta_0^u\rangle\right] \\
&= E\left[\langle q_T(\xi_T^u, x), \exp(\theta\Phi(x))\rangle\right].
\end{aligned}$$

Lemma 5.2 *We have the following backward recursion for the process* β

$$\begin{aligned}
\beta_k^u(x_k) = \int_{\mathbb{R}^n} \Bigg[& \frac{\phi_{k+1}(y_{k+1} - C_{k+1}(u)x)}{\phi_{k+1}(y_{k+1})} \psi_{k+1}(x - A_k(u)x_k - B_k(u)) \\
& \times \beta_{k+1}^u(x) \exp\left(\theta\left\{x' M(u)x + \langle m(u), x\rangle + N(u)\right\}\right)\Bigg] dx. \qquad (5.9)
\end{aligned}$$

Proof

$$\begin{aligned}
\beta_k^u(x_k) &= E\left[\Lambda_{k+1,T}^u D_{k+1,T}^u \exp\theta\Phi(x_T) \mid x_k, \mathscr{Y}_T\right] \\
&= E\left[E\left[\frac{\phi_{k+1}(y_{k+1}-C_{k+1}(u)x_{k+1})\,\psi_{k+1}(x_{k+1}-A_k(u)x_k-B_k(u))}{\phi_{k+1}(y_{k+1})\,\psi_{k+1}(x_{k+1})}\right.\right. \\
&\qquad \times \exp\left(\theta\left\{x'_{k+1}M(u)x_{k+1}+\langle m(u),x_{k+1}\rangle+N(u)\right\}\right) \\
&\qquad \left.\left.\times \Lambda_{k+2,T}^u D_{k+2,T}^u \exp(\theta\Phi(x_T)) \,\middle|\, x_k, x_{k+1}, \mathscr{Y}_T\right] \,\middle|\, x_k, \mathscr{Y}_T\right] \\
&= E\left[\frac{\phi_{k+1}(y_{k+1}-C_{k+1}(u)x_{k+1})\,\psi_{k+1}(x_{k+1}-A_k(u)x_k-B_k(u))}{\phi_{k+1}(y_{k+1})\,\psi_{k+1}(x_{k+1})}\right. \\
&\qquad \times \beta_{k+1}^u(x_{k+1}) \\
&\qquad \left.\times \exp\left(\theta\left\{x'_{k+1}M(u)x_{k+1}+\langle m(u),x_{k+1}\rangle+N(u)\right\}\right) \,\middle|\, x_k, \mathscr{Y}_T\right].
\end{aligned}$$

Integrating with respect to the density of x_{k+1} and using the independence assumption under P gives the result. □

Consider the unnormalized Gaussian densities $\tilde{\beta}_k^u(x,z)$ given by

$$\tilde{\beta}_k^u(x,z) = \tilde{Z}_k^u \exp\left(-\tfrac{1}{2}(x-\gamma_k(z))' S_k^{-1}(x-\gamma_k(z))\right).$$

We put $\tilde{\beta}_T^u(x,z) = \delta(x-z)$, $S_T = 0$, and $\gamma_T = z$. Then γ_k, S_k^{-1} and $\tilde{Z}_k$ are given by the following backward recursions.

$$\boxed{\begin{aligned}
S_k^{-1} &= A'_k\left[Q_{k+1}^{-1} - Q_{k+1}^{-1}\overline{S}_{k+1}Q_{k+1}^{-1}\right]A_k, \\
(S_k^{-1}\gamma_k) &= A'_k Q_{k+1}^{-1}\overline{S}_{k+1}\Big[C'_{k+1}R_{k+1}^{-1}y_{k+1} + Q_{k+1}^{-1}B_k \\
&\qquad + \theta m + (S_{k+1}^{-1}\gamma_{k+1}) - \overline{S}_{k+1}^{-1}B_k\Big], \\
\tilde{Z}_k &= \tilde{Z}_{k+1}\left|Q_{k+1}\right|^{-1/2}\left|\overline{S}_{k+1}\right|^{1/2} \\
&\qquad \times \exp\left[-\tfrac{1}{2}\left\{d_{k+1} - \gamma'_{k+1}S_{k+1}^{-1}\gamma_{k+1}\right\}\right].
\end{aligned}} \tag{5.10}$$

Here

$$\boxed{\begin{aligned}
\overline{S}_{k+1} &= \left[C'_{k+1}R_{k+1}^{-1}C_{k+1} + Q_{k+1}^{-1} + S_{k+1}^{-1}\right]^{-1}, \\
d_{k+1} &= 2y'_{k+1}R_{k+1}^{-1}C_{k+1}\overline{S}_{k+1} \\
&\qquad \times \left[C'_{k+1}R_{k+1}^{-1}y_{k+1} + Q_{k+1}^{-1}B_k + S_{k+1}^{-1}\gamma_{k+1} + \theta m\right] \\
&\qquad + 2\gamma'_k\Sigma_{k+1}^{-1}\overline{\Sigma}_{k+1}\left[Q_{k+1}^{-1}B_k + S_{k+1}^{-1}\gamma_{k+1} + \theta m\right] \\
&\qquad + B'_k\left[Q_{k+1}^{-1} + Q_{k+1}^{-1}\overline{S}_{k+1}Q_{k+1}^{-1}\right]B_k + \theta^2 m'\overline{S}_{k+1}m - 2\theta N
\end{aligned}}$$

again exhibiting the recursive nature of (5.10). Furthermore,

$$\beta_k^u(x) = \int_{\mathbb{R}^n} \tilde{\beta}_k^u(x,z)\exp(\theta\Phi(z))\,dz.$$

Remarks 5.3 β_k^u is the adjoint process and again it is determined by the finite-dimensional parameters γ, S and Z which satisfy the reverse-time, algebraic, recursions (5.10).

■

Dynamic Programming

We have noted that the information state $q_k(x)$ is determined by the finite dimensional parameters

$$\xi_k^u = (\hat{x}_k(u), \Sigma_k(u), Z_k(u)).$$

Given ξ_k^u a control u_k and the new observation y_{k+1}, the Equations (5.8) determine the next value

$$\xi_{k+1}^u = \xi_{k+1}(\xi_k^u, u_k, y_{k+1}).$$

Suppose at some intermediate time k, $0 < k < T$, the information state ξ_k is $\xi = (\hat{x}, \Sigma, Z)$.

The value function for this control problem is

$$V(k,\xi) = \inf_{u \in U(k,T-1)} E\left[\langle q_k^u, \beta_k^u\rangle \mid q_k = q_k(\xi)\right].$$

We can now establish the analog of Theorems 2.5 and 2.6.

Theorem 5.4 *The value function satisfies the recursion:*

$$V(k,\xi) = \inf_{u \in U(k,k)} E\left[k+1, V(\xi_{k+1}(\xi,u,y_{k+1}))\right] \tag{5.11}$$

and $V(T,\xi) = \theta\langle q_T(\xi), \exp\theta\Phi\rangle$.

Proof Now

$$V(k,\xi) = \inf_{u \in U(k,T-1)} E\left[\langle q_k^u(\xi), \beta_k^u\rangle \mid \xi_k = \xi\right].$$

From (5.10) β_k is given by a backward recursion from β_{k+1}, that is, we can write $\beta_k = \beta_k^u(\beta_{k+1}^v)$. Therefore,

$$\begin{aligned}
V(k,\xi) &= \inf_{u \in U(k,k)} \inf_{v \in U(k+1,T)} E\left[\left\langle q_k^u(\xi), \beta_k^u(\beta_{k+1}^v)\right\rangle \mid \xi_k = \xi\right] \\
&= \inf_{u \in U(k,k)} \inf_{v \in U(k+1,T)} E\left[E\left[\left\langle q_{k+1}(\xi_{k+1}^u), \beta_{k+1}^v\right\rangle \mid \mathcal{Y}_{k+1}, \xi_k = \xi\right] \mid \xi_k = \xi\right] \\
&= \inf_{u \in U(k,k)} E\left[\inf_{v \in U(k+1,T)} E\left[\left\langle q_{k+1}(\xi_{k+1}^u), \beta_{k+1}^v\right\rangle \mid \mathcal{Y}_{k+1}, \xi_k = \xi\right] \,\middle|\, \xi_k = \xi\right] \\
&= \inf_{u \in U(k,k)} E\left[V(k+1, \xi_{k+1}(\xi,u,y_{k+1}))\right].
\end{aligned}$$

The interchange of conditional expectation and minimization is justified by application of the lattice property of the controls (Elliott 1982b). □

Write $U_s(k,k)$ for the set of control processes on the time interval $k,\dots,k$ which are adapted to the filtration $\sigma\{\xi_j : k \le j \le k\}$. We call such controls separated.

Theorem 5.5 (Verification) *Suppose $u^* \in U_s(0,T-1)$ is a control which for each $k=0,\dots,T-1$, $u_k^*(\xi_k)$ achieves the minimum in* (5.11). *Then $u^* \in U(0,T-1)$ and is an optimal control.*

Proof Write

$$\overline{V}(k,\xi;u) = E\left[\langle q_k^u(\xi), \beta_k^u\rangle \mid \xi_k = \xi\right].$$

We shall show that

$$V(k,\xi) = \overline{V}(k,\xi;u^*), \qquad \text{for each } k=0,\dots,T. \tag{5.12}$$

For $k=T$ (5.12) is clearly satisfied. Suppose (5.12) is true for $k+1,\dots,T$. Then

$$\begin{aligned}
\overline{V}(k,\xi;u^*) &= E\left[E\left[\left\langle q_{k+1}(\xi_{k+1}^{u^*}), \beta_{k+1}^{u^*}\right\rangle \mid \xi_k = \xi, \mathscr{Y}_{k+1}\right] \mid \xi_k = \xi\right] \\
&= E\left[\overline{V}\left(\xi_{k+1}^{u^*}(\xi), k+1, u^* \in U(k+1,T-1)\right)\right] \\
&= E\left[V\left(k+1, \xi_{k+1}^{u^*}(\xi)\right)\right] \\
&= V(k,\xi).
\end{aligned}$$

This gives (5.12).

Putting $k=0$ we see

$$\overline{V}(0,\xi;u^*) = V(0,\xi) \le \overline{V}(0,\xi;u)$$

for any $u \in U(0,T-1)$. That is, u^* is an optimal control. □

Remark 5.6 This result shows the optimal policy u^* for the risk-sensitive control problem is a separated policy, in that it is a function of the information state ξ. ■

11.6 Risk-Sensitive LQG Control

In this section we present the case of discrete-time risk-sensitive linear-quadratic-Gaussian (LQG) control. This is a specialization of the previous section, to the case where the signal model is linear in both the control and state variables; further, the quadratic exponential cost is risk sensitive with weights independent of the control. This is further specialization presented to achieve a controller that is finite-dimensional in both the dynamic programming solution as well as the information state. For notational simplicity the signal model and index matrices are assumed time invariant.

As the risk-sensitive parameter θ approaches 0 the problem reduces to the classical LQG situation.

Measure Change

The situation is a specialization of that of Section 11.5. We again initially consider a measure P under which there are two sequences of independent random variables, x_k and y_k. These take values in $\mathbb{R}^n$ and $\mathbb{R}^m$, and have densities $\phi = N(0,Q)$ and $\psi = N(0,R)$, respectively. The control parameter u takes values in the real interval $U \subset \mathbb{R}$.

Let A be in $\mathbb{R}^n \times \mathbb{R}^n$, B be in $\mathbb{R}^n$, and C be in $\mathbb{R}^m \times \mathbb{R}^m$. Now, for any control u, define

$$\Lambda_{k+1} = \prod_{k=0}^{k} \frac{\psi(x_{k+1} - Ax_k - Bu_k)\,\phi(y_{k+1} - Cx_{k+1})}{\psi(x_{k+1})\,\phi(y_{k+1})}.$$

Then Λ_k is an $\mathscr{B}_k$ martingale and $E[\Lambda_k] = 1$. A new probability measure can be defined by putting $(dP^u/dP)|_{\mathscr{B}_k} = \Lambda_k$.

Define the processes $v_k = v_k^u$ and $w_k = w_k^u$ by

$$\begin{aligned} v_{k+1} &= x_{k+1} - Ax_k - Bu_k, \\ w_k &= y_k - Cx_k. \end{aligned}$$

Then under P^u, v_k and w_k are two sequences of independent, normally distributed random variables with densities $\phi = N(0,Q)$ and $\psi = N(0,R)$, respectively. Therefore, under P^u, we have the linear system as signal model.

Signal Model

$$x_{k+1} = Ax_k + Bu_k + v_{k+1}, \tag{6.1}$$

$$y_k = Cx_k + w_k. \tag{6.2}$$

Cost

Consider the following specific case of the cost function defined in Section 11.5.

$$V_1(u) = E\left[\Lambda_T \exp\left\{\theta \sum_{k=1}^{T}\left[x_k' M x_k + m' x_k + u_{k-1}' N u_{k-1}\right] + \Phi(x_T)\right\}\right]. \tag{6.3}$$

Finite-Dimensional Information States

The information state is

$$q_k^u(x)\,dx = Z_k(u)\exp\left\{-\tfrac{1}{2}(x - \hat{x}_k(u))'\Sigma_k^{-1}(x - \hat{x}_k(u))\right\}. \tag{6.4}$$

Theorem 6.1 *$\hat{x}_k(u)$, Σ_k^{-1} and $Z_k(u)$ are given by the following algebraic recursions:*

$$
\begin{aligned}
\left(\Sigma_{k+1}^{-1}\hat{x}_{k+1}(u)\right) &= C'R^{-1}y_{k+1} + \theta m + Q^{-1}A\bar{\Sigma}_k\left(\Sigma_k^{-1}\hat{x}_k(u)\right) \\
&\quad + \left(I - Q^{-1}A\bar{\Sigma}_k A'\right)Q^{-1}Bu_k, \\
\Sigma_{k+1}^{-1} &= -2\theta M + C'R^{-1}C + \left(I - Q^{-1}A\bar{\Sigma}_k A'\right)Q^{-1}, \\
Z_{k+1}(u) &= Z_k(u)\,|Q|^{-1/2}\left|\bar{\Sigma}_k\right|^{1/2} \\
&\quad \times \exp\left(-\tfrac{1}{2}\left[b_k(u) - \hat{x}_{k+1}'(u)\Sigma_{k+1}^{-1}\hat{x}_{k+1}(u)\right]\right),
\end{aligned}
\tag{6.5}
$$

where

$$
\begin{aligned}
\bar{\Sigma}_k &= \left(A'Q^{-1}A + \Sigma_k^{-1}\right)^{-1}, \\
b_k(u) &= -2\theta u_k'\left(N + B'Q^{-1}B\right)u_k + \hat{x}_k'(u)\Sigma_k^{-1}\hat{x}_k(u) \\
&\quad - \left(u_k'B'Q^{-1}A - \hat{x}_k'(u)\Sigma_k^{-1}\right)\bar{\Sigma}_k\left(u_k'B'Q^{-1}A - \hat{x}_k'(u)\Sigma_k^{-1}\right)'.
\end{aligned}
$$

Note Σ_k is independent of u.

Dynamic Programming

Theorem 6.2 *The control law minimizing* (6.3), *for the system* (6.1)–(6.2), *is given by the information-state estimate feedback law*

$$
u_k^* = -K_k^a\Sigma_k^{-1}\hat{x}_k - K_k^d, \tag{6.6}
$$

where

$$
\begin{aligned}
K_k^a &= \sigma_k^{-1}\left[h_k'S_{k+1}^a\left[I + f_kS_{k+1}^a\right] - B'\right]d_k, \\
K_k^d &= \sigma_k^{-1}h_k'\left(I + S_{k+1}^a f_k\right)\left(\theta S_{k+1}^a m + S_{k+1}^{b\prime}\right); \\
S_k^a &= \bar{\Sigma}_k + d_k'S_{k+1}^a\left[I + f_kS_{k+1}^a\right]d_k - K_k^{a\prime}\sigma_kK_k^a, \\
S_k^d &= \left(\theta m'S_{k+1}^a + S_{k+1}^d\right)\left[I + f_kS_{k+1}^a\right]d_k - K_k^{b\prime}\sigma_kK_k^a,
\end{aligned}
\tag{6.7}
$$

and

$$
\begin{aligned}
d_k &= Q^{-1}A\bar{\Sigma}_k, \\
f_k &= C'\left(R - CS_{k+1}^aC'\right)^{-1}C, \\
h_k &= \left(I - Q^{-1}A\bar{\Sigma}_kA'\right)Q^{-1}B, \\
\sigma_k &= 2\theta N - B'\left(I - Q^{-1}A\bar{\Sigma}_kA'\right)Q^{-1}B + h_k'\left(S_{k+1}^a\left[I + f_kS_{k+1}^a\right]\right)h_k.
\end{aligned}
$$

Proof Complete the square. □

Tracking

In this section we present results for the case of tracking a desired trajectory $\hat{y}$ in the risk-sensitive LQG framework. We consider the linear dynamics given in Section 11.6, and define a new cost function for the tracking problem:

$$V_1(u) = E\left[\Lambda_T \exp\left\{\theta \sum_{k=1}^{T} \left[x_k' M x_k + (y_k - \tilde{y}_k)' \tilde{M} (y_k - \tilde{y}_k) + u_{k-1}' N u_{k-1}\right] + \Phi(x_T)\right\}\right]. \tag{6.8}$$

The relevant variation of Theorem 6.1 is now:

Theorem 6.3 *The parametrizations of the information state* (6.8), *viz.* $\hat{x}_k(u)$, R_k^{-1} *and* $Z_k(u)$ *are given by the following algebraic recursions:*

$$\boxed{\begin{aligned}
\left(\Sigma_{k+1}^{-1}\hat{x}_{k+1}(u)\right) &= C'R^{-1}y_{k+1} + Q^{-1}A\overline{\Sigma}_k\left(\Sigma_k^{-1}\hat{x}_k(u)\right) \\
&\quad + \left(I - Q^{-1}A\overline{\Sigma}_k A'\right)Q^{-1}Bu_k, \\
\Sigma_{k+1}^{-1} &= -2\theta M + C'R^{-1}C + \left(I - Q^{-1}A\overline{\Sigma}_k A'\right)Q^{-1}, \\
Z_{k+1}(u) &= Z_k(u)\left|Q\right|^{-1/2}\left|\overline{\Sigma}_k\right|^{1/2} \\
&\quad \times \exp\left(-\frac{1}{2}\left[b_k(u) - \hat{x}_{k+1}'(u)_{k+1}^{-1}\Sigma_{k+1}^{-1}\hat{x}_{k+1}(u)\right]\right),
\end{aligned}} \tag{6.9}$$

where

$$\boxed{\begin{aligned}
\overline{\Sigma}_k &= \left(A'Q^{-1}A + \Sigma_k^{-1}\right)^{-1}, \\
b_k(u) &= -2\theta u_k'\left(N + B'Q^{-1}B\right)u_k + \hat{x}_k'(u)\Sigma_k^{-1}\hat{x}_k(u) \\
&\quad - \left(u_k'B'Q^{-1}A - \hat{x}_k'(u)\Sigma_k^{-1}\right)\overline{\Sigma}_k\left(u_k'B'Q^{-1}A - \hat{x}_k'(u)\Sigma_k^{-1}\right)' \\
&\quad - 2\theta\left(y_{k+1} - \tilde{y}_{k+1}\right)'\tilde{M}\left(y_{k+1} - \tilde{y}_{k+1}\right).
\end{aligned}}$$

Dynamic Programming

Theorem 6.4 *The control law minimizing* (6.8) *for the system* (6.2) *is given by*

$$\boxed{u_k^* = -K_k^a\Sigma_k^{-1}\hat{x}_k - K_k^d,} \tag{6.10}$$

where

$$\boxed{\begin{aligned}
K_k^a &= \sigma_k^{-1}\left[h_k' S_{k+1}^a\left[I + f_k S_{k+1}^a\right] - B'\right] d_k,\\
K_k^d &= \sigma_k^{-1}\left[h_k'\left(I + S_{k+1}^a f_k\right) S_{k+1}^{b\prime} - 2\theta h_k' S_{k+1}^a C' R^{-1}\delta_k^{-1}\tilde{M}\right];\\
S_k^a &= \overline{\Sigma}_k + d_k' S_{k+1}^a\left[I + f_k S_{k+1}^a\right] d_k - K_k^{a\prime}\sigma_k K_k^a,\\
S_k^d &= S_{k+1}^d\left[I + f_k S_{k+1}^a\right] d_k - 2\theta\tilde{M}\delta_k^{-1} R^{-1} C S_{k+1}^a d_k - K_k^{b\prime}\sigma_k K_k^a,
\end{aligned}} \tag{6.11}$$

and

$$\boxed{\begin{aligned}
d_k &= Q^{-1} A\overline{\Sigma}_k,\\
f_k &= C'\left(R\delta_k^{-1} R\right)^{-1} C,\\
h_k &= \left(I - Q^{-1} A\overline{\Sigma}_k A'\right) Q^{-1} B,\\
\sigma_k &= 2\theta N - B'\left(I - Q^{-1} A\overline{\Sigma}_k A'\right) Q^{-1} B\\
&\qquad + h_k'\left(S_{k+1}^a\left[I + f_k S_{k+1}^a\right]\right) h_k,\\
\delta_k &= \left(R^{-1} - R^{-1} C S_{k+1}^a C' R^{-1} - 2\theta\tilde{M}\right).
\end{aligned}}$$

11.7 Problems and Notes

Problems

1. Write the proof of Theorem 2.2.
2. Write the proof of the verification Theorem 2.6.
3. Derive the algebraic forward and backward recursions (5.8) and (5.10).
4. Work out the results of Section 11.5 on the finite-dimensional case replacing the dynamics of x (5.1) by

$$x_{k+1} = A_k(u)\, x_k + B_k(u) + w_{k+1},$$

   where w is the noise in the observation process y.
5. Complete the proofs of Theorems 6.2 and 6.4.

Notes

The risk-sensitive problem was first formulated and solved in Jacobson (1973), in the linear/quadratic context with complete state information. Also, the connection with dynamic games was made explicit in this paper. The partially observed problem, in the linear/quadratic discrete-time context, was first solved by Whittle (1981), making use of a certainty equivalence principle. The continuous-time linear/quadratic partially observed risk-sensitive problem was observed by Bensoussan and van Schuppen (1985), using an equivalent problem with complete "state" information expressed in terms of a Kalman filter, modified to include the cost function.

The nonlinear partially observed risk-sensitive problem was solved by James (1992). These results form the basis of Sections 11.1–11.3. In addition, a partially observed dynamic game was solved, and related to the risk-sensitive problem via the asymptotic analysis described in Section 11.2.

In these papers, a version of "filtering" is developed for partially observed deterministic systems in which the "expectation" operation is replaced by a supremum operation. This worst-case, min-max approach can be interpreted as a dynamic game against "nature." It provides a natural framework for robust or H^∞ control.

The nonlinear partially observed risk-sensitive problem were also considered by Whittle (1990a) (and an approximate certainty equivalence principle when the noise is small). A good general reference for risk-sensitive stochastic control is Whittle (1991).

Bensoussan and Elliott (1995) discussed a finite-dimensional risk-sensitive problem with quasi-linear dynamics and cost; the results of Section 11.5 are a discrete-time version of their results. A specialization of these results gives the partially observed problem in the linear exponential quadratic discrete-time context. These results are derived in Collings, Moore and James (1994) and presented in Section 11.6 for comparison with the work of Whittle (1981). Risk-sensitive filtering problems have been studied in Dey and Moore (1995a,b), and Moore, Elliott and Dey (1995).

Chapter 12
Continuous-Time HMM Control

12.1 Introduction

In this chapter the control of HMM in continuous time is discussed. The first situation considered is that of a continuous-time finite-state Markov chain which is not observed directly; rather, there is a conditional Poisson process whose rate, or intensity, is a function of the Markov chain. An equation for the unnormalized distribution q of the Markov chain, given the observations, is derived. The control problem is then formulated in separated form with q as the new state variable. A minimum principle is obtained.

Secondly, the case when the counting observation process is related to the jumps of the Markov chain is discussed. In this situation there is non-zero correlation between jumps in the observation process and jumps of the Markov chain, so new features arise. However, again a minimum principle is obtained together with explicit formulae for the adjoint process.

The optimal control of a Markov chain observed in Gaussian noise is treated in Section 12.3. Equations for the adjoint process are again derived.

The chapter closes with a discussion of the optimal control of a linear stochastic system whose coefficients are functions of a discrete-state Markov chain.

12.2 Robust Control of a Partially Observed Markov Chain

Introduction

In a seminal paper, (Clark 1978), Martin Clark showed how the filtered dynamics giving the optimal estimate of a Markov chain observed in Gaussian noise can be expressed using an ordinary differential equation. The method uses a gauge transformation and, in effect, solves the Wonham-Zakai equation using variation of constants. Such results offer substantial benefits in filtering and in control, often simplifying the analysis and in some settings providing numerical benefits; see, for example, Malcolm, Elliott and van der Hoek (to appear). These results have not hitherto been exploited for the optimal control of such processes, including risk sensitive control..

Here, we consider the optimal control of a partially observed Markov chain. The innovation in the discussion here is that the robust information state filtering dynamics of Clark are used. This gives rise to a simplified and more useful adjoint process for a simplified minimum principle. Also, simplified dynamic programming partial differential equations are developed. The results of the paper apply to signal models with observations observed through Brownian motion and/or counting processes. The new results also apply to the risk sensitive cost functions of filtering and control.

Background material for stochastic control of partially observed Markov chains optimizing terminal costs can be found in (Elliott 1976, 1992; Elliott and Moore 1992a). There is also an extensive literature which deals with optimal stochastic control of partially observed Markov processes. Many of the techniques developed in this literature apply to the Markov chain case here, but in more general settings Fubini's Theorem may not apply as here, so the results here are of independent interest since they often 'go further' than in a more general setting. Examples of this more general literature are (Bismut 1978a; Davis 1979; Fleming 1969; Fleming and Pardoux 1982; Haussmann 1981b; Fleming and Rischel 1975; Fleming and Soner 1975; Wong and Hajek 1985). For risk sensitive estimation and control, a number of results are developed in (Dey and Moore 1995c; James and Elliott 1996; Malcolm et al. to appear), but these fall short of exploiting the robust state estimates of Clark in the optimal control formulations.

The HMM model and Wonham-Zakai filter

States

Suppose the state process $X = \{X_u | 0 \le u \le t\}$ is a finite state Markov process. Without loss of generality the state space of X can be taken to be the set of unit vectors $\{e_1, e_2, \dots, e_n\}$, $e_i = (0, \dots, 1, 0, \dots, 0)' \in \mathbb{R}^n$. A key property is that any nonlinear function of X is linear in X as $f(X) = \langle f, X \rangle$. Here $f = [f(e_1), \dots, f(e_n)]'$, prime denotes transposition and $\langle \, , \rangle$ denotes the scalar product in $\mathbb{R}^n$.

Suppose X is defined on a probability space $(\Omega, \mathscr{F}, P)$ and write $p_t^i \triangleq P(X_t = e_i)$. The set of control parameters is a convex set $U \subset \mathbb{R}^K$. Suppose there is a time dependent family of generators A_u^t, which also depend on the control parameter $u_t \in U$, so that the probability distribution $p_t = \left(p_t^1, p_t^2, \dots, p_t^n\right)'$ satisfies the Kolmogorov forward equation $\dfrac{dp_t}{dt} = A_t^u p_t$.

We suppose the so-called rate matrix A is measurable in $t \in [0, T]$ and differentiable in $u \in U$.

Observations

The observation process $y = \{y_s | s \ge 0\}$ is a process of the form

$$y_t = \int_0^t H_s X_s ds + b_t \in \mathbb{R}^m,$$

where the observations are taken, without loss of generality, as linear in X_s. Here the components of the observations are independent and are either observed through Brownian motion or are a counting process. Write

$$\begin{aligned}\mathscr{F}_t &:= \sigma\{X_s|0 \le s \le t\},\ \mathscr{Y}_t := \sigma\{y_s|0 \le s \le t\},\\ \mathscr{G}_t &:= \sigma\{X_s, y_s|0 \le s \le t\},\end{aligned} \tag{2.1}$$

where $\{\mathscr{F}_t\}$, $\{\mathscr{Y}_t\}$ and $\{\mathscr{G}_t\}$ are, respectively, the corresponding right continuous, complete filtrations.

Suppose the index set $\{1,2\ldots,m\}$ is decomposed as $S \cup \overline{S}$ where $S := \{1,2\ldots,k\}$ and $\overline{S} := \{k+1,\ldots,m\}$. Then on $(\Omega, \mathscr{F}, P)$, we suppose that for $i \in S$ then $\{b^i|i \in S\}$ is a standard Brownian motion, and for $i \in \overline{S}$, $\{y^i|i \in \overline{S}\}$ is a counting process with compensator $\int_0^t h^i_s X_s ds$ where h^i_s is the ith row of H_s.

That is, for $i \in \overline{S}$ then $b^i_t := y^i_t - \displaystyle\int_0^t h^i_s X_s ds$ is a martingale under P.

Feedback Controls

The set $\mathscr{U}$ of admissible control functions will be $\{\mathscr{Y}_t\}$-predictable functions $u : [0,T] \times \Omega \to U$.

State Equations

For $u \in \mathscr{U}$ suppose the corresponding state process is $X^u = \{X^u_t|0 \le t \le T\}$ where

$$\begin{aligned} dX^u_t &= A^u_t X^u_t dt + dW_t,\\ dy_t &= H_t X^u_t dt + db_t.\end{aligned}$$

Measure Change

Suppose that under a reference probability measure $\overline{P}$, the observations $\{y^i_s|0 \le s \le T, i \in S\}$ are standard Brownian motions, $\{y^i_s|0 \le s \le T, i \in \overline{S}\}$ are standard Poisson processes and the Markov chain X^u has the same dynamics, which are independent of y.

Then under $\overline{P}$, $\{y^i_t|i \in S\}$ and $\{(y^i_t - t)|i \in \overline{S}\}$ are $\{\mathscr{Y}_t\}$ martingales.

For $u \in \mathscr{U}$, define $\Lambda^u = \{\Lambda_t|0 \le t \le T\}$ by:

$$\begin{aligned}\Lambda^u_t = \exp\Big(&\int_0^t \sum_{i\in S} h^i_s X^u_s dy^i_s - \frac{1}{2}\int_0^t \sum_{i\in S} h^i_s h^{i'}_s ds\\ &+ \int_0^t \sum_{i\in\overline{S}} (1 - h^i_s X^u_s) ds\Big) \prod_{i\in\overline{S}} \prod_{0<r\le t} h^i_r X^u_r \triangle y^i_r,\end{aligned}$$

where h^i_s is the ith row of H_s, y^i_s is the ith element of y_s, and $\triangle y^i_r$ represent jumps in y^i_r at discrete times r. Then, using Ito calculus,

$$\Lambda_t^u = 1 + \int_0^t \Lambda_s^u \sum_{i\in S} h_s^i X_s^u dy_s^i + \int_0^t \Lambda_{s-}^u \sum_{i\in \overline{S}} (h_s^i X_s^u - 1) d(y_s^i - s),$$

$$d\Lambda_t^u = \Lambda_t^u \sum_{i\in S} h_t^i X_t^u dy_t^i + \Lambda_{t-}^u \sum_{i\in \overline{S}} (h_{t-}^i X_{t-}^u - 1) d(y_t^i - t).$$

Therefore $\Lambda^u \geq 0$ is a martingale on $(\Omega, \{\mathscr{Y}_t\}, \overline{P})$, with $\overline{E}\left[\Lambda^u = 1\right]$.

Define a measure P^u, by setting $\frac{dP^u}{d\overline{P}}\Big|_{\mathscr{G}_t} = \Lambda_t^u$. Then, under P^u, we have via Girsanov's Theorem that, $b_t^i = y_t^i - \int_0^t h_s^i X_s^u ds$, is a standard Brownian motion for $\{i \in S\}$, and a martingale for $\{i \in \overline{S}\}$. Further, under P^u, X^u has the same dynamics. From Bayes' Theorem,

$$E^u\left[X_t^u \mid \mathscr{Y}_t\right] = \frac{\overline{E}\left[\Lambda_t^u X_t^u \mid \mathscr{Y}_t\right]}{\overline{E}\left[\Lambda_t^u \mid \mathscr{Y}_t\right]}. \tag{2.2}$$

Information State

The information state is defined as,

$$q_t^u := \overline{E}\left[\Lambda_t^u X_t^u \mid \mathscr{Y}_t\right] \in \mathbb{R}^n.$$

We see that $\langle q_t^u, \underline{1}\rangle = \overline{E}\left[\Lambda_t^u \mid \mathscr{Y}_t\right]$, where $\underline{1} = (1,1,\dots,1)' \in \mathbb{R}^n$.

The Wonham-Zakai filter

For completeness, we include the Wonham-Zakai filter, which along with the Kalman filter, is finite dimensional and optimal for a broad signal model class of interest in many applications. The information state q_t^u is derived via a filter in Wong and Hajek (1985). A verification exploits the fact that W is a martingale, and Fubini's Theorem is applied to condition on $\mathscr{Y}_t$ under $\overline{P}$, where y^i is Brownian or Poisson. The filter is:

$$q_t^u = q_0 + \int_0^t A_s^u q_s^u ds + \int_0^t \sum_{i\in S} H_s^i q_s^u dy_s^i + \int_0^t \sum_{i\in \overline{S}} (H_s^i - I_n) q_{s-}^u (dy_s^i - ds). \tag{2.3}$$

Here each $H_s^i := diag\{h_s^i\}$ is the diagonal matrix comprised of the elements of h_s^i. This equation is a stochastic equation which is not very suitable for implementation. Alternatively, we can consider the associated transition matrix $\Phi_{t,0}^u$, where

$$d\Phi_{t,0}^u = dA_t^u \Phi_{t,0}^u dt + \sum_{i\in S} H_t^i \Phi_{t,0}^u dy_t^i$$

$$+\sum_{i\in\overline{S}}(H_t^i - I_n)\Phi_{t-,0}^u(dy_t^i - dt), \;\; \Phi^u(0,0) = I_n.$$

Then,

$$q_t^u = \Phi_{t,0}^u q_0.$$

Robust Clark filter

We now describe the robust form of the Wonham-Zakai filter (2.3) due to Martin Clark (Clark 1978) for the Brownian noise observations case and Elliott (Elliott 1992) for the Poisson noise case.

A Gauge Transformation Variable

A key is to work with the diagonal matrix $\varepsilon = \left(\varepsilon_s | 0 \le s \le t\right)$,

$$\varepsilon_t := \exp\left(\sum_{i\in S}\left[H_t^i y_t^i - \frac{1}{2}\langle H_t^i, H_t^i\rangle t\right] + \sum_{i\in\overline{S}}\left[(H_t^i - I_n)t + y_t^i \ln(H_t^i)\right]\right).$$

The associated differential, can be calculated using Ito calculus, as

$$d\varepsilon_t = \sum_{i\in S} H_t^i \varepsilon_t dy_t^i + \sum_{i\in\overline{S}}(H_t^i - I_n)\varepsilon_t(dy_t^i - dt).$$

Notice that ε_t is a martingale.

Robust Information State

For $u \in \mathscr{U}$, consider the vector-valued robust information state process $\overline{q}^u = \left\{\overline{q}_t^u | 0 \le t \le T\right\} \in \mathbb{R}^n$, which is defined by the readily implemented ordinary differential equation:

$$\frac{d\overline{q}^u}{dt} = \varepsilon_t^{-1} A_t^u \varepsilon_t \overline{q}_t^u, \quad \overline{q}_0^u = q_0 = p_0. \tag{2.4}$$

Note that the coefficients of the right hand side, $\varepsilon_t^{-1} A_t^u \varepsilon_t \overline{q}_t^u$, are functions of the observations y_t and controls u_t. Moreover, for $u \in \mathscr{U}$, consider the matrix process $\overline{\Phi}^u = \left\{\overline{\Phi}_{t,s}^u | 0 \le s \le t \le T\right\}$ where

$$\frac{d\overline{\Phi}_{t,s}^u}{dt} = \varepsilon_t^{-1} A_t^u \varepsilon_t \overline{\Phi}_{t,s}^u, \quad \overline{\Phi}_{s,s}^u = \underline{I}, \;\; \overline{q}_t^u = \overline{\Phi}_{t,0}^u q_0. \tag{2.5}$$

Consider also the matrix process $\overline{\Psi}_{t,s}^u$ defined by setting

$$\frac{d\overline{\Psi}_{t,s}^u}{dt} = -\overline{\Psi}_{t,s}^u \varepsilon_t^{-1} A_t^u \varepsilon_t, \quad \overline{\Psi}_{s,s}^u = \underline{I}.$$

Then

$$\frac{d}{dt}\left(\overline{\Psi}^u_{t,s}\overline{\Phi}^u_{t,s}\right) = -\overline{\Psi}^u_{t,s}\varepsilon_t^{-1}A^u_t\varepsilon_t\overline{\Phi}^u_{t,s} + \overline{\Psi}^u_{t,s}\varepsilon_t^{-1}A^u_t\varepsilon_t\overline{\Phi}^u_{t,s} = 0.$$

Consequently $\overline{\Phi}^u_{t,s}$ is non-singular and

$$\overline{\Psi}^u_{t,s} = (\overline{\Phi}^u_{t,s})^{-1}. \tag{2.6}$$

Relationship to the Wonham-Zakai filter

By differentiating it is immediately checked that $\varepsilon_t\overline{q}^u_t$ satisfies (2.3), so by uniqueness,

$$q^u_t = \varepsilon_t\overline{q}^u_t, \quad q^u_t = \varepsilon_t\overline{\Phi}^u_{t,0}q_0, \quad \Phi^u_{t,0} = \varepsilon_t\overline{\Phi}^u_{t,0}.$$

Optimal Control Formulation

We now discuss the optimal control using the robust dynamics for $\overline{q}^u_t$ of (2.4). Suppose in the first instance that the cost is an integral cost together with a terminal cost.

Cost Function

For a control $u \in \mathscr{U}$, the expected cost is,

$$J(u) = E^u\left[\langle c, X^u_T\rangle + \int_0^T \langle \underline{g^u_s}, X^u_s\rangle ds\right],$$

where without loss of generality, the terminal cost and kernel of the integral term are linear in the discrete states X.

Cost Function in Terms of Information State

Applying Bayes Theorem (2.2), noting that $\overline{E}\left[\Lambda_t\right] = 1$,

$$\begin{aligned} J(u) &= \overline{E}\left[\langle c, \Lambda^u_T X^u_T\rangle + \int_0^T \langle \underline{g^u_s}, \Lambda^u_s X^u_s\rangle ds\right], \\ &= \overline{E}\left[\langle c, \overline{E}\left[\Lambda^u_T X^u_T \mid \mathscr{Y}_T\right]\rangle + \int_0^T \langle \underline{g^u_s}, \overline{E}\left[\Lambda^u_s X^u_s \mid \mathscr{Y}_s\right]\rangle ds\right], \\ &= \overline{E}\left[\langle c, q^u_T\rangle + \int_0^T \langle \underline{g^u_s}, q^u_s\rangle ds\right]. \end{aligned} \tag{2.7}$$

The control problem is now: Minimize over $u \in \mathscr{U}$ the cost $J(u)$ of (2.7) subject to the dynamics (2.3).

Cost Function in Terms of Robust Information State

Noting that $q_t^u = \varepsilon_t \overline{q}_t^u$, the control task can be reformulated: Minimize over $u \in \mathscr{U}$ the cost

$$J(u) = \overline{E}\left[\langle \varepsilon_T c, \overline{q}_T^u \rangle + \int_0^T \langle \varepsilon_s \underline{g}_s^u, \overline{q}_s^u \rangle ds\right], \tag{2.8}$$

where $\overline{q}_t^u$ satisfies the ordinary differential equation (2.4).

Cost Function in Terms of Augmented Robust Information State

For convenience, we define augmented matrices as follows:

$$\begin{aligned} \widetilde{A}_t^u &:= \begin{bmatrix} A_t^u & 0 \\ \underline{g}_t^{u\prime} & 0 \end{bmatrix}, \quad \widetilde{\varepsilon}_t := \begin{bmatrix} \varepsilon_t & 0 \\ 0 & 1 \end{bmatrix}, \\ \widetilde{H}_t &:= \begin{bmatrix} H_t & 0 \\ 0 & 0 \end{bmatrix}, \quad \widetilde{q}_t^u := \begin{bmatrix} \overline{q}_t^u \\ \overline{\rho}_t^u \end{bmatrix}, \quad \widetilde{c} := \begin{bmatrix} c \\ 1 \end{bmatrix}. \end{aligned} \tag{2.9}$$

Here $\overline{\rho}_t^u := \int_0^t \langle \varepsilon_s \underline{g}_s^u, \overline{q}_s^u \rangle ds$. Our subsequent focus is the control task to minimize over $u \in \mathscr{U}$ the cost function,

$$J(u) = \overline{E}\left[\langle \widetilde{\varepsilon}_T c, \widetilde{q}_T \rangle\right], \tag{2.10}$$

subject to the augmented robust information state differential equation,

$$\frac{d\widetilde{q}_t^u}{dt} = \widetilde{\varepsilon}_t^{-1} \widetilde{A}_t^u \widetilde{\varepsilon}_t \widetilde{q}_t^u, \quad \widetilde{q}_t^u = \widetilde{q}_0, \tag{2.11}$$

with associated transition matrix differential equation,

$$\frac{d\widetilde{\Phi}_{t,s}^u}{dt} = \widetilde{\varepsilon}_t^{-1} \widetilde{A}_t^u \widetilde{\varepsilon}_t \widetilde{\Phi}_{t,s}^u, \quad \widetilde{\Phi}_{s,s}^u = \underline{I}, \quad (\widetilde{\Phi}_{t,s}^u)^{-1} =: \widetilde{\Psi}_{t,s}^u. \tag{2.12}$$

Dynamic Programming in Terms of Robust Information States

We first define a value function:

$$V_t(\widetilde{q}, y) = \inf_{u \in \mathscr{U}} \overline{E}\left[\langle \widetilde{\varepsilon}_T \widetilde{c}, \widetilde{q}_T \rangle | \widetilde{q}_t = q, y_t = y\right]. \tag{2.13}$$

The dynamic programming equation for our control task is:

$$\begin{aligned} &\frac{\partial V_t(\widetilde{q}, y)}{\partial t} + \inf_{u_t \in \mathrm{U}} \left(\frac{\partial \mathrm{V}_t(\widetilde{\mathrm{q}}, \mathrm{y})}{\partial \widetilde{\mathrm{q}}} \widetilde{\mathrm{A}}_t^{\mathrm{u}} \widetilde{\mathrm{q}}_t \right) + \frac{1}{2} \sum_{\mathrm{i,j} \in \mathrm{S}} \frac{\partial^2 \mathrm{V}_t(\widetilde{\mathrm{q}}, \mathrm{y})}{\partial \mathrm{y}^{\mathrm{i}} \mathrm{y}^{\mathrm{j}}} \\ &\quad + \sum_{j \in \overline{S}} [V_t(\widetilde{q}, y + e_j) - V_{t-}(\widetilde{q}, y)] = 0, \; V_T(\widetilde{q}, y) = \langle \widetilde{\varepsilon}(y) \widetilde{c}, \widetilde{q} \rangle. \end{aligned} \tag{2.14}$$

Here e_j is the unit vector in $\mathbb{R}^m$ with unity in the jth entry and zeros elsewhere. This is the Hamilton-Jacobi-Bellman nonlinear second order parabolic partial differential equation, which is solved backwards in time. Minimizing the term $\frac{\partial V_t}{\partial \widetilde{q}_t} \widetilde{A}_t^u \widetilde{q}_t$ gives an optimal control law of the form $u_t^* = f(\widetilde{q}_t, y_t)$, which is a robust information state and output Markov feedback law. Equation (2.14) may not posses a smooth solution, but a solution can be defined in a viscosity sense. It appears more attractive to implement than earlier versions involving the non-robust information states q_t^u and differentiation with respect to these states. Certainly the Ito term involving the second derivative is simpler here.

One can extract a minimum principle from the term involving the infimum operation. Our approach will proceed along a different path and return to this feature.

Minimum Principle in Terms of Robust Information States

Suppose there is an optimal control $u^* \in \mathscr{U}$. Write $\widetilde{q}^*$ for $\widetilde{q}^{u^*}$, $\widetilde{\Phi}^*_{t,s}$ for $\widetilde{\Phi}^{u^*}_{t,s}$ *etc.* Consider any other control $v \in \mathscr{U}$. For $\theta \in [0,1]$ write,

$$u_\theta(t) = u^*(t) + \theta\big(v(t) - u^*(t)\big) \in \mathscr{U}, \;\; J(u_\theta) \geq J(u^*). \tag{2.15}$$

Consequently, differentiating in θ and letting $\theta = 0$ we have

$$\left.\frac{\partial J(u_\theta)}{\partial \theta}\right|_{\theta=0} \geq 0. \tag{2.16}$$

Standard results on differentiation with respect to a parameter give a preliminary lemma:

Lemma 2.1 *Write $\widetilde{q}_t(\theta)$ for $\widetilde{q}_t^{u_\theta}$. Then $z_t := \left.\frac{\partial \widetilde{q}_t(\theta)}{\partial \theta}\right|_{\theta=0}$ exists and is the unique solution of the ordinary integral equation*

$$\begin{aligned} z_t := \left.\frac{\partial \widetilde{q}_t(\theta)}{\partial \theta}\right|_{\theta=0} &= \int_0^t \widetilde{\varepsilon}_s^{-1} \frac{\partial \widetilde{A}_s^*}{\partial u_s}(v_s - u_s^*)\widetilde{\varepsilon}_s \widetilde{q}_s^* ds \\ &\quad + \int_0^t \widetilde{\varepsilon}_s^{-1} \widetilde{A}_s^* \widetilde{\varepsilon}_s z_s ds. \end{aligned} \tag{2.17}$$

Lemma 2.2 *Writing*

$$\eta_t := \int_0^t \widetilde{\Psi}^*_{s,0} \widetilde{\varepsilon}_s^{-1} \frac{\partial \widetilde{A}_s^*}{\partial u_s}(v_s - u_s^*)\widetilde{\varepsilon}_s \widetilde{q}_s^* ds, \tag{2.18}$$

we have

$$z_t = \widetilde{\Phi}^*_{t,0} \eta_t. \tag{2.19}$$

Proof Equivalently we prove: $dz_t = d\widetilde{\Phi}^*_{t,0}.\eta_t + \widetilde{\Phi}^*_{t,0}.d\eta_t$. This is immediate by differentiation of $\widetilde{\Phi}^*_{t,0}$ from (2.5) and η_t from (2.18), and using (2.6). □

Lemma 2.3 *For the controls* (2.15), *the cost function is,*

$$J(u_\theta) = \overline{E}\big[\langle \widetilde{\varepsilon}_T c, \widetilde{q}_T(\theta)\rangle\big],$$

and cost function sensitivity is,

$$\frac{\partial J(u_\theta)}{\partial \theta}\bigg|_{\theta=0} = \overline{E}\big[\langle \widetilde{\varepsilon}_T \widetilde{c}, \widetilde{\Phi}^*_{T,0}\eta_T\rangle\big].$$

Proof Follows from the differentiation of (2.10), and application of (2.17) (2.19). □

Lemma 2.4 *Defining, for* $0 \le t \le T$,

$$M_t := \overline{E}\big[\widetilde{\Phi}^{*'}_{T,0}\widetilde{\varepsilon}_T\widetilde{c} \mid \mathscr{Y}_t\big],$$

then M is a square integrable vector-valued martingale under P and the filtration $\mathscr{Y}_t$. Moreover, M has a martingale representation

$$M_t = \overline{E}\big[\widetilde{\Phi}^{*'}_{T,0}\widetilde{\varepsilon}_T\widetilde{c}\big] + \sum_{i\in S}\int_0^t \gamma^i_s dy^i_s + \sum_{i\in \overline{S}}\int_0^t \gamma^i_s (dy^i_s - ds),$$

where each γ^i is a $\{\mathscr{Y}_t\}$-predictable n-dimensional vector process such that,

$$\int_0^T \overline{E}\big[|\gamma^i_s|^2\big]ds < \infty.$$

Proof See Elliott (Elliott 1982b). □

Theorem 2.5

$$\frac{\partial J(u_\theta)}{\partial \theta}\bigg|_{\theta=0} = \int_0^T \overline{E}\left[\langle M_s, \widetilde{\Psi}^*_{s,0}\widetilde{\varepsilon}_s^{-1}\frac{\partial \widetilde{A}^*_s}{\partial u_s}(v_s - u^*_s)\widetilde{q}^*_s\rangle\right]ds.$$

Proof Since η is continuous, $d\langle M_t, \eta_t\rangle = \langle dM_t, \eta_t\rangle + \langle M_t, d\eta_t\rangle$, therefore

$$\begin{aligned}\langle M_T, \eta_T\rangle = & \int_0^T \langle M_s, \widetilde{\Psi}^*_{s,0}\widetilde{\varepsilon}_s^{-1}\frac{\partial \widetilde{A}^*_s}{\partial u_s}(v_s - u^*_s)\widetilde{\varepsilon}_s\widetilde{q}^*_s\rangle ds \\ & + \sum_{i\in S}\int_0^T \langle \gamma^i_s, \eta_s\rangle dy^i_s + \sum_{i\in\overline{S}}\int_0^T \langle \gamma^i_s, \eta_s\rangle(dy^i_s - ds).\end{aligned}$$

Therefore, as the last two terms are martingales,

$$\begin{aligned}\frac{\partial J(u_\theta)}{\partial \theta}\bigg|_{\theta=0} &= \overline{E}\big[\langle M_T, \eta_T\rangle\big], \\ &= \int_0^T \overline{E}\big[\langle M_s, \widetilde{\Psi}^*_{s,0}\widetilde{\varepsilon}_s^{-1}\frac{\partial \widetilde{A}^*_s}{\partial u_s}(v_s - u^*_s)\widetilde{\varepsilon}_s\widetilde{q}^*_s\rangle\big]ds.\end{aligned}$$

□

Theorem 2.6 *The the optimal control value $u_s^*(w)$ satisfies, a.e. in s and in ω, in the following,*

$$\langle M_s, \widetilde{\Psi}^*_{s,0}\widetilde{\varepsilon}_s^{-1}\frac{\partial \widetilde{A}_s^*}{\partial u_s}(v_s - u_s^*)\widetilde{\varepsilon}_s\widetilde{q}_s^*\rangle \geq 0. \tag{2.20}$$

Proof For any $s \in [0,T]$, and $\varepsilon > 0$, any $A \in \mathscr{Y}_s$ and $v \in \mathscr{U}$ choose

$$v_r = u_r^*, \quad \text{for } 0 \leq r \leq s \text{ and } s+\varepsilon < r \leq T$$

and

$$v_r(w) = v, \quad \text{for } s < r \leq s+\varepsilon \text{ and } w \in A,$$
$$v_r(w) = u_r^*(w), \quad \text{for } s < r \leq s+\varepsilon \text{ and } w \notin A.$$

Then Theorem 2.5 implies, with $\mathbf{1}_A$ denoting the indicator function,

$$\overline{E}\Big[\int_s^{s+\varepsilon} \langle M_r, \widetilde{\Psi}^*_{r,0}\widetilde{\varepsilon}_r^{-1}\frac{\partial \widetilde{A}_r^*}{\partial u_r}(v_r - u_r^*)\widetilde{\varepsilon}_r\widetilde{q}_r^*\rangle dr\mathbf{1}_A\Big] \geq 0.$$

Consequently, the optimal control value $u_s^*(w)$ satisfies, a.e. in s and in ω

$$\langle M_s, \widetilde{\Psi}^*_{s,0}\widetilde{\varepsilon}_s^{-1}\frac{\partial \widetilde{A}_s^*}{\partial u_s}(v_s - u_s^*)\widetilde{\varepsilon}_s\widetilde{q}_s^*\rangle \geq 0.$$

□

For the case when testing the Minimum Principle at time s, in the usual case when y, u is measurable over only the interval $[0,\ s]$ with $s < T$, then one needs a martingale representation in order to determine M_s. This is in itself an interesting task requiring the solution of a partial differential equation in terms of an adjoint process.

Adjoint Process

First define an adjoint process,

$$\pi_t := \overline{E}\big[\widetilde{\Phi}^{*'}_{T,t}\widetilde{\varepsilon}_T\widetilde{c} \mid \mathscr{Y}_t\big] = \widetilde{\Psi}^{*'}_{t,0}M_t, \;\; M_t = \widetilde{\Phi}^{*'}_{t,0}\pi_t. \tag{2.21}$$

Backwards Partial Parabolic Differential Equation for Adjoint Process

Theorem 2.7 *The adjoint process $\pi_t(\widetilde{q}, y_t)$ satisfies the backwards nonlinear second order partial parabolic equation*:

$$\frac{\partial \pi_t}{\partial t} + \frac{\partial \pi_t}{\partial \widetilde{q}_t}\widetilde{\varepsilon}_t^{-1}\widetilde{A}_t^*\widetilde{\varepsilon}_t\widetilde{q}_t + \frac{1}{2}\sum_{i,j\in S}\frac{\partial^2 \pi_t}{\partial y_t^i \partial y_t^j}$$
$$+ \sum_{j\in\overline{S}}[\pi_t(\widetilde{q}_t, y_t + e_j) - \pi_t(\widetilde{q}_t, y_t)] + \widetilde{\varepsilon}_t\widetilde{A}_t^{*'}\widetilde{\varepsilon}_t^{-1}\pi_t = 0, \;\; \pi_T = \widetilde{\varepsilon}_T\widetilde{c}. \tag{2.22}$$

Proof Notice that π_t depends only on y_t and the control u_t^*, which is Markov. That is, π_t is a function of $y_t, \widetilde{q}_t^*, t$. Earlier work defined such an adjoint, but in terms of the information state q_t. Thus,

$$d\pi_t = \left(\frac{\partial \pi_t}{\partial t} + \frac{\partial \pi_t}{\partial \widetilde{q}_t} \widetilde{\varepsilon}_t^{-1} \widetilde{A}_t^* \widetilde{\varepsilon}_t \widetilde{q}_t + \frac{1}{2} \sum_{i,j \in S} \frac{\partial^2 \pi_t}{\partial y_t^i y_t^j} \right) dt + \sum_{i \in S} \frac{\partial \pi_t}{\partial y_t^i} dy_t^i$$
$$+ \sum_{j \in \overline{S}} \left[\pi_t(\widetilde{q}_t, y_t + e_j) - \pi_t(\widetilde{q}_t, y_t) \right] dy_t^j, \quad \pi_T = \widetilde{\varepsilon}_T \widetilde{c}. \tag{2.23}$$

We now exploit the fact that M_t is a martingale. Applying the Ito rule to the the product $M_t = \widetilde{\Phi}_{t,0}^{*'} \pi_t$, which depends only on $(\widetilde{q}_t, y_t, t)$ gives,

$$\gamma_t^i = \widetilde{\Phi}_{t,0}^{*'} \frac{\partial \pi_t}{\partial y_t^i}, \quad \text{for } i \in S. \tag{2.24}$$

The dt term has coefficient $\widetilde{\Phi}_{t,0}^{*'}$ times the left hand side of (2.22), and since M is a martingale, is the zero process. Consequently, the theorem is established. □

Forwards Integral Equation

The equation (2.22) leads to the simplification of (2.23) as,

$$d\pi_t = \frac{\partial \pi_t}{\partial y_t} - \widetilde{\varepsilon}_t \widetilde{A}_t^{*'} \widetilde{\varepsilon}_t^{-1} \pi_t dt, \quad \pi_T = \widetilde{\varepsilon}_T \widetilde{c}. \tag{2.25}$$

Integrating leads to the result:

Theorem 2.8 *The adjoint process π_t satisfies a forward integral equation*:

$$\pi_t = \overline{E}\left[\widetilde{\Phi}_{T,0}^{*'} \widetilde{\varepsilon}_T \widetilde{c} \right] + \int_0^t \frac{\partial \pi_s}{\partial y_s} dy_s - \int_0^t \widetilde{\varepsilon}_s \widetilde{A}_s^{*'} \widetilde{\varepsilon}_s^{-1} \pi_s ds. \tag{2.26}$$

It remains a challenge to solve such equations with respect to (2.22) and (2.25), even knowing the optimal controls u^*.

Minimum Principle in Terms of the Adjoint Process

From Theorem 2.6 and (2.21) we have,

Theorem 2.9 *The the optimal control value $u_s^*(w)$ satisfies, a.e. in s and in ω, the following,*

$$\left\langle \pi_s, \widetilde{\varepsilon}_s^{-1} \frac{\partial \widetilde{A}_s^*}{\partial u_s} (v_s - u_s^*) \widetilde{\varepsilon}_s \widetilde{q}_s^* \right\rangle \geq 0,$$

where π_s is given by the backward second order nonlinear partial parabolic differential equation (2.22).

Connection to Dynamic Programming

A connection between the adjoint process π_t and the value function V_t of dynamic programming for the optimal control is almost immediate. Substituting $\widetilde{q}_T = \widetilde{\Phi}_{T,t}\widetilde{q}_t$ in the definition for V_t in (2.13), then

$$\frac{\partial V_t(\widetilde{q}_t, y_t)}{\partial \widetilde{q}_t} = \overline{E}\big[\langle \widetilde{\varepsilon}_T \widetilde{c}, \widetilde{\Phi}^*_{T,t} \frac{\partial \widetilde{q}_t}{\partial \widetilde{q}_t} \rangle | \widetilde{q}_t, y_t\big] = \overline{E}\big[\langle \widetilde{\varepsilon}_T \widetilde{c}, \widetilde{\Phi}^*_{T,t} \rangle | \widetilde{q}_t, y_t\big] = \pi_t. \tag{2.27}$$

Following a partial differentiation of the dynamic programming equation with respect to $\widetilde{q}_t$, we recognize that the minimization operation in this differentiated dynamic programming is identical to the Minimum Principle of this section.

Results for Risk-Sensitive Estimation and Control

Risk Sensitive Control

Suppose there is a risk sensitive generalization of our cost function with risk sensitivity parameter $0 \leq \mu$: Consider,

$$J^\mu(u) = \overline{E}\left[\exp\left(\mu\langle c, q_T^u\rangle + \int_0^T \mu\langle \underline{g_s^u}, q_s^u\rangle ds\right)\right]. \tag{2.28}$$

To apply the results developed so far in this paper, first define a modified rate matrix:

$$A_t^\mu(u) := A_t^u + \mu\Gamma_t^u, \;\; \Gamma_t^u := diag\{g_t^u(e_1), \ldots, g_t^u(e_n)\}. \tag{2.29}$$

Also, define

$$G_t^u := \exp\left(\mu\int_0^t g_r^u(X_r)dr\right), \;\; c^\mu := \exp(\mu c). \tag{2.30}$$

Now introduce a risk sensitive information state,

$$q_t^\mu(u) := \overline{E}[\Lambda_t^u G_t^u X_t^u | \mathscr{Y}_t]. \tag{2.31}$$

Following the derivations of the Wonham-Zakai filter, it is shown in James and Elliott (1996) that, a cost function dependent, risk sensitive filter, generating $q_t^\mu(u)$ is,

$$\begin{aligned} dq_t^\mu(u) &= A_t^\mu(u) q_t^\mu(u)dt + \sum_{i\in S} H_t^i q_t^\mu(u)dy_t^i \\ &\quad + \sum_{i\in \overline{S}} (H_t^i - I_n) q_{t-}^\mu(u)(dy_t^i - dt), \;\; J^\mu(u) = \overline{E}[\langle q_T^\mu(u), c^\mu\rangle]. \end{aligned} \tag{2.32}$$

The formulas for the Robust Clark filter, the Risk Sensitive Dynamic Programming equations and the Risk Sensitive Minimum Principle, are derived as previously but using the superscript μ notation in lieu of the over tilde notation. The results are not stated separately since this notation change is trivial.

Risk Sensitive Filtering

The notion of a risk sensitive filtering is related to risk sensitive control. For simplicity of presentation, assume here that we are not seeking any optimal control for the signal generating project but only an optimal 'control' which is the desired risk sensitive filtered estimate. The cost function for the risk sensitive filtering is in fact a special case of a risk sensitive control. For the risk sensitive filtering task, we view the optimal risk sensitive state estimate, $\widehat{X}_t \in \triangle_n$ of X_t, where $\triangle_n$ denotes the n-dimensional simplex, as a 'control' variable which only enters the cost function and not the signal model. Clearly, discrete state estimates here are nonnegative probabilities which add to unity.

Typically, for filtering, as opposed to smoothing, the cost function does not penalize any future costs or any terminal cost, but penalizes the exponential of the integrated quadratic estimation error up until the present time.

For the derivation of the appropriate risk sensitive information state, apply the equations for the risk sensitive information state derived for the the control situation of this section with the notation adjustments, recalling that $\langle X, X\rangle = \langle \mathbf{1}, X\rangle = 1$,

$$\begin{aligned}
u_t &\Rightarrow \widehat{X}_t,\\
\langle g_t^u, X_t\rangle &\Rightarrow \frac{1}{2}\|\widehat{X}_t - X_t\|^2 = \left\langle \frac{1}{2}(\widehat{X}_t'\widehat{X}_t - 1)\mathbf{1} - \widehat{X}_t, X_t\right\rangle,\\
A_t^\mu(u) &\Rightarrow \widehat{A}_t^\mu := A_t + \mu\widehat{\Gamma}_t,\\
&\quad \widehat{\Gamma}_t := \operatorname{diag}\left\{\left\langle \frac{1}{2}(\widehat{X}_t'\widehat{X}_t - 1)\mathbf{1} - \widehat{X}_t, e_1\right\rangle, \dots, \left\langle \frac{1}{2}(\widehat{X}_t'\widehat{X}_t - 1)\mathbf{1} - \widehat{X}_t, e_n\right\rangle\right\}.
\end{aligned}$$

The filter equations (2.32), with these notational adjustments give the risk sensitive information state as,

$$\widehat{q}_t^\mu := \overline{E}[\Lambda_t\widehat{G}_t X_t|\mathscr{Y}_t], \quad \widehat{G}_t := \exp\left(\int_0^t \frac{\mu}{2}\|\widehat{X}_s - X_s\|^2 ds\right),$$

where, rather than calculate this via the Wonham-Zakai version,

$$\begin{aligned}
d\widehat{q}_t^\mu &= \widehat{A}_t^\mu \widehat{q}_t^\mu dt + \sum_{i\in S} H_t^i \widehat{q}_t^\mu dy_t^i\\
&\quad + \sum_{i\in \overline{S}} (H_t^i - I_n)\widehat{q}_{t-}^\mu (dy_t^i - dt),
\end{aligned}$$

it is preferable to calculate this via the robust Clark version,

$$\begin{aligned}
\widehat{q}_t^\mu &= \varepsilon_t \overline{q}_t^\mu,\\
\frac{d\overline{q}_t^\mu}{dt} &= \varepsilon_t^{-1}\widehat{A}_t^\mu \varepsilon_t \overline{q}_t^\mu, \quad \overline{q}_0^\mu = q_0 = p_0.
\end{aligned}$$

Now an optimal risk sensitive state estimate $\widehat{X}_t$ can be obtained via the infinite state associated dynamic programming equations solved forward in time, but here we introduce a simplifying assumption, used in earlier discrete-time formulations (Dey and Moore 1995c), so that the risk sensitive estimator is finite dimensional:

Assumption

Risk sensitive state estimates are calculated at the current time assuming they are known prior to the present time. That is $\widehat{X}_t^{\mu} \in \mathscr{Y}_t$.

Under this assumption, we define the estimate and formally simplify its formulation as:

$$\begin{aligned}
\widehat{X}_t^{\mu} &:= \underset{\widehat{X}\in\Delta_n}{\operatorname{argmin}} \overline{E}\left[d\left(\Lambda_t \widehat{G}_t\right)|\mathscr{Y}_t\right], \\
&= \underset{\widehat{X}\in\Delta_n}{\operatorname{argmin}} \overline{E}\left[d\Lambda_t \widehat{G}_t + \frac{\mu}{2}\Lambda_t \|X_t - \widehat{X}\|^2 \widehat{G}_t dt|\mathscr{Y}_t\right], \\
&= \underset{\widehat{X}\in\Delta_n}{\operatorname{argmin}} \overline{E}\left[\sum_{j=1}^{n}\langle X_t, e_j\rangle(\Lambda_t^{-1} d\Lambda_t + \frac{\mu}{2}\|X_t - \widehat{X}\|^2 dt)\Lambda_t \widehat{G}_t|\mathscr{Y}_t\right], \\
&= \underset{\widehat{X}\in\Delta_n}{\operatorname{argmin}} \sum_{j=1}^{n}\left(\Lambda_t^{-1} d\Lambda_t|_{X=e_j} + \frac{\mu}{2}\|e_j - \widehat{X}\|^2 dt\right)\overline{E}\left[\Lambda_t \widehat{G}_t\langle X_t, e_j\rangle|\mathscr{Y}_t\right], \\
&= \underset{\widehat{X}\in\Delta_n}{\operatorname{argmin}} \sum_{j=1}^{n}\|e_j - \widehat{X}\|^2\langle e_j, \widehat{q}_t^{\mu}\rangle, \\
&= \underset{\widehat{X}\in\Delta_n}{\operatorname{argmin}}\left(\|\widehat{X}\|^2\langle \mathbf{1}, \widehat{q}_t^{\mu}\rangle - 2\langle \widehat{X}, \widehat{q}_t^{\mu}\rangle\right), \\
&= \widehat{q}_t^{\mu}\langle \mathbf{1}, \widehat{q}_t^{\mu}\rangle^{-1}.
\end{aligned}$$

Thus, the risk sensitive information state in this case, when normalized is the desired risk sensitive state estimate.

12.3 The Dependent Case

Again our processes are assumed to be defined on a probability space $(\Omega, \mathscr{F}, P)$. In this section we discuss a conditional Markov process, X, as defined in Section 12.1. However, our observations are now given by the process $\mathscr{J}$ which counts the total number of jumps of X. From Appendix B,

$$\mathscr{J}_t = \int_0^t c(X_r, u)\, dr + \mathscr{Q}_t, \tag{3.1}$$

where $c(X_r, u) = -\sum_{i=1}^{N}\langle X_r, e_i\rangle a_{ii}(r, u)$.

Here the noises in the signal and observation processes are correlated, $[X^u, \mathscr{J}]_t \neq 0$. The control problem we wish to consider is that of choosing $u \in \underline{U}$ so that the expected cost

$$J(u) = E\left[\langle \ell, X_T^u\rangle\right]$$

is minimized. We have, therefore, a signal given by (2.2) and an observation process given by (3.1). Because the signal and observation processes are correlated we shall derive the Zakai equation using semimartingale methods. Write $\{\mathscr{Y}_t\}$ for the right-continuous complete filtration generated by $\mathscr{J}$. If $\{\phi_t\}$, $t \geq 0$, is any process write $\hat{\phi}$ for the $\mathscr{Y}$-optional projection of ϕ. Then $\hat{\phi}_t = E[\phi_t \mid \mathscr{Y}_t]$ a.s. Similarly, write $\overline{\phi}$ for the $\mathscr{Y}$-predictable projection of ϕ. Then $\overline{\phi}_t = E[\phi_t \mid \mathscr{Y}_{t-}]$ a.s. From Elliott (1982b, Theorem 6.48), for almost all ω, $\hat{\phi} = \overline{\phi}$ except for countably many values of t. Therefore,

$$\int_0^t \overline{c}(X_r,u)\,dr = \int_0^t \hat{c}(X_r,u)\,dr = \int_0^t \hat{c}(X_{r-},u)\,dr.$$

Write $\hat{p}_t = \hat{X}_t = E[X_t \mid \mathscr{Y}_t]$ so $\hat{p}_0 = E[X_0] = p_0$ say. Now $c(X_r,u) = \langle a(r,u), X_r\rangle$, where $a(r,u) = -\left(a(r,u)_{11}, \ldots, a(r,u)_{NN}\right)'$, so

$$\hat{c}(X_r,u) = \left\langle a(r,u), \hat{X}_r\right\rangle = \langle a(r,u), \hat{p}_r\rangle.$$

For the vector $c(X_r,u)X_r = \operatorname{diag}(a(r,u))X_r$ we have $\widehat{c(X_r)X_r} = \operatorname{diag}(a(r,u))\,\hat{p}_r$. The innovation process associated with the observations is

$$\tilde{\mathscr{Q}}_t = \mathscr{J}_t - \int_0^t \overline{c}(X_r,u)\,dr = \mathscr{J}_t - \int_0^t \hat{c}(X_r,u)\,dr.$$

See Brémaud (1981). Application of Fubini's theorem shows that $\tilde{\mathscr{Q}}$ is a $\{\mathscr{Y}_t\}$ martingale. Therefore,

$$\mathscr{J}_t = \int_0^t \hat{c}(X_r,u)\,dr + \tilde{\mathscr{Q}}_t. \tag{3.2}$$

Similarly, Fubini's theorem shows that the process

$$\tilde{V}_t := \hat{p}_t - p_0 - \int_0^t A(s,u)\,\hat{p}_{r-}\,ds$$

is a square-integrable $\{\mathscr{Y}_t\}$ martingale. Consequently, $\tilde{V}$ can be represented as a stochastic integral

$$\tilde{V}_t = \int_0^t \gamma_r\,d\tilde{\mathscr{Q}}_r.$$

Therefore,

$$\hat{p}_t = p_0 + \int_0^t A(r,u)\,\hat{p}_{r-}\,dr + \int_0^t \gamma_r\,d\tilde{\mathscr{Q}}_r. \tag{3.3}$$

The problem now is to find an explicit form for γ.

Theorem 3.1

$$\begin{aligned}\gamma_r &= I(\langle \hat{p}_{r-}, a(r,u)\rangle \neq 0)\,\langle \hat{p}_{r-}, a(r,u)\rangle^{-1} \\ &\quad\times \left[\operatorname{diag}(a(r,u))\,\hat{p}_{r-} - \langle \hat{p}_{r-}, a(r,u)\rangle + A(r,u)\,\hat{p}_{r-}\right].\end{aligned} \tag{3.4}$$

Proof The product $\hat{p}_t \mathscr{J}_t$ is calculated two ways. First consider, using the Ito product rule in Appendix A,

$$X_t \mathscr{J}_t = \int_0^t X_{r-} \left(d\mathscr{Q}_r + c\left(X_{r-}\right) dr \right) + \int_0^t \mathscr{J}_{r-} \left(A\left(r,u\right) X_{r-} + dV_r \right) + \left[X, \mathscr{J} \right]_t .$$

Now X and $\mathscr{J}$ jump at the same times, at which $\Delta \mathscr{J} = 1$, so

$$\begin{aligned} \left[X, \mathscr{J} \right]_t &= \sum_{0<r\leq t} \Delta X_r \Delta \mathscr{J}_r = \sum_{0<r\leq t} \Delta X_r \\ &= X_t - X_0 = \int_0^t A\left(r,u\right) X_{r-}\, dr + V_r . \end{aligned}$$

So

$$X_t \mathscr{J}_t = \int_0^t \left(X_{r-} c\left(X_{r-}\right) + \mathscr{J}_{r-} A\left(r,u\right) X_{r-} + A\left(r,u\right) X_{r-} \right) dr + \text{martingale}. \tag{3.5}$$

Taking the $\mathscr{Y}$-optional projection of each side of (3.5)

$$\hat{p}_t \mathscr{J}_t = \int_0^t \left(\operatorname{diag}\left(a\left(r,u\right)\right) \hat{p}_{r-} + \mathscr{J}_{r-} A\left(r,u\right) \hat{p}_{r-} + A\left(r,u\right) \hat{p}_{r-} \right) dr + \text{martingale}. \tag{3.6}$$

However, from (3.2) and (3.3)

$$\hat{p}_t \mathscr{J}_t = \int_0^t \hat{p}_{r-} c\left(X_{r-}\right) dr + \int_0^t \hat{p}_{r-}\, d\tilde{\mathscr{Q}}_r + \int_0^t A\left(r,u\right) \hat{p}_{r-} \mathscr{J}_{r-}\, dr + \int_0^t \gamma_r \mathscr{J}_{r-}\, d\tilde{\mathscr{Q}}_r + \left[\hat{p}, \mathscr{J} \right]_t .$$

Now

$$\begin{aligned} \left[\hat{p}, \mathscr{J} \right]_t &= \sum_{0<r\leq t} \Delta \hat{p}_r \Delta \mathscr{J}_r = \sum_{0<r\leq t} \gamma_r\, d\mathscr{J}_r \\ &= \int_0^t \gamma_r\, d\mathscr{J}_r = \int_0^t \gamma_r\, d\tilde{\mathscr{Q}}_r + \int_0^t \gamma_r c\left(X_{r-}\right) dr. \end{aligned}$$

Therefore,

$$\hat{p}_t \mathscr{J}_t = \int_0^t \left(\hat{p}_{r-} c\left(X_{r-}\right) + A\left(r,u\right) \hat{p}_{r-} \mathscr{J}_{r-} \right) dr + \text{ martingale}. \tag{3.7}$$

The bounded variation processes in (3.6) and (3.7) must be equal, so (3.4) follows. □

Note for any set $B \in \mathscr{Y}_s$, $E\left[I(B)\int_s^t d\mathscr{J}_r\right] = E\left[I(B)\int_s^t c(X_{r-})\,dr\right]$, so γ_r can be taken to be 0 on any set where $c(\hat{X}_{r-}) = 0$.

Suppose there is a constant $\alpha > 0$ such that $-a_{ii}(r,u) > \alpha$ for all i and $r \geq 0$. Then $c(X_r)^{-1} = \langle a(r,u), X_r\rangle^{-1} < \alpha^{-1}$ for all $r \geq 0$. Define the martingale Λ by

$$\Lambda_t = 1 + \int_0^t \Lambda_{r-}\left(c(X_{r-})^{-1} - 1\right) d\mathscr{Q}_r \tag{3.8}$$

and introduce a new probability measure $\overline{P}$ on $\{\Omega, \mathscr{G}\}$ by $E\left[d\overline{P}/dP \mid \mathscr{G}_t\right] = \Lambda_t$.

Then it can be shown that under $\overline{P}$ the process $\mathscr{J}$ is a standard Poisson process, and in particular $\overline{Q} = \mathscr{J}_t - t$ is a martingale. Conversely we can define the $(\overline{P}, \mathscr{F})$ martingale

$$\overline{\Lambda}_t = 1 + \int_0^t \overline{\Lambda}_{r-}\left(c(X_{r-})^{-1} - 1\right) d\overline{\mathscr{Q}}_r. \tag{3.9}$$

Then $\Lambda_t \overline{\Lambda}_t = 1$. To obtain the Zakai equation we take $\overline{P}$ as the reference probability measure and compute expectations under $\overline{P}$. Write $\Pi\left(\overline{\Lambda}_t\right)$ for the $\mathscr{Y}$-optional projection of $\overline{\Lambda}$ under $\overline{P}$. Then for each $t \geq 0$, $\Pi\left(\overline{\Lambda}_t\right) = \overline{E}\left[\overline{\Lambda}_t \mid \mathscr{Y}_t\right]$ a.s. It can be shown that $\Pi\left(\overline{\Lambda}_t\right) = 1 + \int_0^t \Pi\left(\overline{\Lambda}_{r-}\right)\left(\hat{c}(X_{r-}) - 1\right) d\overline{\mathscr{Q}}_r$. By Bayes' rule, for any $\mathscr{Y}_t$-measurable random variable ϕ

$$\hat{p}_t = E\left[p_t \mid \mathscr{Y}_t\right] = \frac{\overline{E}\left[\overline{\Lambda}_t p_t \mid \mathscr{Y}_t\right]}{\overline{E}\left[\overline{\Lambda}_t \mid \mathscr{Y}_t\right]} := \frac{q_t}{\Pi\left(\overline{\Lambda}_t\right)}.$$

Calculating the product $\Pi\left(\overline{\Lambda}_t\right)\hat{p}_t$ we obtain the Zakai equation for q_t:

$$q_t(u) = p_0 + \int_0^t A_r(u)\, q_{r-}(u)\, du + \int_0^t B_r(u)\, q_{r-}(u)\, d\overline{\mathscr{Q}}_r, \tag{3.10}$$

where

$$\begin{aligned} B_t(u) &= \left(\operatorname{diag}(a(t,u)) - I + A_t(u)\right) \\ &= \begin{pmatrix} -1 & a_{01}(t,u) & \dots & a_{0N}(t,u) \\ a_{10}(t,u) & -1 & \dots & a_{1N}(t,u) \\ \vdots & \vdots & \ddots & \vdots \\ a_{N0}(t,u) & a_{N1}(t,u) & \dots & -1 \end{pmatrix}. \end{aligned}$$

The expected cost if $u \in \underline{U}$ is used is

$$\begin{aligned} J(u) &= E[\langle \ell, X_T^u\rangle] = \overline{E}\left[\overline{\Lambda}_T^u \langle \ell, X_T^u\rangle\right] = \overline{E}\left[\left\langle \ell, \overline{\Lambda}_T^u X_T^u\right\rangle\right] \\ &= \overline{E}\left[\left\langle \ell, \overline{E}\left[\overline{\Lambda}_T^u X_T^u \mid \mathscr{Y}_T\right]\right\rangle\right] = \overline{E}[\langle \ell, q_T^u\rangle]. \end{aligned}$$

The control problem has, therefore, been formulated in separated form: find $u \in \underline{U}$ which minimizes

$$J(u) = \overline{E}\left[\langle \ell, q_T^u \rangle\right],$$

where for $0 \leq t \leq T$, $q_t(u)$ satisfies the dynamics (3.10).

Differentiation

Notation 3.2 *For $u \in \underline{U}$ write $\Phi^u(t,s)$ for the fundamental matrix solution of*

$$d\Phi^u(t,s) = A_t(u)\Phi^u(t-,s)\,dt + B_t(u)\Phi^u(t-,s)(d\mathcal{J}_t - dt) \tag{3.11}$$

with initial condition $\Phi^u(s,s) = I$, the $N \times N$ identity matrix.

Note that $A_t(u) - B_t(u) = \operatorname{diag}(1 + a_{ii}(t,u))$ and write

$$D^u(s,t) = \operatorname{diag}\left(\exp \int_s^t (1 + a_{ii}(r,u))\,dr\right).$$

Then if $T_n \leq t < T_{n+1}$,

$$\begin{aligned} \Phi^u(t,0) &= D^u(t,T_n)(I + B_{T_n}(u_{T_n}))D^u(T_n,T_{n-1}) \\ &\quad \times \left[(I + B_{T_{n-1}}(u_{T_{n-1}})) \times \cdots \times (I + B_{T_1}(u_{T_1}))\right] D^u(T_1,0). \end{aligned} \tag{3.12}$$

The matrices $D^u(s,t)$ have inverses

$$\operatorname{diag}\left(\exp - \int_s^t (1 + a_{ii}(r,u))\,dr\right);$$

we make the following assumption:

For $u \in \underline{U}$ and $t \in [0,T]$ the matrix $(I + B_t(u_t))$ is nonsingular.

The matrix Φ is the analog of the Jacobian in the continuous case. We now derive the equation satisfied by the inverse Ψ of Φ.

Lemma 3.3 *For $u \in \underline{U}$ consider the matrix Ψ^u defined by the equation*

$$\begin{aligned} \Psi^u(t,s) = {} & I - \int_s^t \Psi^u(r-,s)A_r(u)\,dr - \int_s^t \Psi^u(r-,s)B_r(u)\,d\overline{\mathcal{Q}}_r \\ & + \int_s^t \Psi^u(r-,s)B_r^2(u)(I + B_r(u))^{-1}\,d\mathcal{J}_r. \end{aligned} \tag{3.13}$$

Then $\Psi^u(t,s)\Phi^u(t,s) = I$ for $t \geq s$.

Proof Recall

$$\Phi^u(t,s) = I + \int_s^t A_r(u)\Phi^u(r-,s)\,dr + \int_s^t B_r(u)\Phi^u(r-,s)\,d\overline{\mathcal{Q}}_r. \tag{3.14}$$

Then by the product rule

$$\begin{aligned}\Psi\Phi &= I + \int_s^t \Psi A \Phi dr + \int_s^t \Psi B \Phi d\overline{\mathscr{Q}}_r \\ &\quad - \int_s^t \Psi A \Phi dr - \int_s^t \Psi B \Phi d\overline{\mathscr{Q}}_r + \int_s^t \Psi B^2 (I+B)^{-1} \Phi d\mathscr{J}_r \\ &\quad - \int_s^t \Psi B^2 \Phi d\mathscr{J}_r + \int_s^t \Psi B^2 (I+B)^{-1} B \Phi d\mathscr{J}_r \\ &= I,\end{aligned}$$

as the integral terms cancel. □

We shall suppose there is an optimal control $u^* \in \underline{U}$. Write q^* for q^{u^*}, Φ^* for Φ^{u^*}, and so on. Consider any other control $v \in \underline{U}$. Then for $\theta \in [0,1]$,

$$u_\theta(t) = u^*(t) + \theta(v(t) - u(t)) \in \underline{U}.$$

Because $U \subset \mathbb{R}^k$ is compact, the set $\underline{U}$ of admissible controls can be considered as a subset of the Hilbert space $H = L^2\left[\Omega \times [0,T] : \mathbb{R}^k\right]$. Now

$$J(u_\theta) \geq J(u^*). \tag{3.15}$$

Therefore, if the Gâteaux derivative $J'(u^*)$ of J, as a functional on the Hilbert space H, is well-defined, differentiating (3.15) in θ, and evaluating at $\theta = 0$, implies

$$\langle J'(u^*), v(t) - u^*(t)\rangle \geq 0$$

for all $v \in \underline{U}$.

Lemma 3.4 *Suppose $v \in \underline{U}$ is such that $u_\theta^* = u^* + \theta v \in \underline{U}$ for $\theta \in [0,\alpha]$. Write $q_t(\theta)$ for the solution $q_t(u_\theta^*)$ of* (3.10). *Then $z_t = \partial q_t(\theta)/\partial\theta|_{\theta=0}$ exists and is the unique solution of the equation*

$$\begin{aligned}z_t &= \int_0^t \left(\frac{\partial A}{\partial u}(r,u^*)\right) v_r q_{r-}^* dr + \int_0^t A_r(u^*) z_{r-} dr \\ &\quad + \int_0^t \left(\frac{\partial B}{\partial u}(r,u^*)\right) v_r q_{r-}^* d\overline{\mathscr{Q}}_r + \int_0^t B_r(u^*) z_{r-} d\overline{\mathscr{Q}}_r.\end{aligned} \tag{3.16}$$

Proof $q_t(\theta) = p_0 + \int_0^t A_r(u^* + \theta v) q_{r-}(\theta) dr + \int_0^t B_r(u^* + \theta v) q_{r-}(\theta) d\overline{\mathscr{Q}}_r$. The stochastic integrals are defined pathwise, so differentiating under the integrals gives the result. □

Comparing (3.14) and (3.16) we have the following result by variation of constants.

Lemma 3.5 *Write*

$$\begin{aligned}\eta_{0,t} = & \int_0^t \Psi^*(r-,0)\left(\frac{\partial A}{\partial u}(r,u^*)\right) v_r q^*_{r-}\, dr \\ & + \int_0^t \Psi^*(r-,0)\left(\frac{\partial B}{\partial u}(r,u^*)\right) v_r q^*_{r-}\, d\overline{\mathscr{Q}}_r \\ & - \int_0^t \Psi^*(r-,0)\left(I+B_r(u^*)\right)^{-1} B_r(u^*)\left(\frac{\partial B}{\partial u}(r,u^*)\right) v_r q^*_{r-}\, d\mathscr{J}_r.\end{aligned} \tag{3.17}$$

Then $z_t = \Phi^*(t,0)\,\eta_{0,t}$.

Proof Using the differentiation rule

$$\Phi^*(t,0)\,\eta_{0,t} = \int_0^t \Phi^*_- \cdot d\eta + \int_0^t d\Phi^* \eta_- + [\Phi,\eta]_t\,.$$

Because $\Phi^*_-\Psi^*_- = I$, therefore

$$\begin{aligned}\Phi^*(t,0)\,\eta_{0,t} = & \int_0^t \left(\frac{\partial A}{\partial u}(r,u^*)\right) v_r q^*_{r-}\, dr \\ & + \int_0^t \left(\frac{\partial B}{\partial u}(r,u^*)\right) v_r q^*_{r-}\, d\overline{\mathscr{Q}}_r \\ & - \int_0^t \left(I+B_r(u^*)\right)^{-1} B_r(u^*)\left(\frac{\partial B}{\partial u}(r,u^*)\right) v_r q^*_{r-}\, d\mathscr{J}_r \\ & + \int_0^t A_r(u)\,\Phi^*(r-,0)\,\eta_{0,r-}\, dr \\ & + \int_0^t B_r(u)\,\Phi^*(r-,0)\,\eta_{0,r-}\, d\overline{\mathscr{Q}}_r \\ & + \int_0^t B_r(u)\left(\frac{\partial B}{\partial u}(r,u^*)\right) v_r q^*_{r-}\, d\mathscr{J}_r \\ & - \int_0^t B_r(u)\,(I+B_r(u^*))^{-1} B_r(u^*)\left(\frac{\partial B}{\partial u}(r,u^*)\right) v_r q^*_{r-}\, d\mathscr{J}_r.\end{aligned}$$

Now the $d\mathscr{J}$ integrals sum to 0, showing that $\Phi^*\eta$ satisfies the same Equation (3.17) as z. Consequently, by uniqueness, the result follows. □

Corollary 3.6

$$\left.\frac{dJ}{d\theta}(u^*_\theta)\right|_{\theta=0} = \overline{E}\left[\langle \ell, \Phi^*(T,0)\,\eta_{0,T}\rangle\right].$$

Proof $J\left(u^*_\theta\right) = \overline{E}\left[\langle \ell, q_T(\theta)\rangle\right]$. The result follows from Lemmas 3.4 and 3.5. □

Write $\Phi^*(T,0)'$ for the transpose of $\Phi^*(T,0)$ and consider the square integrable, vector martingale

$$\mathscr{M}_t := \overline{E}\left[\Phi^*(T,0)'\,\ell \mid \mathscr{Y}_t\right].$$

Then $\mathscr{M}_t$ has a representation as a stochastic integral

$$\mathscr{M}_t = \overline{E}\left[\Phi^*(T,0)'\ell\right] + \int_0^t \gamma_r \, d\overline{\mathscr{Q}}_r,$$

where γ is a predictable $\mathbb{R}^N$-valued process such that

$$\int_0^T \overline{E}\left[\gamma_r^2\right] dr < \infty.$$

Under a Markov hypothesis γ will be explicitly determined below.

Definition 3.7 The adjoint process is

$$p_t := \Psi^*(t,0)' \mathscr{M}_t.$$

Theorem 3.8 *The derivative at* u_θ^* *of the cost is given by*:

$$\begin{aligned}\left.\frac{dJ(u_\theta^*)}{d\theta}\right|_{\theta=0} = \int_0^T \overline{E}\Bigg[\Bigg\langle p_{r-}, \Bigg\{\left(\frac{\partial A}{\partial u}(r,u^*)\right) \\ - (I + B_r(u^*))^{-1} B_r(u^*)\left(\frac{\partial B}{\partial u}(r,u^*)\right)\Bigg\} v_r q_{r-}^*\Bigg\rangle \\ + \left\langle \gamma_r, \Psi^*(r-,0)(I + B_r(u^*))^{-1}\left(\frac{\partial B}{\partial u}(r,u^*)\right) v_r q_{r-}^*\right\rangle\Bigg] dr.\end{aligned} \tag{3.18}$$

Proof First note that

$$\begin{aligned}&\langle \mathscr{M}_T, \eta_{0,T}\rangle \\ &= \int_0^T \left\langle \mathscr{M}_{r-}, \Psi^*(r-,0)\left(\frac{\partial A}{\partial u}(r,u^*)\right) v_r q_{r-}^*\right\rangle dr \\ &\quad + \int_0^T \left\langle \mathscr{M}_{r-}, \Psi^*(r-,0)\left(\frac{\partial B}{\partial u}(r,u^*)\right) v_r q_{r-}^*\right\rangle d\overline{\mathscr{Q}}_r \\ &\quad - \int_0^T \left\langle \mathscr{M}_{r-}, \Psi^*(r-,0)(I + B_r(u^*))^{-1} B_r(u^*)\left(\frac{\partial B}{\partial u}(r,u^*)\right) v_r q_{r-}^*\right\rangle d\mathscr{J}_r \\ &\quad + \int_0^T \langle \gamma_r, \eta_{0,r-}\rangle \, d\overline{\mathscr{Q}}_r + \int_0^T \left\langle \gamma_r \Psi^*(r-,0)\left(\frac{\partial B}{\partial u}(r,u^*)\right) v_r q_{r-}^*\right\rangle d\mathscr{J}_r \\ &\quad - \int_0^T \left\langle \gamma_r, \Psi^*(r-,0)(I + B_r(u^*))^{-1} B_r(u^*)\left(\frac{\partial B}{\partial u}(r,u^*)\right) v_r q_{r-}^*\right\rangle d\mathscr{J}_r.\end{aligned} \tag{3.19}$$

Taking expectations under P, we have

$$\left.\frac{dJ\left(u_{\theta}^{*}\right)}{d\theta}\right|_{\theta=0} = \overline{E}\left[\left\langle \ell, \Phi^{*}(T,0)\,\eta_{0,T}\right\rangle\right]$$
$$= \overline{E}\left[\left\langle \Phi^{*}(T,0)'\,\ell, \eta_{0,T}\right\rangle\right]$$
$$= \overline{E}\left[\left\langle \mathscr{M}_T, \eta_{0,T}\right\rangle\right].$$

Combining the last two terms in (3.19) and using the fact that $\mathscr{J}_t - t$ is a $\overline{P}$ martingale, this is

$$= \int_0^T \overline{E}\Bigg[\left\langle p_{r-}, \left(\frac{\partial A}{\partial u}(r,u^*)\right) v_r q_{r-}^{*}\right\rangle$$
$$- \left\langle p_{r-}, \left(I + B_r(u^*)\right)^{-1} B_r(u^*)\left(\frac{\partial B}{\partial u}(r,u^*)\right) v_r q_{r-}^{*}\right\rangle$$
$$+ \left\langle \gamma_r, \Psi^{*}(r-,0)\left(I + B_r(u^*)\right)^{-1}\left(\frac{\partial B}{\partial u}(r,u^*)\right) v_r q_{r-}^{*}\right\rangle\Bigg] dr.$$

□

Now consider perturbations of u^* of the form

$$u_{\theta}(t) = u^{*}(t) + \theta\left(v(t) - u(t)\right)$$

for $\theta \in [0,1]$ and any $v \in \underline{U}$. Then as noted above

$$\left.\frac{dJ\left(u_{\theta}\right)}{d\theta}\right|_{\theta=0} = \left\langle J'(u^*), v(t) - u^{*}(t)\right\rangle \geq 0.$$

Expression (3.18) is, therefore, true when v is replaced by $v - u^*$ for any $v \in \underline{U}$, and we can deduce the following minimum principle:

Theorem 3.9 *Suppose $u^* \in \underline{U}$ is an optimal control. Then* a.s. *in w and* a.e. *in t*

$$\left\langle p_{r-}, \Delta_1(r,u^*)\, q_{r-}^{*}\right\rangle + \left\langle \gamma_r, \Delta_2(r,u^*)\, q_{r-}^{*}\right\rangle \geq 0. \tag{3.20}$$

Here

$$\Delta_1(r,u^*) = \left\{\left(\frac{\partial A}{\partial u}(r,u^*)\right) - \left(I + B_r(u^*)\right)^{-1} B_r(u^*)\left(\frac{\partial B}{\partial u}(r,u^*)\right)\right\}\left(v_r - u_r^{*}\right),$$

$$\Delta_2(r,u^*) = \Psi^{*}(r-,0)\left(I + B_r(u^*)\right)^{-1}\left(\frac{\partial B}{\partial u}(r,u^*)\right)\left(v_r - u_r^{*}\right).$$

The Equation for the Adjoint Process

The process p is the adjoint process. However, (3.20) also contains the integrand γ. In this section we shall obtain a more explicit expression for γ in the case when u^* is Markov, and also derive forward and backward equations satisfied by p.

We assume that the optimal control u^* is a Markov feedback control. That is,

$$u^* : [0,T] \times \mathbb{R}^N \to U \tag{3.21}$$

so that $u^*\left(s, q^*_{s-}\right) \in U$. Note that if u_m is a Markov control, with a corresponding solution $q_t(u_m)$ of (3.10), then u_m can be considered as a stochastic open loop control $u_m(\omega)$ by setting

$$u_m(\omega) = u_m\left(s, q^*_{s-}(u_m)(\omega)\right).$$

This means the control u_m "follows" the "left limit" of its original trajectory $q_s(u_m)$ rather than any new trajectory.

Lemma 3.10 *Write δ for the predictable "integrand" such that*

$$\Delta p_t = p_t - p_{t-} = \delta_t \Delta \mathscr{J}_t, \qquad i.e., \; p_t = p_{t-} + \delta_t \Delta \mathscr{J}_t.$$

Furthermore, write

$$\begin{aligned} q_{t-} &= q, \\ B_{t-}(u^*(t-,q)) &= B^*(q_{t-}) = B^*(q), \end{aligned}$$

and

$$B_t(u^*(t,q_t)) = B^*(q_t).$$

Then

$$\delta_t(q) = (I + B^*((I+B^*(q))q))^{-1} p_{t-}((I+B^*(q))q) - p_{t-}(q). \tag{3.22}$$

Proof Let us examine what happens if there is a jump at time t; that is, suppose $\Delta \mathscr{J}_t = 1$. Then from (3.10)

$$q_t = (I + B^*(q))q.$$

By the Markov property and from (3.12) and Definition 3.7,

$$\begin{aligned} p_t &= E\left[D^*(t,T_k)\left(I + B'_{T_k}(u^*)\right) \cdots D^*(T,T_M)\ell \mid \mathscr{Y}_t\right] \\ &= p_t(q_t) \\ &= p_t((I+B^*(q))q) \\ &= \left(I + B^*(q_t)'\right)^{-1} p_{t-}((I+B^*(q))q), \end{aligned}$$

where T_M is the last jump time before T. The result follows. □

Heuristically, the integrand δ assumes there is a jump at t; the question of whether there is a jump is determined by the factor $\Delta \mathscr{J}_t$.

Theorem 3.11 *Under assumption* 3.21 *and with* δ_t *given by* (3.22)

$$\gamma_r = \Phi^*(r-,0)'\left(\left(I + B_r'(u^*)\right)\delta_r + B_r'(u^*)\, p_{r-}\right). \tag{3.23}$$

Proof $\Phi^*(t,0)' p_t = \mathscr{M}_t = \overline{E}\left[\Phi^*(T,0)'\ell \mid \mathscr{Y}_t\right] = \overline{E}\left[\Phi^*(T,0)'\ell\right] + \int_0^t \gamma_r\, d\overline{\mathscr{Q}}_r$. However, if u^* is Markov the process q^* is Markov, and, writing $q = q_t^*$, $\Phi = \Phi^*(t,0)$,

$$\begin{aligned}\overline{E}\left[\Phi^*(T,0)'\ell \mid \mathscr{Y}_t\right] &= \overline{E}\left[\Phi'\Phi^*(T,t)'\ell \mid q, \Phi\right]\\ &= \Phi'\overline{E}\left[\Phi^*(T,t)'\ell \mid q\right].\end{aligned}$$

Consequently, $p_t = \overline{E}\left[\Phi^*(T,t)'\ell \mid q\right]$ is a function only of q, so by the differentiation rule:

$$\begin{aligned}p_t = p_0 &+ \int_0^t \frac{\partial p_{r-}}{\partial q}\left(Aq_{r-}\,dr + Bq_{r-}\,d\overline{\mathscr{Q}}_r\right) + \int_0^t \frac{\partial p_{r-}}{\partial r}\,dr\\ &+ \sum_{0<r\le t}\left(p_r - p_{r-} - \frac{\partial p_{r-}}{\partial q}Bq_{r-}\Delta\mathscr{J}_r\right)\\ = p_0 &+ \int_0^t \left[\frac{\partial p_{r-}}{\partial q}\left(Aq_{r-} - Bq_{r-}\right) + \delta_r\right]dr + \int_0^t \delta_r\, d\overline{\mathscr{Q}}_r + \int_0^t \frac{\partial p_{r-}}{\partial r}\,dr.\end{aligned}$$

Evaluating the product:

$$\begin{aligned}\mathscr{M}_t &= \Phi^*(t,0)' p_t\\ &= p_0 + \int_0^t \Phi^*(r-,0)'\left[\frac{\partial p_{r-}}{\partial q}\left(Aq_{r-} - Bq_{r-}\right) + \delta_r\right]dr\\ &\quad + \int_0^t \Phi^*(r-,0)'\frac{\partial p_{r-}}{\partial r}\,dr + \int_0^t \Phi^*(r-,0)'\delta_r\, d\overline{\mathscr{Q}}_r\\ &\quad + \int_0^t \Phi^*(r-,0)'A' p_{r-}\,dr + \int_0^t \Phi^*(r-,0)'B' p_{r-}\,d\overline{\mathscr{Q}}_r\\ &\quad + \int_0^t \Phi^*(r-,0)'B'\delta_r\,d\overline{\mathscr{Q}}_r + \int_0^t \Phi^*(r-,0)'B'\delta_r\,dr.\end{aligned} \tag{3.24}$$

However, $\mathscr{M}_t$ is a martingale, so the sum of the dr integrals in (3.24) must be 0, and

$$\gamma_r = \Phi^*(r-,0)'\left(\delta_r + B_r'(u_r^*)\,\delta_r + B_r'(u_r^*)\,p_{r-}\right).$$

□

Theorem 3.12 *Suppose the optimal control* u^* *is Markov. Then* a.s. *in* ω *and* a.e. *in* t, u^* *satisfies the minimum principle*

$$\left\langle p_{r-}, \frac{\partial A}{\partial u}(r,u^*)(v_r - u_r^*)\,q_{r-}^*\right\rangle + \left\langle \delta_r, \frac{\partial B}{\partial u}(r,u^*)(v_r - u_r^*)\,q_{r-}^*\right\rangle \ge 0. \tag{3.25}$$

Proof Substituting γ from (3.23) into (3.20), and noting $B(I+B)^{-1}-(I+B)^{-1}B=0$, the result follows. (Substituting for B and δ gives an alternative form.) □

We now derive a forward equation satisfied by the adjoint process p:

Theorem 3.13 *With δ given by* (3.22)

$$p_t = \overline{E}\left[\Phi^*(T,0)'\ell\right] - \int_0^t A_r'(u_r^*)\,p_{r-}\,dr - \int_0^t \left(I+B_r'(u_r^*)\right)\delta_r\,dr + \int_0^t \delta_r\,d\mathscr{J}_r. \tag{3.26}$$

Proof $p_t = \Psi^*(t,0)'\mathscr{M}_t$ and from (3.13) this is

$$\begin{aligned}
&= \overline{E}\left[\Phi^*(T,0)'\ell\right] - \int_0^t A'\Psi^{*\prime}M\,dr - \int_0^t B'\Psi^{*\prime}M\,d\overline{\mathscr{Q}}_r \\
&\quad + \int_0^t \left(I+B'\right)^{-1}B'^2\Psi^{*\prime}M\,d\mathscr{J}_r + \int_0^t \Psi^{*\prime}\gamma_r d\overline{\mathscr{Q}}_r \\
&\quad - \int_0^t B'\Psi^{*\prime}\gamma_r\,d\mathscr{J}_r + \int_0^t \left(I+B'\right)^{-1}B'^2\Psi^{*\prime}\gamma_r\,d\mathscr{J}_r \\
&= \overline{E}\left[\Phi^*(T,0)'\ell\right] - \int_0^t A'p_{r-}\,dr - \int_0^t B'p_{r-}\,d\overline{\mathscr{Q}}_r \\
&\quad + \int_0^t \left(I+B'\right)^{-1}B'^2 p_{r-}\,d\mathscr{J}_r \\
&\quad + \int_0^t \left(\left(I+B'\right)\delta_r + B'p_{r-}\right)d\overline{\mathscr{Q}}_r \\
&\quad - \int_0^t \left(I+B'\right)^{-1}B'\left(\left(I+B'\right)\delta_r + B'p_{r-}\right)d\mathscr{J}_r \\
&= \overline{E}\left[\Phi^*(T,0)'\ell\right] - \int_0^t A'p_{r-}\,dr + \int_0^t \left(I+B'\right)\delta_r\,d\overline{\mathscr{Q}}_r - \int_0^t B'\delta_r\,d\mathscr{J}_r
\end{aligned}$$

and the result follows. □

However, an alternative backward equation for the adjoint process p is obtained from the observation that the sum of the bounded variation terms in (3.24) must be identically zero. Therefore, we have the following result:

Theorem 3.14 *With δ given by* (3.22) *the Markov adjoint process $p_t(q)$ is given by the backward equation*

$$\boxed{\frac{\partial p_t}{\partial t} + \frac{\partial p_t}{\partial q}\cdot\left(A^*(q)' - B^*(q)'\right)q + A^*(q)'p_t + \left(I+B^*(q)'\right)\delta_t = 0} \tag{3.27}$$

with the terminal condition

$$p_T = \ell.$$

12.4 Hybrid Conditionally Linear Process

Introduction

The filtering problem, where the state and observation processes are linear equations with Gaussian noise, has as its solution the celebrated result of Kalman and Bucy. For the related partially observed, linear quadratic control problem the separation principle applies, and the optimal control can be described explicitly as a function of the filtered state estimate.

Suppose, however, the coefficients in the linear dynamics of the state process are functions of a noisily observed Markov chain. Both the filtering problem and the related quadratic control problem are now nonlinear, and explicit solutions are either difficult to find or of little practical use. The approximation proposed below is to consider the coefficients in the linear dynamics to be functions of the filtered estimate of the Markov chain. In this way a conditional Kalman filter can be written down. These dynamics lead us to consider a conditionally linear, Gaussian control problem. By adapting techniques of Bensoussan (1982), a minimum principle and a new equation for the adjoint process are obtained.

Dynamics

Consider again a system whose state is described by two quantities, a vector $x \in \mathbb{R}^d$ and a component σ which can take a finite number of values from a set $S = \{s_1, s_2, \dots, s_N\}$. (The value x can be thought of as describing the location, velocity, etc., of an object; σ might then describe its orientation or some other operating characteristic.)

If σ evolves as a Markov process on S we can, without loss of generality, consider the corresponding process described by X evolving on the set $\{e_1, \dots, e_N\}$. Write X_t for the state of this process at time t and $p_t = E[X_t]$. Suppose the generator of the Markov chain is the Q matrix $Q(t) = (q_{ij}(t))$, $1 \le i, j \le N$, so that p_t satisfies the forward equation

$$\frac{dp_t}{dt} = Q(t)\, p_t. \tag{4.1}$$

It follows from (4.1) that on the family of σ-fields generated by X_t the process V_t is a martingale, where

$$V_t = X_t - X_0 - \int_0^t Q(s) X_s\, ds. \tag{4.2}$$

Suppose X is observed only through the noisy process z, where

$$z_t = \int_0^t \Gamma(s, X_s)\, ds + v_t. \tag{4.3}$$

Here v is a Brownian motion independent of V. Write $\{\mathscr{Z}_t\}$ for the right-continuous complete family of σ-fields generated by z and $\hat{X}_t$ for the $\mathscr{Z}$-optional projection of X, so that

$$\hat{X}_t = E[X_t \mid \mathscr{Z}_t] \qquad \text{a.s.}$$

Write $\Delta(s)$ for the vector $(\Gamma(s,e_1),\ldots,\Gamma(s,e_N))$ and $\operatorname{diag}\Delta(s)$ for the diagonal matrix with diagonal $\Delta(s)$.

With an innovation process $\hat{v}_t$ given by $d\hat{v}_t = dz_t - \langle \Delta(t), \hat{X}_t \rangle dt$ it is shown in, for example Elliott (1982b), that the equation for the filtered estimate $\hat{X}$ is

$$\hat{X}_t = \hat{X}_0 + \int_0^t Q(s)\hat{X}_s\, ds + \int_0^t \left(\operatorname{diag}\Delta(s) - \langle \Delta(s), \hat{X}_s \rangle I\right)\hat{X}_s\, d\hat{v}_s. \tag{4.4}$$

Here $\langle , \rangle$ denotes the inner product in $\mathbb{R}^N$ and I is the $N\times N$ identity matrix. Equation (4.4) provides a recursive expression for the best least squares estimate $\hat{X}$ of X given the observations z.

Suppose now the x component of the state is described by the equation

$$dx_t = A(X_t)x_t\, dt + \rho_t dX_t + B(X_t)\, dw_t. \tag{4.5}$$

Here $x \in \mathbb{R}^d$, $w_t = (w_t^1,\ldots,w_t^n)$ is an n-dimensional Brownian motion independent of V and v, and $A(X_t)$, $B(X_t)$ and ρ_t are, respectively, $d\times d$, $d\times n$, and $d\times N$ matrices.

Suppose the x process is observed through the observations of y, where

$$dy_t = Hx_t\, dt + G\, d\beta_t. \tag{4.6}$$

Here $y \in \mathbb{R}^p$, $\beta_t = (\beta_t^1,\ldots,\beta_t^m)$ is an m-dimensional Brownian motion independent of V, v and w and H, (resp. G), is a $p\times d$ (resp. nonsingular $p\times m$) matrix.

Now the y observations also provide information about X, so that altogether we have the states x and X given by (4.5) and

$$X_t = X_0 + \int_0^t Q(s)X_s\, ds + V_t, \tag{4.7}$$

with observations given by (4.3) and (4.6). Write $\{\overline{\mathscr{Y}}_t\}$ for the right continuous, complete filtration generated by y and z, and denote by a bar the $\overline{\mathscr{Y}}$-optional projection of a process so that, for example,

$$\overline{X}_t = E\left[X_t \mid \overline{\mathscr{Y}}_t\right] \qquad \text{a.s.}$$

Define the innovation processes v^*, β^* by

$$\begin{aligned} dv_t^* &= dz_t - \langle \Delta(t), \overline{X}_t \rangle dt, \\ d\beta_t^* &= G^{-1}(dy_t - H\overline{x}_t\, dt). \end{aligned}$$

For vectors $x = (x_1, \ldots, x_d)' \in \mathbb{R}^d$ and $y = (y_1, \ldots, y_n)' \in \mathbb{R}^n$, write xy' for the $d \times n$ matrix (a_{ij}), $a_{ij} = x_i y_j$. Then the filtered estimate of $\binom{x_t}{X_t}$ is given by

$$\begin{aligned} d\begin{pmatrix} \overline{x}_t \\ \overline{X}_t \end{pmatrix} &= \begin{pmatrix} \overline{A(X_t)\,x_t} \\ \mathscr{Q}_t \overline{X}_t \end{pmatrix} dt + \begin{pmatrix} \rho_t \mathscr{Q}_t \overline{X}_t \\ 0 \end{pmatrix} dt \\ &+ \Big(\overline{(x_t, X_t)'\,(Hx_t, \langle \Delta(t), X_t \rangle)'} \\ &- (\overline{x}_t, \overline{X}_t)'\,(H\overline{x}_t, \langle \Delta(t), \overline{X}_t \rangle)' \Big) \begin{pmatrix} G^{-1} & 0 \\ 0 & I \end{pmatrix} \begin{pmatrix} d\beta_t^* \\ dv_t^* \end{pmatrix}. \end{aligned}$$

This is a nonlinear equation. However, the approximation we shall make is to suppose that most of our information about X comes from the observations of z and that we can replace X by $\hat{X}$ in (4.5), where $\hat{X}$ is given by (4.4). Note that $\hat{X}$ is independent of w and β. We can, therefore, state the following result:

Lemma 4.1 *Suppose the state x_t is approximated by $\tilde{x}_t$ where*

$$d\tilde{x}_t = A(\hat{X}_t)\tilde{x}_t\,dt + \rho_t\,d\hat{X}_t + B(\hat{X}_t)dw_t. \tag{4.8}$$

Here $\hat{X}$ is given by (4.4). *Suppose $\tilde{x}$ is observed through the process $\tilde{y}$ where*

$$d\tilde{y}_t = H\tilde{x}_t\,dt + G\,d\beta_t. \tag{4.9}$$

Write $\{\mathscr{Y}_t\}$ for the right-continuous, complete filtration generated by $\tilde{y}$ and $\hat{x}$ for the $\mathscr{Y}$-optional projection of x, so that $\hat{x}_t = E[x_t \mid \mathscr{Y}_t]$ a.s. Then

$$\boxed{\begin{aligned} d\hat{x}_t &= A(\hat{X}_t)\hat{x}_t\,dt + \rho_t\,d\hat{X}_t + P_t H (GG')^{-1}\,d\hat{\beta}_t, \\ \hat{x}_0 &= Ex_0, \end{aligned}} \tag{4.10}$$

where

$$G \cdot d\hat{\beta}_t = dy_t - H\hat{x}_t\,dt \tag{4.11}$$

and P_t is the matrix solution of the Riccati equation

$$\boxed{\begin{aligned} \dot{P}_t &= B(\hat{X}_t)B(\hat{X}_t)' - P_t H' (GG')^{-1} H P_t + A(\hat{X}_t)P_t + P_t A(\hat{X}_t), \\ P_0 &= cov\,x_0. \end{aligned}} \tag{4.12}$$

Proof The proof is left as an exercise. The Kalman filter applies. □

Equations (4.4), (4.10), (4.11), and (4.12) therefore give a finite-dimensional filter for $\hat{x}_t$, which is a conditionally Gaussian random variable given $\hat{X}$ and $\mathscr{Y}_t$.

Note that

$$\begin{aligned} A(\hat{X}_t) &= \sum_{i=1}^{N} A(e_i)\langle e_i, \hat{X}_t \rangle, \\ B(\hat{X}_t) &= \sum_{i=1}^{N} B(e_i)\langle e_i, \hat{X}_t \rangle. \end{aligned}$$

Remark 4.2

$$\begin{aligned} d(\bar{x}_t - \hat{x}_t) = &\big(\overline{A(X_t)x_t} - A(\hat{X}_t)\hat{x}_t\big)dt + \rho_t \mathscr{Q}_t(\overline{X}_t - \hat{X}_t)dt \\ &+ \big(\overline{x(Hx)'} - \bar{x}(H\bar{x})'\big)G^{-1}d\beta_t^* \\ &+ \big(\overline{x\langle\Delta(t), X_t\rangle'} - \bar{x}\langle\Delta(t), \overline{X}_t\rangle'\big)dv_t^* + P_t H(GG')^{-1}d\hat{\beta}_t. \end{aligned}$$

Therefore, with tr denoting the trace of a matrix

$$\begin{aligned} d(\bar{x}_t - \hat{x}_t)^2 = &\, 2(\bar{x}_t - \hat{x}_t)d(\hat{x}_t - \hat{x}_t) \\ &+ tr\big(\overline{x(Hx)'} - \bar{x}(H\bar{x})'\big)(G'G)^{-1}(\overline{Hxx'} - H\overline{xx}')dt \\ &+ tr\big(\overline{\langle\Delta(t), X\rangle x'} - \langle\Delta(t), \overline{X}\rangle\bar{x}'\big)\big(\overline{x\langle\Delta(t), X\rangle'} - \bar{x}\langle\Delta(t), \overline{X}\rangle'\big)dt \\ &+ tr P_t H(GG')^{-2}H'P_t' \cdot dt \\ &+ tr P_t H(GG')^{-1} \cdot G'\big(\overline{x(Hx)'} - \bar{x}(H\bar{x})'\big)dt. \end{aligned}$$

Taking expectations the martingale terms disappear and, under integrability or boundedness conditions on the coefficient matrices, an estimate of order $o(t)$ for $E(\bar{x}_t - \hat{x}_t)^2$ can be obtained. However, this does not appear too useful.

■

Hybrid Control

Suppose the state equation for x now contains a control term, so that

$$dx_t = A(X_t)x_t\,dt + \rho_t dX_t + C_t u(t)\,dt + B(X_t)\,dw_t. \tag{4.13}$$

The observation process is again y, where

$$dy_t = Hx_t\,dt + G\,d\beta_t. \tag{4.14}$$

Assume the control parameter u takes values in some space $\mathbb{R}^k$ and the admissible control functions are those which are predictable with respect to the right-continuous, complete filtration generated by y and $\hat{X}$. C_t is a $d \times k$ matrix.

Suppose the control $\{u_t\}$ is to be chosen to minimize the cost

$$V_1(u) = E\left[\int_0^T (x_t'D_t x_t + u_t'R_t u_t)\,dt + x_T'Fx_T\right]. \tag{4.15}$$

Here Q_t, R_t, and F are matrices of appropriate dimensions and R_t is non-singular. Then (4.3), (4.7), (4.13), (4.14), and (4.15) describe a nonlinear partially observed stochastic control problem whose solution is in general difficult. To obtain a related completely observed problem the approximation we propose is that X_t is replaced by its filtered estimate $\hat{X}_t$ in (4.13) giving a process $\tilde{x}$, where

$$d\tilde{x}_t = A(\hat{X}_t)\tilde{x}_t\,dt + \rho_t\,d\hat{X}_t + C_t u(t)\,dt + B(\hat{X}_t)dw_t. \tag{4.16}$$

The observation process is now $\tilde{y}$, where

$$d\tilde{y}_t = H\tilde{x}_t\,dt + G\cdot d\beta_t \tag{4.17}$$

and the admissible controls are the predictable functions with respect to the right-continuous, complete filtrations generated by $\tilde{y}$ and z.

The cost is taken to be

$$J(u) = E\left[\int_0^T \left(\tilde{x}_t' D_t \tilde{x}_t + u_t' R_t u_t\right)dt + \tilde{x}_T' F \tilde{x}_T\right]. \tag{4.18}$$

Equations (4.16), (4.17), and (4.18) describe a partially observed, linear, quadratic Gaussian control problem which is parametrized by $\hat{X}_t$, a process which is independent of w and β. However, we cannot apply the separation principle, as in Davis (1977), because the coefficients in (4.16) are functions of $\hat{X}$. The usual form of the separation principle involves the solution of a Riccati equation solved backward from the final time T, and we do not know the future values of X. We therefore proceed as follows to derive a minimum principle satisfied by an optimal control. We are in effect considering a completely observed optimal control problem with state variables $\hat{X}$ and $\hat{x}$, where $\hat{X}$ is given by

$$\hat{X}_t = \hat{X}_0 + \int_0^t Q(s)\hat{X}(s)\,ds + \int_0^t \Pi(s)\hat{X}(s)\,d\hat{v}_s \tag{4.19}$$

and

$$\hat{x}_t = m_0 + \int_0^t A(\hat{X}_s)\hat{x}_s\,ds + \int_0^t \rho_s\,d\hat{X}_s + \int_0^t C_s u_s\,ds + \int_0^t P_s H\left(GG'\right)^{-1} d\hat{\beta}_s. \tag{4.20}$$

Here $\Pi(s) = \operatorname{diag}\Delta(s) - \langle \Delta(s), \hat{X}_s\rangle I$ and $m_0 = Ex_0$. Note from (4.12) that the covariance P_t depends on $\hat{X}$. In terms of $\hat{x}$ and P the cost corresponding to control $\{u_t\}$ is (Davis 1977),

$$J(u) = E\left[\int_0^T \left(\hat{x}_t' D_t \hat{x}_t + u_t' R_t u_t\right)dt + \hat{x}_T' F \hat{x}_T + \int_0^T tr(P_t D_t)\,dt + tr(P_T F)\right].$$

The last two terms do not depend on the control, so we shall consider a problem with dynamics given by (4.19) and (4.20), and a cost corresponding to a control u given by

$$J(u) = E\left[\int_0^T \left(\hat{x}_t' D_t \hat{x}_t + u_t' R_t u_t\right)dt + \hat{x}_T' F \hat{x}_T\right]. \tag{4.21}$$

Write $\{\tilde{\mathscr{Y}}_t\}$ for the right-continuous filtration generated by $\tilde{y}$ and z. Write $L_Y^2[0,T] = \{u(t,\omega) \in L^2([0,T]\times\Omega; dt\times dP, \mathbb{R}^k)$ such that for a.e. t, $u(t,\cdot) \in$

$L^2(\Omega, \tilde{\mathscr{Y}}_t, P, \mathbb{R}^k)\}$. Assume U is a compact, convex subset of $\mathbb{R}^k$. Then the set of admissible controls is the set

$$\underline{U} = \left\{u \in L^2_Y[0,T] : u(t,\omega) \in U \text{ a.e. a.s.}\right\}.$$

Suppose there is an optimal control u^*. We shall consider perturbations of u^* of the form $u_\theta(t) = u^*(t) + \theta(v(t) - u^*(t))$ where v is any other admissible control and $\theta \in [0,1]$. Then

$$J(u_\theta) \geq J(u^*).$$

Following and simplifying techniques of Bensoussan (1982), our minimum principle is obtained by investigating the Gateaux derivative of J as a functional on the Hilbert space $L^2_Y[0,T]$. Write $\hat{x}^*$ for the trajectory corresponding to the optimal u^*. Then

$$d\hat{x}^*_t = A(\hat{X}_t)\hat{x}^*_t\,dt + \rho_t d\hat{X}_t + C_t u^*_t\,dt + P_t H\left(GG'\right)^{-1} d\hat{\beta}_t.$$

Given any sample path $\hat{X}$, $\hat{X}_t$ will be considered as a time-varying parameter. Write $\Phi(\hat{X},t,s)$ for the matrix solution of the equation

$$\frac{d}{dt}\Phi(\hat{X},t,s) = A(\hat{X}_t)\Phi(\hat{X},t,s)dt$$

with initial condition $\Phi(X,s,s) = I$.

Lemma 4.3 *Suppose $v \in \underline{U}$ is such that $u^*_\theta = u^* + \theta v \in \underline{U}$ for $\theta \in [0,\alpha]$. Write $\hat{x}^\theta$ for the solution of* (4.20) *associated with u^*_θ. Then $\psi_t = \partial \hat{x}^\theta_{0,t}/\partial\theta|_{\theta=0}$ exists a.s. and*

$$\psi_t = \Phi(\hat{X},t,0)\int_0^t \Phi(\hat{X},s,0)^{-1} C_s v_s\,ds. \tag{4.22}$$

Proof The estimated state is given by

$$\begin{aligned}\hat{x}^\theta_{0,t} &= x_0 + \int_0^t A(\hat{X}_s)\hat{x}^\theta_s\,ds + \int_0^t \rho_s\,d\hat{X}_s \\ &\quad + \int_0^t C_s(u^*_s + \theta v_s)\,ds + \int_0^t P_s H\left(GG'\right)^{-1} d\hat{\beta}_s.\end{aligned} \tag{4.23}$$

From the result of Blagovescenskii and Freidlin (1961), on the differentiability of solutions of stochastic differential equations with respect to a parameter, (4.23) can be differentiated to give

$$\psi_t = \int_0^t A(\hat{X}_s) z_s\,ds + \int_0^t C_s v_s\,ds. \tag{4.24}$$

The solution of (4.24) is then given by (4.22). □

Notation 4.4 *Consider the martingale*

$$\mathscr{M}_t = E\left[2\int_0^T \hat{x}^{*\prime}_s D_s \Phi(\hat{X},s,0)ds + 2\hat{x}^{*\prime}_T F\Phi(\hat{X},T,0) \mid \overline{\mathscr{Y}}_t\right]$$

and write

$$\begin{aligned}
\xi_t &= \mathscr{M}_t - 2\int_0^t \hat{x}_s^{*\prime} D_s \Phi(\hat{X},s,0)\,ds, \\
p_t &= \xi_t \cdot \Phi(\hat{X},s,0)^{-1}, \\
\eta_{0,t} &= \int_0^t \Phi(\hat{X},s,0)^{-1} C_s v_s\,ds.
\end{aligned} \tag{4.25}$$

Then there are square-integrable processes γ and λ such that the martingale M has a representation as a stochastic integral

$$\begin{aligned}
\mathscr{M}_t = E\left[2\int_0^T \hat{x}_s^{*\prime} D_s \Phi(\hat{X},s,0)\,ds + 2\hat{x}_T^{*\prime} F \Phi(\hat{X},T,0)\right] \\
+ \int_0^t \gamma_s\,d\hat{\beta}_s + \int_0^t \lambda_s\,d\hat{v}_s.
\end{aligned}$$

Lemma 4.5 *The derivative of the cost is given by*

$$\left.\frac{dJ\left(u_\theta^*\right)}{d\theta}\right|_{\theta=0} = E\left[\int_0^T \left(p_s C_s v_s + 2u_s^{*\prime} R_s v_s\right) ds\right].$$

Proof From (4.21) the cost is

$$J\left(u_\theta^*\right) = E\left[\int_0^T \left(\hat{x}_s^{\theta\prime} D_s \hat{x}_s^{\theta} + u_{\theta s}^{*\prime} R_s u_{\theta s}^{*}\right) ds + \hat{x}_T^{\theta\prime} F \hat{x}_T^{\theta}\right].$$

Therefore, differentiating we see

$$\left.\frac{dJ\left(u_\theta^*\right)}{d\theta}\right|_{\theta=0} = E\left[2\int_0^T \left(\hat{x}_s^{*\prime} D_s z_s + u_s^{*\prime} R_s v_s\right) ds + 2\hat{x}_T^{*\prime} F z_T\right]. \tag{4.26}$$

Using the above notation

$$\begin{aligned}
\psi_t &= \Phi(\hat{X},t,0)\eta_{0,t}, \\
\xi_T &= 2\hat{x}_T^{*\prime} F \Phi(\hat{X},T,0)
\end{aligned}$$

so

$$\begin{aligned}
\xi_T \eta_{0,T} &= 2\hat{x}_T^{*\prime} F \psi_T \\
&= \int_0^T \xi_s \Phi(\hat{X},s,0)^{-1} C_s v_s\,ds - 2\int_0^T \hat{x}_s^{*\prime} D_s \Phi(\hat{X},s,0)\eta_{0,s}\,ds \\
&\quad + \int_0^T \gamma_s \eta_{0,s}\,d\hat{\beta}_s + \int_0^T \lambda_s \eta_{0,s}\,d\hat{v}_s.
\end{aligned}$$

Substituting in (4.26)

$$
\begin{aligned}
\left.\frac{dJ\left(u_{\theta}^{*}\right)}{d\theta}\right|_{\theta=0} &= E\left[\xi_T \eta_{0,T} + 2\int_0^T \hat{x}_s^{*\prime} D_s \Phi\left(\hat{X}, s, 0\right) \eta_{0,s}\, ds + 2\int_0^T u_s^{*\prime} R_s v_s\, ds\right] \\
&= E\left[\int_0^T \xi_s \Phi\left(\hat{X}, s, 0\right)^{-1} C_s v_s\, ds + 2\int_0^T u_s^{*\prime} R_s v_s\, ds\right] \\
&= E\left[\int_0^T \left(p_s C_s v_s + 2u_s^{*\prime} R_s v_s\right) ds\right].
\end{aligned}
$$

□

Now take v to be of the form $v - u^*$ so that $u_\theta = u^* + \theta\left(v - u^*\right) \in \underline{U}$. Applying Proposition 4.5 to $J\left(u_\theta\right)$ we have the following result.

Corollary 4.6 *The optimal control satisfies the minimum principle*

$$
\boxed{p_s C_s u_s^* + 2u_s^{*\prime} R_s u_s^* = \min_{v \in U}\left(p_s C_s v + 2u_s^{*\prime} R_s v\right) \text{ a.e. a.s.}} \tag{4.27}
$$

Proof u^* is optimal so $dJ(u_\theta)/d\theta|_{\theta=0} \geq 0$, that is for any other admissible control v

$$
E\left[\int_0^T \left(p_s C_s\left(u_s^* - v_s\right) + 2u_s^{*\prime} R_s\left(u_s^* - v_s\right)\right) ds\right] \geq 0.
$$

v can equal u^* except on an arbitrary set of the form $A \times [s, s+h]$, $A \in F_s$. Therefore, a.e. dt and a.s. dP,

$$
p_s C_s\left(u_s^* - v_s\right) + 2u_s^{*\prime} R_s\left(u_s^* - v_s\right) \geq 0,
$$

where the adjoint variable p is given by (4.26). □

Remark 4.7 From Haussmann (1981a) we know the optimal control u^* is feedback, in the sense that at time t it is a function of the states $\hat{x}_t$ and $\hat{X}_t$. However, to avoid derivatives of u^* we suppose u^* always follows the trajectories of $\hat{x}^*$ and $\hat{X}$, even if these trajectories are perturbed. By the Markov property we, therefore, have that p_t is a function of $x = \hat{x}_t$ and $\phi = \hat{X}_t$. Writing $\Phi = \Phi\left(\hat{X}, t, 0\right)$ and $y = 2\int_0^t \hat{x}_s^{*\prime} D_s \Phi\left(\hat{X}, s, 0\right) ds$ we have that $\Psi(t, x, y, \Phi) = p_t(x, \phi) \cdot \Phi + y = \mathcal{M}_t$, a martingale.

■

If we write down the Ito expansion of Ψ the sum of the terms integrated with respect to time must be zero. After division by Φ we have the following equation satisfied by the adjoint process $p = p(t, x, \phi)$.

Lemma 4.8 *Denote the Hessian of p with respect to x [resp. X] by $\partial^2 p/\partial x^2$ [resp. $\partial^2 p/\partial \phi^2$] and write*

$$
\begin{aligned}
\Gamma_t &= p_t H\left(GG'\right)^{-1} + \rho_t \Pi(t) \hat{X}_t, \\
\Lambda_t &= \Pi(t) \hat{X}_t,
\end{aligned}
$$

(4.28) is in fact

$$\frac{\partial p}{\partial t} + pA(\hat{X}_t) + \frac{\partial p}{\partial x} \cdot \left(A(\hat{X}_t)\hat{x}_t + \rho_t Q\hat{X}_t + C_t u_t^* \right)$$
$$+ \frac{\partial p}{\partial \phi} \cdot Q\hat{X}_t + \frac{1}{2} \sum_{i,j=1}^{d} \frac{\partial^2 p}{\partial x_i \partial x_j} \Gamma_t(i)\Gamma_t(j) + \frac{1}{2} \sum \frac{\partial^2 p}{\partial x_i \partial x_j} \Lambda_t(i)\Lambda_t(j)$$

and $tr\left(\Gamma_t'(\partial^2 p/\partial x^2)\Gamma_t\right)$ *[resp.* $tr\left(\Lambda_t'(\partial^2 p/\partial \phi^2)\Lambda_t\right)$*] for the vector with components* $tr\left(\Gamma_t'(\partial^2 p_i/\partial x^2)\Gamma_t\right)$ *[resp.* $tr\left(\Lambda_t'(\partial^2 p_i/\partial \phi^2)\Lambda_t\right)$*].*

Then

$$\boxed{\begin{aligned}&\frac{\partial p}{\partial t} + pA(\hat{X}_t) + \frac{\partial p}{\partial x} \cdot \left(A(\hat{X}_t)\hat{x}_t + \rho_t Q\hat{X}_t + C_t u_t^* \right)\\ &\quad + \frac{\partial p}{\partial \phi} \cdot Q\hat{X}_t + \frac{1}{2} tr\left(\Gamma_t' \frac{\partial^2 p}{\partial x^2} \Gamma_t \right) + \frac{1}{2} tr\left(\Lambda_t' \frac{\partial^2 p}{\partial \phi^2} \Lambda_t \right) = 0,\end{aligned}} \tag{4.28}$$

with terminal condition $p(T, x, \phi) = 2xF$.

12.5 Problems and Notes

Problems

1. Show that under the probability measure $\overline{P}$ defined in Section 12.2 the process $\mathcal{J}$ which counts the number of jumps of the Markov chain is a standard Poisson process (see Appendix A).
2. Using the innovations approach derive the filtered estimate (4.4).
3. Derive the Kalman filter and the Riccati equation given by (4.10) and (4.12).

Notes

The optimal control of a continuous time partially observed Markov chain observed through Brownian noise or by a counting process is derived using the robust filtering equations introduced by Clark. New simpler dynamic programming results are presented using the robust information states of the Clark filters. Also, a minimum principle is derived in terms of new equations for an adjoint process in terms of the robust information states. Generalizations of the formulations to the case of risk sensitive optimal filtering and control are immediate. Perhaps the improvement to filtering applications of the Clark results are more dramatic than for control as presented here, but this can be a subject of further exploration.

In this chapter the control of HMM in continuous time is discussed. The first situation considered is that of a continuous-time finite-state Markov chain which is not observed directly; rather, there is a conditional Poisson process whose rate, or intensity, is a function of the Markov chain. An equation for the unnormalized distribution q of the Markov chain, given the observations, is derived. The control problem is then formulated in separated form with q as the new state variable. A minimum principle is obtained.

Secondly, the case when the counting observation process is related to the jumps of the Markov chain is discussed. In this situation there is non-zero correlation between jumps in the observation process and jumps of the Markov chain, so new features arise. However, again a minimum principle is obtained together with explicit formulae for the adjoint process.

The optimal control of a Markov chain observed in Gaussian noise is treated in Section 12.3. Equations for the adjoint process are again derived.

The chapter closes with a discussion of the optimal control of a linear stochastic system whose coefficients are functions of a discrete-state Markov chain.

Equation (4.28) is a backward parabolic equation. Bismut (1978b) considers a forward equation, with a terminal condition, for the adjoint process.

Other work discussing similar situations to that of Section 12.4 includes Fragoso (1988), Yang and Mariton (1991), and Mariton (1990).

Appendix A
Basic Probability Concepts

Definitions

Let Ω be a set of points. A nonempty class $\mathscr{B}$ of subsets of Ω is called a *σ-field* if $\mathscr{B}$ is closed under complementation and countable unions. The sets $B \in \mathscr{B}$ are said to be *measurable*.

A function P: $\mathscr{B} \to [0,1]$ is called a *probability measure*

1. If $P(\Omega) = 1$;
2. if B_k is a countable sequence of disjoint sets in $\mathscr{B}$, then $P(\bigcup B_k) = \sum P(B_k)$.

The triple $(\Omega, \mathscr{B}, P)$ is *called a probability space*. A set $B \in \mathscr{B}$ is sometimes referred to as an event.

If $X : \Omega \to \mathbb{R}$ is a function we let $\sigma(X) = \sigma(\{\omega \mid X(\omega) \le x\}, x \in \mathbb{R})$. This is the smallest σ-field containing all subsets of the form $\{\omega : X(\omega) \le x\}$. If X is $\mathscr{B}$-measurable, that is, if $\sigma(X)$ is in $\mathscr{B}$, then we call X a random variable.

For $C \in \mathscr{B}$ we define $I(C)$, the *indicator function* of C, as follows

$$\begin{aligned} I(C)(\omega) &= 1 \qquad \text{if } \omega \in C \\ &= 0 \qquad \text{otherwise.} \end{aligned}$$

A random variable is called simple if there are finitely many real numbers $x_1, \ldots, x_k$ such that $\sum P(X = x_i) = 1$. Then $B_i = \{\omega : X(\omega) = x_i\} \in \mathscr{B}$ and $X = \sum_{i=1}^{k} x_i I(B_i)$.

The integral of a simple random variable is defined as

$$\int X\, dP = \sum_i x_i P(X = x_i) = \sum x_i P(B_i).$$

If X is a nonnegative random variable, its integral is defined as

$$\int X\, dP = \lim_{k \to \infty} \int X_k\, dP,$$

where $\{X_k\}$ is a sequence of simple random variables which increases pointwise to X. The existence of such an increasing sequence of functions is established in, for example, Shiryayev (1986). Furthermore, the limit of the sequence of integrals is independent of the increasing sequence of functions. If the above limit is finite, X is said to be integrable. If both its positive and negative parts, $X^+ = XI(X \geq 0)$ and $X^- = -XI(X < 0)$ are integrable, X is said to be *integrable* and we define

$$\int X\,dP = \int X^+\,dP - \int X^-\,dP.$$

The expected value of X is written $E(X)$, and is by definition

$$E(X) = \int_\Omega X\,dP = \int_{\mathbb{R}} x dF(x).$$

Here $F(x) = P(X \leq x)$ is the usual cumulative distribution function of X. For a simple random variable X, then

$$E(X) = \sum_i x_i P(X = x_i).$$

For $C \in \mathscr{B}$, we write

$$\int_C X\,dP = \int I(C) X\,dP.$$

A sequence of random variables $\{X_k\}$ is said to converge *almost surely* $(a.s)$ if

$$P\left[\lim_{k\to\infty} X_k(\omega) \text{ exists and is finite}\right] = 1.$$

The value $\{X_k\}$ converges to X in L_1 if $E[|X_k - X|] \to 0$ as $k \to \infty$. The value $\{X_k\}$ converges to X *in probability* if for each $\varepsilon > 0$ the sequence

$$P[|X_k - X| > \varepsilon] \to 0 \qquad \text{as } k \to \infty.$$

Theorem 1.1 (Radon-Nikodym) *If P and $\overline{P}$ are two probability measures on* $(\Omega, \mathscr{B})$ *such that for each $B \in \mathscr{B}$, $P(B) = 0$ implies $\overline{P}(B) = 0$, then there exists a non-negative random variable Λ, such that $\overline{P}(C) = \int_C \Lambda\,dP$ for all $C \in \mathscr{B}$. We write $d\overline{P}/dP|_{\mathscr{B}} = \Lambda$.*

For a proof see Wong and Hajek (1985). The value Λ is the density of $\overline{P}$ with respect to P. When Ω is a finite, or discrete, sample space $\Lambda(\omega) = \overline{P}/P$.

If A, B are two events, then we define the probability of A given B as

$$P(A \mid B) = \frac{P(A \text{ and } B)}{P(B)},$$

provided $P(B) > 0$. Otherwise $P(A \mid B)$ is left undefined. More generally we have the concept of conditional expectation which we now define.

Let $X \in L_1$ and $\mathscr{A}$ be a sub-σ-field of $\mathscr{B}$. If X is nonnegative and integrable we can use the Radon-Nikodym theorem to deduce the existence of an $\mathscr{A}$-measurable random variable, denoted by $E(X \mid \mathscr{A})$, which is uniquely determined except on an event of probability zero, such that

$$\int_A X\, dP = \int_A E[X \mid \mathscr{A}]\, dP \tag{1.1}$$

for all $A \in \mathscr{A}$. For a general integrable random variable X we define $E[X \mid \mathscr{A}]$ as $E[X^+ \mid \mathscr{A}] - E[X^- \mid \mathscr{A}]$. $E(X \mid \mathscr{A})$ is called the *conditional expectation* of X given $\mathscr{A}$. If A is an event, then we write $P(A \mid \mathscr{A})$ as shorthand for $E(I(A) \mid \mathscr{A})$.

The following is a list of classical results. If $\mathscr{A}_1$ and $\mathscr{A}_2$ are two sub-σ-fields of $\mathscr{B}$ such that $\mathscr{A}_1 \subset \mathscr{A}_2$, then

$$E(E(X \mid \mathscr{A}_1) \mid \mathscr{A}_2) = E(E(X \mid \mathscr{A}_2) \mid \mathscr{A}_1) = E(X \mid \mathscr{A}_1). \tag{1.2}$$

If $X, Y, XY \in L_1$, and Y is $\mathscr{A}$-measurable, then

$$E(XY \mid \mathscr{A}) = YE(X \mid \mathscr{A}). \tag{1.3}$$

If X and Y are independent, then

$$E(X \mid \sigma(Y)) = E(X). \tag{1.4}$$

A *stochastic process* is a mathematical model for any phenomenon evolving or varying in time (or space, etc.) subject to random influences (e.g., the stock market price of a commodity observed in time, the distribution of colors or shades in a noisy picture observed in an unordered two-dimensional lattice, etc.) To capture this randomness we list all outcomes into a measurable space $(\Omega, \mathscr{B})$ usually called the *sample space* on which probability measures can be placed. Thus a stochastic process is a mapping $X_{(\text{index})}(\omega)$ from $\Omega \times \{\text{index space}\}$ into a second measurable space, called the *state space* or the range space.

For a fixed simple outcome ω, $X_{(.)}(\omega)$ is a deterministic function describing *one* possible trajectory or path followed by the process starting from some possible initial position.

If the time index is frozen at t, say, then we have a usual random variable $X_t(.)$.

Theorem 1.2 (Kolmogorov's Extension Theorem) *For all $\tau_1, \ldots, \tau_k, k \in \mathbb{N}$ and τ in the time index, let $P_{\tau_1,\ldots,\tau_k}$ be probability measures on $\mathbb{R}^{nk}$ such that*

$$P_{\tau_{\sigma(1)},\ldots,\tau_{\sigma(k)}}(F_1 \times \cdots \times F_k) = P_{\tau_1,\ldots,\tau_k}\left(F_{\sigma^{-1}(1)} \times \cdots \times F_{\sigma^{-1}(k)}\right)$$

for all permutations σ on $\{1, 2, \ldots, k\}$ and

$$\begin{aligned} &P_{\tau_1,\ldots,\tau_k}(F_1 \times \cdots \times F_k) \\ &\quad = P_{\tau_1,\ldots,\tau_k,\tau_{k+1},\ldots,\tau_{k+m}}(F_1 \times \cdots \times F_k \times \mathbb{R}^n \times \cdots \times \mathbb{R}^n) \end{aligned}$$

for all $m \in \mathbb{N}$, and the set on the right-hand side has a total of $k+m$ factors. Then there exists a probability space $(\Omega, \mathscr{F}, P)$ and a stochastic process $\{X_\tau\}$ on Ω into $\mathbb{R}^n$ such that

$$P_{\tau_1,\ldots,\tau_k}(F_1 \times \cdots \times F_k) = P\left[X_{\tau_1} \in F_1, \ldots, X_{\tau_k} \in F_k\right],$$

for all τ_i in the time set $k \in \mathbb{N}$ and all Borel sets F_i.

Suppose, from now on that the index space is either $\mathbb{R}^+$ or $\mathbb{N}$. To keep track, to record, and to benefit from the flow of information accumulating in time and to give a mathematical meaning to the notions of *past*, *present*, and *future* the concept of *filtration* is introduced. For this we equip our sample space $(\Omega, \mathscr{B})$ with a nondecreasing family $\{\mathscr{B}_\tau, \tau \geq 0\}$ of sub-σ-fields of $\mathscr{B}$ such that $\mathscr{B}_\tau \subset \mathscr{B}_{\tau'}$ whenever $\tau \leq \tau'$. We define $\mathscr{B}_\infty = \sigma\left(\bigcup_{\tau \geq 0} \mathscr{B}_\tau\right)$.

When the time index τ is in $\mathbb{R}$ we are led naturally to introduce the concepts of *right-continuity* and *left-continuity* of a filtration. A filtration $\{\mathscr{B}_\tau, \tau \geq 0\}$ is right-continuous if $\mathscr{B}_\tau$ contains events *immediately after* τ, that is $\mathscr{B}_\tau = \bigcap_{\varepsilon > 0} \mathscr{B}_{\tau+\varepsilon}$.

The stochastic process X is *adapted* to the filtration $\{\mathscr{B}_\tau, \tau \geq 0\}$ if for each $\tau \geq 0$ X_τ is a $\mathscr{B}_\tau$-measurable random variable.

Roughly speaking, a stochastic process X is *predictable* if knowledge about the behavior of the process is left-continuous, or, more precisely, if it is measurable with respect to the σ-field generated by the family of all left-continuous adapted stochastic processes. The concept of predictability is more easily understood in the discrete-time situation where X_k is $\mathscr{B}_k$-predictable if X_k is $\mathscr{B}_{k-1}$-measurable. A stochastic process X is *optional* if it is measurable with respect to the σ-field on $\Omega \times \{\text{time set}\}$ generated by the family of all right continuous adapted stochastic processes which have left limits. Note the concepts of optional and predictable σ-fields on $\Omega \times \{\text{time set}\}$ involve measurability concepts in both ω and τ.

The stochastic process X is a *martingale* with respect to the filtration $\{\mathscr{B}_\tau\}$ if it is $\mathscr{B}_\tau$-adapted, $E\left[|X_\tau|\right] < \infty$ for all τ and $E\left[X_\tau \mid \mathscr{B}_{\tau'}\right] = X_{\tau'}$ for all $\tau' \leq \tau$.

The *variation* of a real-valued f over an interval $[a,b]$ is

$$\int_a^b |df| \triangleq \sup_\pi \sum_{k=0}^{n-1} |f(t_{k+1}) - f(t_k)|,$$

where π is the set of all partitions of the interval $[a,b]$.

A stochastic process X is of *integrable variation* if

$$E\left[\int_0^\infty |dX_s|\right] < \infty.$$

Given an adapted stochastic process X, if there exists a right-continuous predictable process with finite variation and left limits A such that $X_t - A_t$ is a martingale, then A is called the *compensator* of X.

A *special semimartingale* is, roughly speaking, the sum of a martingale and a predictable process of bounded variation.

Theorem 1.3 (Girsanov) *Suppose W_t, $t \in [0,T]$, is an m-dimensional Brownian motion on a filtered space $\{\Omega, \mathscr{B}, \mathscr{B}_t, P\}$. Let $f : \Omega \times [0,T] \to \mathbb{R}^m$ be a predictable process such that*

$$\int_0^T |f_t|^2 \, dt < \infty \text{ a.s.}$$

Write

$$\xi_0^t(f) = \exp\left[\sum_{i=1}^m \int_0^t f_s^i dW_s^i - \frac{1}{2}\int_0^t |f_s|^2 \, ds\right]$$

and suppose

$$E\left[\xi_0^T(f)\right] = 1.$$

If $\overline{P}$ is the probability measure on $\{\Omega, \mathscr{B}\}$ defined by $d\overline{P}/dP = \xi_0^T(f)$, then $\tilde{W}_t$ is an m-dimensional Brownian motion on $\left\{\Omega, \mathscr{B}, \mathscr{B}_t, \overline{P}\right\}$, where

$$\tilde{W}_t^i = W_t^i - \int_0^t f_s^i ds.$$

For a proof see Elliott (1982b).

Consider a process X_τ, $\tau \geq 0$, which takes its values in some measurable space $\{E, \mathscr{E}\}$ and which remains at its initial value $z_0 \in E$ until a random time T, when it jumps to a random position z. The underlying probability space can be taken to be

$$\Omega = [0, \infty] \times E.$$

Write

$$P_t^A = P[T > t \text{ and } z \in A], \qquad P_t = P[T > t];$$

then clearly, if $P_t(C) = 0$, then $P_t^A(C) = 0$ for any $C \subset [0, \infty]$ so that there is a Radon-Nikodym derivative $\lambda(A, s)$ such that

$$P_t^A - P_0^A = \int_{]0,t[} \lambda(A,s) \, dP_s.$$

Write

$$\Lambda(t) = -\int_{]0,t[} \frac{dP_s}{P_{s-}}.$$

The pair (λ, Λ) is the *Lévy system* for the jump process. Roughly, $d\Lambda(t) = -dP_t/P_{t-}$ is the probability that the jump occurs at time t, given it has not happened so far, and $\lambda(dx, s)$ is the conditional distribution of the jump position z given the jump happens at time s.

For two semimartingales X_t and Y_t, the *Ito product rule* gives the product as

$$X_t Y_t = \int_0^t X_{s-} dY_s + \int_0^t Y_{s-} dX_s + [X,Y]_t,$$

where

$$[X,Y]_t = \lim_{n\to\infty} (\text{in prob.}) \left\{ X_0Y_0 + \sum_{0\le k<2^n} \left[\left(X_{t(k+1)2^{-n}} - X_{tk2^{-n}} \right) \right.\right.$$
$$\left.\left. \times \left(Y_{t(k+1)2^{-n}} - Y_{tk2^{-n}} \right) \right] \right\}$$

is the *quadratic variation* of X and Y.

If the process $[X,Y]_t$ has a compensator, this is denoted by $\langle X,Y\rangle_t$ and called the *predictable quadratic variation* of X and Y. If the martingale part of the semimartingale is discontinuous, then $[X,Y]_t = X_0Y_0 + \sum_{0<s\le t} \Delta X_s \Delta Y_s$.

Theorem 1.4 (The Ito Rule) *Let f be a twice continuously differentiable function on $\mathbb{R}$ and let X be a real semimartingale. Then $f(X_t)$ is also a semimartingale and*

$$\begin{aligned} f(X_t) &= f(X_0) + \int_0^t f'(X_{s-})\,dX_s \\ &\quad + \sum_{0<s\le t} \left(f(X_s) - f(X_{s-}) - f'(X_{s-})\Delta X_s \right) \\ &\quad + \frac{1}{2}\int_0^t f''(X_s)\,d[X,X]_s^c . \end{aligned}$$

Theorem 1.5 (The Gronwall Inequality) *Suppose that the continuous function $g(t)$ satisfies*

$$0 \le g(t) \le \alpha(t) + \beta \int_0^t g(s)\,ds, \qquad 0 \le t \le T,$$

with $\beta \ge 0$ and $\alpha : [0,T] \to \mathbb{R}$ integrable. Then

$$g(t) \le \alpha(t) + \beta \int_0^t \alpha(s) \exp(\beta(t-s)), \qquad 0 \le t \le T.$$

A proof can be found in Elliott (1982b).

Theorem 1.6 (Jensen's Inequality) *Let $\phi : \mathbb{R} \to \mathbb{R}$ be convex and let X be an integrable random variable such that $\phi(X)$ is integrable. Then for the probability space $(\Omega, \mathscr{B}, P)$ if $\mathscr{A}$ is a subfield of $\mathscr{B}$,*

$$\phi(E[X \mid \mathscr{A}]) \le E[\phi(X) \mid \mathscr{A}].$$

A proof can be found in Elliott (1982b).

Theorem 1.7 (Fubini) *Let $(\Omega_1, \mathscr{B}_1, P_1)$, $(\Omega_2, \mathscr{B}_2, P_2)$ be two complete probability spaces, and let $f \in L^1(\Omega_1 \times \Omega_2, \mathscr{A}_1 \times \mathscr{A}_2, P_1 \times P_2)$. Then for $A_1 \in \mathscr{A}_1, A_2 \in \mathscr{A}_2$,*

$$\int_{A_1\times A_2} f\,d(P_1 \times P_2) = \int_{A_1} \left(\int_{A_2} f\,dP_2 \right) dP_1 = \int_{A_2} \left(\int_{A_1} f\,dP_1 \right) dP_2.$$

(For simple random variables this theorem specializes to the familiar result that expectations and summations can be interchanged.)

A proof can be found in Loève (1978).

Appendix B
Continuous-Time Martingale Representation

For $0 \leq i \leq N$ write $e_i = (0, \ldots, 1, \ldots, 0)'$ for the ith unit (column) vector in $\mathbb{R}^N$. Consider the Markov process $\{X_t\}$, $t \geq 0$, defined on a probability space (Ω, F, P), whose state space is the set

$$S = \{e_1, \ldots, e_N\}.$$

Write $p_t^i = P(X_t = e_i)$, $0 \leq i \leq N$. We shall suppose that for some family of matrices A_t, $p_t = \left(p_t^1, \ldots, p_t^N\right)'$ satisfies the forward Kolmogorov equation

$$\frac{dp_t}{dt} = A_t p_t. \tag{1.1}$$

$A_t = (a_{ij}(t))$, $t \geq 0$, is, therefore, the family of so-called Q-matrices of the process. Because A_t is a Q-matrix

$$a_{ii}(t) = -\sum_{j \neq i} a_{ji}(t). \tag{1.2}$$

The fundamental transition matrix associated with A will be denoted by $\Phi(t,s)$, so with I the $N \times N$ identity matrix

$$\frac{d\Phi(t,s)}{dt} = A_t \Phi(t,s), \qquad \Phi(s,s) = I, \tag{1.3}$$

$$\frac{d\Phi(t,s)}{ds} = -\Phi(t,s) A_s, \qquad \Phi(t,t) = I. \tag{1.4}$$

[If A_t is constant $\Phi(t,s) = \exp(t-s)A$.]

Consider the process in state $x \in S$ at time s and write $X_{s,t}(x)$ for its state at the later time $t \geq s$. Then $E[X_{s,t}(x)] = E_{s,x}[X_t] = \Phi(t,s)x$. Write $\mathscr{F}_t^s$ for the right-continuous, complete filtration generated by $\sigma\{X_r : s \leq r \leq t\}$, and $\mathscr{F}_t = \mathscr{F}_t^0$. A basic result is the following:

Lemma 1.1 $V_t := X_t - X_0 - \int_0^t A_r X_{r^-}\, dr$ *is an* $\{\mathscr{F}_t\}$ *martingale.*

Proof Suppose $0 \le s \le t$. Then

$$\begin{aligned} E\left[V_t - V_s \mid \mathscr{F}_s\right] &= E\left[X_t - X_s - \int_s^t A_r X_{r^-}\, dr \,\middle|\, \mathscr{F}_s\right] \\ &= E\left[X_t - X_s - \int_s^t A_r X_{r^-}\, dr \,\middle|\, X_s\right] \\ &= E\left[X_t - X_s - \int_s^t A_r X_r\, dr \,\middle|\, X_s\right], \end{aligned}$$

because $X_r = X_{r^-} = \lim_{\varepsilon>0,\varepsilon\to 0} X_{r-\varepsilon}$ for each $\omega \in \Omega$, except for countably many r this is

$$\begin{aligned} &= E\left[X_t \mid X_s\right] - X_s - \int_s^t A_r E\left[X_r \mid X_s\right] dr \\ &= \Phi(t,s) X_s - X_s - \int_s^t A_r \Phi(r,s) X_s\, dr = 0 \end{aligned}$$

by (1.3). Therefore,

$$\begin{aligned} X_t &= X_0 + \int_0^t A_r X_r\, dr + V_t \qquad (1.5) \\ &= X_0 + \int_0^t A_r X_{r^-}\, dr + V_t. \end{aligned}$$

□

We now give a martingale representation result.

Lemma 1.2

$$X_t = \Phi(t,0)\left(X_0 + \int_0^t \Phi(r,0)^{-1}\, dV_r\right). \qquad (1.6)$$

Proof The proof follows from (1.5) by variation of constants. Alternatively, differentiate (1.6). □

If x, y are (column) vectors in $\mathbb{R}^N$ we shall write $\langle x, y\rangle = x'y$ for their scalar (inner) product. Consider $0 \le i, j \le N$ with $i \ne j$. Then

$$\begin{aligned} \langle X_{s^-}, e_i\rangle e_j' dX_s &= \langle X_{s^-}, e_i\rangle e_j' \Delta X_s \\ &= \langle X_{s^-}, e_i\rangle e_j' (X_s - X_{s^-}) = I(X_{s^-} = e_i, X_s = e_j). \end{aligned}$$

Define the martingale

$$V_t^{ij} := \int_0^t \langle X_{s^-}, e_i \rangle e_j' \, dV_s.$$

(Note the integrand is predictable.) Then

$$V_t^{ij} = \int_0^t \langle X_{s^-}, e_i \rangle e_j' \, dX_s - \int_0^t \langle X_{s^-}, e_i \rangle e_j' A_s X_{s^-} \, ds$$

and, writing $\mathscr{J}_t^{ij}$ for the number of jumps of the process X from e_i to e_j up to time t, this is

$$\begin{aligned} &= \mathscr{J}_t^{ij} - \int_0^t \langle X_{s^-}, e_i \rangle a_{ji}(s) \, ds \\ &= \mathscr{J}_t^{ij} - \int_0^t \langle X_s, e_i \rangle a_{ji}(s) \, ds, \end{aligned}$$

because $X_s = X_{s^-}$ for each ω, except for countably many s. That is, for $i \neq j$,

$$\mathscr{J}_t^{ij} = \int_0^t \langle X_s, e_i \rangle a_{ji}(s) \, ds + V_t^{ij}.$$

For a fixed j, $0 \leq j \leq N$, write $\mathscr{J}_t^j$ for the number of jumps into state e_j up to time t. Then

$$\mathscr{J}_t^j = \sum_{i=1}^N \mathscr{J}_t^{ij} = \sum_{i=1}^N \int_0^t \langle X_s, e_i \rangle a_{ji}(s) \, ds + V_t^j,$$

where V_t^j is the martingale $\sum_{i=1}^N V_t^{ij}$. Finally, write $\mathscr{J}_t$ for the total number of jumps (of all kinds) of the process X up to time t. Then

$$\mathscr{J}_t = \sum_{j=1}^N \mathscr{J}_t^j = \sum_{i,j=1}^N \int_0^t \langle X_s, e_i \rangle a_{ji}(s) \, ds + Q_t,$$

where Q_t is the martingale $\sum_{j=1}^N V_t^j$. However, from (1.2)

$$a_{ii}(s) = -\sum_{j=1}^N a_{ji}(s)$$

so

$$\mathscr{J}_t = -\sum_{i=1}^N \int_0^t \langle X_s, e_i \rangle a_{ii}(s) \, ds + Q_t. \tag{1.7}$$

Lemma 1.3

$$\langle V, V \rangle_t = \operatorname{diag} \int_0^t A_r X_{r-} \, dr - \int_0^t (\operatorname{diag} X_{r-}) A_r' \, dr - \int_0^t A_r (\operatorname{diag} X_{r-}) \, dr.$$

Proof Recall $X_t \in S$ is one of the unit vectors e_i. Therefore,

$$X_t X_t' = \operatorname{diag} X_t. \tag{1.8}$$

Now by the *product rule*

$$\begin{aligned} X_t X_t' &= X_0 X_0' + \int_0^t X_{r-} (A_r X_{r-})' \, dr \\ &\quad + \int_0^t X_{r-} \, dV_r' + \int_0^t (A_r X_{r-}) X_{r-}' \, dr \\ &\quad + \int_0^t dV_r X_{r-}' + \langle V,V\rangle_t + ([V,V]_t - \langle V,V\rangle_t), \end{aligned}$$

where $[V,V]_t - \langle V,V\rangle_t$ is an $\{\mathscr{F}_t\}$ martingale. However, a simple calculation shows

$$X_{r-} (A_r X_{r-})' = (\operatorname{diag} X_{r-}) A_r'$$

and

$$(A_r X_{r-}) X_{r-}' = A_r (\operatorname{diag} X_{r-})'.$$

Therefore,

$$\begin{aligned} X_t X_t' &= X_0 X_0' + \int_0^t (\operatorname{diag} X_{r-}) A_r' \, dr \\ &\quad + \int_0^t A_r (\operatorname{diag} X_{r-}) \, dr + \langle V,V\rangle_t + \text{martingale}. \end{aligned} \tag{1.9}$$

Also, from (1.8)

$$X_t X_t' = \operatorname{diag} X_t = \operatorname{diag} X_0 + \operatorname{diag} \int_0^t A_r X_{r-} \, dr + \operatorname{diag} V_t. \tag{1.10}$$

The semimartingale decompositions (1.9) and (1.10) must be the same, so equating the predictable terms

$$\langle V,V\rangle_t = \operatorname{diag} \int_0^t A_r X_{r-} \, dr - \int_0^t (\operatorname{diag} X_{r-}) A_r' \, dr - \int_0^t A_r (\operatorname{diag} X_{r-}) \, dr.$$

□

We next note the following representation result:

Remark 1.4 A function of $X_t \in S$ can be represented by a vector

$$f(t) = (f_1(t), \dots, f_N(t))' \in \mathbb{R}^N$$

so that $f(t, X_t) = f(t)' X_t = \langle f(t), X_t \rangle$ where $\langle , \rangle$ denotes the inner product in $\mathbb{R}^N$.

■

We therefore have the following differentiation rule and representation result:

Lemma 1.5 *Suppose the components of $f(t)$ are differentiable in t. Then*

$$f(t,X_t) = f(0,X_0) + \int_0^t \langle f'(r), X_r \rangle dr + \int_0^t \langle f(r), A_r X_{r-} \rangle dr + \int_0^t \langle f(r), dV_r \rangle . \tag{1.11}$$

Here, $\int_0^t \langle f(r), dV_r \rangle$ is an $\mathscr{F}_t$-martingale. Also,

$$f(t,X_t) = \langle f(t), \Phi(t,0) X_0 \rangle + \int_0^t \langle f(t), \Phi(t,r) dV_r \rangle . \tag{1.12}$$

This gives the martingale representation of $f(t,X_t)$.

References

Aggoun, L., Bensoussan, A., Elliott, R. J. and Moore, J. B. (1995). A finite-dimensional quasi-linear risk-sensitive control problem, *Systems and Control Letters* **25**: 151–157.

Aggoun, L. and Elliott, R. J. (1992). Finite dimensional predictors for hidden Markov chains, *Systems and Control Letters* **19**: 335–340.

Aggoun, L. and Elliott, R. J. (1994). Celestial signal estimation, *Stochastic Analysis and Applications* **12**(4): 399–407.

Aggoun, L. and Elliott, R. J. (1995). Finite dimensional models for hidden Markov chains, *Advances in Applied Probability* **27**: 146–160.

Aggoun, L., Elliott, R. J. and Moore, J. B. (1993a). Adjoint process for Markov chains observed in Gaussian noise, *26th IEEE Asilomor Conference on Signal, Systems and Computers*, IEEE Computer Society Press, pp. 396–399.

Aggoun, L., Elliott, R. J. and Moore, J. B. (1993b). A measure change derivation of continuous state Baum-Welch estimators, *Journal of Mathematical Systems, Estimation and Control* **5**: 359–362.

Anderson, B. D. O. (1968). State and parameter estimation in non-linear systems, *Institute of Engineers of Australia. Elecrical Engineering Transactions* pp. 200–204.

Anderson, B. D. O. and Moore, J. B. (1979). *Optimal Filtering*, Prentice Hall, Englewood Cliffs, NJ.

Aström, K. J. (1965). Optimal control of Markov processes with incomplete information, *Journal of Mathematical Analysis and Applications* **10**: 174–205.

Baras, J. S., Bensoussan, A. and James, M. R. (1988). Dynamic observers as asymptotic limits of recursive filters: Special cases, *SIAM Journal on Applied Mathematics* **48**(5): 1147–1158.

Baras, J. S. and James, M. R. (1993). Robust output feedback control for nonlinear discrete-time systems. Preprint.

Basar, T. S. and Bernhard, P. (1991). *H_∞-Optimal Control and Related Minimax Design Problems: A Dynamic Game Approach*, Birkhauser, Boston.

Baum, L. E. and Petrie, E. (1966). Statistical inference for probabilistic functions of finite state Markov chains, *Annals of the Institute of Statistical Mathematics* **37**: 1554–1563.

Baum, L. E., Petrie, E., Soules, G. and Weiss, N. (1970). A maximization technique occuring in the statistical analysis of probabilistic functions of Markov chains, *Annals of the Institute of Statistical Mathematics* **41**(1): 164–171.

Bensoussan, A. (1982). *Lectures on Stochastic Control*, Vol. 972 of *Springer Lecture Notes in Mathematics*, Springer-Verlag, New York.

Bensoussan, A. and Elliott, R. J. (1995). A finite dimensional risk sensitive control problem, *SIAM Journal of Control and Optimization* **33**: 1834–1846.

Bensoussan, A. and van Schuppen, J. H. (1985). Optimal control of partially observable stochastic systems with an exponential-of-integral performance index, *SIAM Journal on Control and Optimization* **23**: 599–613.

Bertsekas, D. P. (1987). *Dynamic Programming: Deterministic and Stochastic Models*, Prentice Hall, Englewood Cliffs, NJ.

Besag, B. (1977). Efficiency of pseudo-likelihood estimation for simple Gaussian fields, *Biometrika* **64**: 616–618.

Besag, B. (1986). On the statistical analysis of dirty pictures, *Journal of the Royal Statistical Society of London. Series B* **B**(48): 259–302.

Biane, P. (1990). Chaotic representation for finite Markov chains, *Stochastics and Stochastics Reports* **30**: 61–68.

Bingham, J. A. C. (1988). *The Theory and Practise of Modem Design*, Wiley, New York.

Bismut, J.-M. (1976). Linear quadratic optimal stochastic control with random coefficients, *SIAM Journal on Control and Optimization* **14**(3): 419–444.

Bismut, J. M. (1978a). An introductory approach to duality in optimal stochastic control, *SIAM Review* **20**(1).

Bismut, J.-M. (1978b). An introductory approach to duality in optimal stochastic control, *SIAM Review* **20**: 62–78.

Blagovescenskii, J. N. and Freidlin, M. I. (1961). Some properties of diffusion processes depending on a parameter, *Soviet Mathematics* **2**: 633–636.

Boel, R. (1976). Discrete time martingales in filtering and control, *Technical report*, Division of Engineering Cybernetics, NTH Trondheim, Norway.

Boel, R. (1987). Modelling, estimation and prediction for jump processes, *in* H. V. Poor (ed.), *Advances in Statistical Signal Processing*, Vol. 1, JAI Press, pp. 53–116.

Boel, R. and Varaiya, P. P. (1977). Optimal control of jump processes, *SIAM Journal on Control and Optimization* **15**: 92–119.

Boel, R., Varaiya, P. P. and Wong, E. (1975). Martingales on jump processes; Part I: Representation results; Part II: Applications, *SIAM Journal on Control* **13**: 999–1061.

Brémaud, P. M. (1981). *Point Processes and Queues*, Springer-Verlag, New York.

Brémaud, P. M. and van Schuppen, J. H. (1976). Discrete time stochastic systems, *Technical report*, Department of Systems Science and Mathematics, Washington University, St. Louis, MO. Unpublished report.

Caines, P. E. (1988). *Linear Stochastic Systems*, Wiley, New York.

Cameron, R. H. and Martin, W. T. (1944). Transformation of Wiener integrals under translations, *Annals of Mathematics* **45**: 386–396.

Campi, M. and James, M. R. (1992). Discrete-time nonlinear risk-sensitive control. Preprint.

Campillo, F. and le Gland, F. (1989). MLE for partially observed diffusions: Direct maximization versus the em algorithm, *Stochastic Processes and Their Applications* **33**(2): 245–274.

Chen, B. and Tugnait, J. K. (1986). An interacting multiple model fixed-lag smoothing algorithm for markovian switching systems, *37th IEEE Conference on Decision and Control*, Tampa Florida USA.

Chen, B. and Tugnait, J. K. (2000). Interacting multiple model fixed-lag smoothing algorithm for markovian switching systems, *IEEE Transactions on Aerospace and Electronic Systems* **36**(1).

Chung, K. L. and Williams, R. J. (1990). *Introduction to Stochastic Integration*, 2nd edn, Birkhauser, Boston.

Chung, S.-H., Krishnamurthy, V. and Moore, J. B. (1991). Adaptive processing techniques based on hidden Markov models for characterising very small channel currents buried in noise and deterministic interferences, *Philosophical Transactions of the Royal Society of London, Series B* **334**: 357–384.

Clark, J. M. (1978). The design of robust approximations to the stochastic differential equations for nonlinear filtering, *in* J. Skwirzynski (ed.), *Communications Systems and Random Process*

Theory, Darlington 1977, Alphen aan den Rijn, Sijthoff and Noorhoff, the Netherlands, pp. 721–734.

Clements, D. and Anderson, B. D. O. (1975). A non-linear fixed lag smoother for finite state Markov processes, *IEEE Transactions on Information Theory* **1**(21): 446–452.

Collings, I. B., Krishnamurthy, V. and Moore, J. B. (1993). Recursive prediction error techniques for adaptive estimation of hidden Markov models, *Proceedings of the 12th World Congress IFAC*, Vol. V, Sydney, Australia, pp. 423–426.

Collings, I. B. and Moore, J. B. (1994a). Adaptive HMM filters for signals in noisy fading channels, *International Conference on Acoustics, Speech and Signal Processing*, Adelaide, Australia, pp. 305–308.

Collings, I. B. and Moore, J. B. (1994b). An HMM approach to adaptive demodulation of QAM signals in fading channels, *International Journal of Adaptive Control and Signal Processing* **8**(5): 457–474.

Collings, I. B., Moore, J. B. and James, M. R. (1994). An information state approach to linear/risk-sensitive/quadratic/Gaussian control, *33rd IEEE Conference on Decision and Control*.

Davis, M. H. A. (1977). *Linear Estimation and Control*, Chapman and Hall, London.

Davis, M. H. A. (1979). Martingale methods in stochastic control, *Stochastic Control Theory and Stochastic Differential Systems*, Springer Verlag, New York, pp. 85–117.

Davis, M. H. A. (n.d.). A note on the Poisson disorder problem, *Banach Centre Publications* **1**: 65–72.

Davis, M. H. A., Kailath, T. and Segall, A. (1975). Non-linear filtering with counting observations, *IEEE Transactions on Information Theory* **21**: 143–150.

Dembo, A. and Zeitouni, O. (1986). Parameter estimation of partially observed continuous time stochastic processes, *Stochastic Processes and Their Applications* **23**: 91–113.

Dempster, A. P., Laird, N. M. and Rubin, D. B. (1977). Maximum likelihood from incomplete data via the EM algorithm, *Journal of the Royal Statistical Society of London. Series B* **39**: 1–38.

Dey, S. and Moore, J. (1995a). Risk-sensitive filtering and smoothing for hidden Markov models, *Systems and Control Letters* **25**(5): 361–366.

Dey, S. and Moore, J. (1995b). Risk-sensitive filtering and smoothing via reference probability methods, *Proceedings of the Amercican Control Conference*, Seattle, pp. 129–133.

Dey, S. and Moore, J. B. (1995c). Risk-sensitive filtering and smoothing for hidden markov models, *Systems and Control Letters* **25**: 361–366.

di Masi, G. B. and Runggaldier, W. J. (1982). On measure transformation for combined filtering and parameter estimation in discrete time, *Systems and Control Letters* **2**: 57–62.

Doyle, J. C., Glover, K., Khargonekar, P. P. and Francis, B. A. (1989). State-space solutions to standard H_2 and H_∞ control, *IEEE Transactions on Automatic Control* **34**(8): 831–847.

Duncan, T. E. (1967). *Probability Densities for Diffusion Processes with Application to Nonlinear Filtering Theory and Detection Theory*, PhD thesis, Stanford Electronics Laboratories, Stanford, CA. 7001-4.

Elliott, R. J. (1976). A stochastic minimum principle, *Bulletin of the American Mathematical Society* **82**(6): 944–946.

Elliott, R. J. (1982a). *The Nonlinear Filtering Equations*, Vol. 43 of *Springer Lecture Notes in Control and Information Sciences*, Springer-Verlag, New York, pp. 168–178.

Elliott, R. J. (1982b). *Stochastic Calculus and Applications*, Vol. 18 of *Applications of Mathematics*, Springer-Verlag, New York.

Elliott, R. J. (1985a). *Smoothing for a Finite State Markov Process*, Vol. 69 of *Springer Lecture Notes in Control and Information Sciences*, Springer-Verlag, New York, pp. 199–206.

Elliott, R. J. (1985b). A special semimartingale derivation of smoothing and prediction equations, *Systems and Control Letters* **6**: 287–289.

Elliott, R. J. (1987). *Viscosity Solutions and Optimal Control*, Pitman Research Notes in Mathematics, Longman Scientific and Technical, London.

Elliott, R. J. (1990a). Filtering and control for point process observations, *in* J. S. Baras and V. Mirelli (eds), *Recent Advances in Stochastic Calculus*, Springer-Verlag, New York, pp. 1–26.

Elliott, R. J. (1990b). Martingales associated with finite Markov chains, *in* E. Cinlar, R. J. Williams and P. Fitzsimmons (eds), *Seminar on Stochastic Processes*, Vol. 1, Birkhauser, Boston, pp. 61–172.

Elliott, R. J. (1992). A partially observed control problem for Markov chains, *Applied Mathematics and Optimization* **25**: 151–169.

Elliott, R. J. (1993a). A general recursive discrete time filter, *Journal of Applied Probability* **30**: 575–588.

Elliott, R. J. (1993b). New finite dimensional filters and smoothers for noisily observed markov chains, *IEEE Transactions on Information Theory* **39**(1): 265–271.

Elliott, R. J. (1993c). New finite dimensional filters and smoothers for noisily observed Markov chains, *IEEE Transactions on Information Theory* **39**: 265–271.

Elliott, R. J. (1994a). Exact adaptive filters for markov chains observed in gaussian noise, *Automatica* **30**(9): 1399–1408.

Elliott, R. J. (1994b). Exact adaptive filters for Markov chains observed in Gaussian noise, *Automatica* **30**: 1399–1408.

Elliott, R. J. (1995). Recursive estimation for hidden Markov models: A dependent case, *Stochastic Analysis and Applications* **13**: 437–460.

Elliott, R. J. and Aggoun, L. (1992). Estimation for hidden Markov random fields, *Journal for Statistical Planning and Inference* **50**: 343–351.

Elliott, R. J. and Aggoun, L. (1994). Estimation for discrete Markov random fields observed in Gaussian noise, *IEEE Transactions on Information Theory* **40**: 1600–1603.

Elliott, R. J. and Aggoun, L. (1995). M. A. P. estimation using measure change for continuous state random fields, *Systems and Control Letters* **26**: 239–244.

Elliott, R. J. and Aggoun, L. (1998). M. A. P. estimation for hidden discrete random fields, *Stochastic Analysis and Applications* **16**. To appear.

Elliott, R. J., Aggoun, L., Bensoussan, A. and Moore, J. B. (1994). Finite-dimensional exponential LQG control, *American Control Conference*.

Elliott, R. J., Dufour, F. and Sworder, D. (1996). Exact hybrid filters in discrete time, *IEEE Transactions on Automatic Control* **41**: 1807–1810.

Elliott, R. J. and Kohlmann, M. (1989). Integration by part, homogeneous chaos expansions and smooth densities, *Annals of Probability* **17**: 194–207.

Elliott, R. J. and Krishnamurthy, V. (1999). New finite-dimensional filters for parameter estimation of discrete-time linear gaussian models, *IEEE Transactions on Automatic Control* **44**(5).

Elliott, R. J. and Malcolm, W. P. (2000). Reproducing gaussian densities and linear gaussian detection, *Systems and Control Letters* **40**: 133–138.

Elliott, R. J. and Malcolm, W. P. (2001). Robust smoother dynamics for poisson processes driven by an itô diffusion, *IEEE Conference on Decision and Control*, Orlando Florida, USA.

Elliott, R. J. and Moore, J. B. (1992a). Adjoint processes for markov chains observed in gaussian noise, *26th Asilomar Conference on Signals Systems and Computing, Vol 1*, California USA, pp. 396–399.

Elliott, R. J. and Moore, J. B. (1992b). Recursive parameter estimation for partially observed Markov chains, *27th IEEE Asilomor Conference on Signal, Systems and Computers*, Vol. 16, IEEE Computer Society Press, Asilomor.

Elliott, R. J. and Moore, J. B. (1993). Discrete time partially observed control, *in* D. Elworthy, W. N. Everitt and E. B. Lee (eds), *Differential Equations, Dynamical Systems and Control Science*, Vol. 32, Marcel Dekker, New York, pp. 481–490. Festschrift in Honor of Lawrence Markus.

Elliott, R. J. and Moore, J. B. (1996). State and parameter estimation for linear systems, *Journal of Mathematical Systems, Estimation and Control* **6**: 125–128.

Elliott, R. J. and Sworder, D. (1992). Control of a hybrid conditionally linear Gaussian process, *Journal of Optimization Theory and Applications* **74**: 75–85.

Elliott, R. J. and Yang, H. (1992). Forward and backward equations for an adjoint process, *Festschrift for G. Kallianpur*, Springer-Verlag, New York, pp. 61–70.

Elliott, R. J. and Yang, H. (1994). How to count and guess well: Discrete adaptive filters, *Applied Mathematics and Optimization* **30**: 51–78.

Fleming, W. H. (1969). Optimal continuous-parameter stochastic control, *SIAM Review* **11**(4).

Fleming, W. H. and McEneaney, W. M. (1991). *Risk Sensitive Optimal Control and Differential Games*, Vol. 184 of *Springer Lecture Notes in Control and Information Sciences*, Springer-Verlag, New York, pp. 185–197.

Fleming, W. H. and Pardoux, E. (1982). Optimal control for partially observed diffusions, *SIAM J Control and Optimization* **20**(2): 261–285.

Fleming, W. H. and Rischel, R. W. (1975). *Deterministic and Stochastic Optimal Control*, Vol. 1, 25 of *Stochastic Modelling and Applied Probability Series*, Springer Verlag.

Fleming, W. H. and Soner, H. M. (1975). *Controlled Markov Processes and Viscosity*, Stochastic Modelling and Applied Probability Series, Springer Verlag.

Fragoso, M. D. (1988). On a partially observed LQG problem for systems with Markovian jumping parameters, *Systems and Control Letters* **10**: 349–356.

Freidlin, M. I. and Wentzell, A. D. (1984). *Random Perturbations of Dynamical Systems*, Springer-Verlag, New York.

Geman, S. and Geman, D. (1984). Stochastic relaxation, Gibbs distributions, and the Bayesian restoration of images, *IEEE Transactions on Pattern Analysis and Machine Intelligence* **6**: 721–741.

Girsanov, I. V. (1960). On transforming a certain class of stochastic processes by absolutely continuous substitution of measures, *Theory of Probability and Applications* **5**: 285–301.

Glover, K. and Doyle, J. C. (1988). State space formulae for all stabilizing controllers that satisfy an H_∞ norm bound and relations to risk sensitivity, *Systems and Control Letters* **11**: 167–172.

Haussmann, U. G. (1981a). On the adjoint process for optimal control of diffusion processes, *SIAM Journal on Control and Optimization* **19**: 221–243, 710.

Haussmann, U. G. (1981b). Some examples of optimal stochastic controls or: The stochastic maximum principle at work, *SIAM Review* **23**(3).

Hijab, O. (1980). *Minimum Energy Estimation*, PhD thesis, University of California, Berkeley.

Hijab, O. (1984). Asymptotic Bayesian estimation of a first order equation with a small diffusion, *Annals of Probability* **12**: 890–902.

Ho, Y.-C. and Lee, R. C. K. (1968). A Bayesian approach to problems in stochastic estimation and control, *IEEE Transactions on Automatic Control* **1**: 333–339.

Jacobson, D. H. (1973). Optimal stochastic linear systems with exponential performance criteria and their relation to deterministic differential games, *IEEE Transactions on Automatic Control* **18**(2): 124–131.

Jacod, J. (1979). *Calcul Stochastique et Problèmes de Martingales*, Vol. 794 of *Springer Lecture Notes in Mathematics*, Springer-Verlag, New York.

James, M. R. (1991). Finite time observer design by probabilistic-variational methods, *SIAM Journal on Control and Optimization* **29**(4): 954–967.

James, M. R. (1992). Asymptotic analysis of nonlinear stochastic risk-sensitive control and differential games, *Mathematics of Control, Signals and Systems* **5**(4): 401–417.

James, M. R. and Baras, J. S. (1988). Nonlinear filtering and large deviations: A PDE-control theoretic approach, *Stochastics* **23**(3): 391–412.

James, M. R., Baras, J. S. and Elliott, R. J. (1993). Risk-sensitive control and dynamic games for partially observed discrete-time nonlinear systems, *IEEE Transactions on Automatic Control* **39**: 780–792.

James, M. R. and Elliott, R. J. (1996). Risk-sensitive and risk neutral control for continuous-time hidden markov models, *Applied Mathematics and Optimization* **34**: 37–50.

James, M. R., Krishnamurthy, V. and Le Gland, F. (1998). Time discretization for continuous time filters and smoothers for hmm parameter estimation, *Transactions on Information Theory* **42**(2): 593–605.

Jazwinski, A. H. (1970). *Stochastic Processes and Filtering Theory*, Academic Press, New York, London.

Kinderman, R. and Snell, L. (1980). *Markov Random Fields and Their Applications*, Vol. 1 of *Contemporary Mathematics*, American Mathematical Society, Providence, R.I.

Kohlmann, M. (1985). A maximum principle and dynamic programming for a partially observed control problem, *Technical report*, Universität Konstanz. Discussion paper.

Kolmogorov, A. N. (1933). Grundbegriffe der wahrsch, *Ergodic Mathematics* **2**(3).

Krasovskii, N. N. and Lidskii, E. A. (1961). Analytical design of controllers in systems with random attributes (Parts 1, 2 & 3), *Automation and Remote Control* **22**: 1020–1025, 1141–1146, 1289–1294.

Krishnamurthy, V. and Moore, J. B. (1993). On-line estimation of hidden Markov model parameters based on the Kullback-Leibler information measure, *IEEE Transactions on Signal Processing* **41**(8): 2557–2573.

Kulhavy, R. (1990). Differential geometry of recursive nonlinear estimation, *11th IFAC World Congress*, Vol. 3, Tallin, pp. 113–118.

Kumar, P. R. and Varaiya, P. P. (1986a). *Stochastic Systems*, Prentice Hall, Englewood Cliffs, NJ.

Kumar, P. R. and Varaiya, P. P. (1986b). *Stochastic Systems: Estimation, Identification, and Adaptive Control*, Prentice Hall, Englewood Cliffs, NJ.

Kushner, H. J. (1972). Necessary conditions for continuous parameter stochastic optimization problems, *SIAM Journal on Control* **10**: 550–565.

Kushner, H. J. and Shwartz, A. (1984). An invariant measure approach to the convergence of stochastic approximations with state dependent noise, *SIAM Journal on Control and Optimization* **22**(1): 13–27.

Lee, M. H., Kolodziej, W. J. and Mohler, R. R. (1985). Stochastic dynamic system suboptimal control with uncertain parameters, *IEEE Transactions on Aerospace and Electronic Systems* **21**: 594–600.

Limebeer, D. J. N., Anderson, B. D. O., Khargonekar, P. P. and Green, M. (1992). A game theoretic approach to H_∞ control for time varying systems, *SIAM Journal on Control and Optimization* **30**: 262–283.

Liptser, R. S. and Shiryayev, A. N. (1977). *Statistics of Random Processes*, Vol. 1, Springer-Verlag, New York.

Ljung, L. (1977). Analysis of recursive stochastic algorithms, *IEEE Transactions on Automatic Control* **22**(4): 551–575.

Ljung, L. and Söderström, T. (1983). *Theory and Practice of Recursive Identification*, MIT Press, Cambridge MA.

Lodge, J. H. and Moher, M. L. (1990). Maximum likelihood sequence estimation of CPM signals transmitted over Rayleigh flat-fading channels, *IEEE Transactions on Communications Systems* **38**(6): 787–794.

Loève, M. M. (1978). *Probability Theory*, Vol. I and II, 4th edn, Springer-Verlag, New York.

Loparo, K. A., Aslanis, J. T. and Hajek, O. (1987). Analysis of switched systems in the plane (Parts 1 & 2), *Journal of Optimization Theory and Applications* **52**: 365–394, 395–427.

Malcolm, W. P. and Elliott, R. J. (1999). A general smoothing equation for poisson observations, *IEEE Conference on Decision and Control*, Phoenix USA.

Malcolm, W. P., Elliott, R. J. and van der Hoek, J. (to appear). A deterministic discretisation-step upper bound for state estimation via clark transformations, *Journal of Applied Mathematics and Stochastic Analysis* .

Mariton, M. (1990). *Jump Linear Systems in Automatic Control*, Marcel Dekker, New York.

McGarthy, T. P. (1974). *Stochastic Systems and State Estimation*, Wiley, New York.

Moore, J. B. and Weiss, H. J. (1979). Recursive prediction error methods for adaptive estimation, *IEEE Transactions on Systems, Man, and Cybernetics* **9**(4): 197–205.

Moore, J., Elliott, R. and Dey, S. (1995). Risk-sensitive generalizations of minimum variance estimation and control, *Journal of Mathematical Systems, Estimation and Control* . Accepted.

Mortensen, R. E. (1966). *Optimal Control of Continuous-Time Stochastic Systems*, PhD thesis, Electronics Research Laboratories, University of California, Berkeley. ERL-66-1.

Polyak, B. T. and Juditsky, A. B. (1992). Acceleration of stochastic approximation by averaging, *SIAM Journal on Control and Optimization* **30**(4): 838–855.

Qian, W. and Titterington, D. M. (1990). Parameter estimation for hidden Gibbs chains, *Statistics and Probability Letters* **10**: 49–58.

Rabiner, L. R. (1989). A tutorial on hidden Markov models and selected applications in speech recognition, *Proceedings of the IEEE* **77**: 257–286.

Rhee, I. and Speyer, J. L. (1991). A game theoretic approach to a finite-time disturbance attenuation problem, *IEEE Transactions on Automatic Control* **36**: 1021–1032.

Ripley, B. D. (1988). *Statistical Inference for Spatial Processes*, Cambridge University Press, New York.

Rudin, W. (1966). *Real and Complex Analysis*, McGraw-Hill.

Segall, A. (1976a). Dynamic file assignment in computer networks (Part 1), *IEEE Transactions on Automatic Control* **21**: 161–173.

Segall, A. (1976b). Recursive estimation from discrete time point processes, *IEEE Transactions on Information Theory* **22**: 422–431.

Segall, A. (1977). Optimal control of noisy finite state Markov processes, *IEEE Transactions on Automatic Control* **2**: 179–186.

Shiryayev, A. N. (1986). *Probability Theory*, Springer-Verlag, New York.

Sorenson, H. W. and Alspach, D. L. (1971). Recursive bayesian estimation using gaussian sums, *Automatica* **7**(4): 465–479.

Stratonovich, R. L. (1960). Conditional Markov processes, *Theory of Probability and Applications* **5**: 156–178.

Titterington, D. M. (1984). Recursive parameter estimation using incomplete data, *Journal of the Royal Statistical Society of London. Series B* **46**(2): 257–267.

van Schuppen, J. H. (1977). Filtering, prediction and smoothing for counting process observations, a martingale approach, *SIAM Journal on Applied Mathematics* **32**: 552–570.

Viterbi, A. J. (1969). Error bounds for the white gaussian and other very noisy memoryless channels with generalized decision regions, *IEEE Transactions on Information Theory* **IT**(2).

Weinstein, E., Feder, M. and Oppenheim, A. V. (1990). Sequential algorithms for parameter estimation based on the Kullback-Leibler information measure, *IEEE Transactions on Acoustics, Speech, and Signal Processing* **38**(9): 1652–1654.

Whittle, P. (1981). Risk-sensitive linear/quadratic/Gaussian control, *Advances in Applied Probability* **13**: 764–777.

Whittle, P. (1990a). A risk-sensitive maximum principle, *Systems and Control Letters* **15**: 183–192.

Whittle, P. (1990b). *Risk-Sensitive Optimal Control*, Wiley, New York.

Whittle, P. (1991). A risk-sensitive maximum principle: The case of imperfect state observation, *IEEE Transactions on Automatic Control* **36**: 793–801.

Wong, E. and Hajek, B. (1985). *Stochastic Processes in Engineering Systems*, Springer-Verlag, New York.

Wonham, W. M. (1965). Some applications of stochastic differential equations to optimal nonlinear filtering, *SIAM Journal on Control* **2**: 347–369.

Yang, C. and Mariton, M. (1991). A new approximation for the partially observed jump linear-quadratic problem, *International Journal of Science* **22**: 775–781.

Yao, Y. C. (1985). Estimation of noisy telegraph process: Nonlinear filtering versus nonlinear smoothing, *IEEE Transactions on Information Theory* **31**: 444–446.

Zakai, M. (1969). On the optimal filtering of diffusion processes, *Zeitschrift fur Wahrscheinlichkeitstheorie und verwandte Gebiete* **11**: 230–243.

Zeitouni, O. and Dembo, A. (1988). Exact filters for the estimation of the number of transitions of finite-state continuous time Markov processes, *IEEE Transactions on Information Theory* **34**: 890–893.

Author Index

Subject Index

Stochastic Modelling and Applied Probability
formerly: Applications of Mathematics

1 Fleming/Rishel, **Deterministic and Stochastic Optimal Control** (1975)
2 Marchuk, **Methods of Numerical Mathematics** (1975, 2nd ed. 1982)
3 Balakrishnan, **Applied Functional Analysis** (1976, 2nd ed. 1981)
4 Borovkov, **Stochastic Processes in Queueing Theory** (1976)
5 Liptser/Shiryaev, **Statistics of Random Processes I: General Theory** (1977, 2nd ed. 2001)
6 Liptser/Shiryaev, **Statistics of Random Processes II: Applications** (1978, 2nd ed. 2001)
7 Vorob'ev, **Game Theory: Lectures for Economists and Systems Scientists** (1977)
8 Shiryaev, **Optimal Stopping Rules** (1978, 2008)
9 Ibragimov/Rozanov, **Gaussian Random Processes** (1978)
10 Wonham, **Linear Multivariable Control: A Geometric Approach** (1979, 2nd ed. 1985)
11 Hida, **Brownian Motion** (1980)
12 Hestenes, **Conjugate Direction Methods in Optimization** (1980)
13 Kallianpur, **Stochastic Filtering Theory** (1980)
14 Krylov, **Controlled Diffusion Processes** (1980, 2008)
15 Prabhu, **Stochastic Storage Processes: Queues, Insurance Risk, and Dams** (1980)
16 Ibragimov/Has'minskii, **Statistical Estimation: Asymptotic Theory** (1981)
17 Cesari, **Optimization: Theory and Applications** (1982)
18 Elliott, **Stochastic Calculus and Applications** (1982)
19 Marchuk/Shaidourov, **Difference Methods and Their Extrapolations** (1983)
20 Hijab, **Stabilization of Control Systems** (1986)
21 Protter, **Stochastic Integration and Differential Equations** (1990)
22 Benveniste/Métivier/Priouret, **Adaptive Algorithms and Stochastic Approximations** (1990)
23 Kloeden/Platen, **Numerical Solution of Stochastic Differential Equations** (1992, corr. 2nd printing 1999)
24 Kushner/Dupuis, **Numerical Methods for Stochastic Control Problems in Continuous Time** (1992)
25 Fleming/Soner, **Controlled Markov Processes and Viscosity Solutions** (1993)
26 Baccelli/Brémaud, **Elements of Queueing Theory** (1994, 2nd ed. 2003)
27 Winkler, **Image Analysis, Random Fields and Dynamic Monte Carlo Methods** (1995, 2nd ed. 2003)
28 Kalpazidou, **Cycle Representations of Markov Processes** (1995)
29 Elliott/Aggoun/Moore, **Hidden Markov Models: Estimation and Control** (1995, corr. 3rd printing 2008)
30 Hernández-Lerma/Lasserre, **Discrete-Time Markov Control Processes** (1995)
31 Devroye/Györfi/Lugosi, **A Probabilistic Theory of Pattern Recognition** (1996)
32 Maitra/Sudderth, **Discrete Gambling and Stochastic Games** (1996)
33 Embrechts/Klüppelberg/Mikosch, **Modelling Extremal Events for Insurance and Finance** (1997, corr. 4th printing 2008)
34 Duflo, **Random Iterative Models** (1997)
35 Kushner/Yin, **Stochastic Approximation Algorithms and Applications** (1997)
36 Musiela/Rutkowski, **Martingale Methods in Financial Modelling** (1997, 2nd ed. 2005)
37 Yin, **Continuous-Time Markov Chains and Applications** (1998)
38 Dembo/Zeitouni, **Large Deviations Techniques and Applications** (1998)
39 Karatzas, **Methods of Mathematical Finance** (1998)
40 Fayolle/Iasnogorodski/Malyshev, **Random Walks in the Quarter-Plane** (1999)
41 Aven/Jensen, **Stochastic Models in Reliability** (1999)
42 Hernandez-Lerma/Lasserre, **Further Topics on Discrete-Time Markov Control Processes** (1999)
43 Yong/Zhou, **Stochastic Controls. Hamiltonian Systems and HJB Equations** (1999)
44 Serfozo, **Introduction to Stochastic Networks** (1999)
45 Steele, **Stochastic Calculus and Financial Applications** (2001)
46 Chen/Yao, **Fundamentals of Queuing Networks: Performance, Asymptotics, and Optimization** (2001)
47 Kushner, **Heavy Traffic Analysis of Controlled Queueing and Communications Networks** (2001)
48 Fernholz, **Stochastic Portfolio Theory** (2002)
49 Kabanov/Pergamenshchikov, **Two-Scale Stochastic Systems** (2003)

Stochastic Modelling and Applied Probability
formerly: Applications of Mathematics

50 Han, **Information-Spectrum Methods in Information Theory** (2003)
51 Asmussen, **Applied Probability and Queues** (2nd ed. 2003)
52 Robert, **Stochastic Networks and Queues** (2003)
53 Glasserman, **Monte Carlo Methods in Financial Engineering** (2004)
54 Sethi/Zhang/Zhang, **Average-Cost Control of Stochastic Manufacturing Systems** (2005)
55 Yin/Zhang, **Discrete-Time Markov Chains** (2005)
56 Fouque/Garnier/Papanicolaou/Sølna, **Wave Propagation and Time Reversal in Randomly Layered Media** (2007)
57 Asmussen/Glynn, **Stochastic Simulation: Algorithms and Analysis** (2007)
58 Kotelenez, **Stochastic Ordinary and Stochastic Partial Differential Equations: Transition from Microscopic to Macroscopic Equations** (2008)
59 Chang, **Stochastic Control of Hereditary Systems and Applications** (2008)

Printing: Krips bv, Meppel, The Netherlands
Binding: Stürtz, Würzburg, Germany